APPLIED MYCOLOGY AND BIOTECHNOLOGY

VOLUME 1

AGRICULTURE AND FOOD PRODUCTION

APPLIED MYCOLOGY AND BIOTECHNOLOGY

VOLUME 1

AGRICULTURE AND FOOD PRODUCTION

Edited by

George G. Khachatourians
Department of Applied Microbiology & Food Sciences
College of Agriculture
University of Saskatchewan
Saskatoon, SK, Canada

Dilip K. Arora
Department of Botany
Banaras Hindu University
Varanasi, India

2001
ELSEVIER
Amsterdam - London - New York - Oxford - Paris - Shannon - Tokyo

ELSEVIER SCIENCE B.V.
Sara Burgerhartstraat 25
P.O. Box 211, 1000 AE Amsterdam, The Netherlands

First edition 2001

Library of Congress Cataloging in Publication Data
A catalog record from the Library of Congress has been applied for.

ISBN: 0 444 50657 8

♾ The paper used in this publication meets the requirements of ANSI/NISO Z39.48-1992 (Permanence of Paper).
Printed in The Netherlands.

Preface

Fungi, over the course of history, have been a continuous source of great benefit and risk for human life and significant source of experiential and experimental knowledge. Fungi have been used deliberately and, on many occasions, accidentally in pre- and post-harvest agriculture and food production. The development of the discipline of mycology, and its integration with many other interdisciplinary areas has contributed to the field of biotechnology. The subjects and disciplinary areas captured by the title Applied Mycology and Biotechnology is therefore timely. With the advice of our editorial board and concurrence of Elsevier Science, our publishers, we are presenting to you the reader, the most stimulating synthesis of rapidly growing research interests and publications of scholars in this field. In this inaugural volume we have a contextual coverage from current knowledge of principles and techniques to general process applications in the agri-food sector.

The interdisciplinary and complex nature of the subject area combined with the need to consider the social, economical and industrial perspectives require a new focus in this area. The surge of research and development activity in applied mycology and fungal biotechnology relates to the need and utility of fungi in many contexts. These contexts are wide in scope, and include agriculture, animal and plant health, biotransformation of organic or inorganic matter, food safety, composition of nutrients and micronutrients, and human and animal infectious disease. We hope that our readers will share our belief that with these rapid developments, the need for this book is justifiable.

It is generally accepted that knowledge based goods and services will have a crucial role in meeting the challenges of the 21st century and microorganisms and fungi in particular will continue to play a most significant role. We have therefore begun our inaugural volume with a balanced treatment of principles, biotechnological manipulations and applications of major groups of fungi in agriculture and food. The next volume will deal with various specific applications of mycology and fungal biotechnology to food production and processing.

As a professional reference, this book is targeted towards agri-food producer research establishments, government and academic units. Equally useful should this volume be for teachers and students both in undergraduate and graduate studies in departments of food science, food technology, food engineering, microbiology, applied molecular genetics and, of course, biotechnology.

In a field where the turnover of literature is less than 2 years, we hope this compilation is only a beginning as we continue with the preparation of the next volume. Together, these volumes should help us arrive at comprehensive, in depth information on Applied Mycology and Biotechnology. With citation referring to original publications and patent literature, we hope this will serve as a useful reference for knowledgeable veterans and beginners as well as for those crossing disciplinary boundaries and getting into the exciting field of biotechnology.

George G. Khachatourians, Ph.D.
Dilip K. Arora, Ph.D.

Editorial Board for Volume 1

Editors

Associate Editors

Contents

Preface v

Editorial Board for Volume 1 vii

Applied mycology and biotechnology for agriculture and foods 1
G.G. Khachatourians and D.K. Arora

Filamentous fungi – growth and physiology 13
R.W.S. Weber and D. Pitt

Metabolic regulation in fungi 55
G.A. Marzluf

Protein secretion by fungi 73
J.F. Peberdy, G.L.F. Wallis and D.B. Archer

Significance of fungal peptide secondary metabolites in the agri-food industry 115
D.G. Panaccione and S.L. Annis

Plant antifungal peptides and their use in transgenic food crops 145
A.E. Woytowich and G.G. Khachatourians

Clustered metabolic pathway genes in filamentous fungi 165
J.W. Cary, P.-K. Chang and D. Bhatnagar

Molecular transformation, gene cloning, and gene expression systems for filamentous fungi 199
S.E. Gold, J.W. Duick, R.S. Redman and R.J. Rodriguez

Aspergillus nidulans as a model organism for the study of the expression of genes encoding enzymes of relevance in the food industry 239
A.P. MacCabe, M. Orejas and D. Ramón

Detection of food-borne toxigenic molds using molecular probes 267
M.E. Boysen, A.R.B. Eriksson and J. Schnürer

Strain improvement in filamentous fungi – an overview 289
K.M.H. Nevalainen

Fungal solid state fermentation – an overview 305
M.K. Gowthaman, C. Krishna and M. Moo-Young

Role of fungal enzymes in food processing 353
R.K. Saxena, R. Gupta, S. Saxena and R. Gulati

Production of organic acids and metabolites of fungi for food industry 387
N.A. Sahasrabudhe and N.V. Sankpal

Index of Authors 427

Keyword Index 429

Applied Mycology and Biotechnology
Volume 1. Agriculture and Food Production
G.G. Khachatourians and D.K. Arora (editors)

Applied mycology and biotechnology for agriculture and foods

George G. Khachatourians[a] and Dilip K. Arora[b]

[a]Department of Applied Microbiology & Food Sciences, University of Saskatchewan, Saskatoon, SK, S7N 5A8, Canada

[b] Department of Botany, Banaras Hindu University, Varanasi 221 005 , India

In this chapter we highlight recent milestones of basic and applied mycology. The cross connection between applied mycology and biotechnology will give rise to newer applications of applied mycology to agriculture and food. Needless to say, developments in ancillary fields should also continue make significant impact on developments of products and processes for agriculture and food. Disciplinary crossovers of fungal genomics and genes, their regulation, expression, and engineering will have a strong impact in dealing with mycotoxins and with the rational use of mycopesticides in production agriculture. We are confident that applied mycology will continue to be an important driver of world agriculture- food and consequently trade in the years to come.

1. APPLIED MYCOLOGY FOR AGRICULTURE AND FOOD

Our relationship with agriculture and food has been a long and enduring one. Knowledge and use of plant and animal diversity helps sustain human life. In terms of abundance of species, after 6 million insects, fungi with 1.5 million species represent the next major player in the evolutionary drama of the planet. We survive in the diversity of insects, fungi and plants, in our interlocked community of organisms. The world population is expected to increase by 2 billion in the next quarter century. In context, life will be a fine balance in terms of food, space, and species interactions.

The importance of the kingdom of fungi, relates to its interactions on an equal basis with members of plants, animals and prokaryotic microorganisms. Fungal diversity, whether in structure, survival in the extremes of environment, modes of growth and proliferation, production and secretion of extracellular enzymes, peptides and secondary metabolites is consequential to agriculture and food. While the molecular biologists of the day work with a few well studied fungi such as *Saccharomyces cerevisiae*, *Aspergillus nidulans* and *Neurospora crassa* there are 72,000 species waiting in line to be studied (1).

Through advances in mycology we understand fungal organismic structure, physiology, genetic endowment, and primary and secondary metabolism and their purposes (see; chapters by Weber and Pitt; Peberdy, Wallis and Archer; and

Marzluff, in this volume). Advances in genetics and molecular biology and collective strength of the associated tools of analytical chemistry, biology and information technology allows us to probe into the mystery of the fungal world. As we have a growing base of some 10,000 fungal species in various collections and the public domain literature we have some of the relevant knowledge to cross connect to biotechnology for exploitation of fungi for manufacturing of products and deployment of processes of benefit agriculture and food. The knowledge base in mycology is in a good position to lead the new biotechnology of agriculture and food. But this knowledge is trailing so far as its use in the prevention of mycotic infections of animals, humans, foods and plant products are concerned.

2. APPLIED MYCOLOGY AND BIOTECHNOLOGY

Whether through traditional or present day biotechnological routes, fundamental discoveries of mycology and their applications have made a substantial impact in agriculture and food. Today mycology is advancing more rapidly than the past decade. We have significant new understandings of the processes involved in production and post-production agriculture and new products and those that are aided or deterred by fungi. By all criteria these trends are expected to continue.

Historically applications of mycology in agriculture and food have been exploited by many countries and ethnic groups and are now being enhanced by several tools and concepts of biotechnology. In recent years much is emerging in the developed countries of the world that serves as new learning opportunities in application of mycology for the needs of agriculture, food and environment. In general, whether in seed- or plant- development, pre- and post- harvest food crop management, transformation and value differentiation of food commodities, increased production efficiency, increased value-in-use of animal and plant food and non-food markets, mycology has a deciding role.

A large proportion of these innovations come from the new biotechnology by renewed assessment of fungal physiology, biochemistry and genetics in order to determine methods and options for manipulations at the molecular level. Outside the natural sciences, agricultural biotechnology of today also requires the convergence several disciplines from production strategies, process engineering, commerce and international law. Indeed it is the entire process of science, investment, inventions and innovations that are the interdisciplinary and transdisciplinary characteristics of the new biotechnology (2).

Advances in ingredient sub-disciplines of mycology as in the past will remain the drivers of applied agricultural research. With new interests hopefully there will be major investment focused on generating discoveries and their applications towards both conventional and biotechnology oriented useful products and processes or services.

In the sections to follow we will present certain highlights of basic and applied mycology that are exciting developments and which impact the cross connection with biotechnology for applications to agri-food. Needless to say specific developments in fields ancillary to mycology should continue to be of significant impact on the new applied mycology and biotechnology products and processes.

3. GENOMES, GENES, AND ENGINEERING

The evolutionary trend in genetic research has moved from genes, to genomics the science of studying the genome of organisms. There are three major elements, of genomics; structure, evolution and functionality. Today the genomes of half a dozen fungi are being aggressively examined. During a meeting in 1996 a group of mycologists initiated an interest in genomics, of two filamentous fungi (3) and the establishment of Filamentous Fungal Genome Centers. Already annually organized meetings on the genomics of several fungi have become established (3-5). The information on fungal genomics is accumulating and can be updated from various institutions further genomics hold direct and substantial economic ties with many industries (5).

Research in the area of functional genomics includes, molecular and structural biology, bioinformatics, combinatorial chemistry, proteomics, high throughput technologies, model fungi, transgenics and differential gene expression (see; chapters by Bhatnagar; Nevalainen; and Gold, Duick, Redman and Rodriguez in this volume). Both researchers and companies are using functional genomics to determine gene function and transfer of genetic information to particular dimensions of products and processes.

The technology push from the human genome project has had an impact on the development of a generic technology for genetic analysis, including those of fungi and bacteria. With new generations of analytical instruments and systems which speed up gene sequencing and biochip technology 100 to 200 analysis a day can be performed. Just a decade ago one to five such analyses could have been performed using conventional technology. A single nucleotide polymorphisms (SNPs) which is a single base pair mutation can be detected in amplicons ranging from 70-700 base pairs in size. Such measurements are possible with better than 90% sensitivity by e.g. Varian Inc. (Palo Alto, CA) advance technologies. Biochip technologies, such as that from Gene Logic (Gaithersburg, MD) have developed from porous glass chip with one million micro channels of 10 micron size running in 3D to analyze cRNA or cDNA and immunoassays.

Chromosomal rearrangements, recombinations and reorganizations could have very important consequences on fungal pathogens of plants and insects and their roles in applied mycology as fungal biopesticides. Fungi are unique eukaryotic organisms so far as their genomes and extrachromosomal elements are concerned. Whether sexual or asexual, fungal genomic plasticity is amazing. In some cases transposition of DNA sequences through various transposons are widely distributed and generally are found in multiple copies per plant pathogenic fungal genomes. A recent example is *Polyt1,* a member of hAT family of the transposable elements, shown to be active in the genome of the phytopathogenic fungus, *Fusarium oxysporum* (6) and the entomopathogenic fungus *Beauveria nivea* (7, 8). In *B. nivea* the transposon has sequence homology with the *restless* transposon and interestingly enough there may be only one copy in the genome (7, 8). In other cases, mechanisms such as repeat-induced point mutations or RIPs are frequently meitotically unstable creating chromosomal rearrangements. Some of these rearrangements determine purposes of genes and even entire chromosomes in phytopathogenecity of fungi (9, 10). In other fungi, such as *Penicillium chrysogenum*, short specific DNA sequences are involved in tandem reiterations leading to amplification of cluster of the penicillin biosynthetic genes and therefore genetic variability for environmental adaptation (11). These rearrangements add to plasticity of fungal genomes creating chromosome length polymorphism (CLP) as determined by electrophoretic karyotyping analysis of fungal chromosomes.

Chromosomal translocations and rearrangements occur in both natural and laboratory derived (i.e., through mutations, transformations and protoplast fusions) isolates of many fungi. Without regards for the causes, chromosomal rearrangements have been reported for *A. nidulans, Absidia glauca, Acremonium chrysogenum, Ascobolus immersus Candida albicans, Clasosporium ulvum, Cochiliobolus heterostrophus, Colletotrichum gloesporiodies, Coprinus cinereus, Erysiphe graminis, Fusarium oxysporium, Histoplasma capsulatum, Leptosphaeria maculans, Magnaportthe grisea, Magnoporthe grisea, Melamspora lini, N. crassa, Nectaria haematococca, Ophiostoma ulmi, Penicillium chrysogenum, Podospora anserina, Septoria tritici, Saccharomyces cerevisiae Schizosaccharomyces pombe, Septoria nodurum, Sordaria macrospora, Tolypocladium inflatum, Trichoderma reesei,* and *Ustelago hordei.* An important facet of these rearrangements is acquisition of a complete change in traits of cruicial impact in agriculture and food e.g., host pathogenicity or pathotype.

Extraordinary strides have been made to perfect the science of *in vitro* genetic engineering of fungi. At the molecular level there is a tremendous contrast between bacterial and fungal genetic engineering. In the latter case techniques have been much more difficult. This is in part due to difficulty of isolation of DNA, choice for vectors or multiple choices and post transformation stability of constructs and clones. We are still to enjoy the convenience of working with the *E. coli* DH5α version in any fungus. Developments such as *Pichia pastoris* vectors and expression systems (12) which have been commercialized for the production of foreign proteins (Invitrogen Corp.) represent very positive development in the field. In the near future perhaps commercial, transformation-ready fungal cells or propagules will be available. Successful examples of fungal genetic engineering include incorporation of single or multiple heterologous genes to provide simple or multiple expression of genes.

4. FROM REGULATION OF EXPRESSION TO GENE CLUSTERS

Introduced genes must be expressed at the appropriate time to be effective. Understanding of the regulation of gene expression is critical. Recent work has identified genetic elements involved in light regulation of gene expression. In addition, studies of so-called signal sequences associated with chromosomal genes whose products are transported either outside the cell or to organelles within the cell have advanced.

From the classic studies of Beadle and Tatum, in *N. crassa* we have come to understand how metabolic pathways and gene clusters are organized and expressed (13). From the organization and regulation of genes involved in L-proline catabolism, where the clustering was observed, we now recognize many more in *A. nidulans* and other filamentous fungi. Gene clusters in fungi show close linkage of a few or several genes involved in common pathways. Two broad classes of dispensable metabolic pathways are those involved in catabolism of low

Mw nutrients (e.g. nitrate, ethanol, or L-proline) and those involved in natural product synthesis (e.g. antibiotics penicillin and mycotoxins, trichothecenes, aflatoxin, sterigmatocystein, and melanin). These gene clusters which can have as many as two dozen genes and occupy 60kb of DNA, contribute to fungal survival and ecology through their shared features (13). Best studied in this context are the shared gene clusters of *A. nidulans* and *A. parasiticus* for sterigmatocystein and aflatoxin biosynthesis (see chapter by Bhatnagar).

5. FUNGAL THREATS AND BENEFITS FOR AGRICULTURE AND FOOD

5.1. Fungal threats to agriculture and food

History has several episodes of recording the negative social impact of fungi, the great potato famine of Ireland; or ergotism as Hudler's book Magical Mushrooms, Mischievous Molds indicates (14). Plant pathogenic fungi have reduced or threatened availability and safety of food. It is estimated that over 400 fungi can be considered potentially toxigenic of which about 20 are confirmed producers of mycotoxins (15). Food crops and their products and feedstuff contaminated with single or multiple toxigenic fungi are contaminated with toxic metabolite(s).

Human and animal exposure to these metabolites results in well known toxopathological manifestations and death. Diametrically opposite to singular toxins in the environment are multiple and often structurally different mycotoxins. Even at sub-threshold levels, multiple mycotoxin, by their interactions with multiple sites and targets often produce devastating synergistic effects on living cells and whole animals (16-19). Traditionally, prevention of the contamination is the first and most important strategy. In its absence, destruction of fungal propagules or spores, inactivation and decontamination of mycotoxin(s), or inhibition of absorption of mycotoxin in consumed foods or feeds in the digestive tract or skin is the option. The idea that grains contaminated with mycotoxins could be used for ethanolic fermentation during which mycotoxins will be biodegraded has merit (20). In such events, not only improvements of ethanolic fermentation have been observed, but also, levels of mycotoxins in feedstuff are significantly reduced (20,21). A more rational choice is engineering of mycotoxin resistance in cereal and other food crop plants (22). So far however, breeding of corn and cereal grain plants for resistance has been attempted but remain unsuccessful. This is possibly due to multiple modes of action of some of these toxins and hence polygenic nature of resistance (23-25). Perhaps with the isolation of target specific genes a better and fuller resistance could be achieved.

The other problem of food threat to public health comes from the fungal alkaloids leading to ergotism. The fungus, *Claviceps purpurea* is prevalent in the cool climates where rye is grown and *Clavicepes africana* which in the last few years has spread through sorghum from Brazil to USA, Australia and Africa (26). Sorghum is the fifth most important cereal crop in the world. A more recent publication (27) brings to our attention that a large number of *Fusarium* species (*F. solani, F. oxysporum, F. moniliforme, F. proliferatum, F. subglutinans,* and *F. chlamidosporum*) isolated from blood, autopsies or biopsies of organs, cerebrospinal-, bronachioalveolar lavage-, and peritoneal-fluid and, wounds of patients in Japan, a majority (29 out of 37) were mycotoxin producers. Further, all 18 of *F. solani* from this collection were cyclosporin A producers. Therefore a high degree of vigilance and equally high degree of interest in strategies for management of this fungal spread in our environments is needed. At the same time plant breeding and genetic engineering options for combating these issues of public health remains urgently needed.

Whether as phytopathogens damaging the plants in the fields, or post harvest damage of the stored grains, fruits and vegetables, fungi wreak havoc. Whether we look at the aflatoxins, trichothecenes and fuminosins, mycologists face a multiplicity of challenges. From the perspective of fungal food spoilage, the presence of particular fungi and their genetic and epigenetic factors, association with particular foods or commodities, environmental and storage conditions are the key issues. But so are the diagnostics and intervention through sanitization of the foods. Both classical and molecular schemes may be used for examining the indicators of food and feed spoilage (28-30).

DNA probes and monoclonal antibodies are proving useful for diagnosis of fungi or their metabolites-proteins in foods (29, also see; chapter by Boysen, Eriksson and Schnurer in this volume). DNA probes are also useful for diagnostics of the occurrence and spread of fungal species in foods are being used commercially. With the exception of the aflatoxins, trichothecenes etc., the identification of either the genes or mycotoxins on food commodities, in food systems and feeds has not been commercialized for rapid and user friendly, cost-effective adoption. Advances in high-throughput screening (HTS) which is currently the main domain of activity for drug discovery, may be adopted for toxicology of fungal metabolites and their presence in foods.

5.2. Fungi in pest biocontrol

A variety of fungi act as myco-herbicides, insecticides and fungicides. Several fungi have been used commercially for years as pest control agents (31). Applied mycology and biotechnology have improved the efficacy, production and formulation of these fungi. In certain cases the physiological data for the early stages of pathogenesis and identification of gene products and sequencing data are being considered for construction of hyper-virulent biocontrol agents (31,32). There is likely be a large number of additional fungal pathogens of insects and weeds to be discovered and developed to augment the current inventory of biocontrol agents.

We have learned a great deal about pathogenesis beyond the traditional 'spray and count the dead ones'. After the early stages of attachment, establishment and bonding of the germinating fungal spore on insect or plant surfaces much more molecular events are unfolded than that described in the textbooks of last 20 years. Both in phyto- and entomo-pathogenic fungi, the presence of hydrophobins, lectins, organic acids, host surface catabolic enzymes (proteases, chitinases, cutinases, lipases) and invasive (mechanical) forces play a role in both appressorium formation and hyphal peg penetration (31, 33- 36). Appressoria become structurally bonded to the host surface area by glue like substances and in plants, maturation occurs through the deposition of melanin. It is now estimated that the force exerted by the appressoria of *M. girsea* and *Colletotrichum graminicola* are about 8 and 17 micronewtons respectively (37,

38). This magnitude of force exerted over the surface area of a fungal hyphal tip of about one micrometers squared is thought to be sufficient to breach the cuticle and epidermal cell wall of plants (39). It is for certain, that these findings and those reported in this volume by Panaccione and Anis and Woytowich and Khachatourians can be applied to new concepts to promote or prevent fungal interactions with plant or insect hosts.

5.3. Food biotechnology products and processes

Research and development work on food products and processes is less advanced than that in the plant and animal area. For the most part, research to decrease process costs is just beginning. Undoubtedly, biotechnology will enable improvement in important consumer and health associated aspects of food. These may include longer shelf life, improved appearance, improved flavor, and increased perceived healthfulness of the food among others. The light-beer example noted earlier is one of the few completed products or processes in the food area.

Industrial yeasts are involved in the production of many beverages, foods, and some industrial products (40). The edible products, cheese and bread, and the potable alcohol products, beer, wine and spirits, which are over 1.5 billion liters per year all depend on yeast based fermentation (41). Greater volume than potable alcohol is the industrial and fuel grade ethanol which is produced at 24 billion liters per year (41).

Molecular genetic engineering techniques have been developed in recent years to enable the engineering of industrial yeasts. One example is the engineering of multi-ploidy in yeasts for the production of beer with reduced calories by reduction in residual starch. Normal yeasts are unable to completely convert starch to alcohol because of their inability to degrade the starch beyond its branch points. A debranching-enzyme gene has been engineered into the industrial yeast so that it can completely convert starch to alcohol, resulting in the production in a single natural step, light beer. In another case, an industrial yeast has been genetically engineered with the incorporation of a gene to enable lactose utilization, a by-product of cheese manufacture, in the production of high concentrations of ethanol, eliminating a potential pollutant and producing a useful source of energy. Other genes could be incorporated into industrial yeasts to improve the efficiencies of manufacture of cheese, bread, wine, and spirits and to make other useful products.

Besides the genetic engineering aspect of *S. cerevisiae*, fermentation of variety of substrates into alcohol has two additional loci for manipulative improvement for yield or quality of products. Ingledew (41) makes the analogy that the black box in fermentation is the manner in which interdisciplinary sciences of biochemistry, microbiology and process engineering cross cut. He argues that the 'art' of fermentation, a traditional trial and error based experience has given way to the 'science' of the alcohol fermentation. During the past 15 and the previous 100 years of the 'science' of fermentation we have learned about multifactorial issues of fermentation. The yield reducing factors are yeast growth, interfering by-product formation, contaminant bacteria and stuck fermentation (42). Interestingly enough stuck fermentation, i.e., extremely low rate of sugar utilization in nutrient limited fermentation media, can be manipulated by non-genetic means (for a full discussion see, 42). New technology development in fermentation can be through genetics or epi-genetics; either of these are valid and powerful in their contribution to agri-food industries.

In production of certain foods, food additives and ingredients, another facet of biotechnology, fermentation technology, enters the picture. While a number of fungi can be grown in liquid cultures many more depend on solid state fermentation (see chapter by Gowthaman, Krishna and Moo-Young). A number of products such as food enzymes, flavoring agents or fermented foods, various fermented beverages, and organic acids bring values of many kinds to foods (40; also see; chapter by Nevalainen, Sankpal and Sahasrabhude, Ramon, Saxena, and Revuelta in this volume).

6. APPLIED MYCOLOGY AND WORLD TRADE

Many vegetables, fruits, and seeds lose their nutritive and other qualitative values due to loss of moisture, infection with spoilage microorganisms, and senescence. These wastages occur during transport, handling and redistribution. Loss of shelf life alone, e.g. due to lack of refrigeration, is a major contributor to limited market expansion of foods. Saprophytic and pathogenic fungi are major determinants of fruit and vegetable freshness and safety. Application of antifungal peptides and antimicrobial peptides (see chapter by Woytowich and Khachatourians, in this volume) could significantly change this situation. Certain developmental genes of many fungi are expressed at defined times during growth and differentiation. As the knowledge of the regulation of fungal gene expression advances, it is expected that strains will be designed for expression of commodities of high impact in world trade. For example gene regulators that will cause expression of plant protectants at the desired time, control of growth and development of plants of ethnobotanical importance, and alteration of the composition of the harvested product are a few examples and major opportunities in fungal biotechnology for application in trade.

Ancient cultures and aboriginal people and societies based on their traditional experiences and wisdom, used many fungi for food and health uses. Indeed it is expected that transmission and decoding of such knowledge which has been refereed to ethnomycology and its confirmation, extension and utilization should offer new options for nutraceutical and pharmaceuticals (43).

Many fungi, whether single celled or filamentous have had a long relationship with human culture, food production and therefore trade. 'Generally regarded as safe' or GRAS status, has become the hallmark of United States Food and Drug Administration to indicate utility of those fungi which for millennia have had beneficial effects for the agri-food industries. In many countries, exotic or imported commodities and live material have a prerequisite quarantine requirement. With greater knowledge of the biohazards of certain fungi the World Trade Organization meeting in 1994, adopted the Agreement on Sanitary and Phytosanitary (SPS) measures (44). This and the Food and Agriculture Organization's International Plant Protection Convention (IPPC) agreement, have placed increased emphasis on science based phytosanitary regulations. As Palm (44) points out, the cut flower industry's intense competition to export flowers across the globe brings with it a danger.

What is clear is that as we move through the 21-st century, the prospects for grain storage, transportation and processing will change by the same forces that are impacting on all other facets of economy (45). Compared to a government only mode, social, environmental and economic changes will be responsibilities for everyone, whether in public or private sector. Applied mycology and biotechnology approaches can ensure that many agriculture based commodities are free from mycotoxins, mold allergens, and other problems of quality loss during storage.

7. CONCLUSIONS

Products and processes derived from the research and development efforts in applied mycology and biotechnology are growing. It is expected that these efforts should provide new inputs for agriculture and food. On the output side, it is increased productivity, higher value added and improved quality and shelf life of food products, and pest protection of crops whether for food or for non-food markets are expected. In this article several examples were highlighted and more can be found in the other chapters of this volume. These products and processes for food crops and agricultural practice are not only challenging for the scientific community interested in mycology but also for public health and commerce. From various estimates, the values of sales of mycology-based products run into $25 to $100 billion, projected for the year 2010, certainly not a insignificant figure.

From the scientific perspective, the progress made in transferring the basic knowledge to applied science and technology development, and its commercialization during the last five years has been impressive. With the ever crosscutting aspect of a holistic biotechnology education, that is integration of science, commerce, law, and end user-consumer position, this field will provide a strong leadership for new products and processes for agriculture and food. Applied mycology and its future contributions are much more advanced than most recognize, but surely in time we shall see.

REFERENCES

1. D.L. Hawksworth, Mycol. Res. 95 (1991) 641.
2. P. Phillips, and G.G. Khachatourians, (eds.), Innovation, inventions and investment- Canola as a case of knowledge led growth in agrifood industry John Wiley Press, 580pp.*(in prep'n)* 2000.
3. J.W. Benett, Fungal Genet. Biol. 21 (1997) 3.
4. L. Hamer, Fungal Genet. Biol. 21 (1997) 8.
5. R.A. Prade, Fungal Genet. Biol. 25 (1998) 76.
6. E. Gomez-Gomez, N. Anaya, M.I.G. Roncero and C. Hera, Fungal Genet. Biol. 27 (1997) 167.
7. U. Kuck, S. Jacobson and F. Kempken, In P.D. Bridge, Y. Couteaudier and J.M. Clarkson (eds.), Molecular variability of fungal pathogens, CAB International, Wallingford, 1998. pp73-82.
8. F. Kempken, S. Jacobson and U. Kuck, Fungal Genet. Biol. 25 (1998) 110.

9. V.P, Miao, S.F. Covert, and H.D. VanEtten, Science 254 (1991) 1773.
10. J.H. Ahn and J.D. Walton, Plant Cell 8 (1996) 887.
11. F. Fierro and J.F. Martin, Crit. Rev. Microbiol. 25 (1999) 1.
12. J.M. Cregg and D.R. Higgins, Can. J. Bot. 73 (1995) S891.
13. N.P. Keller and T.H. Hohn, Fungal Genet. Biol. 21 (1997) 17.
14. G.W. Hudler, Magical Mushrooms, Mischievous Molds, Princeton Univ. Press. Princeton, 1998.
15. W. De Koe, In Cereal science and technology: Impact on changing Africa, (eds.), J.R.N. Taylor, P.R. Randall and V.H. Viljoen, ZSIR, Pretoria, pp. 807-822 (1993).
16. H.A. Koshinsky and G.G. Khachatourians, Nat. Toxins 1 (1992) 38.
17. H.A. Koshinsky and G.G. Khachatourians, In Y.H. Hui (Ed. in chief), Handbook of foodborne diseases Vol. II pp.463-520 Marcel Dekker Inc., New York 1994.
18. H.A. Koshinsky, A.L. Woytowich and G.G. Khachatourians, In Foodborne disease handbook. Diseases caused by viruses, parasites, and fungi (ed.), Y.H. Hui, J.R. Gorham, K.D. Murrell and D.O. Cliver, Marcel Dekker, Inc., New York, pp. in press 2000.
19. T.J. Jones, Koshinsky, H.A. and G.G. Khachatourians, Nat. Toxins 3 (1995) 104.
20. A. Batanad R. Lasztity, Trends Food Sci. Technol.,10 (1999) 223,
21. H.A. Koshinsky, R.H. Cosby and G.G. Khachatourians, Biotechnol. Appl. Biochem. 16 (1992) 275.
22. T. Medianer, Plant Breeding, 116 (1997).201.
23. G.G. Khachatourians, Can. J. Physiol. Pharmacol., 68 (1990) 1004.
24. H.A. Koshinsky, K.T. Schappert and G.G. Khachatourians, Cur. Genet. 13 (1988) 363.
25. A.E. Woytowich, H.A. Koshinsky and G.G. Khachatourians, Plant Physiol. 114 (1997) 176.
26. R. Bandyopadhyay, D.E. Fredrickson, N.W. McLaren, G.N. Odvody, and M.J. Ryley, Plant Dis. 82 (1998) 356.
27. Y. Sugiura, J.R. Barr, D.B. Barr, J.W. Brock, C.M. Eie, Y. Ueno, D.G. Patterson Jr., M.E. Potter and E. Reiss, Mycol. Res. 103 (1999) 1462.
28. J. Schnurer, J. Olsson, and T. Borjesson, Fungal Genet. Biol. 27 (1999) 209.
29. G.G. Khachatourians and D.K. Arora. In Encyclopedia of Food Microbiology (eds.), R.K. Robinson, C.A. Batt and P. Patel, Academic Press, San Diego, 1999.
30. S. Brul and F.M. Klis, Fungal Genet. Biol. 27 (1999) 199.
31. G.G. Khachatourians, E. Valencia and G.S. Miranpuri, In O. Koul (ed.), Advances In Biopesticide Research, Harwood Academic Publ. Netherlands, (2000) in press.
33. D.D. Hegedus M. Gruber, L. Braun, and G.G. Khachatourians, In Handbook of transgenic food crops, G.G. Khachatourians, W-K. Nip, A. McHughen, R. Scorza and Y-H. Hui (eds.), Marcel Dekker, New York. 2000, in press.
33. L.B. Jeffs and G.G. Khachatourians, Can. J. Microbiol. 43 (1997) 23.
34. L.B. Jeffs, I.J. Xavier, R.E. Matai and G.G. Khachatourians, Can. J. Microbiol. 45 (1999) 936.

35. G.G. Khachatourians, In D.K. Arora, K.G. Mukerji and E. Drouhet (eds.), Handbook of applied mycology, Vol. 2, pp. 613-663, Marcel Dekker, New York. 1991.
36. G.G. Khachatourians, In The Mycota, Vol. 6 (eds.), D.H. Howard, and J.D. Miller, Springer-Verlag, Berlin. pp. 331-363. 1996.
37. N.P. Money, Can. J. Bot. 73 (1995) S96.
38. C. Bechinger, K-F. Giebel, M. Schnell, P. Leiderer, H.B. Deising and M. Bastmeyer. Science, 285 (1999) 1896.
39. R. J. Howard, M.A. Ferrari, D.H. Roach, and N.P. Money, Proc. Natl. Acad. Sci. USA., 88 (1991) 11281.
40. Y-H., Hui and G. G. Khachatourians, (eds.), Food biotechnology: Microorganisms, VCH Publ. New York, 937 pp. 1995.
41. W.M. Ingledew, In, The alcohol textbook (3rd Edition), (eds.) K.A. Jacques, T.P. Lyons and D.R. Kelsall, Nottingham Univ. Press. Nottingham. (1999) pp49-87.
42. W.M. Ingledew, In, The yeasts, (ed.), A.H. Rose, Academic Press. New York, (1993) pp245-291.
43. J. Singh and K.R. Aneja, (eds.), From ethnomycology to fungal biotechnology: Exploiting fungi from natural resources for novel products. Kluwer Academic/ Plenum Publ. New York. 293 pp. 1999
44. M.E. Palm, Mycologia 91 (1999) 1.
45. B.R. Champ, Postharvest Newsletter 47 (1998) 8.

Applied Mycology and Biotechnology
Volume 1. Agriculture and Food Production
G.G. Khachatourians and D.K. Arora (editors)

Filamentous fungi – growth and physiology

R. W. S. Weber* and D. Pitt

Washington Singer Laboratories, School of Biological Sciences, University of Exeter, Perry Road, Exeter EX4 4QG, United Kingdom

Growth and secretion are two intrinsically linked and strongly polarized processes in filamentous fungi. Growth is confined to the extreme apex where a small amount of cell wall material is kept in a plastic, deformable state whereas behind the apex, rigidification occurs by cross-linking of chitin and glucan polymers. Cell wall precursors and enzymes involved in wall synthesis and softening, as well as others secreted into the extracellular medium, are transported to the apex as the cargo of secretory vesicles. Long-distance transport from subapical sites of synthesis probably occurs along microtubules whereas the terminal stage of secretion – and with it hyphal morphogenesis – is controlled by actin microfilaments. In mammalian cells, the *trans*-Golgi network is the central vesicular traffic junction in which sorting of lysosomal and regulated secretory proteins is mediated by clathrin-coated buds, whereas bulk flow of constitutive secretion occurs by coat protein-covered buds. Whilst the vesicle coats in fungi seem to be different from those in mammals, the overall routes of membrane cycling appear similar in that excess membrane material is retrieved from the plasma membrane by endocytosis and is directed towards the vacuole, the fungal equivalent of the mammalian lysosome. Vacuoles are formed in a polarized fashion, becoming prominent in mature regions of the hypha. In addition to performing lytic functions, they are also involved in maintaining cellular homeostasis and in nutrient storage. Subapical cells are also the site at which ATP hydrolysis-dependent proton expulsion takes place, thus establishing an electrochemical gradient across the plasma membrane. The return inward movement of protons is harnessed for the active uptake of nutrients against their concentration gradient, which is a typical function of fungi and takes place mainly at the hyphal tip. As a result, an electrical field is generated which is regarded as a consequence rather than a cause of hyphal growth polarity.

---

*Present address: LB Biotechnologie der Universität, Paul-Ehrlich-Str. 23, D-67663 Kaiserslautern, Germany.

1. INTRODUCTION

Filamentous fungi are so called because their vegetative growth unit consists of a hypha which comprises a walled tube extending at one end by the incorporation of new cell wall material. Whilst such apical indeterminate growth is relatively rare among other organisms, *e.g.* in higher plants being confined mainly to root hairs and pollen tubes, hyphae represent the principal tool which has enabled filamentous fungi to assume a key role in most ecosystems (Carlile, 1995). Extension at the apex coupled with apical secretion of exoenzymes, the formation of branches, and nutrient translocation through mature hyphal segments are prerequisites for the efficient colonization of substrata such as soil, degradation even of recalcitrant substrates such as cellulose or lignin, and scavenging for nutrients. Filamentous fungi are such superior colonizers of soil that most terrestrial plants have recruited them as mycorrhizal partners (Smith and Read, 1997). Indeed, mycorrhizal fungi were already associated with the very first terrestrial plants some 400 million years ago (Simon et al., 1993; Remy et al., 1994). The ability of filamentous fungi to secrete vast quantities of extracellular enzymes, a direct consequence of the polarized organization of the hypha (Wessels, 1993), has also rendered these organisms useful tools for biotechnological purposes (Peberdy, 1994).

A second fungal growth form is the yeast state, consisting of ellipsoid or near-spherical cells which multiply by budding or, rarely, fission. Yeasts are found where penetration of the substratum is not required, *e.g.* as saprotrophs on plant surfaces or in the digestive tracts of animals (do Carmo-Sousa, 1969; Carlile, 1995). Further, certain filamentous fungi are capable of switching between the hyphal and yeast states (Gow, 1995a). Such dimorphism is common in certain animal and plant diseases in which it may aid the dispersal of the pathogen. Yeast cells are viewed as insufficiently polarized hyphae (Wessels, 1993) and will be discussed in this context where appropriate.

The hyphal growth form as the basis of nutrition by extracellular digestion has been used as a criterion to classify the fungi as a separate kingdom (Whittaker, 1969). Latterly, detailed genetic analyses have revealed this kingdom to be polyphyletic, resulting in the exclusion of the Oomycota and other groups (Alexopoulos et al., 1996). The occurrence of hyphae in several distinct evolutionary lineages may be seen as independent confirmation of the need of such structures for the associated mode of nutrition. Since Oomycota such as *Pythium*, *Phytophthora*, *Achlya* or *Saprolegnia* display typical hyphal growth and nutrition, are functionally indistinguishable from the true fungi, and have been used as important research tools by mycologists investigating hyphal growth (Bartnicki-Garcia, 1996), results obtained with them have been included in the present review, even though generalizations were approached with due caution with regards to their uncertain taxonomic status.

Hyphae from all groups of fungi share a basic pattern of ultrastructural organization in that their polarized mode of growth is mirrored by a polarized distribution of most organelles (Howard and Aist, 1979; Howard, 1981), as illustrated in Figures 1-4. A cluster of secretory vesicles occupies most of the space in the apical dome of all growing hyphae whereas other organelles are excluded (Figure 1). Extensive sheets of rough endoplasmic reticulum (ER) are often found in subapical regions of the hyphal tip cell, the major zone of biosynthesis. Energy is provided by mitochondria (Figure 1) which are also most abundant in this zone (Weber et al., 1998) and are often located slightly ahead of the ER. From the base of the main biosynthetic zone backwards, nuclei (Figure 2) are frequently and evenly distributed in the tip cell as well as in mature regions. In the septate higher fungi, vacuoles arise at the base of the tip cell (Figure 3) and become prominent further back, displacing most of the cytoplasm in mature hyphal segments (Figure 4). In the aseptate Zygomycota and Oomycota, the pattern of vacuolar distribution along the hyphal length is similar except that septa are usually absent. Behind the tip cell few further signs of differentiation of vegetative hyphae are observed except for the occasional formation of branches following the establishment of new growing tips.

Hence, 'the key to the fungal hypha lies in the apex' (Robertson, 1965). The present article will focus on the consequences of tip cell polarity for growth and physiology of the hypha. An attempt will then be made to integrate the means by which this polarity may be achieved and regulated.

2. CELL WALL SYNTHESIS

The first thorough observations on the growth of vegetative hyphae were carried out over a century ago (Ward, 1888; Reinhardt, 1892) and established that extension of the growing hypha was confined to the apex. More recently, specific radiolabelling assays have confirmed the existence of a gradient of incorporation of new cell wall material, being highest at the extreme apex and decreasing sharply subapically (Bartnicki-Garcia and Lippman, 1969; Gooday, 1971; Katz and Rosenberger, 1971). This gradient was shown to correspond to a superimposed gradient of decreasing wall plasticity from the apex backwards (Gooday and Trinci, 1980). Synthesis of wall components and rigidification of the cell wall therefore represent two closely integrated processes which will be considered below.

2.1. Chemical composition of fungal cell walls

Fungi from different taxa vary considerably in the composition of their cell walls (Wessels and Sietsma, 1981; Ruiz-Herrera, 1992). However, all hyphal walls are constructed to a similar scheme consisting of a structural scaffold of cross-linked fibres embedded in or surrounded by a matrix of gel-like and crystalline components (Sentandreu et al., 1994). In higher fungi (Asco- and Basidiomycota and Fungi Imperfecti), the two most important cell wall polymers are glucans composed of glucose units, and chitin, a polymer of β-(1,4)-linked *N*-acetylglucosamine. Several types of glucan may be found in mature walls of higher fungi; linear α-(1,3)-linked chains (Wessels et al., 1972) or chains

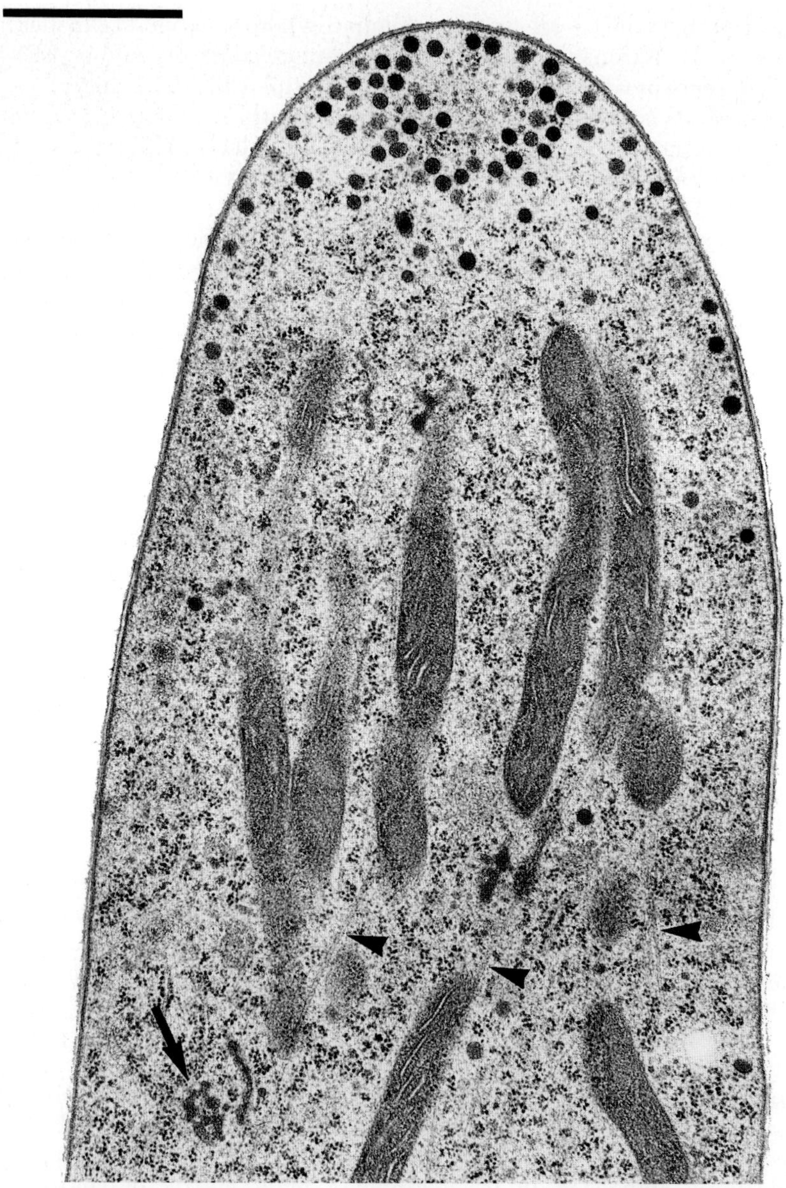

Figure 1. Hyphal tip of *Fusarium acuminatum*. Secretory vesicles are clustered in the apex whereas mitochondria are located subapically, associated with microtubules (arrowheads). A Golgi equivalent is also visible (arrow). Bar, 1 μm. Reproduced from *The Journal of Cell Biology* **87**, 55-64 (1980) by copyright permission of The Rockefeller University Press. Original print kindly supplied by R. J. Howard.

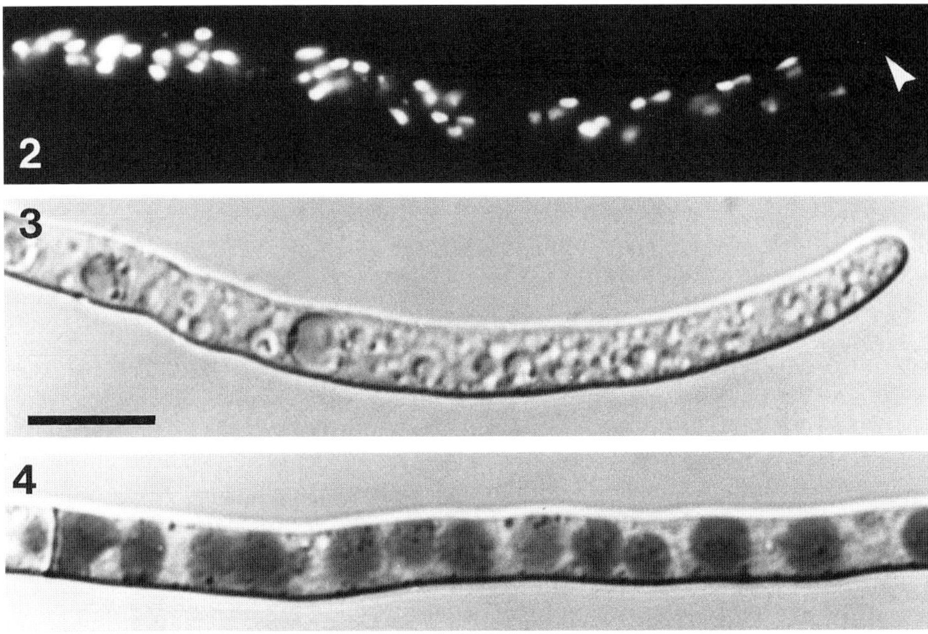

Figures 2-4. Light microscopy of living hyphae of *Botrytis cinerea* (for methods, see Weber et al., 1999). Figure 2. Distribution of nuclei in a tip cell. The position of the apical dome is indicated by an arrowhead. Figure 3. Uptake of Neutral Red into vacuoles in an apical cell. Figure 4. Uptake of Neutral Red into vacuoles in a mature hyphal segment. Bar, 10 μm.

containing a mixture of α-(1,3)- and α-(1,4)-bonds (Bobbitt and Nordin, 1982) are soluble in alkaline solutions (1 M KOH) and constitute approx. 50% of the total cell wall glucan (Sietsma and Wessels, 1994). They often form a crystalline deposit in outer wall layers (Wessels et al., 1972) and are considered to be matrix components. In contrast, the alkali-insoluble glucan fraction of mature hyphal walls consists of highly branched chains with β-(1,3)- and β-(1,6)-linkages (Wessels et al., 1990). These β-glucans are thought to have a structural function by being covalently linked to chitin (Mol and Wessels, 1987; Suarit et al., 1988; Sietsma and Wessels, 1994).

Members of the Zygomycota differ from higher fungi in that much of their chitin is modified by deacetylation, thus producing poly-β-(1,4)-glucosamine (chitosan) (Calvo-Mendez and Ruiz-Herrera, 1987). Chitosan is functionally similar to chitin because it is likewise cross-linked by other polysaccharides, in this case anionic heteropolymers containing glucuronic acid and various neutral sugars. Binding to the cationic chitosan is thought to occur by ionic interactions (Datema et al., 1977a,b).

Oomycota have traditionally been distinguished from the true fungi by the presence of cellulose (β-(1,4)-glucan) instead of chitin in their cell walls (Wessels and Sietsma, 1981), even though some of them are capable of producing chitin under certain conditions (Gay et al., 1993). In any case, cellulose is thought to fulfil an analogous structural role in that it is cross-linked by an alkali-insoluble glucan consisting of β-(1,3)- and β-(1,6)-linkages (Wessels and Sietsma, 1981). An alkali-soluble β-(1,3)-glucan is also present (Wessels and Sietsma, 1981), probably as a matrix component.

Proteins are also found in all fungal cell wall matrices, often in large quantities. Some of them are enzymes involved in cell wall formation or other functions, whilst others are non-catalytic structural proteins (Hunsley and Burnett, 1970; Burnett, 1976). These proteins may bind to each other or to other cell wall components (Marcilla et al., 1991; van Rinsum et al., 1991). Most of them are modified by glycosylation, often very extensively; *e.g.* in the yeast *Saccharomyces cerevisiae*, 90% or more of the total molecular weight of native extracellular proteins may consist of carbohydrate chains (van Rinsum et al., 1991). Mannose is by far the most abundant glycosylation sugar, and hence these glycosylated proteins are often termed mannoproteins. At least in *S. cerevisiae*, mannoproteins rather than polysaccharides are responsible for determining the overall wall porosity and thus limiting the size of molecules capable of diffusing across the cell wall towards or away from the plasma membrane (Zlotnik et al., 1984). It is likely that a similar situation holds for filamentous fungi.

2.2. Biosynthesis of the cell wall

Whereas cell wall proteins are synthesized by cotranslational translocation into the rough endoplasmic reticulum (ER), followed by glycosylation and transport to the plasma membrane *via* vesicular carriers (Peberdy, 1994), chitin is synthesized entirely at the plasma membrane (Sentandreu et al., 1994). Chitin synthase is likely to be a heteropolymeric protein complex (Merz et al., 1999) which is active only when integrated into a lipid membrane (Montgomery and Gooday, 1985). It is delivered in a zymogenic form in specialized structures termed chitosomes (Bartnicki-Garcia et al., 1979). The enzyme accepts its substrate, UDP-*N*-acetylglucosamine, at the cytoplasmic side of the plasma membrane and extrudes a β-(1,4)-linked *N*-acetylglucosamine chain into the nascent cell wall. Fungi may possess several chitin synthase genes which encode enzymes with different cellular functions but have at least a limited ability to compensate for each other such that deletion mutants in all but one chitin synthase gene may still be viable (Bulawa, 1993; Borgia et al., 1996).

In contrast, several different enzymes are required to act in concert to synthesize the structurally complex glucans. Enzymes catalyzing early steps of glucan synthesis are located in the ER (Meaden et al., 1990) and in subsequent stations of the secretory route including the plasma membrane (Roemer and Bussey, 1991; Wessels, 1993). Thus, partially-formed glucan molecules are secreted into the nascent cell wall, along with further glucan synthases which complete glucan cross-linking in the wall (Boone et al., 1990; Hartland et al.,

1991). Nucleotide-glucose seems to serve as the universal substrate at least for chain elongation reactions, UDP-glucose probably being the *in vivo* donor (Sentandreu et al., 1994).

An elegant series of pulse-chase experiments by Wessels and his group, using tritiated glucose or *N*-acetylglucosamine with growing hyphae of the basidiomycete *Schizophyllum commune*, has revealed a plausible course of events leading to cross-linking of chitin and glucan polymers in the nascent wall (Wessels et al., 1990). Immediately after an appropriate pulse, all labelled glucan was found to be alkali-soluble and, likewise, most labelled chitin could be solubilized by chitinase treatment. Later, when the label was located in subapical regions, chitinase-resistant chitin and alkali-insoluble β-(1,3)-glucan with abundant β-(1,6)-cross-links were detected (Wessels et al., 1983; Sietsma et al., 1985). Glucans become alkali-insoluble when they are covalently linked to poly-β-(1,4)-*N*-acetylglucosamine chains which themselves become resistant to chitinase attack only after self-assembly into chitin microfibrils (Sietsma and Wessels, 1994). The delay in the crystallization of these microfibrils may arise partly because the chains contained within must be arranged in an anti-parallel orientation relative to each other, yet they are synthesized and extruded unidirectionally (Sentandreu et al., 1994). Further, binding of small molecules such as Calcofluor White inhibits crystallization (Vermeulen and Wessels, 1986), and similar reversible interactions might regulate chitin polymerization in growing hyphae. This delay in chitin microfibril formation is important because it permits the establishment of cross-linking bonds with glucans, which in turn are thought to be vital for the generation of a tough, turgor-resistant yet elastic mature cell wall.

Much to the frustration of biotechnologists, many fungi release highly viscous glucans into liquid culture media. These are often of the β-(1,3)- and/or β-(1,6)-type (Gorin and Spencer, 1968; Sietsma et al., 1977) and may represent partially-synthesized glucan products which escaped by diffusion before they were firmly incorporated into the nascent cell wall. A famous case is *Botrytis cinerea* (Figures 5 and 6) which can produce glucan slimes capable of ruining botrytified wines (Dubourdieu et al., 1978a,b). Such glucans may be utilized as extracellular carbon reserves because they are degraded when appropriate glucanases are secreted (Dubourdieu and Ribéreau-Gayon, 1980).

From the above it is apparent that cell walls of submerged and aerial hyphae may differ in certain aspects of their carbohydrate composition. This is true also with respect to the protein components of the wall, particularly the hydrophobins, a class of small, usually unglycosylated proteins, some of which are constitutively produced and secreted by the growing apex of submerged as well as aerial hyphae of *S. commune* (Wessels et al., 1991a). From submerged hyphae, they diffuse into the liquid growth medium in a monomeric form whereas on a surface exposed to air, polymerization by non-covalent hydrophobic protein-protein interaction occurs (Wessels, 1997). In such cases, the cell wall is coated by a rodlet layer of hydrophobins forming an extremely water-repellant outer surface. Other hydrophobins are developmentally regulated and have been

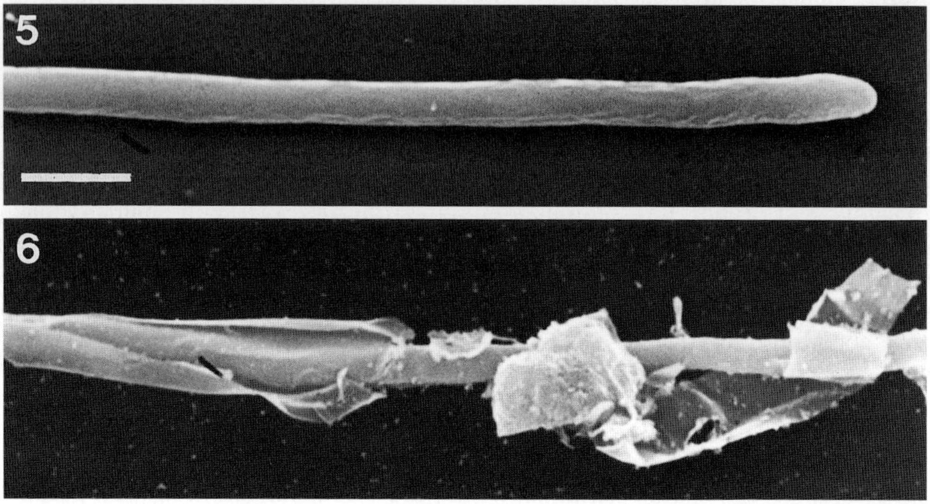

Figures 5 and 6. Comparison of an aerial (Figure 5) with a submerged (Figure 6) hypha of *Botrytis cinerea*. An extracellular slime sheath has been secreted only by the latter. Bar, 10 μm.

shown to be important for the differentiation of hyphae into infection structures such as appressoria (St. Leger et al., 1992; Talbot et al., 1996), the production of dry conidia (Stringer et al., 1991), and the aggregation of hyphae into basidiomycete fruiting bodies (Wessels et al., 1991b; Wessels, 1997).

2.3. The role of lytic enzymes in cell wall formation

From the mechanism of glucan-chitin cross-linking in *S. commune* as described above, hyphal tip extension could be accounted for solely by the presence of a steady-state amount of plastic wall material in the apical dome. However, such a mechanism is unlikely to function on its own, given the fluctuations in the external environment to which the nascent cell wall is exposed. Ward (1888) had already speculated that a balance between biosynthetic and lytic activities might be necessary for apical growth in fungi. A development of this idea, the unitary growth model, has been championed especially by Bartnicki-Garcia (1973, 1996). It suggests that cell wall extension can occur anywhere in the presence of hydrolytic enzymes which soften the wall, synthesizing enzymes which produce new wall material, and turgor pressure which drives expansion of the new wall. Thus, extension of existing apices and formation of new ones by branching can be accommodated in this model.

Fungi do produce chitinases as well as glucanases capable of hydrolysing their own cell wall polymers, and both enzyme types have been found in growing hyphal tips (Notario, 1982; Gooday and Gow, 1990). An apically secreted *endo*-β-

(1,3)-glucanase in *Aspergillus fumigatus* has recently been shown to be capable of *in vivo* hydrolysis of β-glucans which are important as cross-links between chitin fibrils (Fontaine et al., 1997). Two fundamentally different types of chitinolytic enzymes have been characterized in *Mucor* spp. (Rast et al., 1991; Horsch et al., 1997). The first type consists of at least two *endo*-acting chitinases capable of degrading either nascent or more highly polymerized chitin, thereby providing the lytic activity as envisaged by the unitary model. However, a second type of chitin-degrading enzyme, the *exo*-acting *N*-acetylhexosaminidase, has also been found in *Mucor*. Since this enzyme releases *N*-acetylglucosamine, an allosteric activator of chitin synthase, from chitin fragments released by the *endo*-chitinase, it has been proposed to act as a mediator between chitin synthase and *endo*-chitinase, thereby raising the possibility of feedback regulation of the activities of these two enzymes (Horsch et al., 1997; Merz et al., 1999). In true fungi, there is therefore substantial evidence for the role of cell wall hydrolytic enzymes in apical morphogenesis.

Direct proof of such a role has recently been achieved with *Saprolegnia* and *Achlya*. These Oomycota are incapable of adjusting their osmotic pressure in response to hyperosmotic stress (Money, 1994), in contrast to most true fungi (Jennings, 1995; Förster et al., 1998). Hence, the turgor pressure in hyphae of *Achlya* and *Saprolegnia* shows a linear decrease with a correspondingly increasing osmotic strength of the growth medium (Money and Harold, 1992). In this situation substantial hyphal growth was still observed at greatly reduced turgor pressure and even in the absence of measurable turgor (Money and Harold, 1992), due to a softening of the apical cell wall (Kaminskyj et al., 1992a; Money and Harold, 1993; Money, 1997). This in turn was paralleled by an increase in the secretion of *endo*-β-(1,4)-glucanase (Money and Hill, 1997), thought to be involved in breaking glucan cross-links between cellulose microfibrils. In contrast, the secretion of other enzymes such as amylase and protease was not stimulated. It would clearly be of great interest to investigate the hyperosmotic growth response of mutants of true fungi compromised in osmotic adjustment or turgor generation.

Intercalary (non-polarized) growth is a rare phenomenon in fungi, but it does occur at a strikingly fast rate in hyphae located in the expanding stipe of basidiomycete fruit bodies such as the ink-cap, *Coprinus cinereus* (Moore et al., 1979). In this instance, hydrolytic enzymes would seem to be essential to permit the extension of a pre-formed, rigid wall. Indeed, good evidence exists for the involvement of an *endo*-β-(1,3)-glucanase in stipe elongation (Kamada et al., 1985) which presumably acts by breaking the glucan chains interconnecting the chitin microfibrils (Kamada, 1994).

2.4. The Vesicle Supply Centre and the Spitzenkörper

The final transport stage of wall matrix proteins and enzymes involved in the synthesis and lysis of cell wall polymers, as well as any pre-formed glucan material, is known to consist of vesicles which fuse with the plasma membrane (see Figure 1 and Section 3.1.). On the assumption that these vesicles are first collected by a vesicle supply centre (VSC) before being emitted, Bartnicki-Garcia and colleagues created a computer simulation of hyphal growth, the hyphoid model (Bartnicki-Garcia et al., 1989, 1990). Growth can thus be simulated when the VSC is moved forward in a linear fashion. Assuming a constant rate of vesicle emission, a high speed of movement would result in a hyphal tube whereas a slow speed would produce a yeast-like cell. Further, the shape of the apex of the mathematical hypha thus obtained matches exactly that found in most true fungi (Bartnicki-Garcia, 1996).

Structures located in a central position within the apical dome, as assumed by the hyphoid model, had first been observed fortuitously by Brunswik (Brunswik, 1924) who coined the term Spitzenkörper ('apical body'). Subsequently, these structures were found to consist of a dense cluster of vesicles (Figures 1 and 10). However, their functional characterization had to await the advent of computer-enhanced video-microscopy techniques. Using such techniques, the morphological Spitzenkörper was found to behave like a mathematical VSC; it was unfailingly present in growing hyphae yet absent from non-growing ones (López-Franco and Bracker, 1996), and its lateral displacement from the centre of the apical dome preceded a corresponding change in the growth direction of the hypha (Riquelme et al., 1998). Further, time-lapse video microscopy revealed that most hyphal tips do not extend in a linear fashion, but by a fairly regular sequence of pulses which occur at an interval of several seconds. This pulsed growth was found to be caused by bursts of secretion due to fusion events between subapical Spitzenkörper satellites and the main apical Spitzenkörper (López-Franco et al., 1994). Delay in the anterograde movement of a satellite Spitzenkörper by only a few seconds resulted in the formation of a lateral bulge (López-Franco et al., 1995), thus providing evidence for the direct involvement of the Spitzenkörper in cell wall morphogenesis.

These results indicate that a key control element over the rate and direction of hyphal growth must reside inside the hypha, in the apical cytoplasm.

3. THE CYTOSKELETON

There are certain conceptual difficulties (Heath, 1994; Kaminskyj and Heath, 1996) associated with models which place the overall control of hyphal tip growth in an extracellular domain such as the cell wall, be it the model of a steady-state amount of plastic cell wall material at the apex, or the interplay between wall synthesis and lysis as proposed in the unitary model. It would also be difficult to envisage how the apical cell wall can be sufficiently plastic to yield to the turgor pressure which presumably drives hyphal growth in most situations, yet strong enough to resist uncontrolled expansion or rupture – and that in widely differing environmental conditions. Indeed, a wealth of observations points to the cytoplasm within the apical dome of the growing hypha as the key control element of morphogenesis. For instance, hyphal tips subjected to mechanical stress often burst at the base of the extension zone (Reinhardt, 1892; Sietsma and Wessels, 1994), not at the extreme apex which one would expect to be the most plastic region of the wall. The malleable apical cell wall thus seems to be protected by the underlying cytoplasm, which has a growth-restraining function in this case. And yet, the same cytoplasm must be capable of generating force to extend the growing tip in the absence of turgor, at least in *Saprolegnia* and *Achlya* (Section 2.3.). Finally, transport of material from the subapical sites of biosynthesis to the growing apex must be achieved.

In eukaryotes, transport and structural roles at the level of individual cells are commonly carried out by the cytoskeleton. Microtubules (MTs) with microtubule-associated proteins (including the molecular motors dynein and kinesin) as well as actin filaments (AFs) with actin-binding proteins (including myosin motors) have been characterized in filamentous fungi and yeast (Heath, 1994, 1995a; Steinberg, 1998). Intermediate filaments, which fulfil important structural roles in mammalian cells, have also been reported in fungi (Rosa et al., 1990) but are probably of lesser importance, not least because the force of turgor pressure against the rigidified mature cell wall is thought to be sufficient to maintain cell shape (Heath, 1995a). The distribution and putative roles of MTs and AFs in hyphal tip polarity will be explored below.

3.1. Localization of the cytoskeleton in the apical hyphal cell

Since MTs are subjected to extensive depolymerization during incubation of hyphae in chemical fixatives such as formalin or glutaraldehyde, their true abundance was appreciated only after the introduction of the more effective freeze-substitution technique some two decades ago (Howard and O'Donnell, 1987; Bourett et al., 1998). Using this approach, especially when coupled with immunolocalization, MTs have been shown to be arranged parallel to the hyphal growth axis in Oomycota (Heath and Kaminskyj, 1989) as well as true fungi (Howard and Aist, 1979; Howard, 1981; Runeberg et al., 1986), extending from subapical regions into the apical dome, and sometimes even touching the plasma membrane at the extreme apex (Figures 1 and 9). Shorter MTs are also sometimes observed, often in cross-links with AFs (Heath, 1994). Of course, the mitotic spindle during nuclear division is also formed by MTs (Heath, 1995a).

Whilst MTs seem to be localized in a uniform pattern in apical cells of all fungal hyphae, the distribution of actin is considerably more complicated. Much relevant information has been obtained by Heath and colleagues working on *Saprolegnia ferax*. Using the fluorescent actin-binding dye phalloidin-rhodamine, hyphae have been shown to contain a tip-high actin gradient (Heath, 1987; Heath, 1995b). In the apex, actin is present as a dense cap composed of interwoven filaments whereas subapically, discrete cortical actin patches and cables become more prominent (Figure 7). Further, a core of fibrillar actin exists in central regions of growing hyphae (Jackson & Heath, 1993a).

In true fungi, actin is likewise more concentrated in hyphal tips (Cali et al., 1998) and a central actin core has been characterized (Runeberg et al., 1986; Salo et al., 1989), but filamentous caps are not usually observed. Instead, the centre of the Spitzenkörper (Figure 10) is now known to contain a pool of polymerized

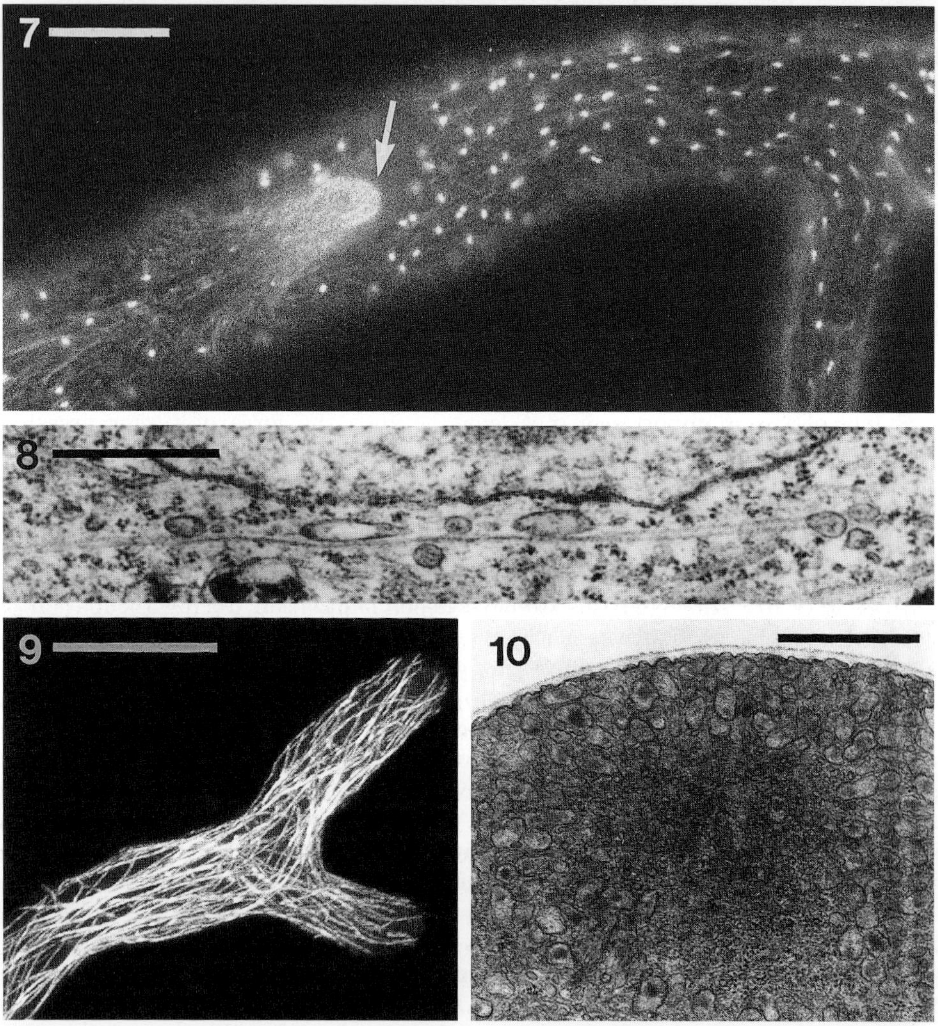

Figures 7-10. The cytoskeleton in filamentous fungi. Figure 7. Actin staining in *Saprolegnia ferax*. Actin fibres and punctate plaques are prominent in a mature hypha whereas a dense actin cap is seen in the the tip of a new lateral branch (arrow). Bar, 10 μm. Original print kindly supplied by I. B. Heath. Figure 8. Association of secretory vesicles with a MT in *Botrytis cinerea*. Bar, 0.5 μm. Figure 9. MTs in a branched hyphal apex of *Rhizoctonia solani* as stained by an α-tubulin antibody. Bar, 10 μm. Reproduced from *Fungal Genetics and Biology* **24,** 3-13 (1998) by copyright permission of *Academic Press*. Original print kindly supplied by R. J. Howard. Fig. 10. The Spitzenkörper of *Botrytis cinerea*, consisting of large secretory vesicles surrounding a core of microvesicles (chitosomes). Bar, 1 μm.

actin (Czymmek et al., 1996; Srinivasan et al., 1996), and actin has also been localized to cortical patches which probably represent the coats of vesicles termed filasomes (Howard, 1981; Bourett and Howard, 1991; Srinivasan et al., 1996). The function of filasomes is unknown; they have been speculated to be of endocytotic origin (Srinivasan et al., 1996) but are unlikely to be directly involved in determining hyphal polarity. As yet, there is no convincing ultrastructural evidence for any sites of direct actin attachment to the plasma membrane in true fungi (Kaminskyj and Heath, 1996), even though their existence may be inferred from physiological studies (see Section 3.2.).

Clearly, more data are required to clarify the pattern of actin localization especially in true fungi. Inconsistencies may arise because AFs are even more labile than MTs and are thus particularly sensitive to depolymerization during experimental manipulation (Heath, 1995a).

3.2. Protection of the growing apex

A pivotal morphogenetic role of apical actin filaments in walled cells was first proposed for pollen tube growth (Picton and Steer, 1982). The model envisaged anchorage of an apical actin cap to the subapical rigid cell wall across the plasma membrane by a rivet-like mechanism. Such anchoring proteins, called integrins, were indeed subsequently shown to exist, first in amoeboid cells in which they provide a direct link between the cytoskeleton and the environment (Hynes, 1992). Integrin homologues were also found in a range of polarized plant cells (Goodner and Quatrano, 1993; Wyatt and Carpita, 1993) and in filamentous fungi (Kaminskyj and Heath, 1995; Corrêa et al., 1996). *Saprolegnia ferax* with its well-characterized actin distribution has provided a particularly amenable system for such studies. Integrin is distributed as patches from the wall extension zone backwards into the rigidified region of the apical cell. These patches were shown to provide a direct link between the actin cytoskeleton, the plasma membrane and the cell wall (Kaminskyj and Heath, 1995). Therefore, the actin cap in *Saprolegnia* is firmly anchored in the rigid subapical cell wall (Kaminskyj and Heath, 1996). The cytoplasmic face of the apical plasma membrane of *Saprolegnia* is further protected by the deposition of another structural protein, spectrin (Heath, 1995b; Kaminskyj and Heath, 1995). The apical dome of *Saprolegnia* is thus constructed on an internal actin-spectrin scaffold which is riveted to the cell wall by integrin. Presumably this scaffold moves forward as soon as the soft anterior wall-integrin connections have hardened (Figure 11).

3.3. Regulation of apical tip extension

In *Saprolegnia*, turgor pressure might be harnessed for tip growth in a controlled fashion by adjustment of the tensile strength of the AFs linking the apical actin cap to the subapical cell wall. In this way, a steady extension of the tip may be achieved whilst at the same time avoiding bursting. Conversely, at reduced turgor pressure, the actin skeleton might employ a different set of molecular motors in order to push the apical cap forward. This, together with the

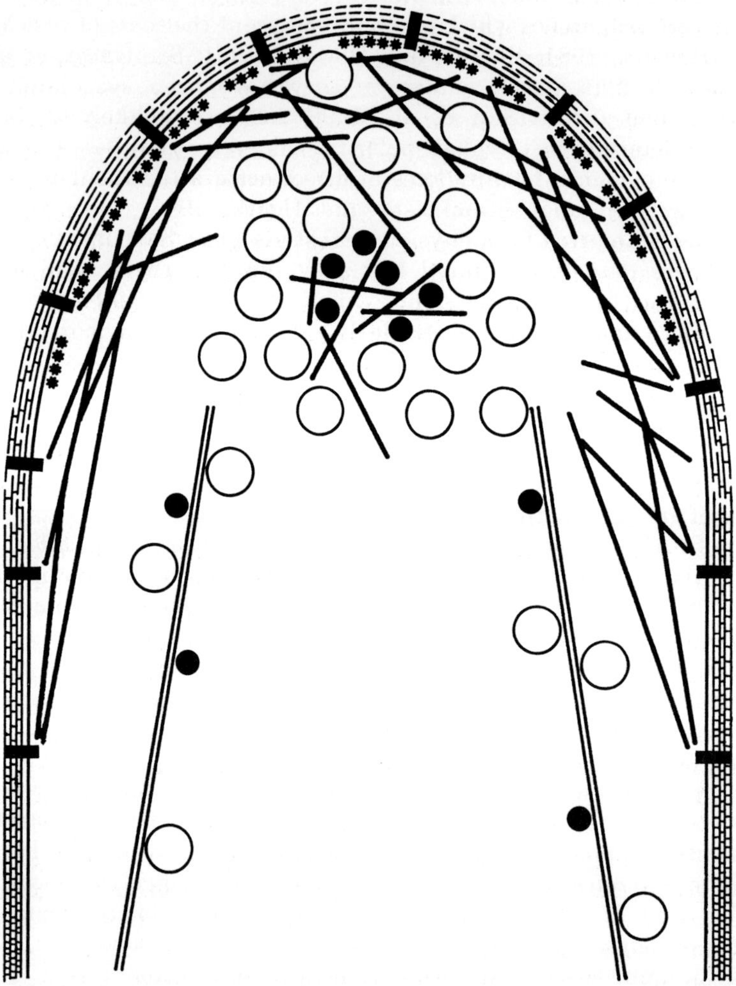

Figure 11. The putative involvement of cytoskeletal elements in hyphal tip growth of fungi. Transport of secretory vesicles (hollow circles) and chitosomes (solid circles) into the apex is mediated by MTs (double lines) whereas vesicle fusion with the plasma membrane is regulated by AFs (thick lines) which form an apical cap and are also found in the Spitzenkörper core. The actin skeleton is anchored to the rigid subapical cell wall by integrin connections (rectangles) which span the plasma membrane. The apex is further strengthened by deposits of spectrin lining the inner surface of the plasma membrane (asterisks). The degree of cell wall rigidification is indicated.

enhanced secretion of wall-softening enzymes (Section 2.3.), might be responsible for achieving growth even in the absence of turgor.

The capacity of actin fibres to contract in response to elevated concentrations of cytoplasmic Ca^{2+}, brought about either directly or *via* its capacity as a second messenger in response to a primary stimulus, is a universal feature among eukaryotes. When such situations are brought about experimentally in fungal hyphae, the cytoplasm of *Saprolegnia* contracts towards the apex at which it remains firmly attached (Jackson and Heath, 1992; Kaminskyj et al., 1992b). Similar observations have been made for several true fungi (McKerracher and Heath, 1987; Jackson and Heath, 1992). In higher fungi, disturbances by irradiation caused a forward-directed contraction of the subapical cytoplasm whilst pulling the Spitzenkörper backwards (López-Franco and Bracker, 1996). The direction of both migratory movements would seem to point at the very area in which, in *Saprolegnia*, the actin cytoskeleton is anchored to the cell wall. Integrin is known to be located in growing apices of higher fungi (Corrêa et al., 1996) as it is in *Saprolegnia*, and may thus be linked to actin in a similar way, even though the distribution of AFs is not as well characterized as in *Saprolegnia*.

The existence of apex-high Ca^{2+} gradients has been demonstrated for a range of fungi (Garrill et al., 1993; Jackson and Heath, 1993b; Gow, 1995b), even though their involvement in hyphal growth is controversial (Harold, 1994). Such gradients are created by the influx of Ca^{2+} through appropriate channels which are concentrated in the apical plasma membrane at least in *Saprolegnia* (Garrill et al., 1993; Jackson and Heath, 1993b). Of particular interest in this respect are the stretch-activated Ca^{2+} channels whose permeability to Ca^{2+} increases when the plasma membrane is under tension (Zhou et al., 1991; Garrill et al., 1993). An elegant if highly speculative model to account for the regulation of tip growth in *Saprolegnia* has been proposed by Heath (1995b) who suggested that the influx of Ca^{2+} through the stretch-activated Ca^{2+} channels might result in an elevated concentration of Ca^{2+} in the apical cytoplasm, which could cause localized contraction of the actin skeleton, which in turn would slow apical extension and at the same time close the Ca^{2+} channels. Gradual constitutive sequestration of Ca^{2+} into its known intracellular stores such as the ER, mitochondria and vacuoles (Pitt and Barnes, 1993; Calvert and Sanders, 1995), all of which are located in subapical or mature regions (see Figures 1, 3 and 4), might reduce the apical Ca^{2+} concentration, leading to a relaxation of the actin skeleton and a resumption of apical extension. Polarity of the hypha would thus be maintained by incorporation of Ca^{2+} channels by secretion into the apical plasma membrane and their confinement there or subapical inactivation or removal (Harold, 1994; Gow, 1995b).

It is impossible at present to assess whether such a model might also apply to true fungi. In *Neurospora crassa* as in other fungi, an apex-high Ca^{2+} gradient is always associated with growth but the stretch-activated Ca^{2+} channels are not, and neither are they concentrated at the tip (Levina et al., 1995; Lew, 1998).

Indeed, it is not yet altogether certain whether Ca^{2+} is an essential nutrient in filamentous fungi (Harold, 1994).

3.3. Vesicle transport to the growing apex

Secretory vesicles are produced in the subapical zone of biosynthesis and must migrate over several micrometres before they fuse with the apical plasma membrane (McClure et al., 1968). Hyphae of several filamentous fungi treated with MT-depolymerizing agents such as MBC (methyl benzimidazole-2-yl-carbamate) or nocodazole, ceased to grow and lost their parallel arrays of MTs. This coincided with the disappearance of the Spitzenkörper, an even redistribution of secretory vesicles throughout the cytoplasm, and displacement of mitochondria from their site of accumulation in the subapical region [Howard and Aist, 1977, 1980; Rupes et al., 1995). In untreated control hyphae, MTs were found to be closely associated with secretory vesicles and mitochondria (Howard, 1981; see also Figures 1 and 8). The involvement of MTs in polarized secretion has been demonstrated also with the secretion of extracellular enzymes, which travel the same route as cell wall material (Hill and Mullins, 1980; Wösten et al., 1991). In the presence of MBC, MTs of *Aspergillus nidulans* were completely depolymerized and secretion of invertase was significantly reduced in intact hyphae, but not in protoplasts, even though invertase synthesis itself was unaffected by MBC in either system (Jochová et al., 1993). The above results imply a role for MTs only in long-range transport in filamentous fungi. Direct evidence for the movement of secretory vesicles along MTs has been obtained by immunolocalization of a kinesin motor with MTs and vesicles in *Neurospora crassa* (Steinberg and Schliwa, 1995). Further, mutant studies have shown that kinesin-type motors are essential for polarized growth in filamentous fungi (Lehmler et al., 1997; Seiler et al., 1997). In marked contrast, in budding cells of *S. cerevisiae*, in which the distance between the sites of vesicle formation and discharge is short, such motors are absent (Steinberg, 1998) and MT depolymerization has no effect on wall growth. Instead, functional MTs are mainly required for nuclear functions (Adams and Pringle, 1984; Huffaker et al., 1988; Makarow, 1988).

In both *S. cerevisiae* and filamentous fungi, actin depolymerization has massive deleterious effects on polarized growth and secretion. In the former, a correlation between AFs and the budding of daughter cells has been obtained especially by analysis of actin mutants (Johnston et al. 1991; Govindan et al., 1995; Cali et al., 1998). In filamentous fungi, the role of actin has been investigated mainly by using actin-depolymerizing agents such as cytochalasins (Betina et al., 1972). In hyphae treated with these compounds, apical growth was disrupted and hyphal tips swelled to yield giant spheres much exceeding the diameter of the original hypha (Betina et al., 1972; Sweigard et al., 1979; Srinivasan et al., 1996].

The conclusion to be drawn from the above results is that MTs deliver secretory vesicles from their subapical sites of synthesis to the apex, whereas the fusion of vesicles with the plasma membrane is controlled by AFs. A somewhat similar pattern of locomotion has been observed in polarized secretion systems in animal cells, such as neuronal axons. Anterograde movement of secretory vesicles from the cell body into the synaptic terminal is constitutive and mediated by MTs (Vale and Goldstein, 1990) whereas the discharge of vesicles in response to appropriate stimuli is governed by Ca^{2+}-dependent actin-myosin interactions in cortical regions of the synapse (Whitaker & Baker, 1983).

In filamentous fungi, such a pattern would tie in with the presence of MTs along the hyphal growth axis and the abundance of actin in the centre of the Spitzenkörper and in the apical cortex (Sections 3.1. and 3.2.). The structural role of the apical actin cap in tip morphogenesis can be reconciled with its additional role in regulating vesicle fusion because the same cytoskeletal elements may bring about different specific movements using different molecular motors. Indeed, a great diversity of such motors has now been described in filamentous fungi (Steinberg, 1998). Naturally, interactions occur between MT and AF components of the cytoskeleton in fungal hyphae, which may obscure the analysis of effects of inhibitors (Torralba et al., 1998). In fact, so intrinsic is this interplay between the two main cytoskeletal elements that the causes and mechanisms of tip polarity are still a matter of great controversy (Heath, 1994; Gow, 1995b).

3.4. Polarity of organelle distribution

Several organelle types other than secretory vesicles are also distributed in a non-random arrangement in fungal hyphae, typically appearing some distance behind the growing apex. By far the greatest body of knowledge exists on the distribution and migration of fungal nuclei which is controlled by MTs (for references, see Heath, 1994; Steinberg, 1998). Nucleus-associated MTs either arise directly on the nuclear membrane from spindle pole bodies, or they run parallel to the hyphal wall and the nucleus is associated laterally with them. As in other eukaryotes, mitosis likewise relies on MTs which form the nuclear spindle; however, fungi are unusual in that the nuclear membrane remains intact throughout the separation of chromosomes. At least in certain Basidiomycota, nuclei are also surrounded by AFs, but instead of contributing to nuclear migration, this actin cage is thought to be involved in monitoring the position of nuclei in response to gravity, thus mediating the gravitropic response that maintains the upright position of mushroom-type fruiting bodies (Monzer, 1996; Moore et al., 1996).

Mitochondrial movement and positioning is determined by AFs in the budding yeast *S. cerevisiae* (Simon et al., 1995) whilst it is dependent on MTs in the fission yeast, *Schizosaccharomyces pombe* (Yaffe et al., 1996). In filamentous fungi, similarly, both AFs (Oakley and Rinehardt, 1985) and MTs (Aist and Bayles, 1991; Steinberg and Schliwa, 1993) have been implicated in mitochondrial movement. Such conflicting results may be explained by the functional redundancy which is often found when transport functions are analyzed by deletion of molecular motors; in many such instances, several different motors are able to perform the equivalent function (Steinberg, 1998).

Elements of the endomembrane system such as the ER and vacuoles are also positioned in a recognizable pattern along the length of the hypha. Little is known about the cytoskeletal factors involved; when hyphae of *S. commune* were subjected to brefeldin A, a compound which inhibits the anterograde membrane flow out of the ER, extensive sheets of ER formed and were associated with MTs (Rupes et al., 1995). This observation, together with the longitudinal alignment of the subapical ER sheets as well as MTs, may be taken as tentative evidence of an association between the ER and MTs. Similarly, vacuole morphology is dependent on functional MTs in *S. cerevisiae* (Guthrie and Wickner, 1988) and has been shown to be determined by MT-associated kinesin in the basidiomycete *Ustilago maydis* (Steinberg et al., 1998).

4. SECRETION AND MEMBRANE TRAFFICKING

Secretion especially of extracellular enzymes is of seminal importance in fungal biotechnology, as demonstrated by the dedication of several chapters of the current volume to aspects of this topic. Protein secretion is initiated in the rough ER where, in a pattern broadly conserved between mammalian cells (Lingappa et al., 1984) and *S. cerevisiae* (Haguenauer-Tsapis and Hinnen, 1984), polypeptide chains are synthesized from their mRNA templates and at the same time translocated into the membrane or lumen of the ER. The affinity of a nascent polypeptide for the ER is determined by a signal peptide, usually situated at the extreme *N*-terminus which comprises the very first amino acids to emerge from the ribosome (Connolly and Gilmore, 1989). From this moment of translocation into the ER until reaching its final destination, the nascent protein remains confined to the lumen or membrane of vesicular structures whose movement determines its fate. In mammalian cells and in *S. cerevisiae*, the secretory route represents the default pathway (Pfeffer and Rothman, 1987), *i.e.* a protein translocated into the rough ER by its signal peptide will be secreted unless it contains further signals which allow it to be recognized and sifted from this bulk flow, usually by receptor-mediated processes.

In the following sections, an attempt will be made to piece together the existing information on routes of membrane traffic in the apical cell of the filamentous fungus hypha, at times borrowing heavily from results obtained for mammalian systems (Figure 12) and *S. cerevisiae*.

4.1. Constitutive transport from ER to Golgi apparatus

In eukaryotes, most proteins that have been at least in transient contact with the ER are glycosylated, unlike cytoplasmic proteins. Glycosylation can be of two types – *N*-glycosylation, in which an $(N\text{-acetylglucosamine})_2(\text{mannose})_9(\text{glucose})_3$ core chain is transferred onto selected asparagine residues, occurs co-translationally as soon as the growing polypeptide contacts the ER lumen (Kornfeld and Kornfeld, 1985). The process of core glycosylation is conserved between mammals, *S. cerevisiae* and filamentous fungi (Lehle, 1981). *N*-glycosylation chains may contribute to the correct folding of the nascent polypeptides in the ER lumen (Olden et al., 1979). In mammals and *S. cerevisiae*, the exposed hydrophobic regions of incorrectly folded proteins are recognized by attachment of molecular chaperones such as protein disulphide isomerase (PDI) and the binding protein BiP, which are themselves inducible by an abundance of malfolded proteins in the ER (Kozutsumi et al., 1988; Tachikawa et al., 1995). They are also inducible by conditions likely to cause abundant malfolding, such as heat-shock or inhibition of *N*-glycosylation (Pouyssegur et al., 1977; Hendrick and Hartl, 1993). Binding of these chaperones to malfolded proteins may assist their folding or, failing that, mark them for retention in the ER (Pelham, 1989). PDI homologues have recently been described also for filamentous fungi (Jeenes et al., 1997).

Whilst the ER-mediated steps of *N*-glycosylation are conserved between different eukaryotic organisms, the second type – *O*-glycosylation – is not. It consists of the addition of a linear chain of mannose moieties to serine or threonine residues which occurs exclusively in the Golgi apparatus of mammalian cells (Johnson and Spear, 1983) whereas in yeast and filamentous fungi, the first sugar is added in the ER (Soliday and Kolattukudy, 1979; Haselbeck and Tanner, 1983). The role of *O*-glycosylation in protein folding is uncertain.

Unless a protein is retained in the ER either selectively by recognition of an appropriate signal sequence, as in the case of ER-resident proteins (Pelham, 1989; Lewis et al., 1990) or by non-specific chaperone binding in the case of malfolded proteins, it will move forward towards the Golgi stack by non-selective bulk flow (Pfeffer and Rothman, 1987). This process has been particularly well described for *S. cerevisiae*. Budding of vesicles from the ER membrane is a constitutive process and is mediated by a specific set of coat proteins (COPs) which assemble as a scaffold on the cytoplasmic ER membrane surface. The unidirectional sequence of steps involved in coat assembly and disassambly is regulated by GTP binding proteins (Rothman and Orci, 1992; Schekman, 1992). Strong similarities exist between COP-coated vesicles involved in ER-to-Golgi transport and those travelling between separate Golgi elements (Rothman and Orci, 1992). The COP coat physically promotes vesicle budding from the donor membrane and is also thought to contain the required targeting information to ensure docking at the target membrane (Pfanner et al., 1990). Subsequent fusion of the vesicle membrane with the target membrane is achieved by disassembly of the COP coat followed by the binding of specific fusion proteins (Waters et al., 1991).

In order to maintain the structural integrity of the ER in the face of a constant flow of membrane material towards the Golgi stack, a retrograde recycling pathway for membrane lipids is required. In the presence of brefeldin A, an inhibitor of COP coat assembly and thus anterograde vesicle movement in mammalian cells (Torii et al., 1995), the vesicle donor compartments swell and tubular continuities between successive Golgi cisternae or between the *cis*-most Golgi compartment and the ER can be observed (Klausner et al., 1992). These membrane tubes, which are maintained by MTs, are believed to be the site of retrograde lipid transport (Lippincott-Schwarz et al., 1990), an altogether different route in form and function from the anterograde one involving vesicular carriers. Brefeldin A has similar effects on filamentous fungi (Rupes et al., 1995; Akashi et al., 1997), indicating that their early constitutive pathway may also be mediated by COP-coated vesicles.

4.2. The Golgi apparatus

It is a well-known paradox that morphologically recognizable Golgi stacks are not commonly found in *S. cerevisiae* and true fungi, even though the latter are among the most potent protein secretors known. Instead, fungi usually contain single flattened cisternae which are thought to represent the morphological equivalent of Golgi stacks (Howard, 1981; see also Figure 1). As an exception to the above, Golgi stacks have been observed in *Schizosaccharomyces pombe* (Ayscough et al., 1993). They are also common in members of the Oomycota (Grove and Bracker, 1970; Nolan and Bal, 1974) – indeed, *Pythium ultimum* was the subject of a classical study which established the eukaryotic secretory route from the ER *via* the Golgi apparatus to the plasma membrane (Grove et al., 1968). Interestingly, both hyphal and yeast phases of the dimorphic fungus *Candida albicans* do produce Golgi stacks when incubated in low concentrations of Brefeldin A (Akashi et al., 1993, 1997). Hence, the curious absence of recognizable Golgi stacks in growing hyphae of filamentous fungi may simply be due to the constantly high efflux of membrane material as a result of very fast anterograde transport under normal conditions.

In *S. cerevisiae*, with the help of analysis of *sec* mutants deficient in specific steps of the secretory route, three functionally distinct Golgi elements – *cis*, medial and *trans* – have been characterized as the compartments in which sequential specific modifications to proteins travelling the secretory route occur (Ballou et al., 1990; Graham and Emr, 1991). Similar functional Golgi compartments are well-known from mammalian cells (Kornfeld and Kornfeld, 1985), but no relevant studies appear to have been performed in filamentous fungi as yet. However, the presence of a functional Golgi apparatus can be inferred from the composition of mature *N*-glycosylation chains on secreted fungal proteins (*e.g.* Rickert and McBride-Warren, 1974; Rudick and Elbein, 1975; Maras et al., 1997) which show extensive modifications from the (*N*-acetylglucosamine)$_2$(mannose)$_8$ chain of *N*-glycosylation which represents the ER exit stage in most eukaryotes (Kornfeld and Kornfeld, 1985). Likewise, post-ER modifications to *O*-glycosylation are common in filamentous fungi (Raizada et al., 1975; Gum and Brown, 1976; Maras et al., 1997).

4.3. The mammalian *trans*-Golgi network (TGN)

In mammalian cells, the main junction in cellular membrane traffic (Figure 12) is located in an array of tubular structures associated with but distinct from the Golgi stack, termed the *trans*-Golgi network, TGN (Griffiths et al., 1985). The TGN produces COP-coated buds giving rise to vesicles which are involved in constitutive secretion and thus represent the final stage of the nonselective bulk flow route originating in the ER (Strous et al., 1983; Burgess and Kelly, 1987).

The same basic molecular machinery is therefore involved in all steps of constitutive secretion from ER through to the plasma membrane (Wilson et al., 1991; Rothman and Orci, 1992). Once formed, constitutive vesicles migrate to the plasma membrane where they are consumed immediately; they typically have a half-life of a few minutes (Wieland et al., 1987).

In addition, the TGN produces a novel type of bud which is coated by a striking polyhedral lattice of coat proteins termed clathrins (Crowther and Pearse, 1981), and their underlying HA-1 adaptors (Pearse and Robinson, 1984, 1990). These adaptors are thought to concentrate membrane-bound receptor proteins into bud regions (Burgess and Kelly, 1987) which, in turn, may selectively accumulate specific luminal proteins (Tooze and Tooze, 1986; Burgess and Kelly, 1987). Clathrin-coated vesicles are involved in regulated secretion, *i.e.* the mass discharge of vesicles following a particular stimulus. This phenomenon is found only in a few mammalian cell types such as endocrine, exocrine and neuron cells (Adelson and Miller, 1985; Buckley and Kelly, 1985).

Furthermore, in virtually all mammalian cells, clathrin-coated buds at the TGN membrane are also involved in selective sorting of lysosomal proteins (Lemansky et al., 1987; Kornfeld and Mellman, 1989). This sorting is mediated by receptors which recognize specific Golgi-mediated modifications – mannose-6-phosphate groups – at the free ends of the *N*-glycosylation chains of luminal enzymes (Kornfeld and Mellman, 1989). Membrane-bound proteins may be recognized by specific signals located in their cytoplasmic regions (Lehmann et al., 1992). After separation from the TGN, the vesicles are diverted into the lysosomal route (see Section 5). The TGN is therefore probably identical with the 'GERL' complex (Golgi-associated ER from which lysosomes arise), a concept formulated earlier (Novikoff and Holtzman, 1970).

One important difference between COP-coated and clathrin-coated vesicles is that in the former, the coat persists until docking at the target membrane has occurred (Pfanner et al., 1990), whereas in the latter the coat usually disassembles soon after the vesicle has left the donor membrane (Pearse and Robinson, 1990). These ultrastructural differences in vesicle surface features may be responsible for different modes of vesicle attachment to the cytoskeleton, resulting in different destinations as well as speed of travel and discharge.

4.4. Vesicle coats in fungi

In mammalian systems, clathrin-coated buds and vesicles are associated with selective transport whereas bulk flow is mediated by COP-coated structures, as shown in the preceding section (Figure 12). Fungi are commonly assumed not to possess a regulated secretory mechanism. However, zoospores of Oomycota such as *Pythium* and *Phytophthora* contain pre-formed peripheral vesicles which synchronously fuse with the plasma membrane during encystment (Hardham et al., 1991). Vesicle fusion is triggered by environmental stimuli and mediated by second messengers, including Ca^{2+} (Griffith et al., 1988) and phosphatidic acid (Zhang et al., 1992). Therefore, this process is a classical case of regulated secretion. Zoospore encystment also occurs in true fungi such as the

Chytridiomycota, even though detailed studies on the mechanisms involved have not yet been carried out.

The above exception apart, filamentous fungi as well as *S. cerevisiae* probably do employ purely constitutive routes to achieve their high secretory activities, in which case any changes in the composition of secreted proteins must be due purely to transcriptional regulation (Peberdy, 1994). In *S. cerevisiae*, the journey from Golgi body to plasma membrane has been estimated to take 3 minutes (Novick et al., 1981), a similar time as in the COP-mediated constitutive transport in mammalian cells. Curiously, when secretory vesicles were isolated from *S. cerevisiae*, they were found to be clathrin-coated (Mueller and Branton, 1984; Casanova et al., 1990) or uncoated (Walworth and Novick, 1987), yet when the yeast clathrin heavy chain gene (*CHC1*) was disrupted, protein export was not impeded (Payne and Schekman, 1985). In fact, clathrin was subsequently shown to play a role in the *retention* of Golgi-resident proteins (Seeger and Payne, 1992a) and in vacuolar protein sorting (Seeger and Payne, 1992b). Whatever the explanation for these observations, they challenge the universality of eukaryotic secretion mechanisms and need to be borne in mind when investigating the corresponding processes in fungi.

Few experimental results are available for filamentous fungi. Clathrin-coated vesicles, similar in size to those of *S. cerevisiae*, were observed in association with subapical Golgi cisternae in *Neurospora crassa* and *Uromyces phaseoli* (That et al., 1987), and the gene encoding γ-adaptin, which is specific to Golgi-associated clathrin-coated buds, has been cloned in *Ustilago maydis* and is required for apical germ tube growth (Keon et al., 1995), but it is unclear whether the Golgi-derived clathrin-coated vesicles are precursors to secretory vesicles or vacuoles.

4.5. Endocytosis in mammalian cells

In mammalian cells, excess membrane material arising from the incorporation of secretory vesicles into the plasma membrane is retrieved by inward budding in a process termed endocytosis (Figure 12). Membrane conductance measurements reveal a wave of inward budding following each burst of regulated secretion (von Grafenstein et al., 1986), and the turnover time for the entire plasma membrane area may be as short as 30-60 minutes (Marsh and Helenius, 1980). Endocytosis is mediated by clathrin-coated pits, initiated by HA-2 adaptors which differ from the HA-1 adaptors involved in coated pit formation at the TGN (Glickman et al., 1989). Various types of molecules are internalized by this route *via* a range of different receptors, including escaped lysosomal enzymes (Geuze et al., 1988; Lehmann et al., 1992) which are bound by the same receptor population that acts in clathrin-coated pits in the TGN (Duncan and Kornfeld, 1988). Endocytosis is accomplished when clathrin-coated vesicles move into the cytoplasm where they shed their coat and fuse with one another to form a tubular network, the early endosome (Kornfeld and Mellman, 1989). At this stage, some receptors may discharge their ligands and return to the plasma membrane (Gruenberg et al., 1989). Others – notably the mannose-phosphate receptors for luminal lysosomal enzymes – move further inwards in discrete uncoated vesicles which ultimately

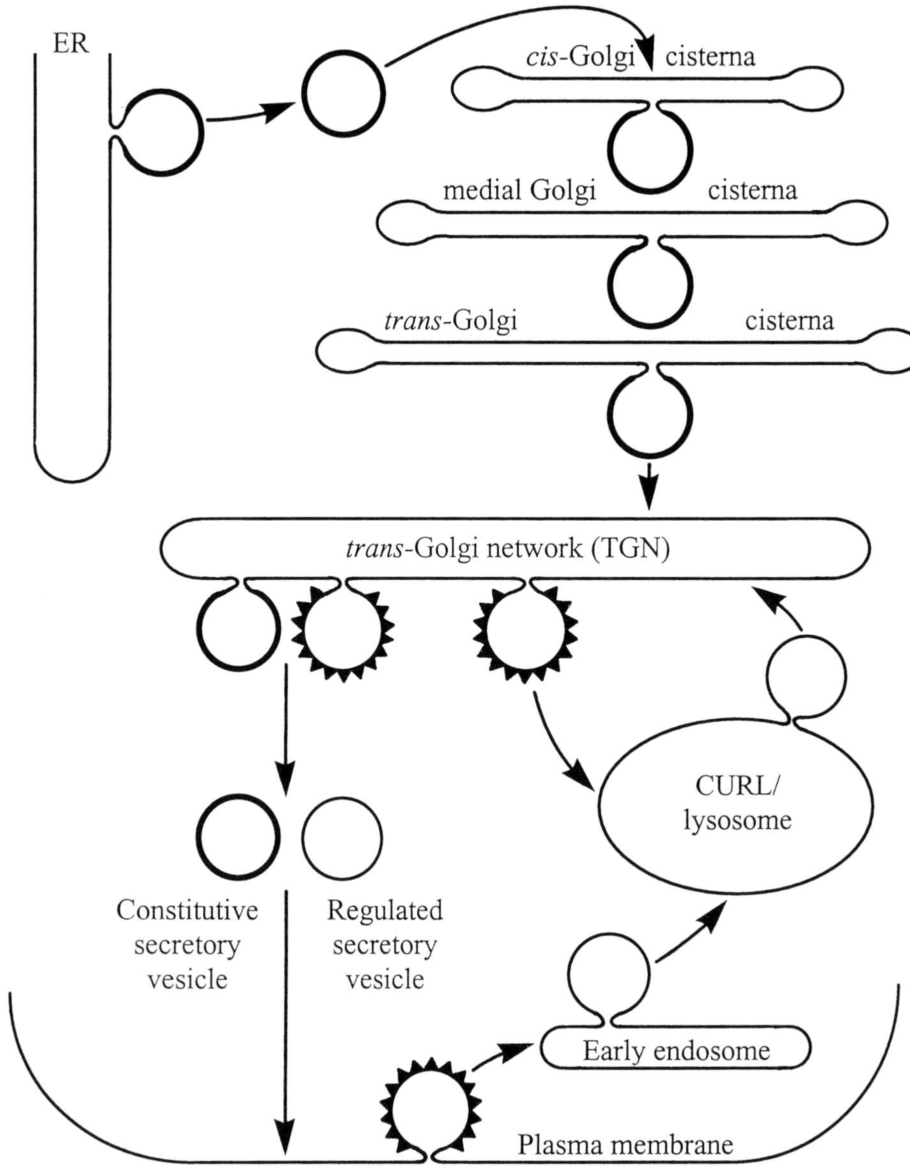

Figure 12. Summary of mammalian vesicular transport routes. Three different carrier types are utilized, *i.e.* COP-coated vesicles (smooth coat), clathrin-coated vesicles (rough coat) which soon shed their coat, and vesicles which appear to be uncoated at the moment of budding.

fuse with the compartment of uncoupling of receptor and ligand, CURL (Geuze et al., 1988; Kornfeld and Mellman, 1989). It is at this stage that the endocytotic and TGN-derived lysosomal transport pathways meet (Griffiths et al., 1988). The CURL has a sufficiently low luminal pH (5.0-5.5) to permit dissociation of lysosomal enzymes from their receptors; the latter are recycled to the TGN by smooth vesicles budding from the CURL, which itself gradually becomes converted into a lysosome as it accumulates endocytotic material (Kornfeld and Mellman, 1989; Schmid et al., 1989).

4.6. Endocytosis in *S. cerevisiae* and filamentous fungi

Any membrane material incorporated into the apex in excess of that required for hyphal extension needs to be retrieved in order to keep the plasma membrane under tension, which may be required for the operation *e.g.* of stretch-activated ion channels (see Section 3.3.). An enormous number of secretory vesicles is continually being incorporated into the growing tip, some 38,000 per minute in a single hypha of *Neurospora* growing at a speed of 36 μm min^{-1} (Collinge and Trinci, 1974). This would correspond to an extension of the plasma membrane area by approx. 1100 μm^2 min^{-1} (assuming a hyphal diameter of 10 μm), but given that secretory vesicles in *N. crassa* have a diameter of approx. 0.1 μm, a total of 1400 μm^2 of membrane surface min^{-1} would be incorporated. Sizeable quantities of excess membrane material may thus have to be retrieved, and endocytosis should be a well-described phenomenon especially in filamentous fungi.

Surprisingly, this is not the case, and even today the mere existence of endocytosis is still doubted by some (Rost et al., 1995; Cole et al., 1997, 1998). However, a recent study on the uptake of fluorescent dyes by germ tubes of *Uromyces fabae* has revealed that the label appeared first in endosome-like structures from which it was either re-secreted through apical secretory vesicles or stored in small subapical vacuoles (Hoffmann and Mendgen, 1998). This pattern is similar in outline to the mammalian membrane cycling described in Section 4.5. (Figure 12). Similar experiments were performed some 20 years ago on spores of a range of fungi, especially *Botrytis cinerea.* Uptake of vital dyes occurred in an energy-dependent manner by fluid-phase endocytosis through coated buds at the plasma membrane; the dye subsequently appeared in small vesicular structures (endosomes) which entered the vacuolar lumen and were digested there, releasing their contents into the vacuolar sap (Wilson et al., 1978, 1980). Similarly, we have observed coated pits in subapical regions of vegetative hyphae of *B. cinerea* (R. W. S. Weber, in preparation). Both the yeast and mycelial forms of *Candida albicans* were also shown to be capable of dye uptake by endocytosis (Basrai et al., 1990).

For *S. cerevisiae*, endocytosis has been readily demonstrated in intact cells and protoplasts by the uptake of enveloped viruses (Makarow, 1985a), α-amylase (Makarow, 1985b), α-type mating factor (Chvatchko et al., 1986) and various fluorescent dyes (de Nobel et al., 1989; Vida and Emr, 1995). In the case of α-factor, uptake was shown to occur by internalization of a specific plasma membrane-located receptor (Jenness et al., 1983). Yeast strains deficient in *CHC1* and devoid of functional clathrin were nonetheless capable of receptor-mediated uptake of α-factor, albeit at a reduced rate, thereby demonstrating that endocytosis, unlike in mammals, is not obligatorily coupled to clathrin-coated buds (Payne et al., 1988). This endocytotic route leads *via* endosomal structures (Vida and Emr, 1995) to the vacuole of *S. cerevisiae* (Dulic and Riezman, 1989).

5. VACUOLES IN FILAMENTOUS FUNGI

In mammalian cells, the lytic system is heterogeneous, consisting of endosomes, primary and secondary lysosomes and residual bodies, whereby lysosomes are essentially identified as membrane-bound organelles which are permanently resident in the cytoplasm, contain a variety of hydrolytic enzymes, and are capable of performing lytic functions (Novikoff and Holtzman, 1970; Holtzman, 1989). Histochemical attempts at applying this lysosome concept to filamentous fungi duly revealed numerous small (<1 μm diameter) organelles evenly distributed throughout hyphae from the apex backwards (Pitt and Walker, 1967; Pitt, 1968; Hislop et al., 1974). Hence, fungi were assumed to contain lysosomes analogous to those in mammalian cells (Wilson, 1973). However, a critical re-evaluation has shown that much of the histochemical reaction product might have been mis-localized into lipid bodies which are abundant in fungal hyphae (Weber et al., 1999), and that the genuine site of activity of hydrolases – notably acid phosphatase, long used as a lysosomal marker enzyme – is found predominantly in large vacuoles (Hänssler et al., 1975, 1977; Weber and Pitt, 1997a). These are identical with the Neutral Red-positive organelles which arise at the base of the apical cell, becoming prominent in intercalary segments (Figures 3 and 4). In *S. cerevisiae*, too, large vacuoles rather than numerous small entities are understood to be the main element of the lytic system (Matile, 1978; Mellman et al., 1986). Under conditions of minimal cytoplasmic disturbance, adjacent vacuoles in hyphae of filamentous fungi may be linked by thin membranous continuities, as revealed by careful studies using appropriate fluorescent dyes (Shepherd et al., 1993a; Rees et al., 1994).

5.1. Vacuoles as elements of the lytic system

If vacuoles are to represent the lytic system in fungi, they must meet several criteria which, in mammals, are fulfilled by lysosomes. Firstly, they should contain hydrolytic enzymes, which has indeed been shown to be the case for the vacuoles of *S. cerevisiae* as well as filamentous fungi (Klionsky et al., 1990; Weber et al., 1999). Secondly, vacuoles should and do act as the sink for the endocytotic route (see Section 4.6.).

Thirdly, lysosomes in mammals are involved in the autophagocytotic uptake and degradation of other organelles (Dunn, 1994). Such autophagic activities have recently been described for *S. cerevisiae* in which they can be stimulated by transferring cells from rich to nutrient-depleted media (Takeshige et al., 1992). In this situation, autophagocytosis in *S. cerevisiae* occurs as a two-step process. First, cisternae limited by a unit membrane form cup-shaped intermediates which ultimately surround a portion of the cytoplasm. Therefore, the cytoplasm becomes enclosed by a double-membrane, the space between the two membranes being the lumen of the former cisternae (Baba et al., 1995). The resulting organelle, termed autophagosome, then becomes associated with the vacuole and its outer membrane fuses with the tonoplast, releasing the inner membrane with the cytoplasm contained therein into the vacuole lumen (Baba et al., 1994). In wild-type cells, this is followed immediately by degradation of the autophagosome, whereas in mutant strains deficient in vacuolar protease activity, intact autophagosomes accumulate in the vacuole (Baba et al., 1994). Interestingly, a constitutive route of delivery of certain vacuolar enzymes to the vacuole has recently been found to utilize the same pathway (Baba et al., 1997). An altogether different entry route into the vacuole of *S. cerevisiae* is taken by redundant proteins and occurs by direct inward budding of the tonoplast (Chiang et al., 1996; Klionsky, 1997).

In filamentous fungi, the few recorded examples of autophagocytosis follow either of the two patterns described above for *S. cerevisiae, i.e. via* autophagosomes (Mims et al., 1995) or without an intermediate stage by a large vacuole directly engulfing adjacent regions of the cytoplasm (Weber et al., 1999). The latter case is morphologically very similar to the events leading to vacuolation in the course of xylem differentiation in higher plants (Buvat, 1977; Marty, 1978). In either case, autophagocytosis is stimulated by nutrient starvation, and evidence of digestion of the engulfed cytoplasm is readily seen. These results therefore strongly implicate vacuoles as the main lytic elements in fungi whereas no such activities have been associated with the smaller, lipid-rich entities formerly thought to be fungal lysosome equivalents.

5.2. The involvement of vacuoles in secretion

An investigation of acid phosphatase secretion in *B. cinerea* revealed the existence of two isoforms, a phosphate-repressible enzyme secreted in great quantities through the growing apex by phosphorus-starved hyphae (Weber and Pitt, 1997a), and a constitutive enzyme of putative vacuolar origin which was secreted by the fusion of vacuoles with lateral (subapical) regions of the plasma membrane (Weber and Pitt, 1997b). Such a secondary secretory route involving protein discharge from vacuoles some distance behind the apex may also act in cellulase secretion in *Trichoderma reesei* (Glenn et al., 1985; Ghosh et al., 1990; Kubicek et al., 1993; Nykänen et al., 1997) and in lignin peroxidase secretion in *Phanerochaete chrysosporium* (Daniel et al., 1989).

Protein secretion has traditionally been postulated to be confined to the growing hyphal tip on the grounds that the pore size in mature hyphal walls would be too small to permit passage of proteins (Wessels, 1993). The mature hyphal wall has been predicted to prevent the exit of any molecules larger than 20 kDa (Chang and Trevithick, 1974), which would include most extracellular proteins (Peberdy, 1994). However, apart from the difficulties involved in making accurate estimates (de Nobel et al., 1989), a crucial point is the nature of the pore size-determining wall component. In *S. cerevisiae*, cell wall porosity is determined not by the polysaccharide components but by the outer mannoprotein layer which can be degraded by proteases (Zlotnik et al., 1984). If this was also the case in filamentous fungi, enzymes secreted by vacuoles could thus still reach the culture fluid with the aid of the co-secreted vacuolar proteases. Whilst the apical route is undoubtedly the major exit route taken by secretory proteins even in the systems described above, it may not be the only such route. The putative role of vacuoles in secretion thus deserves further investigation, especially because clarification is required as to how much of the products of foreign genes finds its way into the vacuolar system and may be degraded there. Several cases have been reported in which secreted heterologous proteins showed evidence of contact with the lytic system prior to secretion, as indicated by protease attack or exposure to a low-pH compartment (Harkki et al., 1989; Berka et al., 1991; Kubicek et al., 1993).

5.3. The role of vacuoles in nutrient storage and transport

The luminal pH of vacuoles in *Neurospora crassa* has been estimated at pH 6.1 (Legerton et al., 1983) and is thus similar to that of *S. cerevisiae* (Preston et al., 1989) but contrasts with the cytoplasmic pH (approx. pH 7.0). Vacuolar acidification is brought about by the activity of a hetero-oligomeric ATPase located in the tonoplast. The inward transport of protons across the tonoplast is driven by ATP hydrolysis at the cytoplasmic side, thereby generating a transmembrane electrochemical potential (Bowman and Bowman, 1986; Kane et al., 1989; Klionsky et al., 1990).

This H^+ gradient is harnessed by fungal vacuoles to fulfil their important roles in cellular homeostasis, whereby the principle of proton-coupled solute transport is the same as that which occurs during nutrient uptake across the plasma membrane, discussed in detail in Section 6.1. One of the best-understood vacuolar functions involves the storage of cationic amino acids, notably arginine, as nitrogen reserves (Davis, 1986; Keenan and Weiss, 1997). Arginine uptake occurs by specific membrane antiport channels in exchange for protons, thus dissipating the H^+ transmembrane gradient (Paek and Weiss, 1989). At nitrogen starvation, the vacuolar arginine pool is mobilized and released into the cytoplasm (Legerton and Weiss, 1984). A similar transfer and storage scheme exists for phosphorus which is accumulated in vacuoles as inorganic phosphate, stored as polyphosphate, and mobilized under conditions of phosphate starvation (Cramer et al., 1980; Cramer and Davis, 1984).

In addition to the storage of nutrients, vacuoles also play a crucial role in maintaining the general ionic balance of the fungal cell. This is particularly well known for Ca^{2+} which is present in low cytoplasmic concentrations (see Section 3.2.) and is removed from the cytoplasm into vacuoles by an H^+ antiport system (Ohsumi and Anraku, 1983; Cornelius and Nakashima, 1987). Interactions may form in the vacuolar lumen between anionic polyphosphate and cations such as arginine, Ca^{2+}, K^+ and Mg^{2+} (Cramer and Davis, 1984; Bücking and Heyser, 1999). Given that the above and many more storage functions of the fungal vacuole are driven by ATP hydrolysis, mutants defective in the vacuolar H^+-ATPase would be expected to show multiple deficiencies. This is indeed the case; in *S. cerevisiae* such mutants are viable under normal growth conditions but are very sensitive to ionic or nutrient imbalances (Stevens, 1992), whereas in *N. crassa*, H^+-ATPase mutants are lethal altogether (Bowman et al., 1992).

Large vacuoles in filamentous fungi may be linked by tubular continuities (Shepherd et al., 1993a; Rees et al., 1994). Much recent work has been carried out on the important ectomycorrhizal fungus *Pisolithus tinctorius* in which these tubes can extend through dolipore septa, thereby connecting vacuoles in adjacent cells (Shepherd et al., 1993b). The presence of phosphate and potassium ions in these vacuoles, together with observations of peristaltic movement of material through the interconnecting tubes (Cole et al., 1998), may point to this system as the morphological manifestation of the well-known phenomenon of long-distance phosphorus transport in mycorrhizal fungi (Timonen et al., 1996). In this way, vacuoles may be involved in homeostasis and transport not only within cells but between adjacent cells of hyphae and, ultimately, the entire mycelium.

6. NUTRIENT UPTAKE

Many filamentous fungi are capable of absorbing organic and inorganic solutes from very dilute extracellular solutions, often concentrating them more than 1,000-fold against their concentration gradient (Griffin, 1994). Such uptake is principally mediated by proteinaceous pores traversing the lipid bilayer of the plasma membrane. These pores are termed channels if they selectively facilitate the entry of a particular solute by diffusion along its electrochemical gradient, whereas they are called porters if energy is spent on actively transporting the solute across the plasma membrane against its gradient (Harold, 1994). Since the effective assimilation of nutrients is one of the hallmarks of the fungal kingdom and is, furthermore, a polarized process, it is considered in this chapter.

6.1. Facilitated diffusion and active uptake

In fungi as in most living organisms, the plasma membrane is the main barrier to diffusion of substances into and out of cells. Passive movement across phospholipid membranes is a slow and non-selective process, probably of minor significance in fungi. Facilitated diffusion is considered to be of more relevance since the rate of diffusion down the gradient of concentration or the electrical potential is attained more rapidly than that predicted from molecular size. Solute movement across the plasma membrane is selective and involves proteinaceous channels (Griffin, 1994). Some of the systems investigated in *S. cerevisiae* (Burger et al., 1959) appear to require the membrane to be in an energized state because they are affected by respiratory inhibitors and uncouplers of oxidative phosphorylation. Whilst the precise nature of energy dependence is unclear, it may involve shuttling movement of the proteinceous carrier, its change in configuration, or the shaping of transient pores (Lagunas, 1993). In fungi, facilitated diffusion is also the basis of thestretch-activated Ca^{2+} channels discussed earlier (Section 3.2.) and provides the main nutrient uptake route when these organisms are grown in rich media, including most media used for routine laboratory culture.

Filamentous fungi are characterized by a pronounced capacity to scavenge certain solutes selectively against their concentration gradient, and to exclude others, by active and ATP-dependent processes which display saturation kinetics typical of enzyme-catalysed reactions. Current views favour a mechanism of indirect consumption of energy, derived from ATP hydrolysis, *via* the depolarization of a transmembrane H^+ gradient. This draws on the principles set out in the chemiosmotic theory of Peter Mitchell in which the generation of an H^+ gradient across the inner mitochondrial membrane by oxidative phosphorylation was postulated to be sufficient to drive the membrane-bound ATPase backwards, leading to ATP synthesis (Mitchell, 1966). When applied to solute transport across the plasma membrane in fungi, it is envisaged that hydrolysis of ATP by H^+-ATPases in the plasma membrane is coupled to the expulsion of protons, generating a transmembrane H^+ gradient (acid outside) which provides the driving force for solute uptake by solute-carrier-H^+ complexes (Garrill, 1995). Elegant experimental demonstrations of such an electrogenic mechanism were provided by Clifford Slayman and colleagues for K^+ and hexose transport in *N. crassa* (Slayman and Slayman, 1968, 1974; Slayman, 1987). Several different carrier types were characterized, *i.e.* uniports and symports, which couple inward H^+ movement with the uptake of uncharged molecules and anions, respectively, and antiports which carry cations inwards whilst anions such as Na^+ are expelled. A large volume of similar experimental evidence has been provided for uniport, symport and antiport mechanisms in other species of fungi for transport of NO_3^- and NH_4^+ (Roos, 1989), amino acids (Gow et al., 1984; Horak, 1986) and other solutes (Garrill, 1995; Jennings, 1995). Charge balance is probably maintained by the opening of K^+ efflux channels (Garrill, 1995). Under conditions of K^+ starvation, *N. crassa* is capable of active uptake of K^+ by *symport* with H^+, leading to an unusually rapid membrane depolarization (Rodriguez-Navarro et al., 1986). This curious finding illustrates the great flexibility and adaptability of uptake systems in filamentous fungi.

Detailed work on the uptake kinetics of hexose sugars suggests complications and variations in these transport phenomena which allow *e.g. N. crassa* to react differentially to high and low sugar concentrations in the environment by using two independent uptake systems. System I comprises a facilitated diffusion mechanism with low affinity (K_m = 8 mM) which operates maximally at relatively high external sugar concentrations whilst the high affinity system II (K_m = 10 μM) is inhibited at such high substrate concentrations and is derepressed only when external sugar is scarce, resulting in the cumulative and energy-dependent uptake of sugar against its gradient (Scarborough, 1970). High-affinity active transport and low-affinity facilitated diffusion systems are present for most solutes in filamentous fungi, whereby the repression of active uptake in nutrient-rich conditions undoubtedly saves metabolic energy (Sanders, 1988), especially when considering that at least one third of the total cellular ATP is spent on establishing the H^+ gradient across the plasma membrane (Gradmann et al., 1978).

Furthermore, if active transport systems operated in an uncontrolled manner even at relatively low external sugar concentrations, detrimental osmotic consequences for cells could arise. In order to accommodate the fact that internal solute levels are maintained within osmotically safe limits irrespective of external concentrations, various theoretical pump and leak mechanisms have been proposed (Eddy, 1982), in addition to the transcriptional repression of active uptake pumps as discussed above. It is generally assumed that filamentous fungi maintain osmotic stability by means of a facilitated leak mechanism through which an excess of accumulated solutes is allowed to escape, thus maintaining a fixed ratio between internal and increasing external concentrations. At high external solute concentrations, slip systems may be activated whereby porters could be modified or uncoupled, and their further synthesis repressed, such that the accumulation ratio declines to 1 and is maintained at this value within physiological external concentrations. An extreme case would be the reversal of uptake into active expulsion in media of very high osmolarity, resulting in an accumulation ratio below 1. Some evidence for a pump/leak system of the slip or exclusion type has been provided for glucose transport in *Penicillium notatum* (Pitt and Barnes, 1987).

6.2. The hypha as an electric field

Nutrient uptake channels and porters in filamentous fungi are thought to be concentrated in the apical cell of a hypha which is the region most likely to be in contact with external nutrients. In the case of active uptake, H^+ ions will enter along with the co-transported nutrients into the apical cytoplasm which thus should be acidified relative to regions further behind. Such tip-acidic internal pH gradients have been described *e.g.* for *Penicillium cyclopium* (Roncal et al., 1993). In contrast, in Chytridiomycota such as *Blastocladiella emersonii*, a finely branched rhizoid system is thought to be the primary site of nutrient uptake (Kropf and Harold, 1982), and protons have been shown to enter through the rhizoid tips (van Brunt et al., 1982), presumably indicating the location of active transport systems (Youatt et al., 1988). An ionic current along the length of a fungal hypha will be established if the porters are spatially separated from the H^+-ATPases which generate the proton gradient, as has been suggested for *Achlya* and *Neurospora* (Kropf et al., 1984; Kropf, 1986; Takeuchi et al., 1988). Indeed, in the latter case, the tip cell has been shown to be altogether devoid of H^+-pumping activity which was strong in subapical segments (Potapova et al., 1988; Harold, 1994). These maturing and often highly vacuolated sections of a hypha, commonly perceived to be metabolically inactive, thus make a direct and substantial contribution to the physiology of the apical cell.

Until recently, transhyphal currents had been thought to be causally involved in hyphal growth polarity (Bartnicki-Garcia, 1973; Kropf et al., 1983, 1984). However, when growing hyphae are transferred into a medium of different composition, the endogenous current may be reversed or abolished altogether whilst hyphal growth carries on unabated (Schreurs and Harold, 1988; Cho et al., 1991). Likewise, the striking observation that hyphae subjected to an external electrical field re-orientate their apices to grow either towards the anode or cathode (McGillivray and Gow, 1986) is best explained by an electrophoretic effect on as yet unidentified plasma membrane proteins which leads to re-orientation of growth (Harold, 1994). At present, the existence of electrical currents in growing hyphae is interpreted to be the result of polarized nutrient uptake and, therefore, a consequence rather than a cause of growth polarity.

7. PERSPECTIVE

Ever since polarized apical growth in filamentous fungi was described over a century ago, mycologists have been investigating this strikingly obvious phenomenon in some detail and have had to realize that its underlying mechanisms are supremely complex and often recalcitrant to experimental investigation. Thus, in our opinion, the physiology of the growing fungal hypha represents the supreme remaining challenge in the whole discipline of mycology. This is surely to a large part due to the intricate interweaving between hyphal tip extension and general physiological processes. Since most organelles are distributed in a highly polarized manner, their physiological roles likewise become manifested in gradients along the hyphal length. For this reason, we have treated hyphal growth and physiology in an integrated manner in the present article.

Polarized growth is due to the polarized secretion of cell wall components and enzymes. Channels and porters involved in nutrient uptake and signalling are transported *via* the same route, as are hydrolytic extracellular enzymes, suggesting a tight link between growth and nutrient uptake in fungi. The transport of secretory vesicles to the apex is mediated by the cytoskeleton which is also involved in maintaining the polarity of distribution of most other organelles such as mitochondria, nuclei, the ER and vacuoles. In more mature hyphal regions or subapical segments of septate fungi, the distribution of organelles becomes non-polar. Nevertheless, the mature hypha has vital contributions to make to the functioning of the growing apex, by fuelling the proton gradient which enables the tip cell to scavenge nutrients *via* active transport, by providing a source of organelles which can migrate through septal pores, and by providing vacuolar storage space. Indeed, the vacuolar system even seems to provide an efficient means of translocation through tubular continuities, thereby co-ordinating the nutrient supply between different growing regions of the fungal mycelium. Further, mature regions are capable of establishing new growing tips, thereby ensuring a uniform density of substrate colonization as the mycelium radiates outwards. Therefore, mature regions must be an integral part in considerations of hyphal growth and polarity.

In most of these fundamental points, the Oomycota are functionally indistinguishable from the true fungi, differing merely in certain means to achieve the same end. For instance, actin microfilaments instead of microtubules seem to be involved in long-distance transport of secretory vesicles; tip growth seems to be regulated by stretch-activated Ca^{2+} channels; and extension of the hyphal tip seems to be independent of turgor pressure. It is as yet impossible to state with confidence whether these differences are greater than those found within the true fungi themselves, and for the time being we consider research performed on Oomycota to be of great relevance to the understanding of fungal physiology, not least because some of the most thoroughly researched bodies of knowledge have been obtained with these organisms.

REFERENCES

Adams, A.E.M. and J.R. Pringle (1984) *J. Cell Biol.* **98:** 934-945.

Adelson, J.W. and P.E. Miller (1985) *Science* **228:** 993-996.

Aist, J.R. and C. Bayles (1991) *Eur. J. Cell Biol.* **56:** 358-363.

Akashi, T., M. Homma, T. Kanbe and K. Tanaka (1993) *J. Gen. Microbiol.* **139:** 2185-2195.

Akashi, T., T. Kanbe and K. Tanaka (1997) *Protoplasma* **197:** 45-56.

Alexopoulos, C.J., C.W. Mims and M. Blackwell (1996) *Introductory Mycology* (4th edition), John Wiley and Sons, New York.

Ayscough, K., N.M.A. Hajibagheri, R. Watson and G. Warren (1993) *J. Cell Sci.* **106:** 1227-1237.

Baba, M., K. Takeshige, N. Baba and Y. Ohsumi (1994) *J. Cell Biol.* **124:** 903-913.

Baba, M., M. Osumi and Y. Ohsumi (1995) *Cell Struct. Funct.* **20:** 465-471.

Baba, M., M. Osumi, S.V.Scott, D.J. Klionsky and Y. Ohsumi (1997) *J. Cell Biol.* **139:** 1687-1695.

Ballou, L., L.M. Hernandez, E. Alvarado and C.E. Ballou (1990) *Proc. Natl. Acad. Sci. USA* **87:** 3368-3372.

Bartnicki-Garcia, S. (1973) In J.M. Ashworth and J.E. Smith (Eds.), *Microbial Differentiation*, Cambridge University Press, Cambridge, pp. 245-267.

Bartnicki-Garcia, S. (1996) In B.C. Sutton (Ed.), *A Century of Mycology*, Cambridge University Press, Cambridge, pp. 105-133.

Bartnicki-Garcia, S. and E. Lippman (1969) *Science* **165:** 302-304.

Bartnicki-Garcia, S., J. Ruiz-Herrera and C.E. Bracker (1979) In J.H. Burnett and A.P.J. Trinci (Eds.), *Fungal Walls and Hyphal Growth*, Cambridge University Press, Cambridge, pp. 149-168.

Bartnicki-Garcia, S., F. Hergert and G. Gierz (1989) *Protoplasma* **153:** 46-57.

Bartnicki-Garcia, S., F. Hergert and G. Gierz (1990) In P.J. Kuhn, A.P.J. Trinci, M.J. Jung, M.W. Goosey and L.G. Copping (Eds.), *Biochemistry of Cell Walls and Membranes in Fungi*, Springer-Verlag, Berlin, pp. 43-60.
Basrai, M.A., F. Naider and J.M. Becker (1990) *J. Gen. Microbiol.* **136:** 1059-1065.
Berka, R.M., K.H. Kodama, M.W. Rey, L.J. Wilson and M. Ward (1991) *Biochem. Soc. Transact.* **19:** 681-685.
Betina, V., D. Miceková and P. Nemec (1972) *J. Gen. Microbiol.* **71:** 343-349.
Bobbitt, T.F. and J.H. Nordin (1982) *J. Bacteriol.* **150:** 365-376.
Boone, C., S.S. Sommer, A. Hensel and H. Bussey (1990) *J. Cell Biol.* **110:** 1833-1843.
Borgia, P.T., N. Iartchouk, P.J. Riggle, K.R. Winter, Y. Koltin and C.E. Bulawa (1996) *Fungal Genet. Biol.* **20:** 193-203.
Bourett, T.M. and R.J. Howard (1991) *Protoplasma* **163:** 199-202.
Bourett, T.M., K.J. Czymmek and R.J. Howard (1998) *Fungal Genet. Biol.* **24:** 3-13.
Bowman, B.J. and E.J. Bowman (1986) *J. Membr. Biol.* **94:** 83-97.
Bowman, B.J., W.J. Dschida and E.J. Bowman (1992) *J. Exp. Biol.* **172:** 57-66.
Brunswik, H. (1924) *Bot. Abh.* **5:** 1-152.
Bücking, H. and W. Heyser (1999) *Mycol. Res.* **103:** 31-39.
Buckley, K. and R.B. Kelly (1985) *J. Cell Biol.* **100:** 1284-1294.
Bulawa, C.E. (1993) *Annu Rev. Microbiol.* **47:** 505-534.
Burger, M., L. Hejmová and A. Kleinzeller (1959) *Biochem. J.* **71:** 233-242.
Burgess, T.L. and R.B. Kelly (1987) *Annu. Rev. Cell Biol.* **3:** 243-293.
Burnett, J.H. (1976) *Fundamentals of Mycology* (2nd edition), E. Arnold, London.
Buvat, R. (1977) *C. R. Hebd. Seances Acad. Sci. Ser. D* **284:** 167-170.
Cali, B.M., T.C. Doyle, D. Botstein and G.R. Fink (1998) *Mol. Biol. Cell* **9:** 1873-1889.
Calvert, C.M. and D. Sanders (1995) *J. Biol. Chem.* **270:** 7272-7280.
Calvo-Mendez, C. and J. Ruiz-Herrera (1987) *Exp. Mycol.* **11:** 128-140.
Carlile, M.J. (1995) In N.A.R. Gow and G.M. Gadd (Eds.), *The Growing Fungus*, Chapman and Hall, London, pp. 3-19.
Casanova, M., A. Parets-Soler, F. Miragall, J.P. Martinez and R. Sentandreu (1990) *Mycol. Res.* **94:** 1026-1030.
Chang, P.L. and J.R. Trevithick (1974) *Arch. Microbiol.* **101:** 281-293.
Chiang, H.L., R. Schekman and S. Hamamoto (1996) *J. Biol. Chem.* **271:** 9934-9941.
Cho, C.-W., F.M. Harold and W.J.A. Schreurs (1991) *Exp. Mycol.* **15:** 34-43.
Chvatchko, Y., I. Howald and H. Riezman (1986) *Cell* **46:** 355-364.
Cole, L., G.J. Hyde and A.E. Ashford (1997) *Protoplasma* **199:** 18-29.
Cole, L., D.A. Orlovich and A.E. Ashford (1998) *Fungal Genet. Biol.* **24:** 86-100.
Collinge, A.J. and A.P.J. Trinci (1974) *Arch. Microbiol.* **99:** 353-368.
Connolly, T. and R. Gilmore (1989) *Cell* **57:** 599-610.
Cornelius, G. and H. Nakashima (1987) *J. Gen. Microbiol.* **133:** 2341-2347.

Corrêa Jr., A., R.C. Staples and H.C. Hoch (1996) *Protoplasma* **194:** 91-102.

Cramer, C.L. and R.H. Davis (1984) *J. Biol. Chem.* **259:** 5152-5157.

Cramer, C.L., L.E. Vaughn and R.H. Davis (1980) *J. Bacteriol.* **142:** 945-952.

Crowther, R.A. and B.M.F. Pearse (1981) *J. Cell Biol.* **91:** 790-797.

Czymmek, K.J., T.M. Bourett and R.J. Howard (1996) *J. Microsc.* **181:** 153-161.

Daniel, G., T. Nilsson and B. Pettersson (1989) *Appl. Environ. Microbiol.* **55:** 871-881.

Datema, R., H. van den Ende and J.G.H. Wessels (1977a) *Eur. J. Biochem.* **80:** 611-619.

Datema, R., J.G.H. Wessels and H. van den Ende (1977b) *Eur. J. Biochem.* **80:** 621-626.

Davis, R.H. (1986) *Microbiol. Rev.* **50:** 280-313.

de Nobel, J.G., C. Dijkers, E. Hooijberg and F.M. Klis (1989) *J. Gen. Microbiol.* **135:** 2077-2084.

do Carmo-Sousa, L. (1969) In A.H. Rose and J.S. Harrison (Eds.), *The Yeasts Vol. 1: Biology of Yeast*, Academic Press, London, pp. 79-105.

Dubourdieu, D. and P. Ribéreau-Gayon (1980) *Compt. Rend. Hebd. Seances Acad. Sci. Ser. D* **290:** 25-28.

Dubourdieu, D., B. Fournet, A. Bertrand and P. Ribéreau-Gayon (1978a) *Compt. Rend. Hebd. Seances Acad. Sci. Ser. D* **286:** 229-231.

Dubourdieu, D., B. Pucheu-Planté, M. Mercier and P. Ribéreau-Gayon (1978b) *Compt. Rend. Hebd. Seances Acad. Sci. Ser. D* **287:** 571-573.

Dulic, V. and H. Riezman (1989) *EMBO J.* **8:** 1349-1359.

Duncan, J.R. and S. Kornfeld (1988) *J. Cell Biol.* **106:** 617-628.

Dunn, W.A. (1994) *Trends Cell Biol.* **4:** 139-143.

Eddy, A.A. (1982) *Adv. Microb. Physiol.* **23:** 1-78.

Fontaine, T., R.P. Hartland, A. Beauvais, M. Diaquin and J.-P. Latgé (1997) *Eur. J. Biochem.* **243:** 315-321.

Förster, C., S. Marienfeld, V.F. Wendisch and R. Krämer (1998) *Appl. Microbiol. Biotechnol.* **50:** 219-226.

Garrill, A. (1995) In N.A.R. Gow and G.M. Gadd (Eds.), *The Growing Fungus*, Chapman and Hall, London, pp. 163-181.

Garrill, A., S.L. Jackson, R.R. Lew and I.B. Heath (1993) *Eur. J. Cell Biol.* **60:** 358-365.

Gay, L., H. Chanzy, V. Bulone, V. Girard and M. Fèvre (1993) *J. Gen. Microbiol.* **139:** 2117-2122.

Geuze, H.J., W. Stoorvogel, G.J.A.M. Strous, J.W. Slot, J.E. Bleekemolen and I. Mellman (1988) *J. Cell Biol.* **107:** 2491-2501.

Ghosh, B.K., T. Ganguli and A. Ghosh (1990) In C.P. Kubicek, D.E. Eveleigh, H. Esterbauer, W. Steiner and E.M. Kubicek-Pranz (Eds.), Trichoderma reesei *Cellulases: Biochemistry, Genetics, Physiology and Application*, Royal Society of Chemistry, Cambridge, 115-138.

Glenn, M., A. Ghosh and B.K. Ghosh (1985) *Appl. Environ. Microbiol.* **50:** 1137-1143.

Glickman, J.N., E. Conibear and B.M.F. Pearse (1989) *EMBO J.* **8:** 1041-1047.
Gooday, G.W. (1971) *J. Gen. Microbiol.* **67:** 125-133.
Gooday, G.W. and A.P.J. Trinci (1980) In G.W. Gooday, D. Lloyd and A.P.J. Trinci (Eds.), *The Eukaryotic Microbial Cell*, Cambridge University Press, Cambridge, pp. 207-251.
Gooday, G.W. and N.A.R. Gow (1990) In I.B. Heath (Ed.), *Tip Growth of Plant and Fungal Cells*, Academic Press, New York, pp. 31-58.
Goodner, B. and R.S. Quatrano (1993) *Plant Cell* **5:** 1471-1481.
Gorin, P.A.J. and J.F.T. Spencer (1968) *Adv. Carbohydr. Chem.* **23:** 367-417.
Govindan, B., R. Bowser and P. Novick (1995) *J. Cell Biol.* **128:** 1055-1069.
Gow, N.A.R. (1995a) In N.A.R. Gow and G.M. Gadd (Eds.), *The Growing Fungus*, Chapman and Hall, London, pp. 403-422.
Gow, N.A.R. (1995b) In N.A.R. Gow and G.M. Gadd (Eds.), *The Growing Fungus*, Chapman and Hall, London, pp. 277-299.
Gow, N.A.R., D.L. Kropf and F.M. Harold (1984) *J. Gen. Microbiol.* **130:** 2967-2974.
Gradmann, D., U.-P. Hansen, W.S. Long, C.L. Slayman and J. Warncke (1978) *J. Membr. Biol.* **39:** 333-367.
Graham, T.R. and S.D. Emr (1991) *J. Cell Biol.* **114:** 207-218.
Griffin, D.H. (1994) *Fungal Physiology* (2nd edition), Wiley-Liss, New York.
Griffith, J.M., J.R. Iser and B.R. Grant (1988) *Arch. Microbiol.* **149:** 565-571.
Griffiths, G., S. Pfeiffer, K. Simons and K. Matlin (1985) *J. Cell Biol.***101:** 949-964.
Griffiths, G., B. Hoflack, K. Simons, I. Mellman and S. Kornfeld (1988) *Cell* **52:** 329-341.
Grove, S.N. and C.E. Bracker (1970) *J. Bacteriol.* **104:** 989-1009.
Grove, S.N., C.E. Bracker and D.J. Morré (1968) *Science* **161:** 171-173.
Gruenberg, J., G. Griffiths and K.E. Howell (1989) *J. Cell Biol.* **108:** 1301-1316.
Gum Jr., E.K. and R.D. Brown Jr. (1976) *Biochim. Biophys. Acta* **446:** 371-386.
Guthrie, B.A. and W. Wickner (1988) *J. Cell Biol.* **107:** 115-120.
Haguenauer-Tsapis, R. and A. Hinnen (1984) *Mol. Cell. Biol.* **4:** 2668-2675.
Hänssler, G., D.P. Maxwell and M.D. Maxwell (1975) *J. Bacteriol.* **124:** 997-1006.
Hänssler, G., D.P. Maxwell, H. Barczewski and E. Bernhardt (1977) *Phytopathol. Z.* **88:** 289-298.
Hardham, A.R., F. Gubler and J. Duniec (1991) In J.A. Lucas, R.C. Shattock, D.S. Shaw and L.R. Cooke (Eds.), *Phytophthora*, Cambridge University Press, Cambridge, pp. 50-69.
Harkki, A., J. Uusitalo, M. Bailey, M. Penttilä and J.K.C. Knowles (1989) *Biotechnol.* **7:** 596-603.
Harold, F.M. (1994) In J.G.H. Wessels and F. Meinhardt (Eds.), *The Mycota Vol. I: Growth, Differentiation and Sexuality*, Springer-Verlag, Berlin, pp. 89-109.
Hartland, R.P., G.W. Emerson and P.A. Sullivan (1991) *Proc. Roy. Soc. Lond. B* **246:** 155-160.
Haselbeck, A. and W. Tanner (1983) *FEBS Lett.* **158:** 335-338.

Heath, I.B. (1987) *Eur. J. Cell Biol.* **44:** 10-16.
Heath, I.B. (1994) In J.G.H. Wessels and F. Meinhardt (Eds.), *The Mycota Vol. I: Growth, Differentiation and Sexuality*, Springer-Verlag, Berlin, pp. 43-65.
Heath, I.B. (1995a) In N.A.R. Gow and G.M. Gadd (Eds.), *The Growing Fungus*, Chapman and Hall, London, pp. 99-134.
Heath, I.B. (1995b) *Can. J. Bot.* **73:** S131-S139.
Heath, I.B. and S.G.W. Kaminskyj (1989) *J. Cell Sci.* **93:** 41-52.
Hendrick, J.P. and F.-U. Hartl (1993) *Annu. Rev. Biochem.* **62:** 349-384.
Hill, T.W. and J.T. Mullins (1980) *Can. J. Microbiol.* **26:** 1132-1140.
Hislop, E.C., V.M. Barnaby, C. Shellis and F. Laborda (1974) *J. Gen. Microbiol.* **81:** 79-99.
Hoffmann, J. and K. Mendgen (1998) *Fungal Genet. Biol.* **24:** 77-85.
Holtzman, E. (1989) *Lysosomes*, Plenum Press, New York.
Horak, J. (1986) *Biochim. Biophys. Acta* **864:** 223-256.
Horsch, M., C. Mayer, U. Sennhauser and D.M. Rast (1997) *Pharmacol. Ther.* **76:** 187-218.
Howard, R.J. (1981) *J. Cell Sci.* **48:** 89-103.
Howard, R.J. and J.R. Aist (1977) *Protoplasma* **92:** 195-210.
Howard, R.J. and J.R. Aist (1979) *J. Ultrastruct. Res.* **66:** 224-234.
Howard, R.J. and J.R. Aist (1980) *J. Cell Biol.* **87:** 55-64.
Howard, R.J. and K.L. O'Donnell (1987) *Exp. Mycol.* **11:** 250-269.
Huffaker, T.C., J.H. Thomas and D. Botstein (1988) *J. Cell Biol.* **106:** 1997-2010.
Hunsley, D. and J.H. Burnett (1970) *J. Gen. Microbiol.* **62:** 203-218.
Hynes, R.O. (1992) *Cell* **69:** 11-25.
Jackson, S.L. and I.B. Heath (1992) *Protoplasma* **170:** 46-52.
Jackson, S.L. and I.B. Heath (1993a) *Protoplasma* **173:** 23-34.
Jackson, S.L. and I.B. Heath (1993b) *Microbiol. Rev.* **57:** 367-382.
Jeenes, D.J., R. Pfaller and D.B. Archer (1997) *Gene* **193:** 151-156.
Jenness, D.D., A.C. Burkholder and L.H. Hartwell (1983) *Cell* **35:** 521-530.
Jennings, D.H. (1995) *The Physiology of Fungal Nutrition*, Cambridge University Press, Cambridge.
Jochová, J., I. Rupes and J.F. Peberdy (1993) *Mycol. Res.* **97:** 23-27.
Johnson, D.C. and P.G. Spear (1983) *Cell* **32:** 987-998.
Johnston, G.C., J.A. Prendergast and R.A. Singer (1991) *J. Cell Biol.* **113:** 539-551.
Kamada, T. (1994) In J.G.H. Wessels and F. Meinhardt (Eds.), *The Mycota Vol. I: Growth, Differentiation and Sexuality*, Springer-Verlag, Berlin, pp. 367-379.
Kamada, T., T. Fujii, T. Nakagawa and T. Takemaru (1985) *Curr. Microbiol.* **12:** 257-259.
Kaminskyi, S.G.W. and I.B. Heath (1995) *J. Cell Sci.* **108:** 849-856.
Kaminskyj, S.G.W. and I.B. Heath (1996) *Mycologia* **88:** 20-37.
Kaminskyj, S.G.W., A. Garrill and I.B. Heath (1992a) *Exp. Mycol.* **16:** 64-75.
Kaminskyj, S.G.W., S.L. Jackson and I.B. Heath (1992b) *J. Microsc.* **167:** 153-168.

Kane, P.M., C.T. Yamashiro, J.H. Rothman and T.H. Stevens (1989) *J. Cell Sci. Suppl.* **11:** 161-178.
Katz, D. and R.F. Rosenberger (1971) *J. Bacteriol.* **108:** 184-190.
Keenan, K.A. and R.L. Weiss (1997) *Fungal Genet. Biol.* **22:** 177-190.
Keon, J.P.R., S. Jewitt and J.A. Hargreaves (1995) *Gene* **162:** 141-145.
Klausner, R.D., J.G. Donaldson and J. Lippincott-Schwartz (1992) *J. Cell Biol.* **116:** 1071-1080.
Klionsky, D.J. (1997) *J. Membr. Biol.* **157:** 105-115.
Klionsky, D.J., P.K. Herman and S.D. Emr (1990) *Microbiol. Rev.* **54:** 266-292.
Kornfeld, R. and S. Kornfeld (1985) *Annu. Rev. Biochem.* **54:** 631-664.
Kornfeld, S. and I. Mellman (1989) *Annu. Rev. Cell Biol.* **5:** 483-525.
Kozutsumi, Y., M. Segal, K. Normington, M.-J. Gething and J. Sambrook (1988) *Nature* **332:** 462-464.
Kropf, D.L. (1986) *J. Cell Biol.* **102:** 1209-1216.
Kropf, D.L. and F.M. Harold (1982) *J. Bacteriol.* **151:** 429-437.
Kropf, D.L., M.D.A. Lupa, J.H. Caldwell and F.M. Harold (1983) *Science* **220:** 1385-1387.
Kropf, D.L., J.H. Caldwell, N.A.R. Gow and F.M. Harold (1984) *J. Cell Biol.* **99:** 486-496.
Kubicek, C.P., R. Messner, F. Gruber, R.L. Mach and E.M. Kubicek-Pranz (1993) *Enzyme Microb. Technol.* **15:** 90-99.
Lagunas, R. (1993) *FEMS Microbiol. Rev.* **104:** 229-242.
Legerton, T.L. and R.L. Weiss (1984) *J. Biol. Chem.* **259:** 8875-8879.
Legerton, T.L., K. Kanamori, R.L. Weiss and J.D. Roberts (1983) *Biochem.* **22:** 899-903.
Lehle, L. (1981) In W. Tanner and F.A. Loewus (Eds.), *Encyclopedia of Plant Physiology, New Series* **13B**, Springer-Verlag, Berlin, pp. 459-483.
Lehmann, L.E., W. Eberle, S. Krull, V. Prill, B. Schmidt, C. Sander, K. von Figura and C. Peters (1992) *EMBO J.* **11:** 4391-4399.
Lehmler, C., G. Steinberg, K.M. Snetselaar, M. Schliwa, R. Kahmann and M. Bölker (1997) *EMBO J.* **16:** 3464-3473.
Lemansky, P., A. Hasilik, K. von Figura, S. Helmy, J. Fishman, R.E. Fine, N.L. Kedersha and L.H. Rome (1987) *J. Cell Biol.* **104:** 1743-1748.
Levina, N.N., R.R. Lew, G.J. Hyde and I.B. Heath (1995) *J. Cell Sci.* **108:** 3405-3417.
Lew, R.R. (1998) *Fungal Genet. Biol.* **24:** 69-76.
Lewis, M.J., D.J. Sweet and H.R.B. Pelham (1990) *Cell* **61:** 1359-1363.
Lingappa, V.R., J. Chaidez, C.S. Yost and J. Hedgpeth (1984) *Proc. Natl. Acad. Sci. USA* **81:** 456-460.
Lippincott-Schwartz, J., J.G. Donaldson, A. Schweizer, E.G. Berger, H.-P. Hauri, L.C. Yuan and R.D. Klausner (1990) *Cell* **60:** 821-836.
López-Franco, R. and C.E. Bracker (1996) *Protoplasma* **195:** 90-111.
López-Franco, R., S. Bartnicki-Garcia and C.E. Bracker (1994) *Proc. Natl. Acad. Sci. USA* **91:** 12228-12232.

López-Franco, R., R.J. Howard and C.E. Bracker (1995) *Protoplasma* **188:** 85-103.
Makarow, M. (1985a) *EMBO J.* **4:** 1855-1860.
Makarow, M. (1985b) *EMBO J.* **4:** 1861-1866.
Makarow, M. (1988) *EMBO J.* **7:** 1475-1482.
Maras, M., X. Saelens, W. Laroy, K. Piens, M. Clayssens, W. Fiers and R. Contreras (1997) *Eur. J. Biochem.* **249:** 701-707.
Marcilla, A., M.V. Elorza, S. Mormeneo, H. Rico and R. Sentandreu (1991) *Arch. Microbiol.* **155:** 312-319.
Marsh, M. and A. Helenius (1980) *J. Mol. Biol.* **142:** 439-454.
Marty, F. (1978) *Proc. Natl. Acad. Sci. USA* **78:** 852-856.
Matile, P. (1978) *Annu. Rev. Plant Physiol.* **29:** 193-213.
McClure, W.K., D. Park and P.M. Robinson (1968) *J. Gen. Microbiol.* **50:** 177-182.
McGillivray, A.M. and N.A.R. Gow (1986) *J. Gen. Microbiol.* **132:** 2515-2525.
McKerracher, L.J. and I.B. Heath (1987) *Exp. Mycol.* **11:** 79-100.
Meaden, P., K. Hill, J. Wagner, D. Slipetz, S.S. Sommer and H. Bussey (1990) *Mol. Cell. Biol.* **10:** 3013-3019.
Mellman, I., R. Fuchs and A. Helenius (1986) *Annu. Rev. Biochem.* **55:** 663-700.
Merz, R.A., M. Horsch, L.E. Nyhlén and D.M. Rast (1999) In P. Jollès and R.A.A. Muzzarelli (Eds.), *Chitin and Chitinases* (in press).
Mims, C.W., E.A. Richardson, R.P. Clay and R.L. Nicholson (1995) *Intern. J. Plant Sci.* **156:** 9-18.
Mitchell, P. (1966) *Biol. Rev.* **41:** 445-502.
Mol, P.C. and J.G.H. Wessels (1987) *FEMS Microbiol. Lett.* **41:** 95-99.
Money, N.P. (1994) In J.G.H. Wessels and F. Meinhardt (Eds.), *The Mycota Vol. I: Growth, Differentiation and Sexuality*, Springer-Verlag, Berlin, pp. 67-88.
Money, N.P. (1997) *Fungal Genet. Biol.* **21:** 173-187.
Money, N.P. and F.M. Harold (1992) *Proc. Natl. Acad. Sci. USA* **89:** 4245-4249.
Money, N.P. and F.M. Harold (1993) *Planta* **190:** 426-430.
Money, N.P. and T.W. Hill (1997) *Mycologia* **89:** 777-785.
Montgomery, G.W.G. and G.W. Gooday (1985) *FEMS Microbiol. Lett.* **27:** 29-33.
Monzer, J. (1996) *Eur. J. Cell Biol.* **71:** 216-220.
Moore, D., M.M.Y. Elhiti and R.D. Butler (1979) *New Phytol.* **83:** 695-722.
Moore, D., B. Hock, J.P. Greening, V.D. Kern, L. Novak Frazer and J. Monzer (1996) *Mycol. Res.* **100:** 257-273.
Mueller, S.C. and D. Branton (1984) *J. Cell Biol.* **98:** 341-346.
Nolan, R.A. and A.K. Bal (1974) *J. Bacteriol.* **117:** 840-843.
Notario, V. (1982) *J. Gen. Microbiol.* **128:** 747-759.
Novick, P., S. Ferro and R. Schekman (1981) *Cell* **25:** 461-470.
Novikoff, A.B. and E. Holtzman (1970) *Cells and Organelles*, Holt, Rinehart and Winston, London.
Nykänen, M., R. Saarelainen, M. Raudaskoski, K.M.H. Nevalainen and A. Mikkonen (1997) *Appl. Environ. Microbiol.* **63:** 4929-4937.
Oakley, B.R. and J.E. Rinehart (1985) *J. Cell Biol.* **101:** 2392-2397.
Ohsumi, Y. and Y. Anraku (1983) *J. Biol. Chem.* **258:** 5614-5617.

Olden, K., R.M. Pratt, C. Jaworski and K.M. Yamada (1979) *Proc. Natl. Acad. Sci. USA* **76:** 791-795.

Paek, Y.L. and R.L. Weiss (1989) *J. Biol. Chem.* **264:** 7285-7290.

Payne, G.S. and R. Schekman (1985) *Science* **230:** 1009-1014.

Payne, G.S., D. Baker, E. van Tuinen and R. Schekman (1988) *J. Cell Biol.* **106:** 1453-1461.

Pearse, B.M.F. and M.F. Robinson (1984) *EMBO J.* **3:** 1951-1958.

Pearse, B.M.F. and M.F. Robinson (1990) *Annu. Rev. Cell Biol.* **6:** 151-171.

Peberdy, J.F. (1994) *Trends Biotechnol.* **12:** 50-57.

Pelham, H.R.B. (1989) *Annu. Rev. Cell Biol.* **5:** 1-23.

Pfanner, N., B.S. Glick, S.R. Arden and J.E. Rothman (1990) *J. Cell Biol.* **110:** 955-961.

Pfeffer, S.R. and J.E. Rothman (1987) *Annu. Rev. Biochem.* **56:** 829-852.

Picton, J.M. and M.W. Steer (1982) *J. Theor. Biol.* **98:** 15-20.

Pitt, D. (1968) *J. Gen. Mirobiol.* **52:** 67-75.

Pitt, D. and J.C. Barnes (1987) *Trans. Brit. Mycol. Soc.* **89:** 359-365.

Pitt, D. and J.C. Barnes (1993) *J. Gen. Microbiol.* **139:** 3053-3063.

Pitt, D. and P.J. Walker (1967) *Nature* **215:** 783-784.

Potapova, T.V., K.B. Aslanidi, T.A. Belozerskaya and N.N. Levina (1988) *FEBS Lett.* **241:** 173-176.

Pouyssegur, J., R.P.C. Shiu and I. Pastan (1977) *Cell* **11:** 941-948.

Preston, R.A., R.F. Murphy and E.W. Jones (1989) *Proc. Natl. Acad. Sci. USA* **86:** 7027-7031.

Raizada, M.K., J.S. Schutzbach and H. Ankel (1975) *J. Biol. Chem.* **250:** 3310-3315.

Rast, D.M., M. Horsch, R. Furter and G.W. Gooday (1991) *J. Gen. Microbiol.* **137:** 2797-2810.

Rees, B., V.A. Shepherd and A.E. Ashford (1994) *Mycol. Res.* **98:** 985-992.

Reinhardt, M.O. (1892) *Jahrb. Wiss. Bot.* **23:** 479-566.

Remy, W., T.N. Taylor, H. Haas and H. Kerp (1994) *Proc. Natl. Acad. Sci. USA* **91:** 11841-11843.

Rickert, W.S. and P.A. McBride-Warren (1974) *Biochim. Biophys. Acta* **336:** 437-444.

Riquelme, M., C.G. Reynaga-Peña, G. Gierz and S. Bartnicki-Garcia (1998) *Fungal Genet. Biol.* **24:** 101-109.

Robertson, N.F. (1965) *Trans. Brit. Mycol. Soc.* **48:** 1-8.

Rodriguez-Navarro, A., M.R. Blatt and C.L. Slayman (1986) *J. Gen. Physiol.* **87:** 649-674.

Roemer, T. and H. Bussey (1991) *Proc. Natl. Acad. Sci. USA* **88:** 11295-11299.

Roncal, T., U.O. Ugalde and A. Irastorza (1993) *J. Bacteriol.* **175:** 879-886.

Roos, W. (1989) *Biochim. Biophys. Acta* **928:** 119-133.

Rosa, A.L., A. Peralta-Soler and H.J.F. Maccioni (1990) *Exp. Mycol.* **14:** 360-371.

Rost, F.W.D., V.A. Shepherd and A.E. Ashford (1995) *Mycol. Res.* **99:** 549-553.

Rothman, J.E. and L. Orci (1992) *Nature* **355:** 409-415.

Rudick, M.J. and A.D. Elbein (1975) *J. Bacteriol.* **124:** 534-541.

Ruiz-Herrera, J. (1992) *Fungal Cell Wall: Structure, Synthesis, and Assembly*, CRC Press, Boca Raton.

Runeberg, P., M. Raudaskoski and I. Virtanen (1986) *Eur. J. Cell Biol.* **41:** 25-32.

Rupes, I., W.-Z. Mao, H. Åström and M. Raudaskoski (1995) *Protoplasma* **185:** 212-221.

Salo, V., S.S. Niini, I. Virtanin and M. Raudaskoski (1989) *J. Cell Sci.* **94:** 11-24.

Sanders, D. (1988) In D.A. Baker and J.L. Hall (Eds.), *Solute Transport in Plant Cells and Tissues*, Longman, Harlow, pp. 106-165.

Scarborough, G.A. (1970) *J. Biol. Chem.* **245:** 1694-1698.

Schekman, R. (1992) *Curr. Opin. Cell Biol.* **4:** 587-592.

Schmid, S., R. Fuchs, M. Kielian, A. Helenius and I. Mellman (1989) *J. Cell Biol.* **108:** 1291-1300.

Schreurs, W.J.A. and F.M. Harold (1988) *Proc. Natl. Acad. Sci. USA* **85:** 1534-1538.

Seeger, M. and G.S. Payne (1992a) *J. Cell Biol.* **118:** 531-540.

Seeger, M. and G.S. Payne (1992b) *EMBO J.* **11:** 2811-2818.

Seiler, S., F.E. Nargang, G. Steinberg and M. Schliwa (1997) *EMBO J.* **16:** 3025-3034.

Sentandreu, R., S. Mormeneo and J. Ruiz-Herrera (1994) In J.G.H. Wessels and F. Meinhardt (Eds.), *The Mycota Vol. I: Growth, Differentiation and Sexuality*, Springer-Verlag, Berlin, pp. 111-124.

Shepherd, V.A., D.A. Orlovich and A.E. Ashford (1993a) *J. Cell Sci.* **104:** 495-507.

Shepherd, V.A., D.A. Orlovich and A.E. Ashford (1993b) *J. Cell Sci.* **105:** 1173-1178.

Sietsma, J.H. and J.G.H. Wessels (1994) In J.G.H. Wessels and F. Meinhardt (Eds.), *The Mycota Vol. I: Growth, Differentiation and Sexuality*, Springer-Verlag, Berlin, pp. 125-141.

Sietsma, J.H., D.M. Rast and J.G.H. Wessels (1977) *J. Gen. Microbiol.* **102:** 385-389.

Sietsma, J.H., A.S.M. Sonnenberg and J.G.H. Wessels (1985) *J. Gen. Microbiol.* **131:** 1331-1337.

Simon, L., J. Bousquet, R.C. Levesque and M. Lalonde (1993) *Nature* **363:** 67-69.

Simon, V.R., T.C. Swayne and L.A. Pon (1995) *J. Cell Biol.* **130:** 345-354.

Slayman, C.L. (1987) *J. Bioenerg. Biomembr.* **19:** 1-20.

Slayman, C.L. and C.W. Slayman (1968) *J. Gen. Physiol.* **52:** 424-443.

Slayman, C.L. and C.W. Slayman (1974) *Proc. Natl. Acad. Sci. USA* **71:** 1935-1939.

Smith, S.E. and D.J. Read (1997) *Mycorrhizal Symbiosis* (2nd edition), Academic Press, San Diego.

Soliday, C.L. and P.E. Kollatukudy (1979) *Arch. Biochem. Biophys.* **197:** 367-378.

Srinivasan, S., M.M. Vargas and R.W. Roberson (1996) *Mycologia* **88:** 57-70.

Steinberg, G. (1998) *Fungal Genet. Biol.* **24:** 161-177.

Steinberg, G. and M. Schliwa (1993) *J. Cell Sci.* **106:** 555-564.

Steinberg, G. and M. Schliwa (1995) *Mol. Biol. Cell* **6:** 1605-1618.
Steinberg, G., M. Schliwa, C. Lehmler, M. Bölker, R. Kahmann and J.R. McIntosh (1998) *J. Cell Sci.* **111:** 2235-2246.
Stevens, T.H. (1992) *J. Exp. Biol.* **172:** 47-55.
St.Leger, R.J., R.C. Staples and D.W. Roberts (1992) *Gene* **120:** 119-124.
Stringer, M.A., R.A. Dean, T.C. Sewall and W.E. Timberlake (1991) *Genes Developm.* **5:** 1161-1171.
Strous, G.J.A.M., R. Willemsen, P. van Kerkhof, J.W. Slot, H.J. Geuze and H.F. Lodish (1983) *J. Cell Biol.* **97:** 1815-1822.
Suarit, R., P.K. Gopal and M.G. Shepherd (1988) *J. Gen. Microbiol.* **134:** 1723-1730.
Sweigard, J.A., S.N. Grove and A.A. Smucker (1979) *J. Cell Biol.* **83:** 307a.
Tachikawa, H., Y. Takeuchi, W. Funahashi, T. Miura, X.D. Gao, D. Fujimoto, T. Mizunaga and K. Onodera (1995) *FEBS Lett.* **369:** 212-216.
Takeshige, K., M. Baba, S. Tsuboi, T. Noda and Y. Ohsumi (1992) *J. Cell Biol.* **119:** 301-311.
Takeuchi, Y., J. Schmid, J.H. Caldwell and F.M. Harold (1988) *J. Membr. Biol.* **101:** 33-41.
Talbot, N.J., M.J. Kershaw, G.E. Wakley, O.M.H. de Vries, J.G.H. Wessels and J.E. Hamer (1996) *Plant Cell* **8:** 985-989.
That, T.C.C.T., K. Hoang-Van, G. Turian and H.C. Hoch (1987) *Eur. J. Cell Biol.* **43:** 189-194.
Timonen, S., R.D. Finlay, S. Olsson and B. Söderström (1996) *FEMS Microbiol. Ecol.* **19:** 171-180.
Tooze, J. and S.A. Tooze (1986) *J. Cell Biol.* **103:** 839-850.
Torii, S., T. Banno, T. Wantanabe, Y. Ikehara, K. Murakami and K. Nakayama (1995) *J. Biol. Chem.* **270:** 11574-11580.
Torralba, S., M. Raudaskoski and A.M. Pedregosa (1998) *Protoplasma* **202:** 54-64.
Vale, R.D. and L.S.B. Goldstein (1990) *Cell* **60:** 883-885.
van Brunt, J., J.H. Caldwell and F.M. Harold (1982) *J. Bacteriol.* **150:** 1449-1461.
van Rinsum, J., F.M. Klis and H. van den Ende (1991) *Yeast* **7:** 717-726.
Vermeulen, C.A. and J.G.H. Wessels (1986) *Eur. J. Biochem.* **158:** 411-415.
Vida, T.A. and S.D. Emr (1995) *J. Cell Biol.* **128:** 779-792.
von Grafenstein, H., C.S. Roberts and P.F. Baker (1986) *J. Cell Biol.* **103:** 2343-2352.
Walworth, N.C. and P.J. Novick (1987) *J. Cell Biol.* **105:** 163-174.
Ward, H.M. (1888) *Ann. Bot.* **2:** 319-382.
Waters, M.G., I.C. Griff and J.E. Rothman (1991) *Curr. Opin. Cell Biol.* **3:** 615-620.
Weber, R.W.S. and D. Pitt (1997a) *Mycol. Res.* **101:** 349-356.
Weber, R.W.S. and D. Pitt (1997b) *Mycol. Res.* **101:** 1431-1439.
Weber, R.W.S., G.E. Wakley and D. Pitt (1998) *Mycologist* **12:** 174-179.
Weber, R.W.S., G.E. Wakley and D. Pitt (1999) *Histochem. J.* **31** (in press).
Wessels, J.G.H. (1993) *New Phytol.* **123:** 397-413.

Wessels, J.G.H. (1997) *Adv. Microb. Physiol.* **38:** 1-45.

Wessels, J.G.H. and J.H. Sietsma (1981) In W. Tanner and F.A. Loewus (Eds.), *Encyclopedia of Plant Physiology, New Series* **13B**, Springer-Verlag, New York, pp. 352-394.

Wessels, J.G.H., D.R. Kreger, R. Marchant, B.A. Regensburg and O.M.H. de Vries (1972) *Biochim. Biophys. Acta* **273:** 346-358.

Wessels, J.G.H., J.H. Sietsma and A.S.M. Sonnenberg (1983) *J. Gen. Microbiol.* **129:** 1607-1616.

Wessels, J.G.H., P.C. Mol, J.H. Sietsma and C.A. Vermeulen (1990) In P.J. Kuhn, A.P.J. Trinci, M.J. Jung, M.W. Goosey and L.G. Copping (Eds.), *Biochemistry of Cell Walls and Membranes in Fungi*, Springer-Verlag, Berlin, pp. 81-95.

Wessels, J.G.H., O.M.H. de Vries, S.A. Ásgeisdottir and J. Springer (1991a) *J. Gen. Microbiol.* **137:** 2439-2445.

Wessels, J.G.H., O.M.H. de Vries, S.A. Ásgeisdottir and F.H.J. Schuren (1991b) *Plant Cell* **3:** 793-799.

Whitaker, M.J. and P.F. Baker (1983) *Proc. Roy. Soc. Lond. B* **218:** 397-413.

Whittaker, R.H. (1969) *Science* **163:** 150-160.

Wieland, F.T., M.L. Gleason, T.A. Serafini and J.E. Rothman (1987) *Cell* **50:** 289-300.

Wilson, C.L. (1973) *Annu. Rev. Phytopathol.* **11:** 247-272.

Wilson, C.L., G.A. Jumper and D.L. Mason (1978) *Phytopathol.* **68:** 1564-1567.

Wilson, C.L., G.A. Jumper and D.L. Mason (1980) *Phytopathol.* **70:** 783-788.

Wilson, D.W., S.W. Whiteheart, L. Orci and J.E. Rothman (1991) *Trends Biochem. Sci.* **16:** 334-337.

Wösten, H.A.B., S.M. Moukha, J.H. Sietsma and J.G.H. Wessels (1991) *J. Gen. Microbiol.* **137:** 2017-2023.

Wyatt, S.E. and N.C. Carpita (1993) *Trends Cell Biol.* **3:** 413-417.

Yaffe, M.P., D. Harata, F. Verde, M. Eddison, T. Toda and P. Nurse (1996) *Proc. Natl. Acad. Sci. USA* **93:** 11664-11668.

Youatt, J., N.A.R. Gow and G.W. Gooday (1988) *Protoplasma* **146:** 118-126.

Zhang, Q., J.M. Griffith and B.R. Grant (1992) *J. Gen. Microbiol.* **138:** 451-459.

Zhou, X.-L., M.A. Stumpf, H.C. Hoch and C. Kung (1991) *Science* **253:** 1415-1417.

Zlotnik, H., M.P. Fernandez, B. Bowers and E. Cabib (1984) *J. Bacteriol.* **159:** 1018-1026.

Applied Mycology and Biotechnology
Volume 1. Agriculture and Food Production
G.G. Khachatourians and D.K. Arora (editors)

Metabolic Regulation in Fungi

G. A. Marzluf

Department of Biochemistry, The Ohio State University, 484 West 12th Avenue, Columbus, Ohio 43210, U.S.A.

Metabolic Pathways in the fungi are subject to a high degree of regulation which frequently occurs at the level of transcription. Expression of structural genes within a control circuit is achieved by a network of interactions between metabolic repressors and inducers with both global-acting regulatory factors and pathway-specific factors. These sophisticated regulatory circuits operate in such a way to allow preferential use of certain compounds to supply a particular metabolite but also enable the cells to efficiently utilize alternative sources when necessary. Positive-acting regulatory proteins as well as negative-acting repressor proteins display sequence-specific DNA binding to elements within promoters. These control elements are arranged in organized arrays to allow multiple controlled responses from a single structural gene. Protein-protein interactions as well as the dynamic turnover of mRNAs and proteins also play a significant role in controlling metabolic pathways.

1. INTRODUCTION

Sophisticated genetic regulatory systems allow the fungi to utilize a diversity of compounds as sources of important metabolites. Accordingly, many different compounds can be used as sources of carbon, nitrogen, sulfur, and phosphorus, and in some cases a single compound provides two or more of these fundamental requirements. Global-acting regulatory circuits that include control genes and metabolites responsible for induction or repression integrate the expression of gene families involved in obtaining these fundamental elements. In addition, the biosynthesis of precursors of macromolecules, e.g. the amino acids, may be subject to very special regulatory mechanisms. Although a great deal of the control occurs at transcription, other types, especially translational regulation also play important roles. Similarly, much of the transcriptional control is positive, although negative-acting factors also exert significant control of metabolic pathways. This contribution will focus on major features of metabolic control in fungi, with emphasis upon that of *Aspergillus nidulans* and *Neurospora crassa*, the two best-studied filamentous fungi, but will also include some aspects of control in the yeast *Saccharomyces cerevisiae*. The conventions now in place to designate genes and their protein products differ for these species; in order to be consistent herein, genes will be presented italicized in small letters, e.g., *nit-2* while the protein encoded by a gene will be represented by capital letters, e.g., NIT2.

2. NITROGEN METABOLIC REGULATION

The fungi can use a surprisingly diverse array of compounds as nitrogen sources and are capable of expressing upon demand the catabolic enzymes of many different pathways. Extensive studies of nitrogen metabolism and its regulation have been conducted with *Saccharomyces cerevisiae, Aspergillus nidulans* and *Neurospora crassa.* Certain nitrogenous compounds - ammonia, glutamine, and glutamate - are preferentially used by these fungi. However, when these primary nitrogen sources are not available or are present at very low concentrations so as to limit growth, many secondary nitrogen sources can be used, e.g., amino acids, proteins, purines, nitrate, and amides. The use of these alternative nitrogen sources is highly regulated and usually requires the *de novo* synthesis of pathway-specific catabolic enzymes and permeases. Their expression is subject to nitrogen metabolic regulation and requires the activation of the structural genes, which are repressed by the presence of preferred nitrogen sources. The *de novo* synthesis of the permeases and catabolic enzymes of a particular catabolic pathway is controlled at the level of transcription and often requires two distinct positive signals: (1) A global signal indicating nitrogen derepression, and (2) a pathway-specific signal which indicates the presence of a substrate or an intermediate of that pathway. This two-step requirement permits the selective expression of just the enzymes of a specific catabolic pathway which allows the utilization of the specific nitrogen source available at the moment. Positive-acting global regulatory genes, *areA* in *Aspergillus* (1,2), *nit-2* in *Neurospora* (3,4), *gln-3* in yeast (5), *nut-1* in *Magnaporthe grisea (6)*, *areA-GF* in *Gibberella fujikuroi* (7), and *nre* in Penicillium (8) specify GATA-type zinc finger transcription factors which activate nitrogen structural genes in the absence of preferred nitrogen sources, i.e., these factors mediate nitrogen catabolite derepression. Regulatory proteins encoded by the pathway-specific control genes allow induction only of the enzymes for the specific pathway.

2.1 NITRATE ASSIMILATION

Nitrate assimilation has been subject to extensive investigation in *A. nidulans* and *N. crassa*. Inorganic nitrate is an excellent nitrogen source for *Aspergillus*, *Neurospora*, and many other fungi, but it is not utilized unless the cells lack the favored nitrogen sources (9-13). Utilization of nitrate requires the *de novo* synthesis of nitrate reductase and nitrite reductase, which is dependent upon both nitrogen derepression and specific induction by nitrate. Mutants which lack nitrate reductase can be readily isolated in many organisms because unlike the wild-type, they are resistant to chlorate. Thus, a simple two-way selection system permits isolation of mutants which lack nitrate reductase (chlorate resistant) and for revertants and suppressor mutants which restore nitrate reductase (use of nitrate). This feature can be exploited to obtain mutants in many different fungal species and has been used to develop transformation systems for various filamentous fungi (14). In addition to mutations in the structural genes which specify nitrate reductase and nitrite reductase, loss of function of molybdenum cofactor genes and of both major and minor regulatory genes lead to the inability to utilize nitrate as a nitrogen source (Table 1).

In *A. nidulans*, *niaD* and *niiA*, the structural genes which encode nitrate reductase and nitrite reductase, respectively, are closely linked but transcribed divergently from a common intergenic control region (12). In *N. crassa*, the structural genes for the two reductases are

Table 1
Aspergillus nidulans and *Neurospora crassa* nitrate assimilatory genes

Function	Genetic loci *Neurospora*	*Aspergillus*
Encodes nitrate reductase	*nit-3*	*niaD*
Encodes nitrate reductase	*nit-6*	*niiA*
Genes which specify a molybdenum cofactor	*nit-1* *nit-7,8,9*	*cnxABC* *cnxE,F,G,H*
Pathway-specific control gene (mediates induction)	*nit-4*	*nirA*
Global regulatory gene	*nit-2*	*areA*
Negative-acting control	*nmr*	*nmrA*

Wild-type strains and nmr mutants can utilize nitrate, but all of the other mutant strains shown above cannot grow on nitrate. Wild-type and all mutants grow readily with ammonium salts or glutamine as nitrogen sources.

unlinked although they are also regulated in a parallel fashion (11,15). Synthesis of messenger RNA for these enzymes in *Aspergillus* and in *Neurospora* requires both nitrogen limitation and nitrate induction (11,12,15). Upon induction and derepression, the synthesis of nitrate reductase mRNA occurs very rapidly and reaches a steady-state level within 15 minutes; this mRNA also turns over rapidly with a half-life of approximately 5 min; the nitrate reductase enzyme itself is subject to turnover (16-18). These features allow a very rapid response to changing environmental nitrogen sources.

The upstream promoter region of the *N. crassa nit-3* gene, which encodes nitrate reductase, possesses a strong binding site for the global-acting NIT2 protein at -180; two additional NIT2 binding sites are located more than 1 Kb upstream. Two binding sites for NIT4, the pathway specific transcription factor, also occur approximately 1 Kb upstream, immediately downstream of the distal NIT2 sites (19). Each of the NIT2 and NIT4 sites contribute to the regulated optimal expression of the nitrate reductase gene.

Regulation of the nitrate assimilatory genes in *A. nidulans* has been explored by mutating potential control sites in the intergenic control region, with ß-galactosidase and ß-glucuronidase serving as reporters for *niaD* (nitrate reductase) and *niiA* (nitrite reductase), respectively (20). The *niiA-niaD* intergenic control region contains four binding sites for NIRA, the pathway-specific factor that mediates nitrate induction, and ten GATA elements which *in vitro* studies identified as binding sites for AREA, the global-acting nitrogen control factor The four NIRA DNA binding sites each conform to a consensus nonpalindromic sequence, **CTCCGHGG,** where H = A, C or T, and all function in nitrate induction of the *niiA* and *niaD* genes(20). However, of the ten AREA binding sites identified by *in vitro* binding studies, only four centrally located sites appear to be physiologically important for nitrogen repression/derepression *in vivo*. These results remind us that a binding site recognized only by its sequence or even identified by *in vitro* DNA binding studies may not have an *in vivo* physiological role in controlling gene expression.

A number of other fungal species can utilize nitrate and possess linked structural genes for nitrate reductase and nitrite reductase which are divergently transcribed and regulated by nitrate induction and nitrogen repression. These include *Aspergillus niger*, *Aspergillus oryzae*, and *Penicillium chrysogenum*. The nitrate reductase genes of *Fusarium oxysporum* and *Leptosphaeria maculans* have also been isolated and sequenced (20). The promoter region of each of these structural genes contains at least one, and usually 3 to 4, putative NIRA-type binding sites. This feature, plus the finding that the *A. nidulans niaD* gene is normally controlled when introduced into several of these other species implies that these fungi possess a similar regulatory mechanism to control nitrate assimilation (20).

2.2 USE OF OTHER SECONDARY NITROGEN SOURCES

Purine metabolism is another example of a highly regulated nitrogen catabolic pathway in *A. nidulans* (21-23) and in *N. crassa* (24-27). The use of purines requires the *de novo* synthesis of a set of catabolic enzymes that occurs only upon nitrogen derepression, mediated by the global-acting AREA protein, and upon induction with uric acid, mediated by *uaY*, a pathway-specific regulatory gene. The *uaY* gene encodes a positive-acting regulatory protein that possesses a GAL4-like Zn_2/Cys_6 DNA binding domain. The UAY protein is required for the expression of at least nine unlinked genes which specify permeases and enzymes that function in the transport and metabolism of purines. The UAY polypeptide binds as a homodimer at promoter elements with a TCGG-N_6-CCGA sequence (21,22,28). This element is required for expression *in vivo* of the *uap* gene, which encodes a specific urate-xanthine permease (22,28).

The utilization by filamentous fungi of various other secondary nitrogen sources, such as acetamide, amino acids, and extracellular proteins is also highly regulated by similar derepression and pathway-specific induction signals as described above. Amino acid transport, L-amino acid oxidase, and phenylalanine-ammonia lyase are examples of activities whose expression in *Neurospora* requires induction, nitrogen derepression, and a functional *nit-2* gene product (29-32).

2.3 GLOBAL ACTING NITROGEN REGULATORY PROTEINS

A global positive-acting regulatory protein in *A. nidulans*, *N. crassa*, *Penicillium chrysogenum*, *Magnaporte grisea,* and numerous other fungi mediates nitrogen repression/derepression (1,3-5,8). These regulatory proteins are related and all members of the GATA family of transcription factors (Table 2). Each of them possess a remarkably similar DNA binding domain which consists of a single Cys_2/Cys_2 type zinc finger motif with a central loop of 17 amino acids and an immediately adjacent basic region and bind to DNA elements which have the core sequence GATA (thus the term

Table 2
Positive and Negative Acting Fungal GATA factors

GATA Factor	Number of Zinc fingers	Regulatory Function (+) or(-)	Species
GLN3	1	(+) Nitrogen	*Saccharomyces cerevisiae*
DAL80	1	(-) Nitrogen	*Saccharomyces cerevisiae*
AF-AREA	1	(+) Nitrogen	*Aspergillus fumigatus*
AREA	1	(+) Nitrogen	*Aspergillus nidulans*
AREA-GF	1	(+) Nitrogen	*Gibberella fujikuroi*
AREA	1	(+) Nitrogen	*Metarhizium anisopliae*
NRE	1	(+) Nitrogen	*Penicillium chrysogenum*
NUT1	1	(+) Nitrogen	*Magnaporthe grisea*
NIT2	1	(+) Nitrogen	*Neurospora crassa*
WC-1	1	(+) Light	*Neurospora crassa*
WC-2	1	(+) Light	*Neurospora crassa*
SRE	2	(-) iron uptake	*Neurospora crassa*
SREP	2	(-) iron uptake	*Penicillium chrysogenum*
URBS1	2	(-) iron uptake	*Ustilago maydis*

GATA factors). The *A. nidulans* AREA, *P. chrysogenum* NRE, and *N. crassa* NIT2 nitrogen regulatory proteins are closely related; AREA is 65% and 42% identical to NRE and NIT2, respectively. Amino acid identity is extremely high in the 50 amino acid sequence which constitutes the DNA binding domain. AREA and NIT2 turn on expression of many unlinked but coregulated genes. Individual structural genes can be expressed at markedly different levels which may at least be partly due to a different organization of AREA or NIT2 recognition elements in the promoters of target genes. The binding sites in the upstream promoter regions of different genes differ markedly in their number, orientation, location, and nucleotide sequence. In nearly all cases, however, AREA or NIT2 cannot alone activate gene expression but requires synergy with pathway-specific regulatory factors as described below. It is also important to recognize that these fungi possess additional GATA-binding factors which serve completely different regulatory functions, e.g., SRE acts in iron regulation and two GATA factors, WC1 and WC2, function together in blue-light signal transduction in *Neurospora crassa* (33,34). A major challenge is to understand how the various GATA factors, with overlapping DNA binding activities, each activates only their own specific set of target genes (Table 2).

2.4 EXPRESSION AND MODULATION OF AREA AND NIT2

Regulatory genes which specify trans-acting factors are themselves frequently subject to a high degree of control, which can include autogenous regulation. Expression of the global acting *areA* regulatory gene is indeed controlled at a number of steps. The *areA*$^{+}$ gene is highly expressed during nitrogen derepression conditions, yielding three different sized mRNAs, of approximately 3.9, 3.6 and 3.2 Kb; *areA* expression is greatly reduced during nitrogen repression (35). The various mRNAs appear to be functionally redundant. Strains with mutations which eliminate any one of these mRNAs still retain normal *areA* function *in vivo*. Thirteen GATA sequences - potential AREA binding sites - occur upstream of the *areA* coding region. The most proximal GATA element is located near the transcription start site for the smallest (3.2 Kb) *areA* mRNA (35). A point mutation which changes the first GATA element greatly reduces or even totally abolishes the 3.2 Kb mRNA, which is also missing in *areA* mutants which lack the AREA protein. These results imply that synthesis of the 3.2 Kb mRNA is controlled by positive autogenous regulation. In contrast, the 5' upstream region of the *N. crassa nit-2* gene lacks GATA sequence elements and its expression does not appear to be subject to autogenous regulation (36).

The *areA* transcript has a long 3' untranslated region (UTR) of 539 bases which contains a perfect direct repeat of 28 nucleotides with a six nucleotide overlap (37). Deletion of one copy of this tandem repeat was without any effect; however, deletion of both copies of the 28 nucleotide repeat yielded a strain that is significantly derepressed for nitrogen related activities. Thus, at least one copy of this element is essential for proper modulation of AREA function (37). This 3' UTR acts at the level of mRNA stability. Turnover of the wild-type areA transcript depends upon the nitrogen status of the cells, and it has a half-life of 40 minutes during nitrogen derepression conditions, but its half-life is only 7 minutes under N-repressed conditions. In contrast, an *areA* transcript which is deleted for the 28 nb tandem repeats has the same half-life, approximately 25 minutes, under both N-repressed and N-derepressed conditions (37). The molecular basis for the differential *areA* transcript stability is unknown, but may involve novel protein factors which recognize the tandem repeat elements. The *N. crassa nit-2* transcript does not contain sequences closely related to these *A. nidulans* nucleotide repeat elements; furthermore, the *nit-2* mRNA appears to be quite stable and is not subject to nitrogen-specific differential turnover.

The sequence of the carboxy terminus of the AREA and NIT2 proteins is conserved and this region plays an important role in nitrogen repression (38,39). When the carboxy C-terminus of the AREA and NIT2 proteins is deleted, they retain strong trans-activation function but their activity becomes largely insensitive to nitrogen catabolite repression. This region appears to interact with the negative-acting NMR protein as described below.

2.5 THE *NMR* REGULATORY PROTEIN AND REPRESSION SIGNALS

The *nmr* gene (nitrogen metabolic regulation) of *N. crassa* and *A. nidulans* is a global negative-acting regulatory gene whose function is required to establish nitrogen catabolite repression. Mutations in the *nmr* gene result in strains derepressed for nitrate reductase and other nitrogen controlled activities in the presence of sufficient ammonia or glutamine to completely repress their expression in *nmr*$^+$ strains (40,41). In the *nmr* mutants, synthesis of nitrate reductase is largely insensitive to nitrogen catabolite repression but still requires induction by nitrate and functional NIT2 and NIT4 proteins (40,42,43). *nmr* might encode a repressor protein or somehow modulate the activity of the positive-acting NIT2 protein. The *N. crassa nmr* gene encodes a protein of 488 amino acids that has no distinctive features such as obvious DNA-binding or protein kinase motifs (44).

Several lines of evidence indicate that the NMR protein functions as a negative regulator by binding to the NIT2 protein and in some way inhibiting the later's trans-activation function, possibly by interfering with DNA binding. A direct interaction between the NMR and NIT2 proteins has been demonstrated by biochemical techniques and supported by genetic analysis (45). Two distinct short regions of the NIT2 protein, both predicted to exist as α-helices, appear to be recognizes by the NMR protein (45). One of these regions is an α-helix within the zinc finger DNA binding domain, and the second is the an α-helix at the carboxy terminus of the protein. Mutant NIT2 proteins with amino acid substitutions in either of these α-helical motifs fail to bind to NMR *in vitro* and display a N-derepressed phenotype *in vivo* (46). These results provide persuasive evidence that the NMR protein exerts a negative regulatory action by binding directly to the NIT2 protein and blocking the trans-activation function of NIT2 during conditions of nitrogen repression. Glutamine appears to be the critical metabolite which exerts nitrogen catabolite repression (10,47). The identity of the element or signal pathway system that senses the presence of repressing levels of glutamine remains a mystery. It is possible that the AREA, NIT2, GLN3 and similar global regulators themselves bind glutamine, or that an accessory protein such as NMR or a complex such as a NIT2/NMR heterodimer recognizes glutamine. On the other hand, as yet unidentified factor(s) may detect glutamine and convey the repression signal to the global activating proteins. An important goal for future research is to identify the signaling system that recognizes and processes environmental nitrogen cues.

2.6 REGULATION OF ALLANTOIN METABOLISM IN YEAST

A complex genetic regulatory system controls allantoin metabolism and the use of other nitrogen sources in *S. cerevisiae* (48). Yeast can use exogenous allantoin as a nitrogen source. Eight structural genes which specify permeases and enzymes involved in this metabolic pathway have been characterized. Some, but not all, of these genes require induction, all are sensitive to nitrogen catabolite repression. Three types of cis-acting elements are found in the 5' promoter region of DAL7 and other inducible genes: UAS_{NTR} (upstream activating nitrogen control sequence), URS (upstream repressing sequence), and UIS (upstream induction sequence).

Five genes, *gln3, dal81, dal82, dal80*, and *ure2* encode regulatory proteins that function to control the utilization of allantoin. Some are pathway specific whereas others are more global in action. GLN3 is a positive-acting GATA-binding protein that is required for expression of all of the allantoin pathway genes and many other genes subject to nitrogen catabolite repression/derepression (5). DAL80 is a negative-acting regulatory protein which binds at URS_{GATA} sites, elements which contain two GATA sequences 15 to 20 bp apart (49,50). The DAL81/UGA35 and DAL82 regulatory proteins are both required for induction of the structural genes served by an UIS element(s). The *dal7* promoter contains UAS_{NTR} sites, recognized by the positive-acting GLN3 protein, and URS sites for the negative-acting DAL80 protein (48). These sites overlap so that GLN3p and DAL80p compete for binding. During nitrogen derepression, the promoter is ready for action but still quiescent. When inducer is present, the UIS is occupied, presumably by DAL81 and DAL82; this causes the balanced competition between GLN3 (positive) and DAL80 (negative) to favor GLN3, leading to enhanced expression of *dal7* and the similarly controlled genes, *dal1, dal2*, and *dur1*.

A surprising result was the finding that mutation of the *ure2* gene results in a loss of nitrogen catabolite repression for some nitrogen catabolic genes, and yet other genes remain subject to nitrogen repression. This suggests that URE2 functions in only one of two branches of the nitrogen regulatory network. URE2 appears to act by way of an interaction with GLN3 (51). A newly discovered regulatory gene, *gat1,* (also called *nil1*) encodes a GATA binding protein with significant homology with GLN3 (52,53). Both NIL1/GAT1 and GLN3 recognize the same GATAAG sites to activate the expression of the *gap1* gene (54). GAT1/NIL1 is a positive activator and is required for full expression of many, but not all, nitrogen related genes. Interestingly, the *dal5, put1*, and *uga4* all require both GLN3 and GAT1/NIL1 for strong expression. The positive effects of GLN3 or GAT1/NIL1 appear to differ depending upon the available nitrogen sources (52). Further work should provide new insight into the presence of dual nitrogen regulatory networks in yeast and explain unexpected results, e.g., the finding that the two GATA-binding activators, GLN3 and GAT1/NIL1, which have similar DNA binding specificity, act together synergistically in some cases, but also can act individually to turn on certain genes.

3. REGULATION OF ACQUISITION OF CARBON SOURCES

Among the multitude of filamentous fungi, the control of carbon metabolism is by far the best understood in *Aspergillus nidulans*. Carbon catabolite repression acts to insure that primary carbon sources, e.g., glucose, are preferentially used. Secondary carbon sources, e.g., ethanol, quinate, proline, acetate, or acetamide are only utilized when primary compounds are absent. Expression of the structural genes which specify the enzymes required to metabolize the alternative carbon sources requires the lifting of catabolite repression and pathway-specific induction (55). The *creA* gene plays the major role in establishing carbon catabolite repression, and, unlike the wild-type, *creA*- mutants express enzymes needed for the utilization of ethanol, proline, acetamide and similar alternative carbon sources in the presence of glucose (55,56).

3.1 CREA IS A NEGATIVE-ACTING REGULATORY PROTEIN

CREA is a sequence-specific DNA binding protein; it contains two Cys2/His2 zinc fingers and binds to a GC-rich consensus sequence, G/CPyGGG/AG, an element similar to that recognized by the well-known mammalian SP1 protein (57-59). The loss of CREA function due to mutation yields a derepressed phenotype, in agreement with its negative action to preclude the expression of various carbon-related genes in the presence of glucose.

In addition to the two zinc finger motifs, CREA contains an alanine rich region, an acid region that may be an activation domain, plus a stretch of 42 amino acids that is identical to that of a homologous protein from *A. niger* and with similarity to the corresponding region of the yeast RGR1 protein of *Saccharomyces cerevisiae* (55,57).

3.2 THE ETHANOL UTILIZATION REGULON

A cluster of genes involved in the utilization of ethanol in *A. nidulans* is subject to

carbon catabolite repression; their expression also requires induction by an alcohol mediated by a pathway-specific transcription factor, ALCR (60,61). This cluster contains *alcR*, which encodes the regulatory protein, *alcA*, which specifies alcohol dehydrogenase, plus two additional genes, *alcX* and *alcM*, whose function is unknown (60). An unlinked gene, *aldA*, which encodes aldehyde dehydrogenase, is controlled in parallel with *alcA*.
ALCR is a DNA-binding protein with a Cys6/Zn2 binuclear zinc cluster. ALCR mediates induction by ethanol or related compounds of each of the *alc* genes and that of *aldA*. The ALCR protein binds to elements located in the promoters of each of these genes; an unusual aspect is that the half sites of the DNA binding element can occur as direct repeats or as inverted repeats (62,63). ALCR is subject to positive autogenous control and binds to elements in its own promoter, thereby, upon induction, elevating its own expression.

3.3 REGULATION OF THE ETHANOL-UTILIZATION GENES

During conditions of carbon catabolite repression, CREA prevents expression of *alcA* and *aldA*, whose products represent the enzymes required for utilization of ethanol. The repression by CREA is two-fold, direct and indirect: (1) CREA binds at elements in the *alcA* and *aldA* promoters and directly inhibits expression of these two structural genes; (2) CREA also binds within the *adhR* promoter and thus inhibits expression of ADHR, the positive-acting factor required to turn on its own expression and that of *alcA* and *aldA* (61).
Significantly, the *alcR* and *alcA* promoter elements for binding of CREA and ALCR overlap so that a competition occurs between these factors when both glucose and ethanol are present, thus allowing some expression of this set of genes. Finally, CREA and the creA DNA binding sites play an important regulatory role even during carbon derepression conditions, and function to limit the level of expression of the *alc* genes; mutational loss of CREA or of its binding sites leads to a significant over-expression of *alcA* and other members of the regulon (55,61).

4. GENES SUBJECT TO DUAL NITROGEN AND CARBON CONTROL

Metabolism of some compounds which can be utilized as both sources of nitrogen and carbon, is subject to dual nitrogen and carbon regulation. Proline can be utilized by *Aspergillus nidulans* as both a nitrogen or carbon source (64,65). A cluster of five genes, *prn A, X, D, B and C*, are responsible for proline metabolism; *prnA* encodes a regulatory protein that mediates proline induction of the structural genes. The function of *prnX* is unknown. Although tightly linked, each gene within the *prn* cluster is expressed individually to yield a monocistronic mRNA (65). The expression of these genes requires induction by proline and is also controlled by both nitrogen and carbon catabolite repression. When proline is present, both glucose and ammonia are required to repress expression their expression. A central control region between *prnD* and *prnB* contains control elements for the PRNA protein and for CREA and AREA, the global-acting factors for carbon and nitrogen control, respectively (66). The *prnB* gene, which encodes the proline permease, is the major target for control, and inducer exclusion is at least in part responsible for regulation of the other *prn* genes of the cluster.

The pathway-specific protein encoded by *prnA* is a DNA binding protein with a GAL4-like Cys_6/Zn_2 domain and a glutamine-rich putative activation domain. Proline, the inducer, activates the PRNA protein, which binds at elements with the sequence CCGG-N_{16}-CCGG (direct repeats of CCGG separated by 16). Nitrogen catabolite derepression of *prnB* requires a functional AREA protein, and two AREA binding sites occur in the *prnD-prnB* intergenic control region. Carbon repression is exerted by the negative-acting CREA protein. The promoter is organized such that upon induction, mediated by PRNA, a limitation for either carbon or nitrogen leads to strong *prnB* gene expression. Control of the *prnB* gene and the entire *prn* cluster represents an elegant example of complex regulation in which multiple signals converge to control expression.

4.1 REGULATION OF ACETAMIDASE EXPRESSION

Another case in which multiple control signals converge upon a single structural gene is the regulation of *amdS*, which encodes acetamidase of *A. nidulans* (67). Acetamide also serves both as a nitrogen and as a carbon source for *Aspergillus*. Expression of *amdS* is highly regulated and requires a derepression signal and at least one of several possible induction signals. Derepression is achieved by limitation for either nitrogen or carbon, mediated by the AREA and CREA proteins, respectively (67). The FACB protein mediates acetate induction of *amdS* and other genes which encode enzymes required for acetate utilization. The FACB protein has a Zn_2/Cys_6 DNA binding motif (68). Expression of *facB* itself is subject to carbon catabolite repression and to acetate induction. The *amdS* gene can also be induced by omega amino acids, e.g., Γ-amino butyrate. Induction is mediated by AMDR, which contains a Zn_2/Cys_6 DNA binding domain (69).

The *amdS* promoter is complex and modular. It contains distinct binding elements for multiple positive and negative acting regulatory proteins, AREA, CREA, FACB, AMDR, and AMDA proteins and also contains a CCAAT sequence which is required for basal expression (70). The CCAAT element is a binding site for the ANCF protein (67). An understanding of the multiple possible DNA-protein and protein-protein interactions which regulate *amdS* will provide insight into the molecular mechanisms by which multiple regulatory factors control expression of a single gene.

4.2 THE SULFUR REGULATORY CIRCUIT

The filamentous fungi *Aspergillus nidulans* and *Neurospora crassa* and the yeast *Saccharomyces cerevisiae* each possess a regulatory circuit which controls the expression of permeases and enzymes which function in the acquisition of sulfur from the environment and its assimilation. When primary sulfur sources are available, catabolite repression prevents the expression of enzymes that are required to utilize secondary sulfur sources. The structural genes which specify these catabolic activities are controlled at the transcriptional level by both positive-acting and negative-acting regulatory factors. In these model organisms, and presumably in most fungi, a regulatory circuit acts to maintain an adequate source of sulfur and to repress the synthesis of various sulfur catabolic enzymes when the cells possess an adequate internal supply of sulfur.

4.3 UTILIZATION OF ALTERNATIVE SULFUR SOURCES

When primary sulfur sources are not available, e.g., methionine, *N. crassa, A. nidulans* and other filamentous fungi can utilize a variety of compounds to supply sulfur.

Choline-0-sulfate occurs widely throughout the plant kingdom and is stored in many fungi, where it may serve as an osmoprotectant and as an internal sulfur source. *Neurospora crassa* readily uses choline-O-sulfate as a secondary sulfur source. Exogenous choline-O-sulfate is transported into the cells as an intact molecule via a specific permease, followed by its hydrolysis by choline sulfatase to yield an internal pool of inorganic sulfate. Expression of choline-0-sulfate permease and choline sulfatase is highly regulated and dependent upon the lifting of sulfur catabolite repression and requires the action of the CYS3 positive regulatory protein. Tyrosine-O-sulfate and other aromatic sulfate esters also serve as secondary sulfur sources. After uptake of tyrosine-O-sulfate, it is hydrolyzed by aryl sulfatase, yielding internal pools of inorganic sulfate and tyrosine (71). Expression of the aromatic sulfate ester transport system and of aryl sulfatase is dependent upon a functional *cys-3*$^{+}$ regulatory gene and is strongly repressed by methionine (71).

N. crassa possesses two distinct sulfate permease species which are encoded by separate, unlinked genes, *cys-13* and *cys-14* (72). These two sulfate transport systems are both subject to sulfur catabolic repression and are positively controlled by the *cys-3* regulatory gene. The *N. crassa cys-14*$^{+}$ gene, the first eukaryotic sulfate transporter gene to be cloned, encodes a protein of approximately 90 kD with 12 putative hydrophobic membrane-spanning domains (73). A number of genes which encode H^{+}/sulfate cotransporters have recently been identified and appear to represent a new superfamily of membrane transport proteins. Smith et al. (74) isolated the yeast *SUL1* gene, and used a sul1 mutant strain to isolate via complementation three sulfate transporter cDNAs derived from the higher plant *Stylosanthes hamata* (75). Two of these, *shst1* and *shst2*, encode distinct high affinity sulfate permeases that are expressed in the roots and presumably function in uptake of inorganic sulfate from the environment. A third gene, *shst3*, specifies a low affinity sulfate transporter that is expressed in leaves and apparently is involved in the internal transport of sulfate between cells within the plant (75). A sulfate transporter has been identified as the product of the gene responsible for the debilitating human genetic disease, diastrophic dysplasia and was identified by comparison to the known sulfate transporters (76). Severe skeletal abnormalities are associated with with genetic disease appear to be impaired sulfation of proteoglycans in cartilage because of a deficiency in sulfate uptake (76). The sulfate transporters of the fungi, higher plants, and mammals show significant homology to each other, and appear to define a new superfamily distinct from other transporters. These findings represent an example in which research with fungi has revealed important concepts in genetic regulation and also contributed to important studies with higher plants and mammals.

4.4 SULFUR REGULATORY GENES

In *Neurospora*, at least three distinct regulatory genes, scon-1, scon-2, and cys-3, regulate the expression of the structural genes which specify sulfur catabolic enzymes (77-79). The scon-1 and scon-2 genes exert negative control over *cys-3*. Loss of *scon-1* or *scon-2* function leads to the constitutive expression of the CYS3 regulatory gene and structural genes of the sulfur circuit. The *scon* genes appear to represent a sequential control network that controls the *cys-3*$^+$ regulatory gene, which in turn, directly controls expression of the structural genes. The *cys-3*$^+$ gene encodes a positive-acting regulatory protein which turns on the expression of cys-14 and ars and, presumably, each of the other coregulated structural genes. The CYS3 protein is a member of the large family of bZip DNA-binding proteins which include the yeast GCN4 protein and the mammalian proteins FOS, JUN, CREB, and C/EBP1 (80-83).

The CYS3 protein binds *in vitro* to multiple sites upstream of the *cys-14*, *ars*, and *scon-2* genes, as well as to two sites in the 5' promoter region of the cys-3 gene itself (80,84-86). CYS3 binding sites have the consensus sequence ATGRYRYCAT, which represents two abutting 5 bp half-sites (87). Natural binding sites differ in their affinity for the CYS3 protein and can deviate slightly from the consensus sequence (87,88).

The sulfur controller-2 (*scon-2*) gene of *N. crassa* plays a integral role in sulfur control by regulating the positive activator CYS3. In *scon-2* mutants, *cys-3* and the sulfur-controlled structural genes such as *ars* and *cys-14* are insensitive to catabolite repression and expressed even in the presence of high levels of sulfur which repress these genes in *scon-2*$^+$ strains. The *scon-2*$^+$ gene encodes a protein which contains in its carboxy-terminal half contains six ß-transducin repeats, each consisting of approximately 40 amino acid residues (86). Proteins of the ß-transducin family regulate gene expression, and although the exact function of the repeat structures is still unclear, they appear to mediate protein-protein interactions.

The *scon-2* gene itself is also highly regulated and its expression only occurs during sulfur-derepressing conditions and requires a functional CYS3 protein. The *scon-2* promoter has several CYS3 binding sites (86). The *scon-1* and *scon-2* gene products appear to act together to prevent *cys-3* expression; the *scon-1* mutant expresses *cys-3*, *ars*, and *cys-14* constitutively.

4.5 OPERATION OF THE *NEUROSPORA* SULFUR CIRCUIT

A number of critical questions must be solved to fully understand fully the action of the sulfur regulatory circuit, e.g., it is important that the *scon-1* gene be isolated and the nature of its product fully characterized. Similarly, it is essential to identify the factor which recognizes the repressing sulfur metabolite and the way the signal is conveyed to the *scon* gene products. Yet, sufficient insight of the sulfur circuit is available to suggest some features of its operation. It is clear that the positive-acting CYS3 protein carries out the final step during sulfur derepression by turning on *ars*, *cys-14*, and other genes which encode sulfur catabolic enzymes. During sulfur repression conditions, the SCON2 ß-transducin protein plays a central role in preventing *cys-3* function; however, the SCON2 protein is present in the greatest amounts, presumably in an inactive state, during sulfur limitation, when CYS3 is actively turning on the structural genes of the circuit. This apparent paradox most likely represents a complex interacting system of positive and negative feedback loops, so that when the cells suddenly experience an abundant sulfur source, the cellular pool of SCON2 can be rapidly converted into an active form to prevent CYS3 function, thus quickly shutting down expression within the entire sulfur circuit. SCON2, or a SCON1-SCON2 heteromultimer may bind directly to the CYS3 protein by virtue of the ß-transducin repeat motifs and promote CYS3 protein turnover or inhibit its function. The loss of CYS3 (or inhibition of its function) will prevent further *cys-3* gene expression which is dependent upon positive autogenous control. The cellular content of *cys-3* mRNA and CYS3 protein will rapidly decrease during sulfur repression due to turnover, and consequently, the family of structural genes which require CYS3 for expression will be silenced (89). Once the sulfur circuit is shut down, a much smaller amount of SCON2 may suffice to hold it in a quiescent state. When the fungal cells that have enjoyed an abundant supply of sulfur experience sulfur starvation, the negative controls preventing *cys-3* expression may be overturned by conversion of SCON1 and SCON2 into inactive forms, and cys-3 transcription is turned on. The resulting CYS3 protein, via positive autogenous control further increases *cys-3* expression, which in turn leads to expression of the entire set of unlinked sulfur-controlled structural genes.

4.6 SULFUR REGULATION IN *ASPERGILLUS NIDULANS* AND YEAST

Four different "sulfur controller" genes, sconA, sconB, sconC, and sconD, have been identified in *A. nidulans* by mutations, each resulting in the loss of methionine repression of sulfur amino acid biosynthetic enzymes, and appear to resemble the *scon-1* and *scon-2* mutants of *Neurospora* and MET30 of yeast (90). The *sconB* gene encodes a protein of the ß-transducin family containing seven WD-40 repeats that span the C-terminal half of the protein (A Paszewski, unpublished data). The SCONB transducin protein of *Aspergillus* is closely related to the *Neurospora* SCON2 protein (74% amino acid identity) and to the *S. cerevisiae* MET30 protein (63% identity). These proteins appear to carry out similar functions in sulfur regulation. The *Aspergillus sconC* gene has also been isolated and encodes a protein that displays 57% homology to the human RNA polymerase II elongation factor-like OCP2 protein (A Paszewski, unpublished results). SCONC and the OCP2 proteins each possess a PEST sequence and a P-loop motif which may be important in their action.

In *Saccharomyces cerevisiae*, the structural genes which encode the enzymes for the biosynthesis of cysteine and methionine are all repressed by high concentrations of S-adenosyl methionine. At least four different regulatory factors participate in controlling this set of structural genes. One factor, MET30 acts negatively and has five ß-transducin repeats with homology to the *Neurospora* SCON2 protein (91). Expression of the entire set of yeast sulfur assmilatory genes is dependent upon MET4, a positive-acting protein, which has a bZip motif with homology to the *N. crassa* CYS3 protein (92,93). A second bZip protein, MET28 is also required for optimal expression of the *met* structural genes, except one, *met25* (94). MET28 binds only very weakly to DNA and lacks intrinsic activation potential; MET28 appears to function in MET28-MET4 heterodimers, formed by interaction of their leucine zipper motifs. Different regions of the bZip domain of MET4 and MET28 have strong homology with the *Neurospora* CYS3 protein. The bZip protein CYS3 of Neurospora appears to fulfill the functions of these two yeast bZip proteins (94). Surprisingly, expression of some yeast *met* genes requires CBF1 (centromere binding factor-1). A heteromeric complex which contains the MET4, MET28 and CBF1 proteins binds at the *met16* promoter (94). Neither MET4 nor MET28 alone can bind the *met16* promoter, but only as a member of the high molecular weight complex containing CBF1. CBF1-dependent changes in the chromatin structure at the *met16* locus are restricted to the immediate region of the CDEI element, indicating that CBF1 does not act by phasing nucleosomes; rather, it stabilizes the DNA binding of MET4 and MET28 (94,95).

The recent advances in our understanding of the operation of the sulfur regulatory circuit in *A, nidulans, N. crassa*, and *S. cerevisiae* are significant. Similarities but also significant differences in sulfur genetic regulation occur in these closely related organisms. Additional work should lead to an appreciation of the precise molecular mechanisms and multiple interactions which govern this important area of cellular metabolism.

5. CONVERGENCE OF MULTIPLE REGULATORY CIRCUITS

Extracellular proteins can serve as a sole source of nitrogen, carbon, or sulfur for *N. crassa*. The expression of a single structural gene that encodes an extracellular alkaline protease is turned on by distinct regulatory factors that signal a limitation for N, C, or S (96). The response to sulfur catabolite derepression is mediated by CYS3, the bZip regulatory protein which controls the entire set of sulfur catabolic genes (81). On the other hand, when sulfur is plentiful, but the cells are starving for carbon, derepression must involve the function of CRE as defined in *A. nidulans*; similarly, when limited only for nitrogen, derepression of the alkaline protease gene is mediated by the nitrogen regulatory factors. Thus, the promoter that serves the protease gene must be extremely complex with elements which allow its activation by signals from the three independent circuits. To add to this complexity, it is noteworthy that synthesis and secretion of the alkaline protease not only requires derepression but also is dependent upon the presence of an extracellular protein; a peptide derived from the external protein is thought to provide an inductive signal (97). Similarly, *Aspergillus nidulans* expresses neutral and alkaline proteases when subjected to multiple derepression states (98). In *N. crassa*, an extracellular alkaline ribonuclease is synthesized upon limitation for either nitrogen, carbon, or phosphorus. Expression of this phosphatase gene requires activation by the NIT2 or NUC1 positive regulatory proteins to respond to nitrogen or phosphorus starvation, respectively (99). The promoters which govern the structural genes encoding these extracellular proteases and nucleases must be unusually complex and contain multiple elements that allow responses to several independent derepression signals and to requisite inductive signals, as well as the ambient pH (100,101).

6. OTHER METABOLIC CONTROL SYSTEMS

There exist other equally complex and exciting systems which regulate other metabolic pathways. Unusual mechanisms occur in many of these other systems, which, regrettably, cannot be addressed here. Among the many areas of metabolic regulation which deserve exploration are the fascinating systems of general and cross-pathway controls of amino acid biosynthesis in the fungi (102), pH-control of gene expression

(103,104), iron metabolism (105,106), and the remarkably complex regulatory network that governs phosphorus metabolism (107,108).

7. ACKNOWLEDGMENT

Research in the author's laboratory has been supported by grant GM-23367 from the National Institutes of Health.

REFERENCES

1. Kudla, B., Caddick, M.X., Langdon, T., Martinez-Rossi, N.M., Bennett, C.F., Sibley, S., Davis, R.W., and Arst, H.N. (1990) EMBO J.9, 1355-1364.
2. Caddick, M.X. (1992) In Molecular Biology of Filamentous Fungi (U. Stahl and P. Tudzynski, Eds.), pp. 141-152. VCH Press, Weinheim.
3. Stewart, V. and Vollmer, S.J. (1986) Gene46, 291-295.
4. Fu, Y.H. and Marzluf, G.A. (1987) Mol. Cell. Biol.7, 1691-1696.
5. Minehart, P.L. and Magasanik, B. (1991) Mol. Cell. Biol.12, 6216-6226.
6. Froeliger, E. and Carpenter, B. (1996) Mol. Gen. Genet.251, 647-656.
7. Tudzynski, B., Homann, V., Feng, B., and Marzluf, G.A. (1999) Mol. Gen. Genet.261, 106-114.
8. Haas, H., Bauer, B., Redl, B., Stoffler, G., and Marzluf, G.A. (1994) Curr. Genet.
9. Cove, D.J. (1979) Biol. Rev.54, 291-327.
10. Premakumar, R., Sorger, G.J., and Gooden, D. (1979) J. Bacteriol.137, 1119-1126.
11. Fu, Y.H. and Marzluf, G.A. (1987) Proc. Natl. Acad. Sci. USA84, 8243-8247.
12. Johnstone, I.L., McCabe, P.C., Greaves, P., Cole, G.E., Brow, M.A., Gurr, S.J., Unkles, S.E., Clutterbach, A.J., Kinghorn, J.R., and Innis, M. (1990) Gene90, 181-192.
13. Crawford, N.M. and Arst, H.N. (1993) Ann. Rev. Genet.27, 115-146.
14. Unkles, S.E., Campbell, E.I., Carrez, D., Grieve, C., Contreras, R., Fiers, W., and van den Hondel, C.A.M. (1989) Gene78, 157-166.
15. Exley, G.E., Colandene, J.D., and Garrett, R.H. (1993) J. Bacteriol.175, 2379-2392.
16. Okamoto, P.M., Garrett, R.H., and Marzluf, G.A. (1993) Mol. Gen. Genet.238, 81-90.
17. Sorger, G.J. and Premakumar, R. (1978) Biochim. Biophys. Acta540, 33-47.
18. Marzluf, G.A. (1981) Microbiol. Revs.45, 437-461.
19. Fu, Y.H., Feng, B., Evans, S., and Marzluf, G.A. (1995) Mol. Microbiol.15, 935-942.
20. Punt, P.J., Strauss, J., Smit, R., Kinghorn, J.R., van den Hondel, C.A., and Scazzocchio, C. (1995) Mol. Cell. Biol.15, 5688-5699.
21. Diallinas, G. and Scazzocchio, C. (1989) Genetics122, 341-350.
22. Suárez, T., Oestreicher, N., Kelly, J., Ong, G., Sankarsingh, R., and Scazzocchio, C. (1991) Mol. Gen. Genet.230, 359-368.
23. Suárez, T., Oestreicher, N., Peñalva, M.A., and Scazzocchio, C. (1991) Mol. Gen. Genet.230, 369-375.
24. Reinert, W.R. and Marzluf, G.A. (1975) Mol. Gen. Genet.139, 39-55.

25. Nahm, B.H. and Marzluf, G.A. (1987) J. Bacteriol.170, 1943-1948.
26. Lee, H.J., Fu, Y.H., and Marzluf, G.A. (1990) Mol. Gen. Genet.222, 140-144.
27. Lee, H.J., Fu, Y.H., and Marzluf, G.A. (1990) Biochem.29, 8779-8787.
28. Suarez, T., de Queiroz, M.V., Oestreicher, N., and Scazzocchio, C. (1995) EMBO J.14, 1453-1467.
29. Facklam, T. and Marzluf, G.A. (1978) Biochem. Genetics16, 343-350.
30. Sikora, L. and Marzluf, G.A. (1982) J. Bacteriol.150, 1287-1291.
31. Sikora, L. and Marzluf, G.A. (1982) Mol. Gen. Genet.186, 33-39.
32. Xiao, X.D. and Marzluf, G.A. (1993) Curr. Genet.24, 212-218.
33. Ballario, P., Vittorioso, P., Magrelli, A., Talora, C., Cabibbo, A., and Macino, G. (1996) EMBO J.15, 1650-1657.
34. Linden, H. and Macino, G. (1997) EMBO J.16, 98-107.
35. Langdon, T., Sheerins, A., Ravagnani, A., Gielkens, M., Caddick, M.X., and Arst, H.N. (1995) Mol. Microbiol.17, 877-888.
36. Chiang, T.Y. and Marzluf, G.A. (1994) Biochem.33, 576-582.
37. Platt, A., Langdon, T., Arst, H.N., Kirk, D., Tollervey, D., Sanchez, J.M., and Caddick, M.X. (1996) EMBO J.15, 2791-2801.
38. Stankovich, M., Platt, A., Caddick, M.X., Langdon, T., Shaffer, P.M., and Arst, H.N. (1993) Mol. Microbiol.7, 81-87.
39. Marzluf, G.A. (1996) In The Mycota: Biochemistry and Molecular Biology (R. Brambl and G.A. Marzluf, Eds.), pp. 357-368. Springer, Berlin.
40. Premakumar, R., Sorger, G.J., and Gooden, D. (1980) J. Bacteriol.144, 542-551.
41. Tomsett, A.B., Dunn-Coleman, N.S., and Garrett, R.H. (1981) Mol. Gen. Genet.182, 229-233.
42. Sorger, G.J., Brown, D., Farzannejad, M., Guerra, A., and Jonathan, M. (1989) Mol. Cell. Biol.9, 4113-4117.
43. Tomsett, A.B. and Garrett, R.H. (1981) Mol. Gen. Genet.184, 183-190.
44. Young, J.L., Jarai, G., Fu, Y.H., and Marzluf, G.A. (1990) Mol. Gen. Genet.222, 120-128.
45. Xiao, X., Fu, Y.H., and Marzluf, G.A. (1995) Biochem.34, 8861-8868.
46. Pan, H.G. and Marzluf, G.A. (1996) Gene
47. Chang, L.W. and Marzluf, G.A. (1979) Mol. Gen. Genet.176, 385-392.
48. Cooper, T.G. (1996) In The Mycota: Biochemistry and Molecular Biology (R. Brambl and G.A. Marzluf, Eds.), pp. 139-169. Springer, Berlin.
49. Cunningham, T.S. and Cooper, T.G. (1993) J. Bacteriol.175, 5851-5861.
50. Cunningham, T.S., Dorrington, R.A., and Cooper, T.G. (1994) J. Bacteriol.176, 4718-4725.
51. Blinder, D., Coschigano, P.W., and Magasanik, B. (1996) J. Bacteriol.178, 4734-4736.
52. Stanbrough, M., Rowen, D.W., and Magasanik, B. (1995) Proc. Natl. Acad. Sci. USA 92, 9450-9454.
53. Coffman, J.A., Rai, R., Cunningham, T., Svetlov, V., and Cooper, T.G. (1996) Mol. Cell. Biol.16, 847-858.
54. Stanbrough, M. and Magasanik, B. (1996) J. Bacteriol.178, 2465-2468.
55. Felenbok, B. and Kelly, J.M. (1996) In The Mycota: Biochemistry and Molecular Biology (R. Brambl and G.A. Marzluf, Eds.), pp. 369-378, Springer, Berlin.
56. Shroff, R.A., Lockington, R.A., and Kelly, J.M. (1996) Can. J. Microbiol.42,

950-959.
57. Dowzer, C.E.A. and Kelly, J.M. (1991) Mol. Cell. Biol.11, 5701-5709.
58. Kulmburg, P., Mathieu, M., Dowzer, C., Kelly, J., and Felenbok, B. (1993) Mol. Microbiol.7, 847-857.
59. Cubero, B. and Scazzocchio, C. (1994) EMBO J.13, 407-415.
60. Fillinger, S. and Felenbok, B. (1996) Mol. Microbiol.20, 475-488.
61. Fillinger, S., Panozza, C., Mathieu, M., and Felenbok, B. (1995) FEBS Lett.368, 547-550.
62. Kulmburg, P., Sequeval, C., Lenouvel, F., Mathieu, M., and Felenbok, B. (1992) Mol. Cell. Biol.12, 1932-1939.
63. Kulmburg, P., Sequeval, C., Lenouvel, F., Mathieu, M., and Felenbok, B. (1992) J. Biol. Chem.267, 21146-21153.
64. Hull, E.P., Green, P.M., Arst, H.N., and Scazzocchio, C. (1989) Mol. Microbiol.3, 553-560.
65. Sophianopoulou, V. and Scazzocchio, C. (1989) Mol. Microbiol.3, 705-714.
66. Sophianopoulou, V., Suárez, T., Diallinas, G., and Scazzocchio, C. (1992) Mol. Gen. Genet.
67. Hynes, M.J. and Davis, M.A. (1996) In The Mycota: Biochemistry and Molecular Biology (R. Brambl and G.A. Marzluf, Eds.), pp. 381-393. Springer, Berlin.
68. Katz, M.E. and Hynes, M.J. (1989) Mol. Cell. Biol.9, 5696-5701.
69. Andrianopoulos, A. and Hynes, M.J. (1990) Mol. Cell. Biol.10, 3194-3203.
70. van Heeswijck, R. and Hynes, M.J. (1991) Nucleic Acids Res.19, 2655-2660.
71. Marzluf, G.A. (1994) Adv. in Genetics31, 187-206.
72. Marzluf, G.A. (1970) Arch. Biochem. Biophys.138, 254-263.
73. Ketter, J.S., Jarai, G., Fu, Y.H., and Marzluf, G.A. (1991) Biochem.30, 1780-1787.
74. Smith, F.W., Hawkesford, M.J., Prosser, I.M., and Clarkson, D.T. (1995) Mol. Gen. Genet.247, 709-715.
75. Smith, F.W., Ealing, P.M., Hawkesford, M.J., and Clarkson, D.T. (1995) Proc. Natl. Acad. Sci. USA92, 9373-9377.
76. Hastbacka, J., Chapelle, A., Mahtani, M.M., Clines, G., Hamilton, B.A., and Lander, E.S. (1994) Cell78, 1073-1087.
77. Marzluf, G.A. and Metzenberg, R.L. (1968) J. Mol. Biol.33, 423-437.
78. Burton, E.G. and Metzenberg, R.L. (1972) J. Bacteriol.109, 140-150.
79. Paietta, J.V. (1990) Mol. Cell. Biol.10, 5207-5214.
80. Fu, Y.H. and Marzluf, G.A. (1990) J. Biol. Chem.265, 11942-11947.
81. Fu, Y.H., Paietta, J.V., Mannix, D.G., and Marzluf, G.A. (1989) Mol. Cell. Biol.9, 1120-1127.
82. Kanaan, M. and Marzluf, G.A. (1991) Mol. Cell. Biol.11, 4356-4362.
83. Kanaan, M., Fu, Y.H., and Marzluf, G.A. (1992) Biochem.31, 3197-3203.
84. Hanson, M.A. and Marzluf, G.A. (1973) J. Bacteriol.116, 785-789.
85. Paietta, J.V. (1992) Mol. Cell. Biol.12, 1568-1577.
86. Kumar, A. and Paietta, J.V. (1995) Proc. Natl. Acad. Sci. USA92, 3343-3347.
87. Li, Q. and Marzluf, G.A. (1996) Curr. Genet.30, 298-304.
88. Marzluf, G.A., Li, Q., and Coulter, K. (1995) Can. J. Bot.73, S167-S172.
89. Tao, Y. and Marzluf, G.A. (1998) J. Bacteriol.180, 478-482.
90. Natorff, R., Balinska, M., and Paszewski, A. (1993) Mol. Gen. Genet.238,

185-192.

91. Thomas, D., Kuras, L., Barbey, R., Cherest, H., Blaiseau, P., and Surdin-Kerjan, Y. (1995) Mol. Cell. Biol.15, 6526-6534.
92. Thomas, D., Jacquemin, I., and Surdin-Kerjan, Y.H. (1992) Mol. Cell. Biol.12, 1719-1727.
93. Kuras, L. and Thomas, D. (1995) Mol. Cell. Biol.15, 208-216.
94. Kuras, L., Cherest, H., Surdin-Kerjan, Y., and Thomas, D. (1996) EMBO J.15, 2519-2529.
95. O'Connell, K.F., Surdin-Kerjan, Y., and Baker, R.E. (1995) Mol. Cell. Biol.15, 1879-1888.
96. Hanson, M.A. and Marzluf, G.A. (1975) Proc. Natl. Acad. Sci. USA72, 1240-1244.
97. Drucker, H. (1973) J. Bacteriol.116, 593-599.
98. Cohen, B.L. (1973) J. Gen. Microbiol.77, 521-528.
99. Lindberg, R.A. and Drucker, H. (1984) J. Bacteriol.157, 375-379.
100. Tilburn, J., Sarkar, S., Widdick, D.A., Espeso, E.A., Orejas, M., Mungroo, J., Penalva, M.A., and Arst, H.N. (1995) EMBO J.14, 779-790.
101. Arst, H.N. (1996) In The Mycota: Biochemistry and Molecular Biology (R. Brambl and G.A. Marzluf, Eds.), pp. 235-240. Springer, Berlin.
102. Sachs, M.S. (1996) In The Mycota: Biochemistry and Molecular Biology (R. Brambl and G.A. Marzluf, Eds.), pp. 315-345. Springer, Berlin.
103. Espeso, E.A., Tilburn, J., Sanchez-Pulido, L., Brown, C.V., Valencia, A., Arst, H.N, and Penalva, M.A. (1997) J. Mol. Biol.274, 266-480.
104. Orejas, M., Espeso, E.A., Tilburn, J., Sakar, S., Arst, H.N., and Penalva, M.A. (1995) Genes and Dev.9, 1622-1632.
105. Voisard, C., Wang, J., McEvoy, J.L., Xu, P., and Leong, S.A. (1993) Mol. Cell. Biol.13, 7091-7100.
106. Zhou, L., Haas, H., and Marzluf, G.A. (1998) Mol. Gen. Genet.259, 532-540.
107. Littlewood, B.S., Chia, W., and Metzenberg, R.L. (1975) Genetics79, 419-434.
108. Kang, S. and Metzenberg, R.L. (1990) Mol. Cell. Biol.10, 5839-5848.

Applied Mycology and Biotechnology
Volume 1. Agriculture and Food Production
G.G. Khachatourians and D.K. Arora (editors)

Protein secretion by fungi

John F Peberdy[a] Gregg L F Wallis[a,b] and David B Archer[c]

[a]Microbiology Division, School of Biological Sciences,
University of Nottingham, University Park,
Nottingham NG7 2RD, UK

[b]Biochemistry Division, School of Biomedical Sciences, The Medical School,
University of Nottingham
Nottingham NG7 2RD, UK

[c]Food Safety Sciences Division
Institute of Food Research
Norwich Research Park
Colney
Norwich NR4 7UA

Dr J F Peberdy
Tel: +44 115 951 3231
Fax: +44 115 951 3274
Email: john.peberdy@nottingham.ac.uk

Protein secretion is a vital process in fungi. For many, the secretion of hydrolytic enzymes provides a crucial step in their nutrition in nature. However, in recent years the list of different types of secreted proteins that have been discovered has extended significantly. These have been shown to have a diversity of functions including toxic molecule transport and control of desiccation. The majority of secreted proteins are glycosylated and our understanding of this aspect of fungal biochemistry has also extended in recent years. This review addresses the process of protein secretion from the cytological, biochemical and genetical standpoints. Advances in technology in many areas of scientific approach have enabled a better and growing understanding of this important process in fungi.

1. INTRODUCTION

Protein secretion is a vital process in all fungi, including yeasts and filamentous fungi. The roles and importance of these secreted proteins show similarities and significant differences in these two growth forms. Secreted proteins are diverse in their function. They include enzymes which play a key role in nutrition and proteins associated with the cell wall, which may have some structural or recognition role or a protective function as in the case of the recently described hydrophobins.

The secretion of enzymes is related to the invasive growth that occurs in both saprotrophic and biotrophic filamentous fungi. The former play a key role in the breakdown of plant wastes especially cellulose and lignocellulose and many fungi are known to produce the enzymes which hydrolyse these polysaccharides. Both facultative and obligate biotrophs are found associated with plant hosts, and the former also with insects, crustaceans and nematodes. Not surprisingly the plant biotrophs also produce cellulases as well as pectinases, whereas the insect and nematode pathogens produce proteases, lipases and chitinases. In laboratory cultures secreted enzymes are normally detected in the culture medium, although whether this occurs in the natural environment is unknown; it is possible that in this situation they are retained within the wall or on its outer surface. Yeasts also secrete enzymes, in some the molecules become part of the fabric of the wall and are thus associated with the hydrolysis of soluble molecules passing through, but there are others, which release their enzymes into the external environment in the same manner as filamentous fungi. Several enzymes secreted by fungi of both groups have been developed for commercial application in a variety of industrial processes (1).

Secreted proteins are also associated with the fungal cell wall; these probably have a diversity of functions, some of which have still to be described. However, there is evidence that some are involved in recognition processes and have a role in interactions between fungi and other organisms. A unique group of small polypeptides, the hydrophobins, are also wall-associated molecules. These may also have several functions, but the most significant appears to be in the protection of aerial reproductive structures from desiccation. Another location for secreted proteins is the periplasmic space, although little is known about this part of the fungal cell.

The delivery of proteins to the surface of the protoplast and the extracellular environment is dependent upon the complex secretory pathway, which exists in these eukaryotic organisms. This is comprised of a series of membrane bounded organelles which are linked by the movement of vesicles between them. These organelles are the sites for post-translational modification of proteins and embodied in them are the mechanisms for sorting and retention which involve molecular signals and transmembrane domains with protein structures (2). The key organelles are well known and include the Endoplasmic Reticulum and a Golgi apparatus. In fungi the former is readily identified, but the latter has a distinctive form. What follows in this review is a synthesis of our present understanding of the key process, which involves these organelles and is a vital one in the existence of fungi.

2. TRANSLOCATION OF PROTEINS INTO THE ENDOPLASMIC RETICULUM

2.1 Protein translocation

Internal structural localization is a defining part of a eukaryotic cell. A major component of this is the ability of individual proteins to be localized in different cellular organelles and compartments. Proteins that reside within the "reticuloendothelial" system (endoplasmic reticulum (ER), Golgi apparatus (GA), lysosomes) and those that are to be secreted from the cell are first inserted into the ER, hence to the GA and then to their final destination. This process is essentially conserved in all eukaryotic cells and is termed protein translocation.

Proteins of the reticuloendothelial system and those to be secreted enter the ER while they are still attached to the ribosome and while they are still being translated (co-translational translocation). They are directed to do so by a hierarchy of signals, that not only direct the protein to the ER system but also control the final destination. It could be thought of as a pass or permit that allows a protein to reside and travel between specified cellular compartments. For example an *N*-terminal sequence is required to direct a nascent polypeptide to the ER membrane, while mature proteins that reside in the ER posses a *C*-terminal tetrapeptide that allows them to return from the GA, and a glycan addition of mannose-6-phosphate is employed as the lysosomal signalling motif (3-6). The cotranslational pathway predominates in higher eukaryotic cells but in *Saccharomyces cerevisiae* a posttranslational translocation also occurs, which does not involve the ribosome nor the signal recognition particle (SRP) and the polypeptide chain is synthesized in the cytosol and then transported to the ER membrane for translocation. It would appear that a signal sequence is also responsible for determination of which pathway a particular protein is translocated by (4,7). If a particular protein is destined for the mitochondria (or chloroplast) it is first synthesized on a cytosolic ribosome and then directed to the organelle with an *N*-terminal sequence of approximately 25 amino acids in length (post-translational translocation, 8). Mitochondrial protein translocation has been reviewed by Pfanner *et al.*, (9) and is beyond the scope of this article.

2.2 Signal or peptide sequence

An *N*-terminal signal or leader sequence/peptide is present on virtually all proteins (in effect nascent polypeptides) that are to be transported co- or post-translationally and is used to target the protein to the ER membrane. This is situated immediately before the ATG and comprises 15-70 amino acids. The N-terminal part is of variable length and carries a net negative charge. This is followed by a hydrophobic core region of 6-15 amino acids that is essential for activity, and is connected to the C-terminal region by a helix-breaking residue (proline, glycine, serine). The C-terminal region contains the cleavage site for signal peptidase activity and generally contains alanine or glycine amino acids at -1 and -3 from the cleavage site (reviewed by 3-5). The absence of an absolute conserved amino acid sequence suggests that the amino acid residues in this sequence are able to form a particular conformation, which is responsible for the recognition of the signal by components of the translocation machinery (the SRP).

Ng et al., (10) have divided proteins destined for the reticuloendothelial system in yeast into three classes, i) SRP-dependant (cotranslational), SRP-independent (post-translational) and iii) proteins that can use both pathways. The hydrophobicity of the signal peptides affects the efficiency of translocation (the longer hydrophobic peptides are more likely to be recognised by the SRP) and has been suggested as the determinant of which pathway is used (7,10). The conformation of the signal peptide may also play a role in this outcome (11). In yeast all essential proteins must be able to use the post-translational pathway if required, since SRP depleted cells are sill viable.

2.3 The translocation process

Following targeting, the polypeptides are translocated through the membrane at specific sites. These are known as translocons. The translocation of proteins into the ER has been mainly studied in yeast and mammalian systems although the process and its component parts are probably conserved in the filamentous fungi. The subject has been adequately reviewed (3-5,12-14), and only a summary will be given here.

The cotranslational targeting process begins when the signal peptide sequence has emerged from the ribosome that initiates translation on the cytosolic mRNA. The SRP attaches to the translated leader sequence and halts translation. The mammalian SRP is an 11S ribonucleoprotein containing 6 functional proteins and 7S RNA, which provides the structural backbone. Five of these proteins have defined sub-functions: recognition of the polypeptide signal sequence conformation (SRP 54), binding to the ribosomal complex, halting translation (SRP 9/14) and then binding to the SRP-receptor (SRP 68/72) (3,5). In contrast, the yeast SRP is an 16S particle consisting of the ScR1 RNA, six polypeptides with similarity to their mammalian counterparts including SRP54p and one additional protein (15-19). The SRP binds to the signal sequence via its methionine-rich M-domain, and also interacts directly with the ribosome; it impedes translation when about 70 amino acids have been translated. The SRP is bound by the SRP-Receptor (on the ER cytosolic face, sometimes called the docking protein) and the ribosome then attaches itself to the membrane surface (at the translocation site). The SRP complex disassociates from both the ribosome and the signal peptide following the hydrolysis of GTP, and once released from the SRP-Receptor is recycled back into the cytosol. The signal peptide should now be in close proximity to the translocon protein complex; Sec61 and the translocating chain associating membrane protein (TRAM). Successful recognition of this sequence leads to entry of the nascent polypeptide into the ER membrane through a hydrophilic pore formed by Sec61. As the growing chain is translated it is inserted into the ER membrane and the signal peptide is then cleaved by a lumenally located signal peptidase, followed by *N*-linked glycosylation at specific sites by the oligosaccharyltransferase. When translation is completed the ribosome and mRNA are released from the ER cytosolic membrane.

The translocon apparatus in yeast consists of about 20 proteins; the SRP receptor, a complex of ER integral membrane proteins (Sec61p, TRAM) the signal peptidase and the oligosaccharyltransferase (see section 3.3). The mammalian SRP-receptor is a dimer comprising a larger α subunit (69 kDa) and a smaller β subunit (30kD). The *N*-terminus of the former is anchored in the ER, while the cytoplasmic domain may function in SRP/7S recognition (3,4,20). The *S. cerevisiae* homologue is 32% identical to the mammalian counter part and is also part of the GTPase super family (21). The binding of the ribosomal/mRNA/nascent polypeptide to the ER membrane is to a protein complex that forms the hydrophilic channel in the membrane through which the nascent polypeptide will pass. This complex was identified by the use of yeast *sec* (secretion) mutants and is called the Sec61 complex and is the central component of the translocon. It consists of 3 transmembrane proteins Sec61αβγ (Sec61p, Sbh1p and Sbh2p, Sss1p, respectively in *S. cerevisiae*) which form the gated hydrophilic membrane channel and are also responsible for ribosomal binding and signal peptide recognition (14,22). The *S. cerevisiae* structural genes that code for these proteins have been identified and cloned; *SEC61* (α subunit; 23), *SEB1/SBH1* and *SEB2/SBH2* (β subunits; 24) and *SSS1* (γ subunit; 25). Homologues of this complex have been found in *Schizosaccharomyces pombe* and *Yarrowia lipolytica* (26). Both the α and γ subunits are essential for translocation and cell viability and the β subunit facilitates co-translational translocation by interacting with the signal peptidase complex (27). Sec61p may also be involved in the transport of misfolded secretory proteins from the ER to the cytosol, for proteolysis (28). There is also evidence of another trimeric complex in yeast, that may only be involved in the co-translocational process (29). TRAM is a major protein that becomes crosslinked to and stimulates translocation of the polypeptide chain, but its functions are only partly understood (27,30). Although this protein has only been found in mammalian cells, there is thought to be a yeast homologue (14). The mammalian signal peptidase is a complex of 5 proteins located on the lumenal side of the membrane which cleaves the entire signal sequence when it has entered the ER (5). It would seem that the yeast peptidase enzyme has a similar complexity (31-347).

The post-translational translocation process operates in all eukaryotes but would appear to be more important in the yeasts and possibly other fungi. In contrast to the co-translational mode, this process is understood better in these lower eukaryotes. The fully synthesized polypetide is held in an unfolded state by cytosolic chaperones, such as the Hsp70 family (35-37), which must be removed by the Dnaj homolog Ydj1p before it can be translocated (36). The post-translational translocon of *S. cerevisiae* consists of a heptameric complex formed by the Sec61 complex and the Sec62-Sec63 complex (25,36). The Sec62-Sec63 consists of 4 other proteins (Sec62p, Sec63p, Sec71p and Sec72p). The signal peptide is recognised by Sec61p and while still bound is able to interact with Sec62p and Sec71p (38). The ER resident chaperones Kar2p (yeast homologue of BiP) (39-41) and Lhs1p (42,43) are also required for translocation and via an interaction with Sec63p (44,45) may be required to 'pull-through' the polypeptide into the ER lumen. Once inside the ER lumen, the signal peptide is removed and the polypeptide is folded and/or glycosylated.

3. PROCESSING OF PROTEINS FOR SECRETION

3.1 Glycosylation

3.1.1 Protein glycosylation

Glycosylation of proteins is a co- and post-translational modification that would appear almost universal for proteins destined for the extracellular environment. This process proceeds as the protein transgresses the secretory pathway. The biological functions of the glycan attached to proteins remains one of biochemistry's unanswered questions, although, it is probably involved in the correct folding and maturation of the protein inside the cell and in increased stability outside the cell (46,47).

3.1.2 *N*-linked protein glycosylation.

The eukaryotic *N*-linked protein glycosylation pathway consists of three distinct phases, which occur in the ER: i) the synthesis of a 14 residue oligosaccharide linked to a lipid carrier-dolichol phosphate (P-Dol), ii) the transfer of this preassembled core oligosaccharide onto selected asparagine residues on a nascent polypeptide as it emerges from the translocon into the ER lumen, and iii) the processing of the *N*-linked oligosaccharide that begins before the newly synthesized protein traverses the secretory pathway into the compartments of the GA. These processes are conserved in nearly all eukaryotes, from yeast to man (48-54), partially so in the archaea (55) and similar reactions also occur in the eubacteria (56-59).

3.1.2 Dolichol pathway

Dolichol phosphate is the membrane bound lipid carrier upon which the core oligosaccharide is assembled. Dolichols are linear, long-chain isoprenoids with a saturated alpha (adjacent to the hydroxyl) isoprene unit (51,59). The biosynthetic pathway of dolichol assembly, initially follows that for sterol and ubiquinone but the pathways diverge after the synthesis of farnesylpyrophosphate which is then polymerised by the action of *cis*-prenyl transferase to give dolichol molecules of differing chain length. In any organism dolichols are present as a limited-family of molecules of differing chain length, which are species specific (59). In *S. cerevisiae* this family is principally composed of 15-16 isoprene units (dolichol-15,16) (60) and *S. pombe* possesses dolichol-16,17 (61). In filamentous fungi, such as *Aspergillus* spp. slightly longer dolichol-20,21 are found (62) which are very similar to those found in mammalian systems (59). The phosphorylation of dolichol to give P-Dol (which is the co-enzymic carrier) is produced by a CTP requiring dolichol kinase. The gene encoding this kinase has been cloned in yeast (*SEC59*; 63), and inactivation of the gene product leads to a rapid decrease in the levels of P-Dol and protein secretion. This suggests that the flux through the dolichol pathway could be regulated by the supply of P-Dol. P-Dol also serves as the donor of mannose and glucose from Man-P-Dol (MPD) and Glc-P-Dol (GPD) respectively, at later steps in the oligosaccharyl assembly. If each of the steps in the dolichol cycle required a separate transferase, there are at least (including the oligosaccharyl transferase) 9 enzymes requiring P-Dol as a donor or acceptor, in the dolichol pathway and also two more responsible for the synthesis of MPD and GPD, emphasising the importance of P-Dol in this biosynthetic pathway and in *N*-linked glycosylation.

The biosynthesis of the P-Dol-linked oligosaccharide begins with the formation of *N*-acetylglucosamine-P-P-Dol by the enzyme *N*-acetylglucosamine phosphate transferase (GPT; 64) This structure is further glycosylated (Fig.1) by the addition of another *N*-acetylglucosamine, nine mannose and three glucose residues to give the complete $Glc_3Man_9GlcNAc_2$ dolichyl pyrophosphate linked oligosaccharide (Fig.2) which is then transferred onto the asparaginyl residue of newly synthesized polypeptides. The two GlcNAc residues and the five innermost mannose residues (residues 1-7, Fig.2; Table 1) are donated by nucleotide sugars (UDP-*N*-GlcNAc and GDP-Man, respectively), while the remaining mannose and glucose residues (residues 8-14, Fig.2; Table1) are donated from the dolchol phosphate sugars, MPD and GPD, respectively. Although, the sugars to form these latter glycosylphophodolichols are also donated by GDP-mannose and UDP-glucose, respectively. The biosynthesis, transportation, cloning of the genes involved and the availability of these nucleotide sugars, in the yeast secretory pathway has been recently reviewed (52, 65), and similar genes are being found and cloned in filamentous fungi (66). The genes encoding those proteins responsible for the biosynthesis of MDP and GDP, Man-P-Dol synthase (DPMS) and Glc-P-Dol (DPGS) synthase, respectively, have been isolated and cloned in fungi (Table 1). The structural gene for DPMS (*DPM1)*

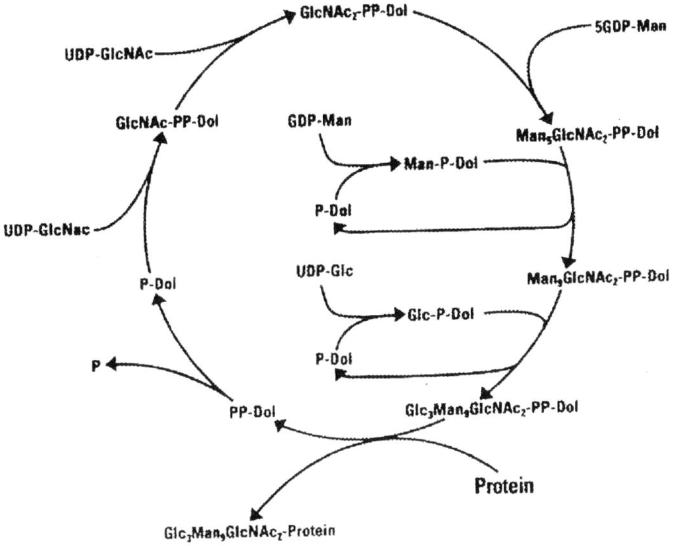

Figure 1. The dolichol cycle
The endoplasmic reticulum localised dolichol cycle pathway leading to *N*-glycosylation of proteins in eucaryotes. The oligosaccharide $Glc_3Man_9GlNAc_2$ linked to and thereby anchored to the ER membrane by dolichyl pyrophosphate is constructed by the stepwise addition of the individual sugars by membrane bound glycosyltransferases. The $Man_5GlcNAc_2$-P-P-Dol intermediate is synthesized on the cytosolic side of the ER membrane, and is then translocated into the ER lumen and then the remaining mannose and glucose residues are added, before the completed 14-residue oligosaccharide is transferred onto a nascent polypeptide, by the oligosaccharytransferase. The P-Dol produced from this reaction and that produced from the donation of mannose and glucose from Man-P-Dol and Glc-P-Dol, respectively, to the growing oligosaccharide, may be recycled.

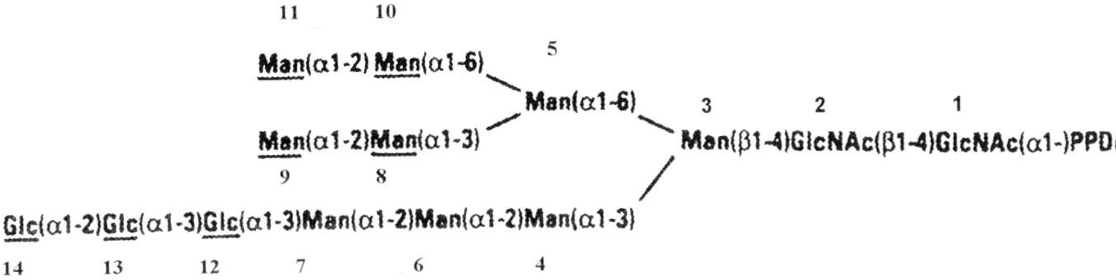

Figure 2
The fully glycosylated core oligosaccharyl diphosphodolichol that is assembled on the lipid carrier and transferred onto polypeptide asparagine residues by oligosaccharyl transferase. The numbers refer to the probable order of addition. The underlined sugars are donated from mannosyl or glucosyl phosphodolichol. (modified from Hemming, (51) with permission.)

Table 1. Dolichol pathway of *N*-linked glycosylation: Cloned genes from yeast and other fungi.

Step[1]	Bond formed	Probable localisation[1]	Gene loci	Lethality	Reference
1	GlcNAc α1-PPDol	Cytosolic	*ALG7*	+	64,78,79
2	GlcNAc β1,4 GlcNAc	Cytosolic	?		
3	Man β1,4 GlcNAc	Cytosolic	*ALG1*	+	83
4	Man α1,3 Man	Cytosolic	*ALG2*?	+	84,85
5	Man α1,6 Man	Cytosolic	*ALG2*?	+	84,85
6	Man α1,2 Man	Cytosolic	?		
7	Man α1,2 Man	Cytosolic	?		
8	Man α1,3 Man	Lumenal	*ALG3*	-	86
9	Man α1,2 Man	Lumenal	*ALG9*	-	52, 87
10	Man α1,6 Man	Lumenal	*ALG12*	-	52
11	Man α1,2 Man	Lumenal	?		
12	Glc α1,3 Man	Lumenal	*ALG6*	-	75, 88
13	Glc α1,3 Glc	Lumenal	*ALG8*	-	89
14	Glc α1,2 Glc	Lumenal	*ALG10*	-	90
	Man β1-PDol	Cytosolic	*DPM1*	+	67-69
	Glc β1-PDol	Cytosolic	*ALG5*	-	76

[1]Probable order of addition of each residue refer to Figure 2.
[2] see 49, 50

encodes a transmembrane protein that is essential in yeast (67), which has also been cloned from *Ustilago maydis* (68) as well as from other eukaryotes (69). DPMS activity has also been found in many other fungi (70-73). MPD (the product of DPMS) is involved in the first step of protein *O*-glycosylation and GPI anchor formation in yeast (74) and probably in other fungi (70, 73). This provides an explanation for the lethal phenotype observed when the gene is disrupted. DPGS is encoded by *ALG5*, a 38kDa transmembrane protein, the loss of which is not lethal but results in an underglycosylation of some proteins (75, 76). No other eukaryotic DPGS encoding genes have been cloned, although the enzyme activity has been detected in microsomes from *Candida albicans* (77) and *Aspergillus niger* (73).

The assembly of the P-Dol-linked oligosaccharide is catalysed by up to 14 separate membrane-bound glycosyltransferases which probably have their active sites facing the cytosolic (steps 1-7) or lumenal (steps 8-15) face of the ER (49,50). The majority of the proteins involved in these reactions have yet to be isolated. However, many of the genes (named the *ALG*, asparagine-linked glycosylation, genes) have been cloned in yeast and other fungi (Table 1). The first enzyme, GPT, has been purified from higher eukaryotic cells (64), and the encoding structural gene (*ALG7*) cloned from a number of fungal sources including *S. cerevisiae* (78), *S. pombe* (79), and *A. niger* (Sørensen and Peberdy, unpublished data). *ALG7* is an essential gene that confers resistance to the antibiotic tunicamycin and is probably an important control point of the dolichol pathway (64, 80). Of the remaining 13 glycosyltransferase enzymes the genes encoding 9 of them have been cloned, and studied in detail (Table 1), principally in yeast, and they provide a good model of the process not only for filamentous fungi but also for eukaryotes in general

The proposed topology of P-Dol-linked oligosaccharide biosynthesis in eukaryotes (49, 50), implies that there are several translocation steps of the oligosaccharylpyrophosphoryldolichol (between the addition of residues 7-8) and of the glycosylphosphodolichols (MPD and GPD) from the cytosolic to the lumenal face of the ER. The mechanisms and the necessary components for this translocation are poorly understood, at present. However, it is known that it is a protein-mediated and not a passive process.

The transcriptional control of the *ALG* genes also remains obscure. Increasing evidence suggests a role for these genes in the yeast cell cycle (80). These genes belong to a class of housekeeping genes, since they are expressed in all cell types, regardless of the physiological and proliferative condition. All of the *ALG* genes studied so far, have multiple potential TATA boxes and GC rich regions, produce unstable transcripts with short half-lives, and the gene transcripts (*ALG7, ALG1* and *ALG2*), show complex decay patterns (80, 81). It is of note that these 'early' genes (whose products operate in the initial phases of the dolichol cycle) are also essential in yeast, while those of the latter ones possessing simple mRNA decay patterns are not (Table 1). It is likely that the levels of the 'early' *ALG* gene transcripts have to be maintained within a narrow range, otherwise the levels of their translational products become deleterious to cell function (80). The inter-relationships between mRNA decay and translational control in eukaryotes have been reviewed (82).

It is a paradox that while the transfer of at least the $Man_5GlcNAc_2$ oligosaccharide is probably essential for viability, the fully glycosylated $Glc_3Man_9GlcNAc_2$ oligosaccharide is the preferred substrate for the oligosaccharyl transferase. This suggests that these other 'outer' sugar residues (Glc_3 and Man_4) must also play a part in the efficient folding, maturation and secretion of the protein.

3.1.3 Oligosaccharyltransferase.

An important, or some would say, pivotal step in the formation of glycoproteins is the 'en-bloc' transfer of the dolichol-linked oligosaccharide precursor onto defined asparagine acceptor sites on the nascent polypeptide. This occurs during the translocation of the polypeptide into the lumen of the rough ER, and is catalysed by a heterooligomeric membrane-bound complex, known as the oligosaccharyltransferase (OST; 53, 91, 92). The preferred donor for this process is the complete dolichol linked 14-residue core oligosaccharide precursor (DolPP-$GlcNAc_2Man_9Glc_3$, Fig.2), although the transfer of only DolPP-$GlcNAc_2$ has been observed (93, 94), suggesting a wide substrate specificity. The presence of the glucose residues seems to enhance glycosyl transfer (95) and in particular the terminal α-1,2 linked residue (90), thus, allowing a more favourable conformation (96) that is recognised by the OST (97) and a consequential lowering of the K_m (93).

It is now well established that the sequon Asn-X-Ser/Thr (N-X-S/T), where X is any amino acid except proline (98) is the potential acceptor site on the newly synthesized polypeptide. However, it has been estimated that 10-30% of potential sequons are either not or only partially glycosylated (53). Thus, other polypeptide-related factors such as the influence of the hydroxy, 'X ' and other flanking amino acids, the local protein conformation and degree of folding and the proximity of the sequon to; transmembrane (potential) domains, the amino or carboxyl terminus, other glycosylated asparagine residues and disulphide-bonded C residues may all influence the glycosylation of a particular sequon by the OST (99). The importance of these in control of oligosaccharide addition by OST is not well understood, at present.

The active OST from *S. cerevisiae* has been purified as a complex of six polypeptide subunits, although an active tetrameric complex has also been described (100-102). To date, genes encoding nine yeast protein subunits have been cloned, and five of them have been shown to be essential (*OST1, OST2, WBP1, SWP1 AND STT3*; 53). All nine polypeptides have at least one predicted membrane spanning domain and Ost1p, Wbp1p and Stt3p are glycoproteins, which raises the intriguing possibility that the complex must be able to process some of its own components (53). A combination of genetic and biochemical evidence has suggested that the OST complex could consist of three subcomplexes; 1) Stt3p-Ost4p-Ost3p, 2) Swp1p-Wbp1p-Ost2p, and 3) Ost1p-Ost5p (103), of which the second could be the catalytic core (53). The substrate binding domains have been proposed to be Wbp1p and Ost1p for binding of the PP-Dol-linked oligosaccharide and recognition of the polypeptide sequon, respectively (102, 104). These component subunits from yeast have a high degree of sequence homology with those from vertebrate, invertebrate and plant systems, which is consistent with the evolutionary conservation of this process (53, 92). However, there has been no reported evidence for the OST in the filamentous fungi, although it is presumed to be present.

3.1.4 ER processing of *N*-linked glycans.

The initial stages of *N*-linked glycoprotein processing are remarkably conserved in all eukaryotes. The processing glycosidases of yeast have been studied in detail and they provide a good model of the process not only for filamentous fungi but also for eukaryotes in general (54, 105).

The processing of *N*-linked glycans appears to commence simultaneously with the polypeptide glycosylation by OST. In the ER, the terminal α1,2-linked glucose (residues 14, Fig.2) is cleaved by the action of glucosidase I (EC 3.2.1.106) and sequentially the remaining two α1,3-linked glucose residues (Fig.2, residues12-13) are removed by glucosidase II. In *S. cerevisiae* the terminal α1,2-linked mannose on the inner (middle) α1,6-arm (residue 9) is then removed by the ER-located processing α1,2-mannosidase (EC 3.2.1.113) and this $Man_8GlcNAc_2$ oligosaccharide is then enlarged by Golgi located mannosyltransferases to give the mature glycoprotein (54, 106, 107). In higher eukaryotes the $Man_8GlcNAc_2$ glycan is further processed to $Man_5GlcNAc_2$ by other ER and Golgi α1,2-mannosidases, which is the substrate for the action of GlcNAc transferase I, the building block for the formation of complex and hybrid glycan structures (91, 108). In *S. pombe* the ER α1,2-mannosidase appears to be absent and $Man_9GlcNAc_2$ is the minimum glycan precursor (109). There is however, a putative glucosidase I gene sequence (Genbank O14255).

The processing α-glucosidases from *S. cerevisiae* have been purified (partially purified for glucosidase II) and the genes coding for them have been cloned. Glucosidase I is a type II transmembrane glycoprotein (possesses a single membrane spanning domain and the N-terminus in the cytosol) of 107 kDa and the cloned gene *CWH41* was isolated during a screen for calcofluor white hypersensitivity. The protein possesses a significant sequence identity to the human glucosidase I (54, 110). Interestingly, no ER retention motif has been found in the sequence of the polypeptide. Glucosidase II has been partially purified from yeast membranes and a putative gene (*ROT2*) has been cloned by homology to its mammalian counterpart (111). This gene encodes a putative protein possessing no transmembrane or ER-retention signal motifs; however, a non-catalytic β-subunit could be responsible for ER localization (111). The catalytic function of both these glucosidases is sensitive to the inhibitor 1-deoxynorijimycin (112). There could be a link between the action of these processing glucosidases and cell wall β1,6 glucan synthesis in yeast (113, 114).

The specific α1, 2 mannosidase has been purified from yeast (115) and the corresponding (non-essential) gene, *MNS1* has been cloned (116). The polypeptide is a type II membrane glycoprotein of 63-66kDa, and its catalytic site probably faces the ER lumen (54). The catalytic domain sequence has about 35% sequence identity to mammalian processing α-mannosidases, and database searches suggest that there is a family of these genes (class 1 α-mannosidases) that has been conserved throughout evolution, although this includes enzymes that also remove the other α1,2 linked mannoses (on both the α1,3 and α1,6 arms of the *N*-linked glycan), not just the specific residue removed by the yeast enzyme to give $Man_8GlcNAc_2$. Neither of these three yeast processing glycosidases (111,117) possess the expected ER retention motifs (118) but the transmembrane domains have been implicated in this process (54).

In filamentous fungi, α1,2 mannosidases have been purified and their genes cloned from *Penicillium citrinum* (*msdC*; 119,120) and from *Aspergillus satoi* (*A. phoenicis*) (*msdS*; 121, 122). In both cases, the coding sequences are similar to each other (ca. 70% identity) and to both yeast and mammalian (ca. 30%) sources and the enzymes are capable of processing the full oligomannose oligosaccharide from $Man_9GlcNAc_2$ to $Man_5GlcNAc_2$. Interestingly, the *P. citrinum* enzyme is secreted into the culture medium, during growth, as well as possessing an internal (ER) localisation. This limited information, in addition to the knowledge available on the mature *N*-glycan structures found, allow us to propose that processing in filamentous fungi is closer to that in higher eukaryotes than that found in the yeasts.

3.1.5 *N*-linked glycan structures

Following processing of the $Glc_3Man_9GlcNAc_2$ oligosaccharide in the ER, the glycoproteins move to the GA where elongation, modification and maturation of both *N*- and *O*-linked glycans by a variety of glycosyltransferase enzymes, occurs. The asparagine- or *N*-linked glycan structures have a common core oligosaccharide of $Man_3GlcNAc_2$, $Man_5GlcNAc_2$ or $Man_{8-9}GlcNAc_2$ and this is usually modified by the addition of many other sugar residues. These additions have allowed a simple classification of *N*-linked glycans into those containing only mannose and *N*-acetyl glucosamine (oligomannose type), those composed of *N*-acetyl galactosamine units and, usually, neuraminic (sialic) acids (complex type) and those containing a mixture of the two others (hybrid type). This classification was introduced on the basis of higher eukaryotic, principally, mammalian glycoprotein structures (123). Although the evidence, apart from that from *S. cerevisiae*, is limited, fungal *N*-linked glycans are primarily of the oligomannose type and oligomannose type glycans that have been modified by the addition of other sugars such as galactose, glucose and galactofuranose and of charged residues such as phosphate and sulphate. No complex or hybrid types are found in fungi. The glycosyltransferases (such as GlcNAc-transferases, β-galactose-transferases and the sialytransferases; 108) required to synthesize these latter glycan types have not been identified in the lower eukaryotes and probably evolved in conjunction with organisms possessing a circulatory system.

As described in the previous section yeast glycoproteins are only trimmed down to $Man_{8-9}GlcNAc_2$. Yeasts elongate their *N*-linked oligosaccharides into two forms; a smaller 'core' series of up to 15 mannose or hexose residues and a much larger 'outer chain' family of up to 200 hexose residues. This outer core which is highly immunogenic, is constructed of a poly-α1,6-linked mannose backbone which is further derivatised in a species dependent manner and which is sometimes capped with; short -α1,2-linked (*S. cerevisiae, Pichia pastoris*) or α1,3-linked (*S. cerevisiae*) mannose residues, mannose linked by a phosphodiester bond (*S. cerevisae* and *P. pastoris*) or -α1,2-linked galactose (*S. pombe*) (50, 105, 124). The mannose is donated from GDP-Man.

S. cerevisiae elongates its core glycans by the addition of α1,2-, α1,3-, and α1,6-linked mannose residues (125) and by the addition of mannose via a phosphodiester linkage. A number of the genes that control this glycoprotein maturation in the Golgi have been isolated and cloned. The *KTR* (killer toxin related) family contains nine members and the *MNN1* (identified by changes in the cell wall mannan composition) family six, separate mannosyltransferases (106, 107). The former have been shown to be type II membrane proteins with variable sequence identity to each other (11-53%), but they do have a common conserved region (107, 126-128). Those that have been fully characterised have α1,2 mannosyltransferase activity. There are also homologues present in *C. albicans* and *S. pombe* genomic sequences. Only one of the *MNN1* family has been characterised so far, a Golgi-localized α1,3-mannosyltransferase (129, 130). Mannosylphosphate modification of two positions on both the inner core and the elongated outer chains are found in *S. cerevisiae* (131). These impart a negative charge to the glycan/glycoprotein. The genes responsible for mannosylphosphate transfer, *MNN4 and MNN6* have been cloned (131). Mannosylphosphate addition to *N*-linked glycans is also found in the higher eukaryotes where it serves as a signal for translocation to the lysosomal compartments (46). This does not appear to be their function in yeasts.

S. pombe adds an α1,2-linked mannose residue followed by an α1,3-linked galactose to its $Man_9GlcNAc_2$ trimmed glycan, and its elongated oligosaccharides contain α1,2-galactose and pyruvylated galactose (pyruvate-2,(4,6)Gal-β1,3,Gal) attached to the poly-α1,6 linked mannose (124, 132, 133). The α1,2 galactosyltransferase (134) and the UDP-Gal transporter (135) have both been identified and cloned. The investigations into protein glycosylation in *P. pastoris* have mainly been using recombinant (mammalian) proteins rather than native glycoproteins. However, from this research it has been deduced that *P. pastoris* can produce core glycans of $Man_{8\text{-}18}GlcNAc_2$ (136). The elongated 1,6-linked mannose chains are not as long as those reported for *S. cerevisiae*, are without α1,3 mannosyl substituents but they do contain phosphodiester linked mannose (124, 137, 138). Pathogenic *Candida* species have received a large amount of study on the structures of their glycans and there involvement in pathogenicity (139-141). *C. albicans* produces many different oligomannose type structures and unusually a poly-β1,2-linked mannose attached to the elongated mannose by phosphodiester linkages (142, 143).

Investigations of *N*-linked glycan structures in filamentous fungi have been relatively sparse and primarily have been conducted on glycoproteins of industrial significance; enzymes such as cellulases, amylases and oxidases, that have usually been produced and secreted by a limited number of fungal species such as *Trichoderma*, *Aspergillus*, and *Fusarium*. Therefore, these results and their conclusions must be viewed with some caution. It is clear that the hypermannosylation and extension of the α1,3 arm by the addition of poly-α1,6-linked mannose residues found in the glycoproteins of yeasts does not occur in filamentous fungi. A summary of the known structures suggests that oligomannose type-glycans of $Man_3GlcNAc_2$ up to $Man_{12}GlcNAc_2$ (144, 145) which can be modified with glucose, mannose, galactose (principally in the furanoic conformation) and mannose linked by a phosphodiester linkage are synthesized in filamentous fungi. Such a range of structures suggests; i) that processing of the core-linked oligosaccharide can continue down to $Man_5GlcNAc_2$ (146,147) and possibly to the trimannosyl core (148), as found in mammalian cells, and the mannosidases to perform this trimming have been identified (119, 121, 149), ii) that incomplete processing of the glucose by glucosidase II occurs but does not appear to affect the subsequent maturation and secretion of the glycoprotein (147, 150, 151), and iii) that maturation of glycoproteins includes the addition of further mannosyl (144, 145; Wallis et al., unpublished observations), galactofuranosyl (145, 150-152; Wallis et al.,unpublished observations) and mannosyl-phosphate residues (147, 153) to the trimmed glycan core, presumably by as yet unidentified Golgi-located glycosyltransferases. In contrast to the situation in *S. pombe*, galactose is generally present in the furanoic (5-membered ring) conformation in filamentous fungi. Galactofuranosyl residues are absent from the higher eukaryotes and appear to be confined to bacteria, fungi and protozoa (154). They have been found as a component of hyphal walls in a number of species; *Aspergillus fumigatus* (155), *Ascobolus furfuracens* (156), *Fusarium sp* (157), *Neurospora crassa* (158), *Penicillium charlesii* (159) and *Trichophyta sp.* (160) and as constituents of *N*-linked glycans (145, 150-152; Wallis et al., unpublished observations). In some recent observations from our laboratory (Wallis, unpublished observations), we have found oligomannose *N*-glycans substituted with up to 10 galactofuranosyl residues on an α-galactosidase secreted by *A. niger*, and have obtained data that suggest these unusual sugars are a universal modification on secreted glycoproteins from *A. niger*.

3.1.6 *O*-glycosylation

Glycans can be attached to the hydroxy amino acids threonine or serine; this post-translational modification of secretory proteins is less conserved amongst eukaryotes than the core processes of *N*-linked glycosylation. The site of attachment appears to favour Thr over Ser (161), similar to the situation in *N*-linked glycosylation whereby Thr is also favoured as the third amino acid in the sequon (99). However, in contrast to the latter, no consensus sequence determining which of the many Thr/Ser in a polypeptide will be glycosylated has yet been found. *O*-linked glycans are generally composed of short <7, unbranched, sugar residues, but there may be as many as 40 separate sites on a given protein, and in many yeast and fungal proteins that may constitute 80-90% of the total glycan. *O*-linked glycosylation begins in the ER and concludes in the GA in lower eukaryotes, but occurs only in the Golgi in higher eukaryotes (50, 105, 161).

As with many biochemical processes, the study and knowledge of *O*-linked glycosylation, is further advanced in *S. cerevisiae* than in other fungi. Protein *O*-glycosylation in this organism begins with the transfer of a single mannose residue from phosphodolichol mannose (MPD) onto the polypeptide. This reaction is catalysed by a family of Dol-P-β-D-Man: protein *O*-α-D-mannosyltransferases (EC 2.4.109, Pmtp) located in the ER (50, 74, 161, 162). The use of glycosylphosphodolichols as the donor for the first *O*-linked residue appears unique to the lower eukaryotes. In nearly all recorded examples the only sugar transferred this way is mannose, however, there are a few reports that suggest that glucose could be too (163). The *PMT* gene family codes for seven potential protein *O*-mannosyltransferases, which share 50-80% sequence identity and identical hydropathy profiles which suggest that they are integral membrane proteins (161, 164, 165). This redundancy suggests that the different Pmt enzymes prefer different protein substrates and/or distinct polypeptide sequences on which to act (163). That such an 'over redundancy' has evolved indicates the importance of *O*-mannosylation in the yeast life cycle (164). There is some evidence for the transcriptional control of the *PMT* genes during the cell cycle (161). In the other yeasts, *PMT1* activity and homologues have also been found; *S. pombe* (Genbank acc. nos. O13898 and O42933) and *C.albicans* (166, 167). The *O*-linked chain is then elongated following translocation of the protein to the GA by the action of mannosyltransferase of the *KTR* and *MNT1* families, to produce a maximum (in *S. cerevisiae*) of five unbranched α-linked mannose residues (50, 107). The second and third mannoses are linked by α1,2 residues catalysed by Ktr1p, Ktr3p and Mnt1p (128) and the final two mannoses are also added by Mnt1p and linked by α1,3 bonds (107, 161). The second mannose can also be modified by the attachment of a mannose via a phosphodiester linkage, similar to the situation in *N*-glycans (131). Therefore in summary, the *O*-linked glycans of *S. cerevisiae* are composed of between one and five α-linked mannose residues and such structures have been found on wall-bound glycoproteins (168). A similar situation is presumed to exist in *P. pastoris* (169, 170). In *C. albicans* the mannose chains are extended to six (139) or seven mannose residues (171). *S. pombe*, is unique among the yeast studied so far as it is able to add α-linked galactose to the first *O*-linked mannosyl residue (132), and the structural gene (*gma12*$^{+}$) that encodes for the α1,2 galactosyltransferase that catalyses this reaction is also involved in galactosylation of *N*-linked glycans (172).

In filamentous fungi, the biosynthetic enzymes (and their structural genes) involved in *O*-linked glycosylation have not received any study. However, the glycan structures have been determined in a limited number of glycoproteins (133, 163, 173, 174). A notable feature of secreted fungal enzymes that hydrolyse large insoluble polysaccharides such as cellobiohydrolase and glucoamylase are that these proteins possess a distinct domain structure; the catalytic domain being connected to a substrate binding domain by a linker domain that has a high proportion of *O*-glycosylated hydroxy amino acids (146, 175). In some proteins up to 40 glycosylated residues may be present in this linker domain (173, 176). These *O*-linked glycans are similar to those found in the yeasts, being principally composed of short chains (1-4) of α-linked mannose, sometimes modified by the addition of glucose and galactose (146, 163, 173, 175). In contrast to this situation, the primary structures of *O*-glycans present on hyphal wall glycoproteins from a *Fusarium* species contained rhamnose, galactofuranose, glucuronic acid and phosphate linked to the *O*-glycosyl mannose, in addition to the usual hexoses (174). We have also found galactofuranoic residues in the *O*-linked glycan (probably terminally attached to mannose) of glucoamylase secreted by *A. niger* (177). When more hyphal wall glycoproteins are analysed, perhaps more of these complex structures will be observed. However, present knowledge suggests that the *O*-linked glycans of filamentous fungi are similar in composition and structure to those of the yeasts than higher eukaryotes.

3.2 Folding of secretory proteins and the unfolded protein response.

Proteins destined for secretion by fungi enter the lumen of the endoplasmic reticulum (ER) where folding occurs prior to translocation to the cell exterior. Protein folding is assisted by molecular chaperones and folding catalysts (foldases) which are resident in the lumen of the ER. The folding process can be disturbed and lead to the accumulation of mis- or partly folded proteins which trigger an ER stress response called the unfolded protein response (UPR). UPR is a stress response with common features across kingdoms (178, 179). Amongst the fungi UPR is best understood in the yeast *Saccharomyces cerevisiae* although research with filamentous fungi and other yeasts has confirmed existence of the UPR.

3.2.1. ER lumen chaperones and foldases

Several fungal genes that encode lumenal chaperones and foldases have been cloned and descriptions of the genes and derived proteins summarised by Gething (180). Chaperones are proteins that assist the correct non-covalent assembly of other polypeptides *in vivo* (181) and the most abundant chaperone in the ER lumen is known as BiP (binding protein) which is an ATP-dependent 70kDa protein found in eukaryotes (182). Genes encoding BiP and related chaperones have now been cloned from several fungi (Table 2). The BiP-encoding gene is essential for protein secretion by *S. cerevisiae*

Table 2. Cloned lumenal chaperones and foldases from fungi.

BiP (binding protein) and related chaperones.

Protein	Organism	Reference
BiP	*Saccharomyces cerevisiae*	238
BiP	*Schizosaccharomyces pombe*	239
BiP	*Kluyveromyces lactis*	240
BiP	*Yarrowia lipolytica*	241
BiP	*Aspergillus niger/awamori*	242, 243
BiP	*Trichoderma reesei*	M. Penttilä, unpublished
BiP	*Neurospora crassa*	244
Lhs1p/(Ssi1p)	*Saccharomyces cerevisiae*	245
Yam 6	*Schizosaccharomyces pombe*	190

Protein disulfide isomerase (PDI) and related proteins

PD1	*Saccharomyces cerevisiae*	194, 195
PD1	*Aspergillus niger*	199, 246
PD1	*Aspergillus oryzae*	247,248
PD1	*Humicola insolens*	249
PD1	*Trichoderma reesei*	M. Penttilä, unpublished
EUGp	*Saccharomyces cerevisiae*	197
TIGAp	*Aspergillus niger*	200
TIGAp	*Neurospora crassa*	200
MPD1p	*Saccharomyces cerevisiae*	250
MPD2p	*Saccharomyces cerevisiae*	250

Other Chaperones

Calnexin	*Saccharomyces cerevisiae*	251
Calnexin	*Schizosaccharomyces pombe*	239
Calnexin	*Aspergillus niger/awamori*	Huaming Wang, unpublished

(183, 184) and its essential role has since been confirmed in other fungi, e.g. *Aspergillus awamori* (185). In common with other proteins resident in the lumen of ER, BiP has a C-terminal sequence of four amino acids which is essential for its retention in the ER. The precise sequences can vary between protein and host organism but HDEL is found in the *S. cerevisiae* BiP protein; other sequences found in fungi are similar, e.g KDEL, SDEL. Being soluble in the lumen, BiP is able to associate with polypeptides during their folding but also specifically interacts with ER membrane-bound components of the translocation system for entry of nascent polypeptides into the lumen and retro-transport of poorly folded proteins from the lumen for degradation. As will be described later, BiP is also a key component in the initiation of the unfolded protein response. The various functions of BiP have largely been established in *S. cerevisiae*.

BiP of *S. cerevisiae* is functionally analogous, and shares substantial sequence identity in its substrate-binding domain to the DnaK protein of *E.coli* which interacts with DnaJ in a specific interaction. DnaJ-like proteins are found in the ER of *S. cerevisiae* (186) and BiP is known to interact specifically with them, e.g. the lumenal co-chaperone Scj1 (187) and the transmembrane Sec63p. Sec63p is an integral component of the protein translocation complex (188, 189), to which BiP binds and interacts with the Sec61p complex to afford ATP-dependent translocation of polypeptides to the ER lumen (190). The translocation channel is gated (perhaps by BiP) and requires a functional signal sequence at the N-terminus of the nascent polypeptide before opening (191). Misfolded secretory proteins are exported from the ER for degradation and this retro-transport occurs also via Sec61p channels involved BiP (192). BiP is a crucial component of the folding of secretory proteins but interacts with other (co-) chaperones and foldases within the ER. Most of these studies have been performed in mammalian cells (180) where the kinetics of association and dissociation of BiP and other lumenal proteins to folding polypeptide appears to depend upon the nature of the polypeptide and physiological condition of the cells.

Protein disulfide isomerase (PDI) is a foldase that catalyses the formation and isomerisation of disulfide bonds in the oxidising (relative to the cytosol) environment of the ER lumen. PDI also has some chaperone activity (193). PDI is an abundant protein in yeast and is the representative of a family of structurally and, probably, functionally related proteins that are characterised by the presence of thioredoxin-like domains separated by similar numbers of amino acids. PDI is essential in *S. cerevisiae* (194. 195) due to its catalysis of non-native disulfide bond shuffling (196). PDI function in *S. cerevisiae* can be, at least partially, replaced by a structurally related protein EUG1p found in yeast (197) and by mammalian PDI or by a related mammalian protein Eep72 (198). Similarly, *A. niger* and *N. crassa* contain structurally-related PDI family proteins (199, 200) although the extent of functional overlap has not been defined. Cloned fungal genes that encode PDI family proteins are listed in Table 2.

The primary functions of PDI, including roles unrelated to the formation of disulfide bonds such as a component of prolyl-4-hydroxylase and the triglyceride transfer protein, have been summarised by Freedman *et al.* (193). The formation and shuffling of disulfide bonds are catalysed by PDI in the relatively oxidising environment of the ER lumen which is buffered primarily by a glutathionine (GSH/GSSG) couple (201). The mechanism for maintenance of the oxidative lumen has not been firmly established but involves a protein Ero1p in *S. cerevisiae* which is associated with the lumenal side of the ER membrane and transcription of the encoding gene is induced by UPR (202, 203).

The complexity of protein folding in the lumen of the ER is such that all components and their interactions are not known. For the purpose of this Section only the key components of the assisted protein folding within the ER can therefore be given. Table 2 lists the cloned fungal genes encoding BiP, PDI family proteins and also calnexin. For information on other lumen chaperones, co-chaperons and foldases, refer to Gething (180). Lumenal peptidyl prolyl isomerases (PPIase) have also been summarised by Dolinsky and Heitmann (204). Calnexin is a membrane-bound calcium-binding chaperone with lectin activity, recognising terminal glucose on the core glycan of glycoproteins (205) although non-glycosylated protein can also be recognised.

Calnexin is a component of the quality control system in the lumen which ensures that only correctly folded proteins, and in this case glycoproteins, are able to progress in the secretory pathway. In mammalian cells and some fungi trimming of terminal glucose units from the core glycan of glycoproteins is catalysed by glucosidases in the lumen and the addition of a single glucose to the core glycan, by a glucosyl transferase (UDP-glucose:glycoprotein glucosyltransferase). The monoglucosylated peptides associate with calnexin and release requires glucosidase II activity. Genes encoding the glucotransferase have been described in *S. pombe* (206) and *A. niger* (J. Lambert and J. Peberdy, unpublished) but the gene appears to be absent from *S. cerevisiae* (207, 208). The gene is essential in *S. pombe* only under conditions of severe ER stress that cause under-glycosylation of secretory proteins, a situation that was shown also to induce transcription of the BiP gene (209). Unfolded peptides are returned to calnexin following glucosylation by the glucosyltransferase whereas fully folded proteins are not glucosylated and progress to the next stage of the secretory pathway (205). Deletion of the calnexin gene is lethal in *S. pombe* (210, 211) but not in *S. cerevisiae* (212). Calreticulin is a mammalian lumenal protein related in sequence (ca. 30% identity) and function to calnexin but it has not yet been reported in any yeast or filamentous fungus.

3.2.2. The unfolded protein response

The accumulation of malfolded proteins in the lumen of the ER leads to the up-regulated transcription of genes encoding ER-resident chaperones and foldases (179, 213). This stress response is called the unfolded protein response (UPR) The promoter of the *S. cerevisiae* KAR2 gene, encoding BiP, contains elements which ensure constitutively high levels of transcription. The promoter also contains a 22bp motif which confers inducibility (which can be over 100-fold) by factors which cause the accumulation of malfolded proteins. The 22bp motif was termed the unfolded protein response element (UPRE) (214) and has since been found in the promoters of other fungal genes subject to induction by UPR. The KAR2 UPRE contains a 10bp core sequence (GGACAGCGTG) which was essential for the UPR although core flanking sequences in the 22bp motif are also important. The essential region of the core element was further refined to the partial palindrome CAGCGTG (215) and this element (as CANNNTG) is found in other UPR-regulated genes (216). Partial conservation of this core element and, to a lesser extent, the flanking regions are also found in mammalian UPR-regulated genes (214) indicating at least some conservation of UPR-across eukaryotes. The UPR motif in mammalian cells has been renamed the ER stress response element (ERSE), $CCAATN_9CCACG$, (217). Although the ERSE has not been found in *S. cerevisiae* UPR-responsive genes, a homologue was found in the *A. niger bipA* gene promoter (217), which is a gene whose transcription is regulated by UPR (185). The *A. niger bipA, pdiA* and *tigA* genes are all UPR-regulated (185, 199, 200) although the motifs responsible for mediating the response have not been functionally determined.

That the UPR-responsive genes share common motifs suggests a common regulator protein which has been identified as a bZIP protein and named Hac1p in *S. cerevisiae* (215, 218). Several components of the UPR have now been identified in *S. cerevisiae*. The mechanism for signalling an accumulation of misfolded proteins in the ER lumen which results in the synthesis of active Hac1p, and the data that account for the model, have been summarised elsewhere (179, 213, 219) and are illustrated in Fig. 1. Ire1p is a transmembrane kinase, probably with direct contact with the nucleus and ER lumen, which mediates a signal indicating the accumulation of malfolded proteins or enhanced synthesis of ER membrane. Ire1p is activated via dimerisation and autophosphorylation, the primary signal is not firmly established but could be a diminished level of a chaperone (e.g. BiP). This would unite the different signals, e.g. malfolded proteins or enhanced ER membrane synthesis, which are two conditions that lower resident free chaperone levels whether by binding to malfolded proteins or by increase in ER lumen volume that occurs during ER membrane proliferation. ER membrane proliferation is induced in *S. cerevisiae* by over-expressing integral membrane proteins and leads to UPR (220). Activated Ire1p has riboendonuclease activity which splices HAC1 precursor mRNA (ligation to the cleaved mRNA requires a tRNA ligase named Rgl1p) and the mature mRNA is translated to give active Hac1p (221) which, in turn, binds to the UPRE and activates transcription of UPR-responsive genes.

Homologues of IRE1 have been found in mammalian cells (222, 223) suggesting that similar UPR mechanisms operate in mammalian cells as they do in *S. cerevisiae*. Homologues of UPR components have not yet been reported in filamentous fungi but the expectation is that they will be found and early indications are that they do exist (D. Jeenes, unpublished information). We have already seen, however, that some divergence is apparent between *S. cerevisiae* and mammalian cells; Hac1p binds to the UPRE in *S. cerevisiae* and a different bZIP protein, ATF6, is the putative ERSE-binding transcription factor in mammalian cells (217). A motif with high identity to the ERSE consensus, found in the *A. niger bipA* promoter, may indicate that filamentous fungi are not identical to yeast in the mechanism of UPR.

Hac1p mediates not only UPR, but also the enhanced synthesis of membrane phospholipids in *S. cerevisiae* by overcoming repression by Opi1p at the UAS_{INO} motif, CANNTG (219). In the mammalian cells, the synthesis of membrane fatty acids and sterols is regulated by the bZIP sterol responsive element (SRE) binding protein (SREBP), the active form of which is a dimerised protein formed following proteolytic cleavage of an ER membrane protein (224). The SREBPs co-ordinate the synthesis of membrane lipids by regulating the transcription of several enzymes, including acetyl-CoA carboxylase, fatty acid synthase and HMG CoA synthase. Putative SRE-1 sites and half sites are found in the *A. nidulans accA* gene (encoding acetyl-CoA carboxylase) although they have not been assessed for mediating the observed induction of *accA* by factors normally associated with UPR (225; J. Morrice, unpublished observation). The

Figure 3. UPR mechanisms in *S. cerevisiae*.
The transmembrane kinase/endoribonuclease Ire1p senses (a) an acccumulation of unfolded proteins in the ER lumen or (b) a requirement for enhanced membrane lipid synthesis. Ire1p dimerises, autophosphorylates and its endoribonuclease activity, together with the ligase Rlg1p, splices the precursor *HAC1* mRNA to produce mature mRNA and its translated product Hac1p. This bZIP transcription factor binds to the UPRE in the promoters of UPR-regulated genes to induce their transcription. Hac1p also relieves the negative transcriptional control exerted by Opi1p to stimulate phospholipid biosynthesis. The scheme was adapted from Cox *et al.* (219). Spatial considerations within the cell are discussed in the text and described also in Pahl and Baeurle (179).

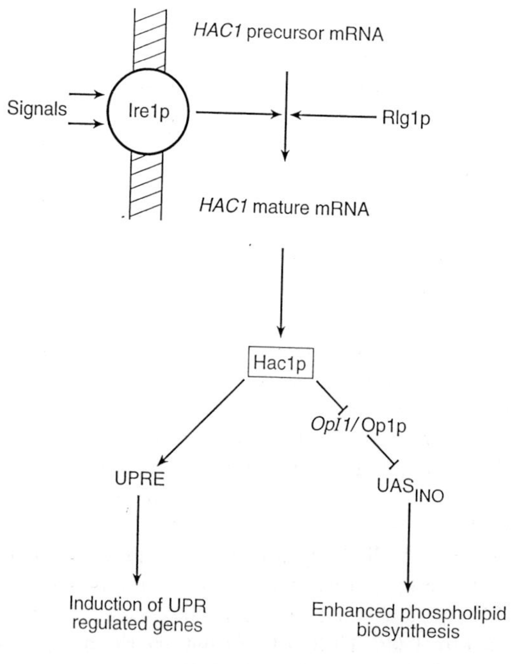

UPR system of mammalian cells is similar to that of *S. cerevisiae* but clearly more complex (217) and it remains to be seen how conserved UPR is within the fungi.

3.2.3. Genetic manipulation of the secretory pathway in fungi

The secreted yields of fungal enzymes can be increased by conventional mutagenesis-based strain improvement strategies. Commercial strains of yeasts and filamentous fungi have been produced in this way for production of enzymes. Heterologous enzymes are also produced by yeasts and filamentous fungi as hosts and, although commercially-acceptable yields of enzymes have been achieved, yields are generally very low compared to yields of native enzymes. Although mutagenesis can also be used to increase secreted yields of heterologous proteins, the identification of the secretory pathway as a significant bottleneck (226-228) has led to attempts at targeted improvements of the secretory pathway with most effort being applied at the protein folding step (229).

The accumulation of malfolded proteins in the ER induces the UPR and, consequently, leads to increased transcription of foldases and chaperones. As described above, the UPR is brought abut experimentally by tunicamycin (inhibits N-glycosylation), reducing agents, inhibit disulfide bond formation) or calcium ionophores (disrupt the calcium-rich environment of the ER lumen). In addition, over-expression of heterologous proteins induces the expression of BiP in *S. cerevisiae* (230, 231) and filamentous fungi (185). Even so, the extractable levels of BiP protein and PDI were reduced under these conditions (231) suggesting that BiP and PDI were sequestered by the malfolded protein. Analogous results were obtained for PDI in *A. niger* (C. Ngiam, unpublished information). Attempts have therefore been made to alter the production of BiP and PDI. Over-expression of BiP and PDI might overcome a possible deficiency. Lowering of BiP levels (below wild-type levels) decreased the secretion of three different heterologous proteins by *S. cerevisiae* but enhanced BiP levels did not significantly affect heterologous protein secretion (232). Rat scFv secretion by *S. cerevisiae* was enhanced over 2 fold when the *S. cerevisiae* BiP was over-expressed tenfold. (233). Thaumatin secretion from *S. cerevisiae* was not increased by enhanced BiP levels (234), but secretion of prochymosin was increased and further increased by disruption of *PMR1* (234) the gene which encodes a membrane-localised ATP-dependent Ca^{++} pump. Yields of thaumatin were extremely low in this work and the effects of BiP or PMR1p levels were less marked at higher prochymosin expression (234, 235). Even if secreted heterologous protein levels are not affected by BiP expression, the ratio of cell-associated protein to secreted protein may be increased by raising BiP levels, as observed in *A. niger* (185).

Over-expression of PDI (by ca. 16 fold) in *S. cerevisiae* enhanced the secreted yields of two heterologous proteins by 4 and 10 fold (236) but enhanced expression of PDI in *A. niger* did not alter the secretion of hen lysozyme by *A. niger* (C. Ngiam, unpublished). Rat scFv secretion by *S. cerevisiae* was enhanced by over-expression of either rat PDI or yeast PDI and co-over-expression of BiP and PDI gave more than an additive increase in scFv secretion (233). Disruption of the gene encoding calnexin in *S. cerevisiae* (which is not lethal) did not alter the secretion of two native enzymes but slightly enhanced (c.a. 30%) the secreted levels of heterologous α_1-antitrypsin (211). Also, the secretion by *S. cerevisiae* of a glycosylated variant of hen egg white lysozyme was enhanced 2-5 fold by disruption of the calnexin-encoding gene (237), but no effect on secretion of prochymosin by calnexin disruption in *A. niger* was observed (Huaming Wang, unpublished information).

Taken together, these studies indicate that genetic manipulation of the folding and quality control aspects of the secretory pathway can alter the secreted yields of heterologous proteins. The effects observed depend upon the particular heterologous protein and fungal species used as expression host. Protein folding and quality control requires several components which interact with the secretory protein and with each other. Without further knowledge of the detail of protein folding it is not possible to precisely predict the outcome of deliberately altering the expression of a gene encoding just one component of the secretory pathway. Even if one bottleneck is overcome, the next will quickly be reached. Even so, the secretory pathway, and the environment of the ER lumen in particular, holds the key to improving the secreted yields of many heterologous protein from fungi. Manipulation of several components of the protein folding system at once may be possible, and co-manipulation of BiP and PDI levels is a beginning. Such an objective may be feasible through the UPR system because the transcription of several components of the folding apparatus are similarly regulated by Hac1p. It must be remembered, however, that this is a stress response and constitutive up-regulation of stress-responsive genes will not provide a healthy cell. This is already confirmed in *S. cerevisiae* where constitutive expression of *HAC1* is toxic and leads to slow growing cells (221). Even so, this work showed that all UPR-responsive genes were co-ordinately up-regulated and demonstrated an approach to control the folding apparatus within a cell, rather than individual components.

4. CELL BIOLOGY OF THE SECRETION PATHWAY

4.1 Organelles of the secretory pathway

The dynamics of intracellular processing and transport of secretory proteins in eukaryotic organisms has engaged the interest and energies of cell biologists for several decades. The opportunity to develop and employ a genetic, ultrastructural and biochemical approach in *S. cerevisiae* has meant that this fungus has been a key model in our understanding of this important cellular process. Ultrastructural investigations, which aided our understanding of the secretory pathway in filamentous fungi, were published some two decades ago (252). In higher eukaryotes the secretory pathway involves two key organelles, the Endoplasmic Reticulum and the Golgi apparatus. In fungi the former is clearly evident, however, the Golgi is differently organised and has the form of dispersed cisternae, and probably organised as a network of membranous tubules dispersed throughout the cytoplasm. Such a structure is more equivalent to the structures found in a mitotic animal cell.

An important feature of the secretory pathway is the vesicular transport system, which is involved in the carriage of proteins from the ER, where they have been processed as described above, to the Golgi and on to the plasma membrane (253). The movement of these vesicles in both yeasts and filamentous fungi is polarised in the direction of the growing point of the cell. This is most pronounced in filamentous fungi because of the elongated filamentous nature of the cells and is manifest intracellularly by a spatial distribution of organelles. The vesicles are most apparent in the hyphae at the apices where they are seen in clusters. It was observations of this arrangement that led to speculation of vesicular involvement in protein secretion, and other processes, more than

20 years ago (254). An important step in the characterisation of vesicles in fungi was the discovery of a protein coat on the vesicles (255). Similar vesicular coating had been reported in mammalian cells (256). The main protein involved is a 180kDa molecule, which has been characterised as clathrin in both types of organisms (255, 257). Observations on thin sections of hyphae of *Neurospora crassa* suggested that the coated vesicles were commonly associated with golgi cisternae (255). This coating might be required for the transportation of vesicles to the growing point of the cell where it is removed prior to fusion of vesicles with the plasma membrane. Not all the vesicles seen the in the sections were coated, and in general these were smaller and interpreted as being chitosomes involved in chitin biosynthesis at the cell wall.

Using phase contrast microscopy the cytoplasm at the hyphal tip is frequently seen to include a refractile or dark phase body – the apical body or Spitzenkorper. This structure is interpreted in transmission electron microscopic sections as the vesicle cluster (258, 259). Clearly these techniques reveal nothing of the dynamics of such a structure, however, recent studies in which video-microscopy was used showed the Spitzenkorper to be a "pleomorphic complex constantly changing size and shape" (260). A co-ordinated role for this vesicle complex in the polarised processes of hyphal tip growth and protein secretion is a reasonable hypothesis that explains the invasive growth characteristic of filamentous fungi. Evidence to confirm this view was made available in studies, which involved the immunostaining of hyphal tips of *A. niger* that were secreting glucoamylase (261). Furthermore mutations that alter growth polarity, e.g. as seen in the *mob* mutant in *Neurospora crassa* which develops bulbous regions in the hyphae, can result in significant increases in protein secretion (262)

4.2 Genetic and Mechanistic features of the secretory pathway

The formation, budding and docking of vesicles from the ER to the Golgi are just some of the aspects of the secretion mechanism that have been dissected genetically in yeast by the isolation of temperature sensitive mutations (263, 264). Two types of vesicles have been identified COPI and COPII (cytosolic **co**at **p**roteins), based on the proteins of a cytosolic origin involved in their formation. Anterograde transport i.e. ER to Golgi, is mediated by COPII vesicles, however, the role of COPI vesicles is less clear. A primary involvement in retrograde transport from the Golgi back to the ER has not been clarified (265). The COPII proteins are used in the formation of vesicles which bud from the ER. Some 4 COPII proteins have been identified to date via the genes (*sec*) which encode them. Two pairs of proteins are formed into heterodimers, this comprise Sec23p-Sec24p and Sec13p-Sec31p (266-268). Other proteins involved in the overall process include Sec12p which is a specific guanine exchange factor which supplies GTP to a GTPase (known as Sar1p) which is utilised in the recruitment of the COPII proteins into the membrane. Reconstitution experiments have demonstrated the role of these proteins in vesicle formation (269). Another Sec protein required for vesicle budding is Sec16p. This binds, at the COOH terminal domain, with Sec23p (270, 271) and another protein located in the ER (271).

COPI proteins have been identified in mammalian vesicles and are therefore not so well characterised genetically. The proteins occur as a stoichiometric set of subunits. Two Sec proteins from yeast, equivalent to two mammalian COP proteins are Sec26p and Sec27p (272, 273).

The fusion of ER generated vesicles with the target Golgi membrane is highly controlled by SNAREs (274). These are protein complexes integral in the structure of the vesicle membrane (vSNARE) and target membrane (tSNARE) (275). In the case of vSNAREs in yeast, four putative proteins, Bet1p, Bos1p, Sec22p and Ykt6p (276, 277), have been implicated in the fusion with the Golgi tSNARE Sed5p (278). Other proteins are also involved in the overall process. Uso1p is required for the assembly of both SNAREs (279) and another, Sly1p, associates with Sed6p, but the significance of this is not known (277). More recently a cytosolic protein, identified as Sec35p, has been shown to be required for association with Uso1p to promote docking of vesicles to the Golgi (280).

The retrograde transport of vesicles from the Golgi to the ER also involves a SNARE mechanism. The ER tSNARE involves three proteins Ufe1p, Sec20p and Tip20p (281), however, there may be some functional conservation in vSNAREs because Sec22p is also believed to function in the retrograde process (281).

Fusion of vesicles with the target membrane results in a vSNARE/tSNARE complex and involves further SNAP proteins. In yeast two, Sec17p and Sec18p, have been identified. The latter is an ATPase and has an N-terminal domain attached to two consensus ATP-binding sites (282). Once the fusion event has occurred the complex is probably no longer significant. This is suggested by the dynamic behaviour of SNAREs which may also be recycled between organelles during vesicle transport (283, 284). Retention recycling of membrane proteins, in and between organelles, is most probably associated to functional domains in the molecule. Thus in Sec12p a transmembrane domain is believed to be the signal which controls retrieval and the amino-terminal cytoplasmic domain includes the site retention signal (285).

Targeting of vesicles from the Golgi to specific docking sites on the plasma membrane is essential for polarised growth which occurs in both the yeast cell and the fungal hypha. Such sites have been located at the growth zone of a budding yeast cell and comprise a protein complex called the exocyst. Two proteins identified in the exocyst are Sec10p and Sec15p (286). Analysis of the former identified two functional domains, the larger amino terminal domain of Sec10p complexes with Sec15p and promotes vesicle fusion with the plasma membrane. Over-expression of this domain blocks exocytosis and results in an accumulation of vesicles; with the smaller carboxy-terminal domain, mophogenetic effects, promoting the elongation of cells, are observed. A third protein Sec3p is also found at sites of exocytosis and has been suggested to interact with actin in the vesicle targeting process (287).

The polarised movement of vesicles in fungal cells is thought to involve the cytoskeleton. Whether both the actin and tubulin components, or only one of them, are involved remains to be resolved. Evidence available to support an involvement has been obtained from two approaches. Studies involving the use of inhibitors such as methyl benzimidazole-2-yl carbamate suggest a role for both microtubules (288-290) and actin (290). Useful data has also been obtained in a video-microscopy study on a kinesin-deficient strain of *Neurospora crassa*. This mutant exhibits extensive distortion to hyphal morphology and concomitant failure to assemble a Spitzenkorper (291). This suggests that the microtubule associated protein, kinesin has the same property of organelle translocation along microtubules fungi, as in mammalian cells. Nevertheless genetic studies on the cytoskeletal elements are required to confirm the role of the microtubules and/or actin.

Studies with inhibitors have also been valuable in other aspects of the secretion process. Brefeldin A is known to inhibit secretion and in several fungal systems has been found to cause gross distortion of the endomembrane system at concentrations which inhibit hyphal growth (292, 293). Experiments with *Candida albicans*, growing in its hyphal form, showed similar effects with the accumulation of abnormal membranous structures in the subapical regions of hyphal cells (294). In *S. cerevisiae*, the inhibitor had the same effect on the Golgi which lead to hypertrophy of the ER connected cis-Golgi network (295)

The diversity of proteins secreted by yeasts has led to the suggestion that distinct pathways are involved in their transport to their location. In *Yarrowia lipolytica*, four distinct secretory pathways have been proposed for proteins used for different purposes, i.e. proteins that are secreted, proteins destined for the plasma membrane, proteins for cell walls and proteins involved in peroxisome biosynthesis (296).

In contrast ultrastructural examinations on appresoria of fungal pathogens show the presence of this organelle (297). It is possible that this difference relates to the differentiation of the hyphal tip for the infection step.

5. THE CELL WALL AND PROTEIN SECRETION

5.1. Mediation of secretion by the cell wall

Proteins delivered by the secretory pathway to the periplasmic space still have to traverse the cell wall. Studies on protoplast secretion by protoplasts (298) suggest that the wall can be a barrier restricting the movement of proteins into the external medium. Clearly there are differences in the walls of different fungi. In filamentous fungi and some yeasts, the wall is a porous structure allowing the free flow of small molecules into and out of the cell, however, even the smallest secreted proteins e.g. hydrophobins, are too large to move out of the cell on a simple diffusion gradient. The accumulation of vesicles at the hyphal tips suggests that protein molecules, which have been processed during their passage from the ER to the tip vesicles, will be discharged at the apex as the vesicles fuse to grow the plasma membrane. The apical tip is also the site of nascent wall synthesis, but in this region it is assumed that the different polymers of the wall, i.e. chitin and ß-glucan are independent layers. In the sub-apical region the polymers become cross-linked by covalent and non-covalent bonding to create a rigid wall. The hyphal tips are therefore very plastic and it can be supposed that proteins released on the plasma membrane surface are carried in the flow of cell wall material eventually reaching the outer surface of the wall where they either remain or are released into the surrounding medium (299).

This model is supported by a variety of studies, which focus on the hyphal wall. Treatment of fungal hyphae with enzyme cocktails used in the isolation of protoplasts confirms the view that proteins are trapped in the more rigid regions of the wall (298). Mutants with a phenotype reflected in a modified wall composition and altered hyphal morphology may either secrete more or less protein than the wild type. Modifications to the cell wall may be significant in hyper-secreting strains as found in some cellulose hyper-secreting mutants of *T. reesei* (300).

5.2. Proteins in the cell wall

Once thought to be part of a fungal cell of little importance, the cell wall is now seen as a highly dynamic cell structure which has significant and important roles in fungal growth, development and in their interactions with other organisms and their environment generally. The wall is also important in secretion, either in aiding the release of proteins to the external environment or as the location of secreted proteins, which have a wall-related function.

The full extent of the types of proteins in fungal cell walls has still to be resolved. The situation is better understood in yeasts than in other fungi. Two types of protein have been identified, GPI-proteins (proteins with a glycosyl phosphatidylinositol anchor; 301) and Pir-proteins (a gene family of **p**roteins with **i**nternal **r**epeats; 302). In *S. cerevisiae* and *C. albicans* these proteins occur as glycosylated and non-glycosylated molecules. The Pir-proteins from *S. cerevisiae* are highly *O*-glycosylated and the mannan chains form cross linkages which are important in anchoring the molecules into the fabric of the wall (303). The glycoproteins are clearly processed through the secretory pathway, as may be the non-glycosylated proteins although this has not been investigated. The glycoproteins are characterised by the high level of mannosylation (known as high molecular weight mannoproteins, HMWM). In *S. cerevisiae* this component, which form radially extending fibrillae on the outer surface of the wall, account for 30-50% of the wall dry weight (304). In *C. albicans* the HMWM are thought to be the major elicitor of immunity to the fungus in patients with Candidiosis infections (305, 306). In *C. albicans* a significant number (estimates range from 20 to 40) of smaller proteins have also been reported (307, 308).

Several studies have shown the dynamic features in wall composition and architecture during cell ageing and morphogenesis, and in response to alterations in environmental conditions. Examples of this are proteins designated Cwp2p (309), G-proteins (310) and a protein called Sed1p (311). Cwp2p in present in the wall of *S. cerevisiae* at all times during the cell cycle, however, it increases in abundance during the G2 phase. At this time the cells show an increase in resistance to small polypeptide inhibitors such as nisin, and in resistance of the walls to zymolyase (309). The G proteins are small GTP-binding proteins which have key signaling functions in morphogenesis; they include Bud2p which is involved in bud site selection, Cdc24p which organises both the bud site and the cell projection at conjugation and Rho1p which regulates the ß-1,3-glucan synthase involved in wall synthesis (310). Sed1p is another protein associated with the integrity of the cell wall; its expression increased in stationary phase cells and enhances their resistance to zymolyase (311).

5.3. Functions of wall bound and secreted proteins

Proteins and glycoproteins in the cell wall have a range of functions. Many are enzymes such as, invertase and acid phosphatase in *S.cerevisiae*, and many more in *C. albicans* (Table 3). In filamentous fungi the situation is less clear as definitive studies are limited. One example can be described. Recent studies with *Aspergillus fumigatus* would tend to suggest otherwise. This pathogen although commonly associated with

Table 3 Periplasmic, Cell wall and Extracellular enzymes in *Candida albicans* (adapted from 362)

Enzyme	Gene	Location	Function	References
Cell wall substrates				
Exo-ß-(1,3)-glucanase	*EXG*	Cell wall, extracellular	Cell wall morphogenesis	340-342
ß-1,3-glucan transferase	*BGL1*	Cell wall	Cell wall metabolism	343
Chitinase	*CHT1-3*	Periplasm, cell wall extracellular	Hydrolytic enzyme, cell wall morphogenesis	344, 345
ß-*N*-acetylglucosamine	*HEX1*	Periplasmic, extracellular	Hydrolytic enzyme, virulence factor?	346, 347
Extracellular substrates				
Secreted aspartyl Proteinase	*SAP1-9*	Extracellular, cell surface	Putative virulence factor	348, 349
Phospholipase A		Cell wall, surface, extracellular	Hydrolytic enzyme	350
Phospholipase B	*PLB1*	Extracellular	Hydrolytic enzyme, putative virulence factor	351
Phospholipase C		Extracellular	Hydrolytic enzyme	352
Lysophospholipase		Cell wall, surface, extracllular	Hydrolytic enzyme	352
Lysopholipase-transcylase		Extracellular	Hydrolytic enzyme, putative virulence factor	353, 354
Esterase		Extracellular	Hydrolytic enzyme	355
Hemolytic factor		Cell wall,extracellular	Hydrolytic enzyme	356
Acid phosphatase		Periplasmic, surface	Hydrolytic enzyme	357
Lipase	*LIP1*	Extracellular	Hydrolytic enzyme	358
Hyaluronidase		Extracellular	Hydrolytic enzyme, virulence factor?	359
Chondroitan sulphatase		Extracellular	Hydrolytic enzyme, virulence factor?	359
Metalloprotease		Cell wall, extracellular?	Hydrolytic enzyme	360
Trehalase		Cell wall, extracellular	Hydrolytic enzyme	361

infections of the lung can be very invasive into deeper body tissues in immunocompromised patients. As the fungus is known to produce a number of proteinases it was reasonably assumed that these were important in tissue invasion. Mutants blocked in the production of these enzymes were, however, found to be still invasive. Analysis of these strains revealed the presence of aspartic and serine proteases in the wall at hyphal tips that could provide a localised mechanism for tissue invasion (312). Enzymes that are secreted into a growth medium can also be found in cell wall material (298), however, whether these can be described as both wall bound and extracellular is an issue for debate.

Glycoproteins at the cell surface are thought to have a role in recognition or as elicitors. Such proteins are now being discovered in fungi, which have some form of association with plants. One such example is the pathogen of tobacco, *Phytopthora parasitica* var. *nicotianae* in which a 34kDa glycoprotein has been found and purified (313). This molecule was shown to be an elicitor when tobacco plants exposed to the molecule through their roots showed an upregulation of lipoxygenase activity and the accumulation of a hydroxyproline-rich glycoprotein. The timing of expression of this glycoprotein, was not determined, however, in *Colletotrichum lindethianum* a proline rich protein, with two putative glycosylation sites, has been found which is expressed in the fungus only when in the host plant (314). It is suggested that the molecule forms cross-linking components in the cell wall. Cell wall proteins with recognition functions have also been described in ectomycorrhizal fungi in studies on the symbiosis of *Psilothus tinctorius* and *Eucalyptus* spp. (315). Dramatic changes in the cell wall proteins of the fungus were seen at the early stages of the interaction, which implied significant spatial and temporal phenomena. Analysis of the protein profile revealed an increase in a 32kDa acidic polypetide and a decrease in a 95kDa glycoprotein.

In the yeast form of *Candida albicans*, the wall glycoproteins have an important role in the adhesion of the fungus to the epithelial cells of the host. The relative roles of the protein and mannan components of the these molecules, in adhesion, is a matter of debate. However, a key factor in this process is the hydrophobicity of the cell surface, with hydrophobic cells adhering more strongly than hydrophilic cells. Hydrophobicity can be a function of either moiety of the molecule. Some evidence suggests that changes in hydrophobicity can result from changes in the outer mannan chains caused by the presence of more phosphodiester linkages (316). However other workers stress the stress the significance of the protein component (317) which can be quite disparate (318).

The discovery of the unique group of proteins, the hydrophobins is one of the major advances in fungal biochemistry in this decade. Since their discovery in 1991 (319) there have been reports for their presence in many fungi (320) which implies they are probably ubiquitous in terrestrial species. These molecules are generally small peptides comprised of *ca* 75-90 amino acids, always including 8 cysteine residues, which contribute to the hydrophobic properties. When released onto the surface of aerial hyphae, fruit bodies and other structures exposed to the atmosphere, the proteins self-assemble to form rodlet structures. It is also possible that hydrophobins have a role in adherence of hyphae to surfaces as they grow over them (321) and are involved in plant fungus interactions both symbiotic (315) and pathogenic (322). When fungi are grown in submerged liquid culture the hydrophobins are released into the growth medium. Molecular analysis of hydrophobin genes has shown that many fungi produce several of these proteins that may be expressed at different stages of development and have different functions. In *Schizophyllum commune* at least four hydrophobins have been identified, SC1, SC3, SC4 and SC6 and their expression shown to be developmentally regulated. SC3 is, so far, unique in that it is the only hydrophobin expressed in the monokaryon of *S. commune*. Synthesis of the protein is suppressed in dikaryons with compatible B mating genes. Expression of SC1, SC4 and SC6 is associated with the dikaryon and fruit body. In these basidiomycete reproductive structures, hydrophobins are important as surface coverings for the fruit body and lining the air cavities in these structures (323). Families of hydrophobins have also been reported in *Agaricus bisporus* (324-4), *Trichoderma reesei* (327-329), and *Coprinus cinereus* (330).

5.4 Glycoproteins in the plasma membrane

The importance of glycoproteins in plasma membranes has arisen from quite divergent work on fungal pathogens of animals and plants. The pathogen of mammals *Pneumocystis carinii* (331) recently designated as a fungus, and the entomopathogen *Entomophaga maimaiga* (332) have both been shown to have glycoproteins, of yet unknown significance in their plasma membranes. In both *Aspergillus fumigatus* and plant pathogens, the plasma membrane is the location for glycoproteins that are homologous to ABC transporters of mammalian cells that are responsible for the efflux of drugs from cells (333). In fungi transcription of the genes encoding these proteins is enhanced in the presence of fungicides suggesting a role in multi-fungicide resistance (334).

6. PHYSIOLOGICAL ASPECTS OF PROTEIN SECRETION IN FUNGI

Despite the industrial importance of enzyme production for commercial application (see Chapters 16 and 17) there is little published information on the physiology of protein secretion in fungi. The position is now changing as the interest to exploit fungi as host for recombinant protein production (see Chapter 6) and data is now becoming available from both batch and continuous culture studies and primarily with transgenic strains. Pluschkell *et al.*, (335) using a batch system showed that glucose oxidase secretion by *Aspergillus niger* followed two phases. The first, rapid phase occurred during the linear growth period, however, not all the enzyme produced was secreted. A second slower phase of secretion followed when almost all the accumulated enzyme appeared. The final amount of secreted enzyme was 90% of the total, it is possible that the cell bound enzyme was trapped in the cell wall.

Trinci and co-workers have made comparisons between batch and continuous cultures using the secretion of glucoamylase by *A. niger* (336, 337) as experimental systems. Comparing batch and continuous cultures it was shown that the specific production rates for glucoamylase were three times greater in continuous culture than in batch culture. Furthermore production in continuous culture increased with dilution rate. The conclusions from this work are perhaps expected, namely the close coupling of specific protein production with the growth rate of the fungal culture.

Attempts to increase protein secretion have also involved the use of detergents in the growth medium. Supplementation of the medium with Tween 80 has been found to increase enzyme release in growing cultures of *Thermomyces languinosus* (338) and in

Trichoderma reesei (339) In contrast Triton X-100 caused a decrease in secretion (338). The underlying mechanisms which cause these effects are unknown.

7. CONCLUDING REMARKS

We have focussed in this review upon the post-ribosomal synthesis and secretion of glycoproteins from yeasts and filamentous fungi. These are vital products of metabolism which have a variety of key roles in the fungal lifestyle. Our understanding of the molecular events involved in this area is increasing rapidly and these advances are being made not only with yeasts such as *S. cerevisiae* and *S. pombe* but with filamentous fungi that are comparatively less amenable to genetic and molecular analysis. There is a degree of commonality in this area across the fungal community that aids research and information derived from studies in yeasts, which help with other fungi. This does not mean, however, that studies of glycoprotein synthesis and secretion by filamentous fungi are necessarily derivative. In comparison with yeasts, filamentous fungi occupy a more diverse range of natural habitats and, consequently, have developed a breadth of ways in which they interact with their ecosystems. Many species are particularly adept at the secretion of enzymes for degredative purposes and this facility has been exploited commercially for the supply of bulk enzymes for a range of biotechnological and industrial applications. Many filamentous fungi are pathogenic and the secretion of enzymes and elicitors are important features of the fungal-host interaction. Thus, the secretion of proteins either by yeasts or filamentous fungi has both common features and aspects peculiar to their morphology and lifestyle of the individual species. Improved understanding of protein secretion by fungi will provide a better basis for their commercial exploitation and also for controlling fungal diseases of plants and animals.

REFERENCES

1. D.B. Archer and D.A. Wood, In N.A.R. Gow and G.M. Gadd (eds) The Growing Fungus, Chapman and Hall, London, 1995, 137-162.
2. H.R.B. Pelham, Cell Struc. Func., 21 (1996) 413.
3. P. Walter and A.E.Johnson, 10 Annu. Rev. Cell. Biol. (1994) 87.
4. T.A. Rapaport, B. Jungnickel and U. Kutay, Annu. Rev. Biochem. 65 (1996a) 271.
5. K. Romisch and A. Corsi, In S.M. Hurtley (ed) Protein Targeting, Protein translocation into the endoplasmic reticulum, IRL Press. Oxford, 1996, 101-122.
6. R.D. Teasdale and M.R. Jackson, Ann. Rev. Cell. Dev. Biol. 12 (1996) 27
7. C. Rothe and L. Lehle, Eur. J. Biochem. 252 (1998) 16.
8. T. Omura, J.Biochem. 123 (1998) 1010.
9. N. Pfanner, E.A. Craig and A. Honlinger, Annu. Rev. Cell Dev. Biol. 13 (1997) 25.
10. D.T.W. Ng, J.D. Brown and P. Walter, J. Cell Biol. 134 (1996) 269.
11. S. Matoba and D.M. Ogrydziak, D.M. J. Biol. Chem. 273 (1998) 18841.
12. T.A. Rapoport, M.M. Rolls and B. Jungnickel, Curr. Opin. Cell Biol. 8 (1996) 499.
13. A. E. Johnson, Trends in Cell Biol. 7 (1997) 90.
14. K-W Kalies and E. Hartman, Eur. J. Biochem. 254 (1998) 1.
15. B.C. Hann and P. Walter, P. Cell 67 (1991) 131.
16. B. Hann, C.J. Stirling and P. Walter, Nature 356 (1992) 532.
17. C.J. Stirling and E.W. Hewitt, Nature 356 (1992) 534.

18. J.D. Brown, B.C. Hann, K.F. Medzihradszky, M. Niwa, A.L. Burlingame and P. Walter, EMBO J. 13 (1994) 4390.
19. M. Sanchez, J.M. Beckerich, C. Gaillardin and A. Dominguez, Gene 203 (1997) 754.
20. J.D. Miller, S. Tajima, L. Lauffer and P. Walter, J. Cell Biol. 128 (1995) 273.
21. S.C. Ogg, M.A. Poritz and P. Walter, Mol. Biol. Cell 3 (1992) 895.
22. M. Pilon, K. Romisch, D. Quach and R. Schekman, Mol Biol. Cell 9 (1998) 3455.
23. C.J. Stirling, J. Rothblatt, M. Hosobuchi, R. Deshaies and R. Schekman, Mol. Biol. ell 3 (1992) 129.
24. J. Toikkanen, E. Gatti, K. Takei, S. Saloheimo, V.M Olkkonen, H. Soderlund, P. DeCamilli and S. Keranen, Yeast 12 (1996) 425.
25. Y. Esnault, M.O. Blondel, R.J. Deshaies, R. Scheckman, and Kepes, F. EMBO J. (1993) 4083.
26. J Broughton, D. Swennen, B.M. Wilkinson, P. Joyet, C. Gaillardin, and C.J Stirling, J. Cell Sci. 110 (1997) 2715.
27. K.U. Kalies, T.A. Rapoport and E. Hartman, J. Cell Biol. 141 (1998) 887.
28. M. Pilon, R. Schekman and K. Romisch, EMBO J. 16 (1997) 4550.
29. R.S. Hegde, S. Voight, T.A. Rapoport and V.R. Lingappa, Cell 92 (1998) 621.
30. K. Finke, K. Plath, S. Panzer, S. Prehn, T.A. Rapoport, E. Hartman, and T. Sommer, EMBO J. 15 (1996) 1482.
31. H. Fang, S. Panzer, C. Mullins, E Hartman and N. Green, J. Biol. Chem. 271 (1996) 16460.
32. H. Fang, H., Mullins, C. and Green, N. J. Biol. Chem. 272 (1997) 13152.
33. C. Mullins, H.A. Meyer, E. Hartman, N. Green, and H. Fang, J. Biol. Chem. 271 (1996) 29094.
34. H.A. Meyer, and E. Hartman, J. Biol. Chem. 272 (1997) 13159.
35. T. Langer, T. and W. Neupert, Curr. Topics Microbiol. Immunol. 167 (1991) 3.
36. S.K. Lyman, and R. Schekman, Experientia 12 (1996) 1042.
37. E. Paunola, T. Stuntio, E. Jamsa, and M. Makarow, Mol. Cell Biol. 9 (1998) 817.
38. K. Plath, W. Mothes, B.M.Wilkinson, C.J. Stirling, C.J. and T.A. Rapoport, Cell 94 (1998) 795.
39. S. Panzner, L. Dreier, E. Harman, S. Kostka, and T.A. Rapoport, Cell 81 (1995) 561.
40. J.L. Brodsky, J. Goeckeler, and R. Scheckman R. Proc. Natl. Acad. Sci USA 92 (1995) 9643.
41. K.E.S. Matlack, K. Plath, B. Misselwitz, and T.A. Rapoport, Science 277 (1997) 938.
42. R.A. Craven, M. Egerton, and C.J. Stirling, EMBO J. 15 (1996) 2640.
43. R.A. Craven, j.R. Tyson, J.R. and C.J. Stirling, Trends Cell Biol. 7 (1996) 277.
44. A.K. Corsi, and R. Schekman, J. Cell Biol. 137 (1997) 1483.
45. S.K. Lyman, and R. Schekman, Cell 88 (1997) 85.
46. H. Lis and N. Sharon, Eur. J. Biochem., 218 (1993) 1.
47. A. Varki, Glycobiol., 3 (1993) 97.
48. W. Tanner, and L. Lehle, L. Biochim. Biophys. Acta , 906 (1987) 81.
49. C. Abeijon, and C.B. Hirschberg, Trends Biochem Sci., 17 (1992) 32.
50. A. Herscovics, and P. Orlean, FASEB J., 7 (1993) 540.
51. F.W. Hemming, In J. Montreuil, J.F.G.Vliegenhart, and H. Schacter (eds.) Glycoproteins. (1995) Elsevier, Amsterdam. pp 127-143.
52. P. Burda, and M. Aebi, Biochim. Biophys. Acta, 1426 (1999) 239.

53. R. Knauer, R. and L. Lehle, Biochim. Biophys. Acta, 1426 (1999) 259.
54. A. Herscovics, Biochim. Biophys. Acta, 1426 (1999) 275.
55. M. Sumper, and F.T. Wieland, In J. Montreuil, J.F.G. Vliegenhart and H. Schacter (eds.), Glycoproteins. (1995) Elsevier, Amsterdam, pp 455-474.
56. T.D. Bugg and P.E. Brandish, FEMS Lett., 119 (1994) 255.
57. C. Whitfield, Trends Microbiol., 3 (1995) 178.
58. S. Moens and J. Vanderleyden, Arch. Microbiol., 168 (1997) 169.
59. S.S. Krag, Biochem. Biophys. Res. Comm., 243 (1998) 1.
60. W.L. Adair and N. Cafmeyer, Arch. Biochem. Biophys., 259 (1987) 589.
61. G.J. Quelhorst, J.S. Pitrowski, S.E. Steffen and S.S. Krag, Biochem. Biophys. Res. Comm. 244 (1998) 546.
62. F.W. Hemming, In Goodwin, T.W. (ed.), Biochemistry of lipids, Ser. 1, Vol. 4, Butterworths, London, (1974) pp 39-97.
63. L. Heller, P. Orlean, and W.L. Adair, Proc. Natl. Acad. Sci 89 (1992) 111.
64. M.A. Lehrman, Glycobiol. 1 (1991) 553.
65. C.B. Hirshberg, P.W. Robbins, and C. Abeijon, Ann Rev. Biochem. 67 (1998) 49.

66. J.S. Kruszewska, M. Saloheimo, M. Penttila, and G. Palamarczyk, Curr. Genet. 33 (1998) 445.
67. P. Orlean, C. Albright and P.W. Robbins, J. Biol. Chem. 263 (1988) 17499.
68. J.W. Zimmerman, C.A. Specht, B.X. Cazares, and P.W. Robbins, Yeast 12 (1996) 765.
69. PA. Colussi, C.H. Taron, J.C. Mack, and P. Orlean, Proc. Natl. Acad. Sci. USA (1997) 7873.
70. J. Kruszeska, R. Messner, C.P. Kubicek, and G. Palmarczyk, J. Gen. Microbiol. 135 (1990) 301.
71. B.L. Arroyo-Flores, C. Calvo-Mendez, A. Flores-Carreon and E. Lopez-Romero, Micorbiol. UK 141 (1995) 2289.
72. B.L. Arroyo-Flores, C. Calvo-Mendez, A. Flores-Carreon and E. Lopez-Romero, Antonie van Leeuw Intl J. Gen. Mol. Microbiol. 73 (1998) 289.
73. G.L.F. Wallis, F.W. Hemming, and J.F. Peberdy, Biochim. Biophys. Acta, 1426 (1999a) 91.
74. P. Orlean, Mol. Cell Biol. 10 (1990) 5796.
75. K.W. Runge, T.C. Huffaker, and P.W. Robbins, J. Biol. Chem. 259 (1984) 412.
76. S.T. Heesen, L. Lehle, A. Weissmann, and M. Aebi, Eir. J. Biochem. 224 (1994) 71.
77. J. Rodriguez Bonilla, L. Vargas Rodriguez, C. Calvo-Mendez, A. Flores-Carreon, and E. Lopez-Romero, Antonie van Leeuw Intl J. Gen. Mol. Microbiol., 73 (1998) 373.
78. K.O. Hartog, and B. Bishop, Nucl. Acids Res., 15 (1987) 36277.
79. J. Zou, S. Scocca, and S.S. Krag, Arch. Biochem. Biophys., 317 (1995) 487.
80. M.A. Kukuruzinska, and K. Lennon-Hopkins, Biochim. Biophys. Acta, 1426 (1999) 359.
81. K. Lennon, R. Pretel, J. Kesselheim, S.T. Heesen, and M.A. Kukuruzinska, Glycobiol., 5 (1995) 633.
82. A. Jacobson, and S.W. Peltz, Ann. Rev. Biochem., 65 (1996) 693.
83. C.F. Albright, and P.W. Robbins, J. Biol. Chem., 265 (1990) 7042.
84. T.C. Huffaker, and P.W. Robbins, Proc. Natl. Acad. Sci. USA, 80 (1983) 7466.

85. B.J. Jackson, C.D. Warren, B.Bugge, and P.W. Robbins, P.W. Arch. Biochem. Bipohys., 272 (1989) 203.
86. M. Aebi, J. Gassenhuber, S.T. Domdey, and S.T. Heesen S.T. Glycobiol., 6 (1996) 439.
87. P. Burda, S.T. Heesen, A. Brachat, A. Wach, A. Duesterhoeft, and M. Aebi, Proc. Natl. Acad. USA, 93 (1996) 7160.
88. G. Reiss, S.T. Heesen, J. Zimmerman, P.W. Robbins, and M. Aebi, Glycobiol., 6 (1996) 493.
89. I. Stagljar, S.T. Heesen, and M. Aebi, Proc. Natl. Acad. Sci., 91 (1994) 5977.
90. P. Burda, and M. Aebi, Glycobiol., 8 (1998) 455.
91. A. Verbert, In Glycoproteins (J. Montreuil, H. Schachter, J.F.G. Vliegenthart eds) Elsevier, (1995) pp145-152.
92. S. Silberstein, and R. Gilmore, FASEB J., 10 (1996) 849.
93. C.B. Sharma, L. Lehle, and W. Tanner, W. Eur. J. Biochem., 116 (1981) 101.
94. R. Cueva, C. Cotano, G. and Larriba, G. Yeast 14 (1998) 773.
95. L. Ballou, P. Gropal, B. Krummel, M. Tammi and C. Ballou, Proc. Natl. Acad. Sci. USA, 83 (1986) 3081.
96. A.J. Petrescu, T.D. Butters, G. Reinkensmeier, S. Petrescu, F.M. Platt, R.A. Dwek and M.R. Wormald, EMBO J., 16 (1997) 4302.
97. E. Alvarado, T. Nukada, T. Ogawa, and C.E. Ballou, Biochemistry, 30 (1991) 881.
98. T. Roitsch and L. Lehle, Eur. J. Biochem. 181 (1989) 525.
99. S.H. Shakin-Esheleman, Trends Glycosci. Clycotechnol., 8 (1996) 115.
100. D.J. Kelleher and R. Gilmore, J. Biol. Chem., (1994) 12908.
101. R. Knauer and L. Lehle, FEBS Lett., 344 (1994) 83.
102. R. Pathak, T.L. Hendrickson and B. Imperiali, Biochemistry, 34 (1995) 4176.
103. D. Karaoglu, D.J. Kelleher and R. Gilmore, J. Biol. Chem., 272 (1997) 32513.
104. Q. Yan, G.D. Prestwich and W.J. Lennarz, J. Biol. Chem. 274 (1999) 5021.
105. L. Lehle and W. Tanner, (1995). In (J. Montreuil, H. Schachter and J.F.G. Vliegenthart eds). Glycoproteins, Elsevier. (1995) pp145-152.
106. N. Dean, biochim. Biophys. Acta, 1426 (1999) 309.
107. M. Lussier, A-M. Sidicu and H. Bussey, Biochim. Biophys. Acta 1426 (1999) 323.
108. H. Schacter In (J. Montreuil, H. Schachter and J.F.G. Vliegenthart eds) Glycoproteins, Elsevier. (1995) pp153-200.
109. F.D. Ziegler, T.R. Gemmill, and R.B. Trimble, J. Biol. Chem., 269 (1994) 12527.
110. P.A. Romero, G.J.P. Dijkgraaf, S. Shahinian, A. Herscovics, and H. Bussey, Glycobiol., 7 (1997) 997.
111. E.S. Trombetta, J.F. Simons, and A. Helenius, J. Biol. Chem. 271 (1996) 27509.
112. Y.T. Pan, and D.A. Elbein, In *Glycoproteins* (J. Montreuil, H. Schachter and J.F.G. Vliegenthart eds) Elsevier. (1995) pp 415-454.
113. C. Abeijon, and L.Y. Chen, mol. Cell Biol., 9 (1998) 2729.
114. J.K. Simons, M. Ebeersold, A. and Helenius, EMBO J. 17 (1998) 396.
115. F.D. Zeigler, and R.B. Trimble, Glycobiol., 1 (1992) 605.
116. A. Camirand, A. Heysen, B. Grondin, and A. Herscovics, J. Biol. Chem. 266 (1991) 15210.

117. J. Burke, F. Lipari, S. Igdoura, and A. Herscovics, Eur. J. Cell Biol., 70 (1996) 298.
118. R.D. Teasdale, and M.R. Jackson, Ann. Rev. Cell Dev. Biol., 12 (1996) 27.
119. T. Yoshida, T. Inoue, and E. Ichishima, Biochem. J., 290 (1993) 349.
120. T. Yoshida, and E. Ichishima, biochim. Biophys. Acta, 1263 (1995) 159.
121. T. Inoue, T. Yoshida and E. Ichishima, Biochim. Biophys. Acta, 1253 (1995) 141.
122. A. Fujita, T. Yoshida, and E. Ichishima, Biochim. Biophys. Acta, 238 (1997) 779.
123. J.F.G. Vliegenthart and J. Montreuil, In *Glycoproteins* (J. Montreuil, H. Schachter and J.F.G. Vliegenthart eds). Elsevier. (91995) pp13-28.
124. T.R. Gemmill and R.B. Trimble biochim. Biophys. Acta 1426 (1999) 227.
125. R.B. Trimble and P.H. Atkinson Glycobiol. 2 (1992) 57.
126. L. Mallet, F. Bussereau and M. Jacquet, Yeast 10 (1994) 819.
127. M. Lussier, A-M. Sidicu, E. Winnett, D.H. Vo, J. Sheraton, A. Dusterhoft, R.K. Storms and H. Bussey, Yeast (1997a) 267.
128. M. Lussier, A-M. Sidicu, F. Bussereau, M. Jaquet and H. Bussey, J. Biol. Chem., 272 (1997b) 15527.
129. C.L. Yip, S.K. Wech, F. Klebl, T. Gilbert, P. Seidel, F.J. Grant, P.J. O'Hara and V.L. MacKay, Proc. Nat. Acad. Sci. USA, 91 (1994) 2727.
130. T.R. Graham, M. Seeger, G.S. Payne, V.L. MacKay, and J. Emr, J. Cell Biol., (1994) 667.
131. Y. Jigami, and T. Odani, Biochim. Biophys. Acta, 1426 (1999) 335.
132. C.E. Ballou, Ballou, L. and G. Ball, Proc. Natl. Acad. Sci. USA, 91 (1994) 9327.
133. T.R. Gemmill and R.B. Trimble, Glycobiol., 8 (1998) 1087.
134. T.G. Chappell, M. Hajibagheri, K. Ayscough, M. Pierce and G. Warren, mol. Cell. Biol., 5 (1994) 519.
135. M. Tabuchi, N. Tanaka, S. Iwahara and K. Takegawa, Biochim. Biophys. Acta 232 (1997) 121.
136. R.G. Miele, S.L. Nilsen, T. Brito, R.K. Bretthaueer and F.J. Castellino, Biotechnol. Appl. Biochem., 25 (1997) 151.
137. R.B. Trimble, P.H. Atkinson, J.F. Tschopp, R.R. Townsend and F. Maley, J. Biool. Chem., 266 (1991) 22807.
138. H.A.Kang, J.H. Sohn, E.S. Choi, B.H. Chung, M.H. Yu and S.K. Rhee, Yeast, 14 (1998) 371.
139. R.P. Podzorski, G.R. Gray and R.D. Nelson, J. Immunol. 144 (1990) 707.
140. R.D. Nelson, N. Shibata, R.P. Podzorski and M.J. Herron, Clin. Microbiol. Rev. 4 (1991) 1.
141. W.L. Chaffin, J.L. LopezRibot, M. Casanova, D. Gozalbo and J.P. Martinez, Microbiol. Mol. Biol. Rev. 62 (1998) 130.
142. A. Suzuki, A Takata, Y. Oshie, A. Tezuka, A. Shibata, H. Kobayashi, Y. Okawa and S. Suzuki, FEBS Lett., 373 (1995) 275.
143. P.A. Trinel, C. Cantelli, A. Bernigaud, T. Jonault and D. Poulain, Microbiol. UK 142 (1996) 2263.
144. Y. Chiba, Y. Yamagata, S. Iljima, T. Nakajima and E. Ichishima, Curr. Microbiol. 27 (1993) 281.
145. I. C. Almeida, D.C.A. Neville, A. Mehlert, A. Treumann, M.A.J. Feguson, J.O. Previato and L.R. Travassos, Glycobiol., 6 (1996) 507.

146. E. Aleshin, C. Hoffman, L.M. Firsov and R.B. Honzatko, J. Mol. Biol. 238 (1994) 575.
147. M. Maras, A. de Bryun, J. Schraml, P. Herdewijn, M. Claeyssens, W. Fiers and R. Contreras, Eur. J. Biochem., 245 (1997) 617.
148. P. Limongi, M. Kjalke, J. Vind, J.W. Tams, T. Johansson and K.G. Welinder, Eur. J. Biochem., 227 (1995) 270.
149. W. Martinet, M. Maras, X. Saelens, W.M. Jou and R. Contreras, Biotechnol. Lett., 20 (1998) 1171.
150. T. Takayanagi, K. Kushida, K. Idonuma and K. Ajisaka, Glycoconj. J., 9 (1992) 229.
151. T. Takayanagi, A. Kimura, S. Chiba and K. Aijisaka, Carbohyd. Res. 256 (1994) 149.
152. M. Ohta, S. Emi, H. Iwamoto, J. Hirose, K. Hiromi, H. Itoh, T. Shin, S. Murao and F. Matsura, Biosci. Biotech. Biochem., 264 (1996) 1123.
153. I.C. Kuan and M. Tien, J. Biol. Chem., 264 (1989) 20350.
154. R.M. Lederkremer and W. Colli, Glycobiol., 5 (1995) 547.
155. J-P. Latge, H. Koybayshi, J-P. Debeaupuis, M. Diaquin, J. Sarfati, J-M. Wieruszeski, E. Parra, J-P. Bouchara and B. Fournet, Infect. Immun., 62 (1994) 5424.
156. J.F. Groisman and R.M. Lederkremer, Eur. J. Biochem., 165 (1987) 327.
157. N. Ramli, H. Shinohara, K. Takegawa and S. Iwahara, J. Ferm. Bioeng., 78 (1994) 341.
158. J.A. Leal, J. Jimenez-Barbero, B. Gomez-Miranda, A. Prieto, J. Domenech and M. Bernabe, Carbohydr. Res., 283 (1996) 215.
159. C.J. Unkefer and J.E. Gander, J. Biol. Chem. 265 (1990) 685.
160. K. Ikuta, N. Shibata, J.S. Blake, M.V. Dahl, R.D. Nelson, K. Hisamichi, H. Kobayashi, S. Suzuki and Y. Okawa, Biochem. J., 323 (1997) 297.
161. S. Strahl-Bolsinger, M. Gentzsch and W. Tanner, Biochim. Biophys. Acta, 1426 (1999) 297.
162. S. Strahl-Bolsinger, T. Immervoll, R. Deutzmann and W. Tanner, Proc. Natl. Acad. Sci. USA, 90 (1993) 9916.
163. A.N. Savelev, E.V. Eneyskaya, L.S. Isaevalvannova, K.A. Shabalin, A.M. Golubev and K.N. Nuestroev, Glycon. J. 14 (1997) 897.
164. G. Gentzsch and W. Tanner, EMBO J., 15 (1996) 5742.
165. G. Gentzsch and W. Tanner, Glycobiol. 7 (1997) 481.
166. A. Weston, P.M. Nassau, C. Henly and M.S. Marriot, Eur. J. Biochem., 215 (1993) 845.
167. C. Timpel, S. Strahl-Bolsinger, K. Ziegelbauer. and J.F. Ernst, J. Biol. Chem., (1998) 20837.
168. M.J. Kuranda and P.W. Robbins, J. Biol. Chem. 266 (1991) 19758.
169. H. Heimo, K. Palmu and I. Suominen, Prot. Exp. Pur. 10 (1997) 70.
170. J.G. Duman, R.G. Miele, H. Liang, D.K. Grella, K.L. Sim, F.J. Castellino and R.K. Bretthuer, Biotechnol. Appl. Biochem., 28 (1998) 39.
171. M.P. Hayette, G. Strecker, D. Faille, D. Dive, D. Camus, D.W.R. MacKenzie and D. Poulain, J. Clin. Microbiol., 30 (1992) 411.
172. M. Kainuma, N. Ishida, T. Yoko-o, S. Yoshoika, M. Takauchi, M. Kawakita and Y. Jigami, Glycobiol., 9 (1999) 133.

173. A. Gunnarsson, B. Svensson, B. Nilsson and S. Svensson, Eur. J. Biochem., 145 (1984) 463.
174. T. Jikibara, K. Tada, K. Takegawa and S. Iwahara, J. Biochem., 111 (1992) 230.
175. M.J. Harrison, A.S. Nouwens, D.R. Jardine, N.E. Zachara, A.A. Gooley, H. Nevalainen and N.H. Packer, Eur. J. Biochem., 256 (1998) 119.
176. K.N. Neustroev, A.S. Krylov, O.N. Abroskina, L.M. Firsov, V.V. Nasonnov and A.Y. Khorlin, Biochem. (Moscow), 56 (1991) 288.
177. G.L.F. Wallis, R.J. Swift, F. W. Hemming, A.P.J. Trinci and J.F. Peberdy, (Submitted, 1999b).
178. C. E. Shama, J. S. Cox and P. Walter, Trends Cell Biol., 4 (1994) 56.
179. H.L. Pahl and P.A. Baeurle, Trends Cell Biol. 7 (1997) 50.
180. M..J. Gething, M.-J. ed. (1997) Guidebook to Molecular Chaperones and Protein-Folding Catalysts. Sambrook and Tooze, Oxford University Press.
181. R.J. Ellis, Phil. Trans. Roy. Soc. B, 329 (1993) 257.
182. M.J. Gething and J. Sambrook, Nature 355 (1992) 33. Protein folding in the cell. Nature 355, 33-45.
183. J.P. Vogel, L.M. Misra and M.D. Rose, J. Cell Biol., 100 (1990) 1885.
184. T. Nguyen, D.T.S. Law and D.B. Williams, Proc. Natl. Acad. Sci. USA, 88 (1991) 1565.
185. P.J. Punt, I.A. van Gemeren, J. Drink-Kuijvenhoven, J.G.M. Hessing, G.M. van Muijlwijk-Hartweld, A. Beijersbergen, C.T. Verrips and C.A.M.J.J. van den Hondel, Appl. Microbiol. Biotechnol. 50 (1998) 447.

186. D.M. Cyr, In: Guidebook to Molecular Chaperones and Protein-Folding Catalysts. (M.-J. Gething, ed), Sambrook and Tooze, Oxford University Press, (1997) pp. 89-95.
187. P.A. Silver, In: Guidebook to Molecular Chaperones and Protein-Folding Catalysts. (M.-J. Gething, ed), , Sambrook and Tooze, Oxford University Press, (1997), pp. 110-112
188. R.J. Deshaies, S. Sanders, D.A. Feldheim and R. Schekman, Nature 349 (1991) 806.
189. J. Brodsky, and R. Schekman, J. Cell Biol., 123 (1993) 1355.
190. R.A. Craven, J.R. Tyson and C.J. Stirling, Trends Cell Biol., 7 (1997) 277.
191. S.K. Lyman and R. Schekman, In: Guidebook to Molecular Chaperones and Protein-Folding Catalysts. (M.-J. Gething, ed), Sambrook and Tooze, Oxford University Press. (1997) pp 506-514.
192. M. Pilon, R. Schekman and K. Römisch, EMBO J., 16 (1997) 4540
193. R.B. Freedman, T.R. Hirst and M.F. Tuite, Trends Biochem. Sci., 19 (1984) 331.
194. R. Farquhar, N. Honey, S.J. Murant, P. Bossier, L. Schultz, D. Montgomery, R.W. Ellis, R.B. Freedman and M.F. Tuite, M.F. Gene 108 (1991) 81.
195. M. LaMantia, T. Miura, H. Tachikawa, H.A. Kaplan, W.J. Lennarz and T. Mizunaga, Proc. Natl. Acad. Sci. USA, 88 (1991) 4453.
196. M.C. Laboissière, S.L. Sturley and R.T. Raines, J. Biol. Chem., 270 (1995) 28006.
197. C. Tachibana and T.H. Stevens, T.H. Mol. Cell Biol., 12 (1992) 4601.

198. R. Gunther, M. Srinivasan, S. Haugejordan, M. Green, I.M. Ehbrecht and H. Küntzel, H. J. Biol. Chem., 268 (1993) 7728.
199. C. Ngiam, D.J. Jeenes and D.B. Archer, Curr. Geent., 31 (1997) 133.
200. D.J. Jeenes, R. Pfaller and D.B. Archer, Gene, 193 (1997) 151.
201. C. Hwang, A.J. Sinsky and H.F. Lodish, Science 257 (1992) 1496.
202. A.R. Frand and C.A. Kaiser, Mol. Cell, 1 (1998) 161.
203. M.G. Pollard, K.J. Travers and J.S. Weissman Mol. Cell., 1 (1998) 171.
204. K. Dolinsky and J. Heitman, In: Guidebook to Molecular Chaperones and Protein-Folding Catalysts. (M.-J. Gething, ed), Sambrook and Tooze, Oxford University Press (1997), pp 359-369.
205. A. Helenius, E.S. Trombetta, D.N. Herbert and J.F. Simons, J.F. Trends Cell Biol., 7 (1997) 193.
206. F.S. Fernandez, M. Jannatipour, U. Hellman, L.A. Rokeach and A.J. Parodi, EMBO J., 15 (1996) 705.
207. F.S. Fernandez, S.E. Trombetta, U. Hellman and A.J. Parodi, J. Biol. Chem., 269 (1994) 30701.
208. C.A. Jacob, P. Burda, S. te Heesen, M. Aebi and J. Roth, Glucobiol., 8 (1998) 155.
209. S. Fanchiotti, F. Fernandez, C. D'Alessio and A.J. Parodi, J. Cell Biol., 143 (1998) 625.
210. M. Jannatipour and L.A. Rokeach, J. Biol. Chem., 270 (1995) 4845.
211. F. Parlatti, D. Dignard, J.J.M. Bergeron and D.Y. Thomas, EMBO J. (1995a) 3064.
212. F. Parlatti, M. Dominguez, J.J.M. Bergeron and D.Y. Thomas, D.Y. J. Biol. Chem., 270 (1995b) 244
213. C. Sidrauski, R. Chapman and P. Walter, Trends Cell Biol., 8 (1998) 245.
214. K. Mori, A. Sant, K. Kohno, K. Normington, M-J. Gething and J.F. Sambrook, EMBO J., 11 (1992) 2583.
215. K. Mori, T. Kawahara, H. Yoshida, H. Yanagi and T. Yura, Genes to Cells 1 (1996) 803.
216. K. Mori, N. Ogawa, T. Kawahara, H. Yanagi and T. Yura, J. Biol. Chem. 273 (1998) 9912.
217. H. Yoshida, K. Haze, H. Yanagi, T. Yura and K. Mori, J. Biol. Chem. 273 (1998) 33741
218. J.S. Cox and P. Walter, Cell, 87 (1996) 39191.
219. J.S. Cox, R.E. Chapman and P. Walter, Mol. Biol. Cell, 8 (1997) 1805.
220. R. Menzel, F. Vogel, E. Kärgel and W-H. Schunk, Yeast, 13 (1997) 1211.
221. T. Kawahara, H. Yanagi, T. Yura and K. Mori, Mol. Biol. Cell, 8 (1997) 1845.
222. X-Z. Wang, H.P. Harding, Y. Zhang, E.M. Jolicoeur, M. Kuroda and D. Ron, EMBO J., 17 (1998) 5708.
223. W. Tirasophon, A.A.Welihinda and R.J. Kaufman, Gen. Devel., 12 (1998) 1812.
224. M.S. Brown and J.L. Goldstein, Cell, 89 (1997) 331.
225. J. Morrice, D.A. MacKenzie, A.J. Parr, and D.B. Archer, Curr. Genet., 34 (1998) 379.
226. J.R. Shuster, Curr. Op. Biol., 2 (1991) 685.
227. D.B. Archer and J.F. Peberdy, CRC Crit. Rev. Biotech., 17 (1997) 273.

228. R.J. Gouka, P.J. Punt and C.A.M.J.J. van den Hondel, Appl. Microbiol. Biotech. 47 (1997) 1.
229. M.F. Tuite and R.B. Freedman, Trends Biotech., 12 (1994) 432.
230. M. Tokunaga, A. Kawamura and K. Kohno, J. Biol. Chem., 267 (1992) 17553.
231. A.S. Robinson and K.D. Wittrup, Biotech. Prog., 11 (1995) 171.
232. A.S. Robinson, J.A. Bockhaus, A.C. Voegler and K.D. Wittrup, J. Biol. Chem., 271 (1996) 10017.
233. E.V. Shuster, R.T. Raines, A. Plückthun and K.D. Wittrup, Nature Biotech., 16 (1998) 733
234. M.M. Harmsen, M.I. Bruyne, H.A. Raué and J. Maat, Appl. Microbiol. Biotech. 46 (1996) 365.
235. R.A. Smith, M.J. Duncan and D.T. Moir, Science 229 (1985) 1219.
236. A.S. Robinson, V. Hines and K.D. Wittrup, Biotechnology 12 (1994) 381.
237. H. Arima, T. Kinoshita, H.R. Ibrahim, H. Azakami and A. Kato, FEBS Lett., 440 (1998) 89.
238. M.D. Rose, L.M. Misra and J.P. Vogel, Cell 57 (1989) 1211.
239. A.L. Pidoux and J. Armstrong, EMBO J., 11 (1992) 1583.
240. M.J. Lewis and H.R.B. Pelham, Nucl Acid Res., 18 (1990) 6438.
241. I.H. Lee and D.M. Ogrydziak, Yeast, 13 (1997) 513.
242. I.A. van Gemeren, P.J. Punt, A. Drint-Kuyvenhoven, M.P. Brockhuijsen, A. van't Hoog, A. Beijersbergen, C.T. Verrips and C.A.M.J.J. van den Hondel, Gene, 198 (1997) 43.
243. M.J. Hijarrubia, J. Casqueiro, S. Gutiérrez, F.J. Fernandez and J.F. Martin, Curr. Genet. 32 (1997) 138.
244. D. Techel, T. Hafker, S. Muschnev, M. Reimann, Y.Z. Li, C. Monnerjahn and L. Rensing, Biochim. Biophys. Acta, 1397 (1998) 21.
245. R.A. Craven, M. Egerton and C.J. Stirling, EMBO J., 15 (1996) 2640.
246. S. Malpricht, A. Thamm and N.Q. Khanh, Biotech. Lett., 18 (1996) 445.
247. C.M. Hjort, A fungal protein disulfide isomerase. Patent WO95/00636 (1995).
248. B.R. Lee, O. Yamada, K. Kitamoto and K. Takahashi, J. Ferm. Bioeng., 82 (1996) 538.
249. T. Kajino, K. Sarai, T. Imaeda, C. Idekoba, O. Asami, Y. Yamada, M. Hirai and S. Udaka, Biosci. Biotech. Biochem., 58 (1994) 1424.
250. H. Tachikawa, Y. Takeuchi, W. Funahashi, T. Miura, X.D. Gao, D. Fujimoto, T. Mizunaga and K. Onodera, FEBS Lett., 369. 212. 216.
251. C. De Virgilio, N. Bückert, J-M. Neuhaus, T. Boller and A. Wiemken, Yeast, 9 (1993) 185.
252. G.W. Gooday, In: Fungal Differentiation- A contemporary synthesis (ed. Smith, J.E.). Marcel Dekker Inc., New York. (1983) pp315-356.
253. N.R. Salama and R. Schekman, Curr. Opin. Cell Biol., 7 (1995) 536.
254. P. Markham, In: The Growing Fungus (N.A.R. Gow and G.M. Gadd eds) Chapman and Hall, London pp 75-98, (1995).
255. T.C. C-T. Thay, K. Hoang-Van, G. Turian, H.C. Hoch, Eur. J. Cell Biol., 43 (1987) 183.
256. T. Kanseki and K. Kadota, J. Cell biol., 42 (1969) 202.
257. B.M.F. Pearse, Proc. Natl. Acad. Sci. USA, 73 (1976) 1255.

258. M. Girbardt, Protoplasma, 67 (19969) 413.
259. S.N. Grove and C.E. Bracker, J. Bacteriol., 104 (1970) 989.
260. R. Lopez Franco and C.E. Bracker, Protoplasma, 159 (1996) 90.
261. H.A.B. Wosten,_S.M. Moukha,_J.H. Sietsma and J.G.H. Wessels, J. Gen. Microbiol., 137 (1991) 2017.
262. I.H. Lee, R.G. Walline and M. Plamamm, Mol. Microbiol. 29 (1998) 209.
263. R. Schekman and L. Orci, Science, 271 (1996) 1526.
264. M.J. Kuehn and R. Schekman, Curr. Opin. Cell Biol. 9 (1997) 477.
265. E.C. Gaynor, T.R. Graham and S.D. Emr, Biochim. Biophys. Acta, 1404 (1998) 33.
266. C. Barlowe, L. Orci, T Yeung, M Hosobuchi, S. Hamamoto, N. Salama, MF Rexach, M. Ravazzola, M. Aherdt and R. Schekman, Cell, 77 (1994) 985.
267. N.R. Salama, J.S. Chuang and R. Schekman, Mol. Cell Biol., 8 (1997) 205.
268. C. Barlowe, Biochim. Biophys. Acta, 1404 (1998) 67.
269. C. Barlowe, FEBS Lett., 369 (1995) 93.
270. P. Espenshade, R.E. Gimeno, E. Holzmacher, P. Yueng and C.A. Kaiser, J. Cell Biol., 31 (1995) 311.
271. R.E. Gimeno, P Espenshade and C.A. Kaiser, J. Cell. Biol., 131 (1995) 325.
272. R. Duden, M. Hosobuchi, S. Hamamoto, M. Winey, B. Byers and R Schekman, J. Biol. Chem. 24 (1994) 486.
273. B. Gerich, L. Orci, H. Tschochner, F Lottspeich, M. Ravazzola, M. Amherdt, F. Wieland and C. Harter, Proc. Natl. Acad. Sci. USA, 92 (1995) 3229.
274. S.R. Pfeffer, Ann Rev. Cell Devel. Biol., 12 (1997) 441.
275. T. Söllner, S.W. Whitehart, M. Brunner, H. Erdjument-Bromage, S. Geromanos, P. Tempest and J.E. Rothman, Nature, 362 (1993) 318.
276. J.P. Lian, S. Stone, Y Jiang, P. Lyons and S Ferreonovick, Nature, 372 (1994) 698.
277. M. Søgaard, K. Tani, R.R. Ye, S. Geromanos, P. Tempst, T. Kirchhausen, J.E. Rothman and T. Söllner, Cell, 78 (1994) 937.
278. K.G. Hardwick and H.R.B. Pelham, J. Cell Biol., 131 (1992) 513.
279. S.K. Sapperstein, W. Lupashin, H.D. Schmidt, M.G. Waters, J. Cell Biol., 132 (1996) 756.
280. S.M. van Rheenan, X.C. Cao, V.V. Lupashin, C. Barlowe and M.G. Waters, J. Cell Biol., 141 (1998) 1107.
281. M.J. Lewis, J.C. Rayner and H.R.B. Pelham, EMBO J., 16 (1997) 3017.
282. J.C. Hay and R.H. Scheller, Curr. Opin. Cell Biol., 9 (1997) 505.
283. W. Ballensiefen, D. Ossipov and H.D. Schmidt, J. Cell Sci., 111 (1998) 1507.
284. S. Wooding and H.R.B Pelham, Mol. Biol. Cell, 9 (1998) 2667.
285. M. Sato, K. Sato and A. Nakano, 134 (1996) 279.
286. D. Roth, W. Guo and P. Novick, Mol. Biol Cell, 9 (1998) 1725.
287. F.P. Finger, T.E. Hughes and P. Novick, Cell, 92 (1998) 559.
288. J. Jochova, I. Rupes and J.F. Peberdy, Mycol. Res., 97 (1993) 23.
289. J.R. Delucas, I.F. Monistrol and F. Laborda, Mycol. Res., 97 (1993) 961.
290. S. Torralba, M. Raudaskoski and A.M. Pedregosa, Protoplasma, 202 (1998) 54.
291. S. Seiler, F.E. Nargang, G. Steinberg and M. Schliwa, EMBO J., 16 (1997) 3025.
292. T.M. Bourett and R.J. Howard, Protoplasma, 190 (1996) 151

293. B. Satait Jeunemaitre, L. Cole, T. Bourett, R. Howard and C. Hawes, J. Microsc. Oxf., 181 (1996)162.
294. T. Akashi T. Kanbe and K. Tanaka, Protoplasma, 197 (1997) 45.
295. A. Rambourg, Y Clermont, C.L. Jackson and F. Kepes, Anat. Rec., 241 (1995) 1.
296. V.I. Titorenko, D.M. Ogrydziak and R.A. Rachubinski, Mol. Cell Biol., 17 (1997) 5210.
297. G.G. Hu and F.H.J. Rijkenberg, J. Phytopath., 146 (1998) 39.
298. M.H. Vainstein and J.F. Peberdy, Mycol. Res., 95 (1991) 1270.
299. J.H. Seitsma, H.A.B. Wosten and J.G.H. Wessels, Can. J. Bot., 73 (1995) S388.
300. H. Nevalainen, I. Lavygina, D. Neethkling and N. Packer, J. Biotech., 42 (1995) 53.
301. P. Orlean, In The Molecular and Cellular Biology of the Yeast *Saccharomyces*, Vol3 Cell Cycle and Cell Biology, (J.R. Pringle, J.R. Broach and E.W. Jones eds), Cold Spring Harbor Laboratory Press, NY (1997) 229-362.
302. A. Toh-E, S. Yasunga, H. Nisogi, K. Tanaka, T. Oguchi and Y. Matsui, Yeast, 9 (1993) 481.
303. V. Mrsa, T. Seidl, M. Gentzch and W. Tanner, Yeast, 13 (1997) 1145.
304. Fleet, G.H. In The Yeasts , eds Rose, A.H. and Harrison, J.S. Vol 4 pp 199-277 Academic Press (1991).
305. M. Casanova, M.L. Gil, L. Cardeñoso, L., J.P. Martínez and R. Sentandreu, Infect. Immun., 57 (1989) 262.
306. M. Casanova, J.P. Martinez and W.L. Chaffin, Infect. Immun., 58 (1990) 3810.
307. M. Casanova, J.L. López-Ribot, C. Monteagudo,A. Llombart-Bosch, R, Sentendreu and J. P. Martínez., Infect. Immun., 60 (1992) 4221.
308. Chaffin, W.L. and Stocco, D.M., Can J. Bot., 29 (1988) 1438.
309. S.K. Dielbandhoesing, H. Zhang, L.H.P. Caro, J.M. van der Vaart, F.M. Klis, C.T. Verripa and S. Brul, Appl. Env. Microbiol., 64 (1998) 4047.
310. E. Cabib, J Drgonova and T. Drgon, Ann. Rev. Biochem., 67 (1998) 307.
311. H. Shimoi, H. Kitagaki, H. Ohmori, Y. Iimura and K. Ito, J. Bacteriol., 180 (1998) 3381.
312. U. Reichard, Mycoses, 41 (1998) 78 1998.
313. N. Sejalon Delmas, F.V. Mateos, A. Bottin, M. Rickauer, R. Dargent and M.T. Esquerre Tugaye, Phytopathology, 87 (1997) 899.
314. S.E. Perfect, R.J. O'Connell, E.F. Green, C. Doering Saad and J.R. Green, Plant J., 15 (1998) 273.
315. D. Tagu and F. Martin, New Phytol., 133 (1996) 73.
316. J. Masouka and K.C. Hazen, Microbiology, 143 (1997) 3015.
317. C. Imbert Bernhard, A. Valentin, M. Mallie and J.M. Bastide, Expt. Mycol., 19 (1995) 247.
318. Y. Fukazawa and K. Kagaya, J. Med. Vet. Mycol., 35 (1997) 87.
319. J.G.H. Wessels, O.M.H. DeVries, S.A. Asgeirsdottir and F.H.J.Schuren, Plant Cell, 3 (1991) 793.
320. M.J. Kershaw and N.J. Talbot, Fun. Genet. Biol., (1998) 18.
321. H.A.B. Wosten, F.H.J. Schuren and J.G.H. Wessels, J.G.H. EMBO J., 13 (1964) 5848.

322. N.J. Talbot, M.J. Kershaw, G.E. Wakley, O.M.H. de Vries, J.G.H. Wessels and J.E. Hamer, Plant Cell, 8 (1996) 985.
323. J.G.H. Wessels, S.A. Asglerdottir, K.U. Birkenhkamp, O.M.H. deVries, L.G. Lugones, J.M.J.Scheer, F.H.J. Schuren, T.A. Schuurs, M.A. van Weeter and H.A.B. Wosten, H.B, Can. J. Bot., 73 (1995) S273.
324. P.W.J. DeGroot, P.J. Schaap, A.S.M. Sonnenberg and J. Visser, J. Mol. Biol. 257 (1996) 1008.
325. L.G. Lugones, J.S. Bosscher, K. Scholtmeyer, O.M.H. de Vries and J.G.H Wessels, Microbiology, 142 (1996) 1321.
326. L.G. Lugones, H.A.B. Wosten and J.G.H. Wessels, Microbiology, 144 (1998) 2345.
327. T. Nakari Setala, N. Aro, N. Kalkkinen, E. Alatalo and M. Penntilä, Eur. J. Biochem., 235 (1996) 248.
328. G. Munoz, T. Nakari Setala, E. Agosin and M. Penntilä, Curr. Genet., 32 (1997) 225.
329. T. Nakari Setala, N. Aro, M. Ilmen, G. Munoz, N. Kalkkinen and M. Penntilä, Eur. J. Biochem., 248 (1997) 415.
330. S.A. Asgeirsdottir, J.R. Halsall and L.A. Casselton, Fun. Genet. Biol., 22 (1997) 54.
331. J. Vasquez, A.G. Smulian, M.J. Linkw and M.T. Cushion, Inf. Immun., 64 (19xx) 290.
332. M.J. Bidochka and A.E. Hajek, Mycol. Res., 100 (1996) 1094.
333. M.B. Tobin, R.B. Peery and P.L. Skatrud, Gene, 200 (1997) 11.
334. G. Del Sorbo, A.C. Andrade, J.G.M. van Nistelrooy, J.A.L. van Kan, E.E. Balzi and M.A. De Waard, Mol. Gen. Genet., 254 (1997) 417.
335. S. Pluschkell, K. Hellmuth and U. Rinas, Biotechnol. Bioeng., 51 (1996) 215.
336. R.J. Swift, M.G. Wiebe, G.D. Robson and A.P.J. Trinci, Fung. Gen. Biol., 25 1998) 1087.
337. J.M. Withers, R.J. Swift, M.G. Wiebe, G.D. Robson, P.J. Punt, C.A.M.J.J. van den Hondel and A.P.J. Trinci, Biotechno. Bioeng., 59 (1998) 407.
338. S. Arnesen, S.H. Eriksen, J. Olsen and B. Jensen, Enz. Microbiol. Technol., 23 (1998) 249.
339. J. Kruszewska, G. Palamarczyk and C.P. Kubicek, J. Gen. Microbiol., 136 (1990) 1293.
340. R.S.M. Chambers, M.J. Broughton, R.D. Cannon, A. Carne, G.W. Emerson and P.A. Sullivan, J. Gen. Microbiol., 139 (1993) 325.
341. J.P. Luna-Arias, E. Andaluz, J.C. Ridruejo, I. Olivero and G. Larriba, Yeast, 7 (1991) 833.
342. L.F. Mackenzie, G.S. Brooke, J.F. Cutfield, P.A. Sullivan and S.G. Withers, J. Biol. Chem., 272 (1997) 3161.
343. R.P. Hartland, G.W.Emeson and P.A. Sullivan, Proc. R. Soc. London Ser. B 246 (1991) 155.
344. G.W. Gooday, W-Y. Zhu and R.W. O'Donnell, FEMS Microbiol. Lett., 100 (1992) 387.
345. K.J. McCreath, C.A. Specht and P.W. Robbins, Proc. Natl. Acd. Sci. USA, 92 (1995) 2544.

346. R.D. Cannon, K. Niimi, H.F. Jenkinson and M.G. Shepherd, J. Bacteriol., 176 (1994) 2640.
347. C. Molloy, M.G. Shepherd and P.A. Sullivan, Exp. Mycol., 19 (1995) 178.
348. A.K. Banerjee, K. Ganesan and A. Datta, J. Gen. Microbiol., 137 (1991) 2455.
349. F. De Bernardis, P. Chiani, Mciccozzi, G. Pellegrini, T. Ceddia, G. DiOffizzi, L. Quinti, P.A. Sullivan and A, Cassone, Infect. Immun. 64 (1996) 466.
350. F. McDonald and F.C. Odds, J. Gen. Microbiol., 129 (1983) 431.
351. Y. Banno, T. Yamada and Y. Nozawa, Sabouraudia, 23 (1985) 47.
352. J. Pontøn and J.M. Jones, infect. Immun., 54 (1986) 864.
353. M. Takahishi, Y. Banno and Y. Nozawa, J. Med. Vet. Mycol., 29 (1991) 193.
354. M. Takahishi, Y. Banno, Y. Shikano, S. Mori and Y. Nozawa, Biochim. Biophys. Acta, 1082 (1991) 161.
355. T. Tsuboi, H. Komatsuzaki and H. Ogawa, Infect. Immun., 64 (1996) 2936.
356. J. M. Manns, D.M. Mosser and H.R. buckley, Infect. Immun. 62 (1994) 5154.
357. F.W. Chattaway, S. Shenolikar and A.J.E. Barlow, J. Gen. Microbiol., 83 (1974) 423.
358. Y. Fu, A.S. Ibrahim, W. Fondi, X. Zhoud, C.F. Ramos and M.A. Ghannoum, Microbiology, 143 (1997) 331.
359. M.T. Shimizu, N.Q. Almeida, Y. Fantinato and C.S. Unterkircher, Mycoses, 39 (1996) 161.
360. B. El Moudni, M.H. Rodier, P. Babin and J.L. Jacquemin, J. Mycol. Med., 7 (1997) 5.
361. M. Molina, R. Cenamor, M. Sánchez and C. Nombela, J. Gen. Microbiol., 133 (1987) 609.
362. W.L. Chaffin, J.L. Lopez-Ribot, M. Casanova, D. Gazalbo and J.P. Martinez, Microbiol. Mol. Biol. Rev., 62 (1998) 130.

Applied Mycology and Biotechnology
Volume 1. Agriculture and Food Production
G.G. Khachatourians and D.K. Arora (editors)

Significance of fungal peptide secondary metabolites in the agri-food industry

Daniel G. Panaccione[a] and Seanna L. Annis[b]

[a]Division of Plant and Soil Sciences, West Virginia University, 401 Brooks Hall, Box 6057, Morgantown, WV 26506-6057 USA

[b]Department of Biological Sciences, University of Maine, 5722 Deering Hall, Orono, ME 04469-5722 USA

Among the vast and varied secondary metabolites that fungi produce are small peptides, synthesized not as a result of translation, but rather by an unusual class of enzymes called nonribosomal peptide synthetases. These peptide secondary metabolites impact the agri-food industry in a numbers of ways. Ergopeptines and other peptides produced by plant pathogens or symbionts poison our food or animal feed. The production of HC-toxin, victorin, and other peptides by plant pathogens affects the yield and genetic constitution of some important crop plants. Still other peptides, such as beauvericin and destruxin, have a positive effect by contributing to the anti-insect arsenals of fungi used as biological control agents. In this chapter we address the nature and biosynthesis of these peptides, the impact of fungal peptides in examples from several agricultural systems, and, finally, speculate on how through biotechnological manipulations we may alter, augment, or diminish the biosynthesis of these peptides for our benefit.

1. CHEMICAL NATURE AND HISTORICAL SIGNIFICANCE

1.1. Ergopeptines

Ergopeptines are small peptides containing lysergic acid and three amino acids that vary between and define the members of the ergopeptine family. The best studied ergopeptine, ergotamine, contains D-lysergic acid, L-phenylalanine, L-proline and L-alanine (Fig. 1). Other ergopeptines differ by one or two amino acid substitutions. Ergopeptines are the most stable and toxic members of the ergot alkaloid family that also contains lysergic acid, simple derivatives of lysergic acid, and 'clavines', some of which are precursors to lysergic acid (1,2). Ergopeptines are produced by several fungi in the Clavicipitaceae (Hypocreales,

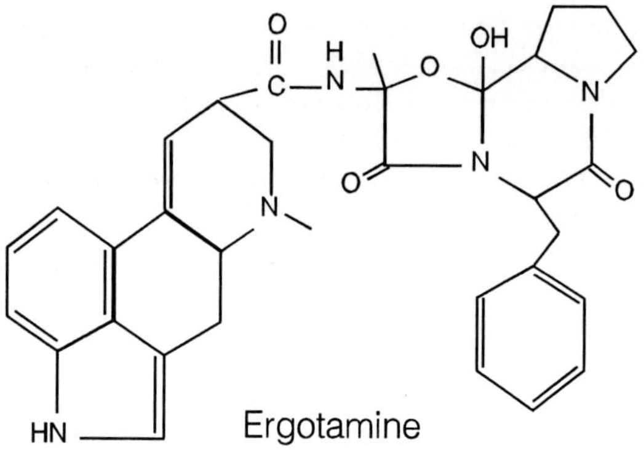

Figure 1. Structure of ergotamine. Other ergopeptines differ by substitution of amino acid components. For example, ergovaline differs from ergotamine by having valine in place of phenylalanine.

Ascomycetes) including the 'ergot' pathogen of rye, *Claviceps purpurea*, and several mutualistic grass endophytes in the genus *Neotyphodium*.

Claviceps purpurea causes a tissue replacement disease of rye in which the developing seed is replaced by a sclerotium, the fungus's overwintering structure. The sclerotium is filled with ergopeptines and other ergot alkaloids. If sclerotia are harvested with the grain and ground into flour, human consumption of large quantities of ergopeptines can result. In years of severe ergot infection 30% of the 'grain' harvested can be ergot sclerotia (3); 1% contamination can lead to a community-wide epidemic of ergotism (4).

Symptoms associated with ergot poisoning include increased blood pressure, hyperthermia, destruction of the nervous system, reduced lactation and reproductive capability, and gangrene of the extremities (5,6). Historically, ergopeptines produced by *C. purpurea* have caused a tremendous amount of human suffering. Diseases known as 'St. Anthony's fire' or 'holy fire' plagued people who relied on large quantities of rye bread for sustenance. Matossian (4) correlated trends in population growth in rye-growing areas of Europe with conditions favorable or unfavorable for the development of ergot on rye and with accounts of ergotism from historical records. Witch persecutions were more frequent in predominantly rye growing areas and during winters following growing seasons that had climatic conditions conducive to the development of ergot (4,7). Historical accounts of 'the great fear' that preceded the French revolution are consistent with symptoms of an epidemic of ergot poisoning; climatic records of the preceding years support this hypothesis (4). Kavaler (8) suggests that the army of the Russian czar, Peter the Great, was stopped from invading the Constantinople in 1722 because his troops were suffering from ergot poisoning.

Knowledge of the fungus and its toxins, a better balanced diet, and mechanical means of removing sclerotia from grain have essentially eliminated ergotism in people in the developed world (7). However, ergopeptines still have a tremendous impact on the agri-food industry, not through human poisoning but through their role as toxins in forage grasses. In the United States, there is considerable current emphasis on ergopeptines and their role in fescue toxicosis, a condition in grazing animals resulting from ingestion of ergopeptines produced by the mutualistic endophyte *N. coenophialum*, commonly found in tall fescue. This topic is explored in detail in section 3.1.

1.2. Peptide phytotoxins

Several nonribosomally synthesized peptides are phytotoxic. HC-toxin and victorin, produced by *Cochliobolus* species, and AM-toxin produced by *Alternaria alternata* 'apple pathotype' are particularly interesting because they are host-selective toxins; their toxicity is restricted to certain genotypes of the host plant. Other phytotoxic peptides, such as destruxin B produced by *Alternaria brassicae* and enniatins produced by several species of *Fusarium,* have more broad-spectrum activity against plant host genotypes.

The two *Cochliobolus* toxins and the fungi that produce them share a similar history. The fungus that we now know as *Cochliobolus carbonum* race 1 was not known prior to its emergence as a leaf spot pathogen of certain maize breeding lines in the late 1930's. A resistance gene, already widespread in other maize breeding lines at the time of emergence of the *C. carbonum*, has controlled the pathogen since then. The discovery of HC-toxin (9) and that a single genetic locus controls its production by Scheffer and colleagues (10) in the 1960's rekindled interest in the organism, though the structure of the toxin would not be elucidated for another 20 years. HC-toxin is a cyclic tetrapeptide composed of D-proline, L-alanine, D-alanine, and the unusual, aliphatic amino acid L-2-amino-9, 10-epoxy-8-oxodecanoic acid (11,12) (Fig. 2). Similar molecules containing substitutions of glycine for D-alanine, or D-hydroxyproline for D-proline can be isolated from HC-toxin producing isolates (13). The single genetic locus controlling HC-toxin production identified by Scheffer and colleagues (10) has proven to be a large tract of DNA containing several genes involved in HC-toxin biosynthesis (14,15). The maize gene conferring resistance to HC-toxin was the first plant disease resistance gene to be cloned (16). The nature of resistance as the enzymatic detoxification of HC-toxin and the distribution of the detoxifying enzyme among other grain crops suggest that similar toxins maybe involved in other plant/pathogen interactions (17). The impact of this toxin and related toxins may go well beyond the brief time *C. carbonum* was an economically important pathogen. These topics are explored in section 3.2.

Cochliobolus victoriae was discovered in the 1940's as a novel pathogen of the oat variety 'Victoria' and its derivatives, which had become widely planted because of their resistance to crown rust (18). The susceptibility of these varieties of oats to *C. victoriae* was due to their extraordinary sensitivity to a

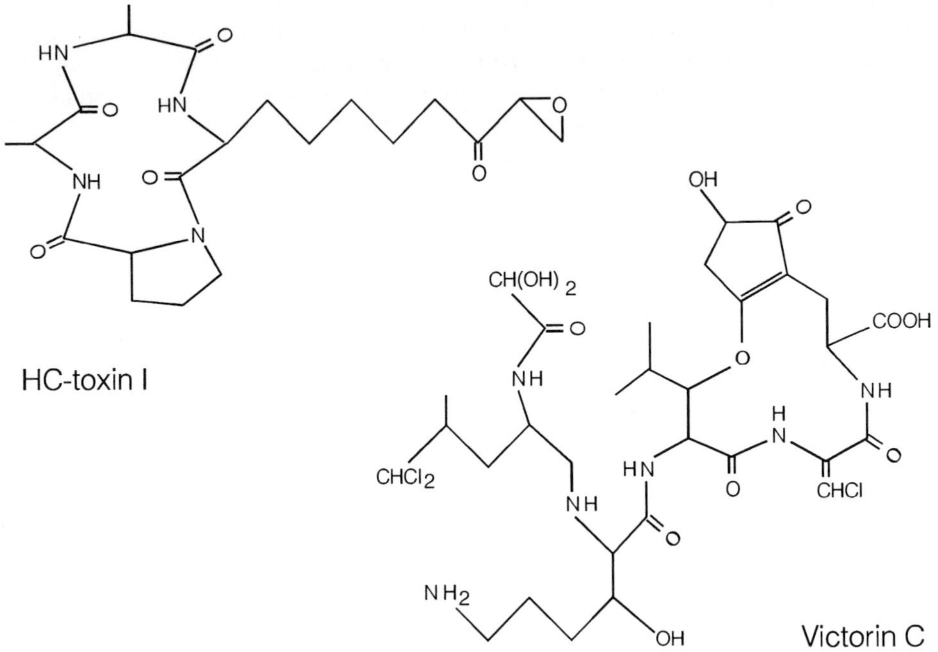

Figure 2. Structures of the major forms of HC-toxin and victorin.

peptide toxin, victorin, produced by the fungus. Victorin is the most phytotoxic compound ever discovered when incubated with sensitive oat genotypes, yet it is virtually harmless to any other plant. Victorin is an unusual chlorinated and partially cyclic pentapeptide that contains, in its most common form, 5,5-dichloroleucine, 3-hydroxylysine, chloroacrylic acid, 3-hydroxyleucine and a novel amino acid called victalanine (19) (Fig. 2). Similar molecules varying in degree of chlorination or hydroxylation of amino acid side chains, or with an amino acid substitution, have been characterized from culture filtrates of *C. victoriae* (20). Victorin and *C. victoriae* are still studied by plant pathologists interested in the molecular mechanisms of plant disease. Moreover, the threat of what victorin can do to a sensitive genotype still impacts the agri-food industry through limitations it places on oat breeding programs. More information on the pathogen and the toxin will be discussed in section 3.2.

Nonribosomally synthesized peptides produced by *Alternaria* species also exhibit phytotoxicity. These are AM-toxin produced by the 'apple pathotype' of *A. alternata*, and destruxin B produced by *A. brassicae*. AM-toxin is a cyclic tetrapeptide containing residues of L-alanine and three nonprotein amino acids: L-2-aminomethoxyphenylvaleric acid; L-2-hydroxyisovaleric acid; and, dehydroalanine (21). This toxin is host-selective and is a critical component of the leaf blotch disease of a few cultivars of apple, including 'Red Gold' (22). It is the only known small peptide toxin produced among the various host-selective toxins produced by different *A. alternata* pathotypes (22).

Destruxin B produced by *A. brassicae* is a cyclic depsipeptide containing D-2-hydroxy-N-methylvaleric acid, L-proline, L-isoleucine, N-methyl-L-valine, N-methyl-L-alanine, and β-alanine (Fig. 3). Destruxin B and other derivatives have been detected in cultures of *A. brassicae* and from leaves infected with the fungus and exhibiting 'black spot disease' symptoms (23,24,25). Destruxins are not considered host-specific because they incite necrosis and chlorosis on a variety of plant species. However, species of *Brassica* are particularly susceptible, requiring 10-fold less destruxin B than other plants to cause the same symptoms (26). Resistance to destruxin B in leaves and pollen from different cultivars and species of *Brassica* is similar to resistance of the plants to infection by *A. brassicae* and can be used to screen for *A. brassicae* resistant plant tissue (27,28). Interestingly, destruxins also are produced by a distantly related fungus, *Metarhizium anisopliae*, that is a pathogen of insects. Destruxins may have a stronger impact on the agri-food industry through their role in controlling insect pests and will be considered from that perspective.

One additional notable family of peptide phytotoxins are the enniatins, cyclohexadepsipeptides that are produced by various plant pathogenic and entomopathogenic species of *Fusarium* (29). Enniatins are toxic to insects and plants and have a role in the virulence of *F. avenaceum* to potato (29,30,31,32), as will be discussed in section 3.2. Most *Fusarium* species produce a complex of different enniatins consisting of a basic unit of an N-methylated branched-chain L-amino acid (valine, leucine or isoleucine) and D-2-hydroxyisovaleric acid, repeated three times to make a hexadepsipeptide (33) (Fig. 3). Enniatin synthetase, purified from *F. oxysporum*, has all the necessary activities to manufacture enniatins (34). This is in contrast to some other fungal toxic peptides that require additional enzymes to activate some of the amino acids or modify the product of the peptide synthetase to its final form. The role of enniatin in the virulence of *Fusarium* to plants will be discussed in section 3.2. Enniatins very closely resemble the insect toxin called beauvericin produced by *Beauveria bassiana* and which is discussed extensively in the following section.

1.3. Insect toxins

Two of the most common entomopathogenic fungi, *Beauveria bassiana* and *Metarhizium anisopliae*, produce cyclic peptides toxins that may be involved in their pathogenicity to insects. Both fungi have a worldwide distribution and wide host ranges (35). In the early 1800's, the silk industry in Europe and Asia was seriously threatened by a mysterious condition called muscardine that caused silkworm larvae to die and then mummify covered in a white material (36). In 1835, Agostino Bassi discovered that *Beauveria bassiana* was the infectious agent causing white muscardine disease of silkworm pupae. This was the first description of a disease of animals caused by a fungus or any microorganism and was likely the first published study supporting the germ theory of disease. (Generally, the first evidence of infectious agents causing disease is credited to Koch for his work on anthrax decades later.) Beauvericin, a cyclic hexadepsipeptide toxin, consists of alternating units of N-methylphenylalanine and 2-hydroxyisovaleric acid (Fig. 3) and was first isolated from culture filtrates of *B. bassiana* that were toxic to brine shrimp (37). Beauvericin is produced by other entomopathogenic fungi, including *Paecilomyces fumoso-roseus* and other species of *Beauveria* (38,39,40). *Beauveria bassiana* also produces the insecticidal analogues of beauvericin, beauvericin A and B, that have 2-hydroxy-3-methylpentanoic acid residues substituted for one and two of the 2-hydroxyisovaleric acid residues, respectively (41).

Beauvericin synthetase has been purified from *B. bassiana* and found to have an enzymatic mechanism very similar to that of enniatin synthetase (40,42). Studies on the specificity of the enzyme showed that phenylalanine can be replaced by a number of other aromatic or aliphatic amino acids. However, valine, one of the amino acids in enniatin was not accepted by the enzyme (40).

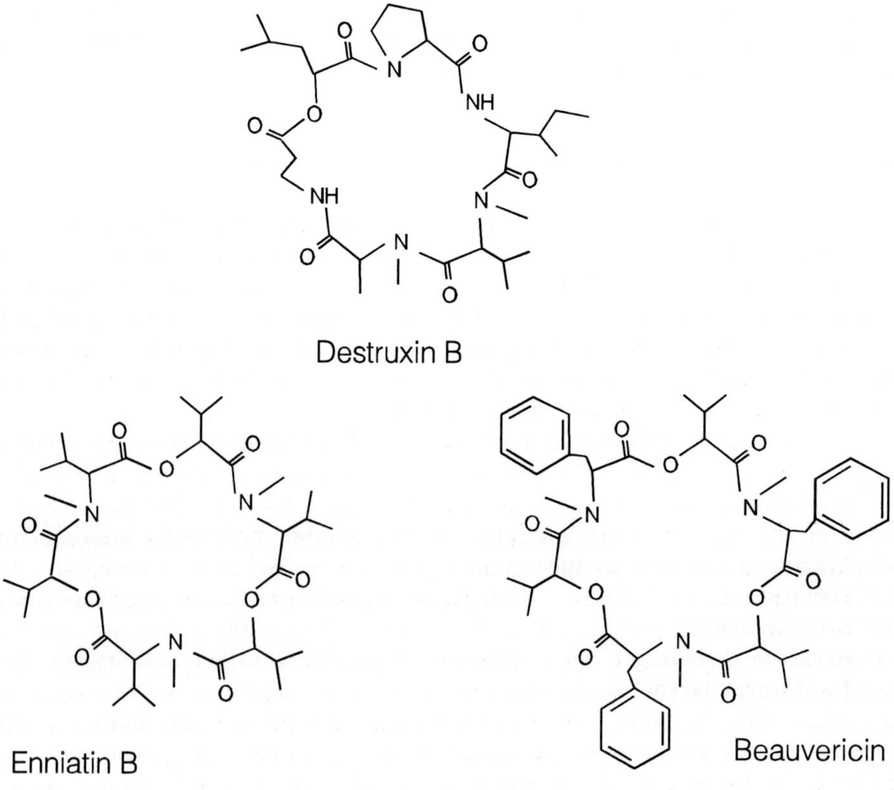

Figure 3. Structures of the major forms of destruxin, enniatin, and beauvericin.

Other minor cyclic tetradepsipeptides, called beauverolides, have been isolated from *B. tenella* and *P. fumoso-roseus*, and although they are not directly insecticidal, they may have an effect upon the immune system of insects (43). Bassianolide, an insecticidal depsipeptide, is also produced by *B. bassiana* and a common pathogen of scale insects, *Verticillium lecanii* (previously called *Cephalosporium lecanii*) (35). Bassianolide is an octadepsipeptide consisting of alternating residues of L-N-methylleucine and 2-hydroxyisovaleric acid residues (44).

Metarhizium anisopliae is also a common entomopathogenic fungus with a wide host range (35). In 1879, Metchnikoff first isolated *M. anisopliae* from an infected beetle and suggested that it may be useful as a biological control agent for insect pests (35). This fungus is commercially available as a control for cockroaches and is being investigated as a possible biocontrol agent of a number of agricultural pests (45). In 1966, Roberts (46) found the hemolymph of insects that were naturally infected with *M. anisopliae* was toxic to other insects and in 1969, he isolated insecticidal cyclic hexadepsipeptides, destruxins, from the culture filtrates of this fungus (47). *Metarhizium anisopliae* produces several forms of destruxin that vary from destruxin B (section 1.2; Fig. 3) by having different 2-hydroxy acid residues in place of 2-hydroxy-N-methylvaleric acid (48). Destruxins are toxic to insect larvae and may have a role in the pathogenicity of *M. anisopliae* to its hosts (35).

The role of beauvericin, bassianolide and destruxin in the interactions of entomopathogenic fungi with their hosts is discussed in section 3.3.

1.4. Other important fungal peptides from nonagricultural systems

Several fungi that are not associated with the agri-food industry also produce nonribosomally synthesized peptides that strongly impact human health as antibiotics, immunosuppressants, or toxins. Some of the more noteworthy examples are mentioned here for general interest and as examples of molecules whose activities may be used or altered in biotechnological manipulations to produce novel peptides as described in section 4.3.

The β-lactam antibiotics, penicillin, cephalosporin and their derivatives, produced by *Penicillium* and *Cephalosporium* spp., among others, have at their core a linear peptide ACV, δ-(L-2-aminoadipyl)-L-cysteinyl-D-valine, that is manufactured by ACV synthetase (42,49). The peptide, ACV, is then modified by various enzymes to produce penicillin, cephalosporin and their derivatives. The profound effect penicillin has had on medical treatment has spurred the search for other fungal antibiotics.

Cyclosporins produced by *Tolypocladium niveum* and *Tolypocladium cylindrosporum* are important immunosuppressive drugs used in human organ-transplant operations and in the treatment of auto-immune disorders (50). Cyclosporins are cyclic peptides consisting of 11 amino acid residues, some of which are modified (42). Interestingly, *T. niveum* and *T. cylindrosporum* are soil fungi that are highly pathogenic to mosquito larvae (35) and cyclosporins produced by these species are toxic to mosquito larvae (51).

Some of the most deadly human poisons known are the amatoxins which are bicyclic peptides produced by certain species of the genera *Amanita*, *Lepiota*, *Conocybe* and *Galerina* (52). The two major groups of amatoxins are the amanitins with at least 9 analogs that have 8 amino acid residues and the phallotoxins with at least 7 analogs that have 7 amino acid residues. *Amanita phalloides* is the main source of fatal mushroom poisonings of humans, and the amanitins are thought to be the main cause of fatalities due to their toxic effects upon liver cells. The peptide α-amanitin is a specific inhibitor of eukaryotic RNA polymerase II and is used in studies of cell physiology and gene expression (53). This toxic peptide has been proposed as a specific treatment for cancer through targeted drug delivery methods (53). There are possible future uses of other amatoxins for cell physiology studies and disease treatments.

2. BIOSYNTHESIS OF FUNGAL PEPTIDE TOXINS

2.1. Peptide synthetases

Fungal peptides are synthesized via a thiotemplate mechanism similar to that first elucidated for the biosynthesis of the bacterial peptide antibiotics, gramicidin and tyrocidine (54). These peptides are assembled on large multifunctional enzymes that perform several sequential steps. Amino acids (or α-hydroxy acids) that will be incorporated into the peptide are first 'charged' by adenylation such that the α phosphate of ATP is attached via a phosphoester linkage to the carboxylic acid functional group, with release of PPi. The charged amino acid, which is still associated with the enzyme, is then covalently bound via a thioester linkage to enzyme bound pantethenate, with concomitant release of AMP. This same series of reactions is catalyzed independently for each of the constituent amino acids by separate modules arranged in series along the peptide synthetase (Fig. 4). Peptide bonds are then formed between the bound amino acids, and the peptide is released via a postulated thioesterase activity. The reversibility of the amino acid adenylation step at equilibrium allows that reaction to serve as the basis of the simple ATP/PP_i exchange assay used to identify peptide synthetases (55). The biochemical details of peptide synthesis have been reviewed recently by Marahiel et al. (56).

2.2. Molecular genetics of fungal peptide biosynthesis

Several genes that encode fungal peptide synthetases have been cloned and sequenced. Among these are the ACV synthetase genes of *Aspergillus nidulans*, *Penicillium chrysogenum*, and *Cephalosporium acremonium* (57,58,59,60), the *HTS1* gene encoding HC-toxin synthetase from *Cochliobolus carbonum* (14,61), the enniatin synthetase gene of *Fusarium scirpi* (62), the cyclosporine synthetase gene of *Tolypocladium niveum* (50), and lysergyl peptide synthetase 1 (LPS1)

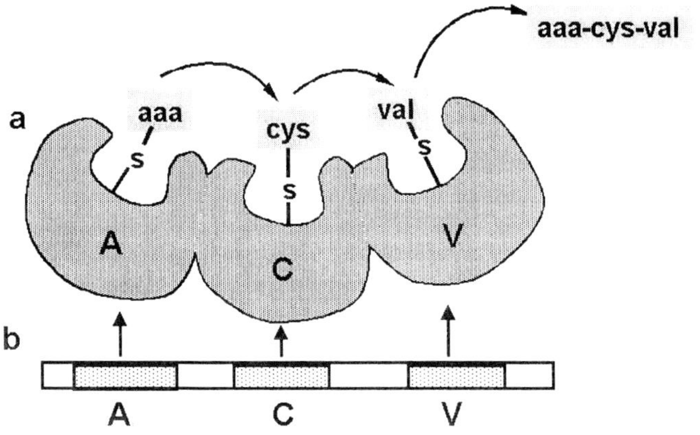

Figure 4. (a) Representation of a three-module peptide synthetase; peptide bonds are forming between the thioesterified amino acids on ACV synthetase. (b) Representation of the ACV synthetase gene with shaded areas encoding their respective peptide synthetase modules.

gene of *Claviceps purpurea* (63). The deduced products of each of these genes consist of a series of peptide synthetase modules occurring at regular intervals along the polypeptide. A module is approximately 650 amino acids long, and contains aminoacyl activation and thiolation domains (56). Without exception in the peptide synthetases studied thus far, the number of modules encoded by the gene corresponds to the number of amino acids adenylated and thioesterified by the gene product. Modifications such as epimerization or N-methylation are catalyzed by additional domains in some modules (56).

Peptide secondary metabolites often occur as families of related molecules, differing by amino acid substitutions. The different peptides in a family are likely generated by a single peptide synthetase as a result of relaxed specificity of amino acid recognition by individual domains. This point is demonstrated by the observations that all forms of HC-toxin are eliminated by knocking out the HC-toxin synthetase genes (14) and all enniatins are eliminated by knocking out the enniatin synthetase gene (62).

3. IMPORTANCE OF FUNGAL PEPTIDES IN SEVERAL AGRICULTURAL SYSTEMS

3.1. Significance of ergopeptines as contaminants of forage grasses

Several species of fungi in the family Clavicipitaceae and their anamorphic relatives have the ability to produce highly toxic ergopeptines alkaloids. These fungi include several species in the genera *Claviceps* and *Neotyphodium* that have had a particularly strong impact on agriculture because they produce ergopeptines within hosts that serve as food for humans and/or livestock. The historical significance of ergotism resulting from human ingestion of alkaloids from *C. purpurea* was described in section 1.1. In modern agriculture, a far greater impact from ergopeptines is associated with animal consumption of forage grasses infected with endophytic symbiotic members of the tribe Balansiae of the Clavicipitaceae (64). The best studied examples of the grass/endophyte symbiosis are tall fescue (*Festuca arundinacea*) infected with *Neotyphodium coenophialum*, and perennial ryegrass (*Lolium perenne*) infected with *Neotyphodium lolii*.

Tall fescue, the predominant perennial forage grass in the mid-Atlantic and southern United States, benefits through its symbiotic association with *N. coenophialum* by increased growth and reproduction as well as through higher levels of tolerance to insect feeding and drought stress (64,65,66). However, endophyte-infected fescue is frequently a poor quality forage for cattle and other grazing animals, resulting in weight loss or reduced weight gain, reduced fertility, reduced lactation, higher body temperature and respiration, and in more severe cases a gangrene-like disease of the animals' limbs, known as fescue foot (65,66).

The mammalian toxicity in endophyte-infected fescue is associated with the production of ergopeptines (in particular ergovaline) by *N. coenophialum* (67,68). The symptoms of fescue toxicosis agree well with the symptoms of ergopeptine poisoning described in section 1.1. The vasoconstrictive activity of ergopeptines is likely responsible for the dry gangrene known as fescue foot and for the sensation that compels affected animals to stand in water. The remaining symptoms of fescue toxicosis, including fertility problems and reduced milk production, could result from the interaction of the ergopeptines with dopamine, norepinephrine, and serotonin receptors (5,6). The construction of near-isogenic strains differing in their ability to produce ergopeptines (as described in section 4.1) would facilitate a direct test of ergopeptines as the critical factor in fescue toxicosis.

Most of the positive agronomic attributes associated with *N. coenophialum* infection are generally attributed to fungal metabolites or processes other than the production and accumulation of ergopeptines. For example, drought tolerance in *N. coenophialum*-infected plants may be affected by decreased stomatal conductance and transpiration, and by the accumulation of compatible solutes, possibly polyols or a class of pyrrolizidine alkaloids called lolines produced by *N. coenophialum* (65,68). Insecticidal and insect feeding deterrence activities appear to be due mainly to the accumulation of lolines and the pyrrolopyrazine alkaloid, peramine. Lolines have broad-spectrum, contact insecticidal activity and also have been shown to deter feeding by important plant pests at concentrations at which they are found in infected grasses (68). Similarly, peramine, which is produced in endophyte-infected perennial ryegrass as well as in endophyte-infected tall fescue, deters feeding of several perennial ryegrass pests at concentrations at which it is found in plants in the field (68,69). Interestingly, peramine appears to be derived from an arginine-proline cyclic dipeptide by methylation and reduction steps (70), though there is as yet no evidence for the activity of a fungal peptide synthetase in its biosynthesis.

Ergopeptines and simpler ergot alkaloids (clavines and lysergic acid derivatives) also have been shown to have some anti-insect activities. Clay and Cheplick (71) found that fall armyworm larvae (*Spodoptera frugiperda*) gained less weight and consumed less plant material when grown on maize leaf disks soaked in solutions of ergopeptines or simpler clavine ergot alkaloids. However, the concentrations of ergot alkaloids required to reduce weight gain and feeding were one to two orders of magnitude greater than those typically encountered in endophyte-infected grasses. Ergopeptines were no more effective than the simpler clavine alkaloids in these assays. Recently, Ball et al. (72) fed adult beetles (*Heteronychus arator*) from the scarab family an artificial, carrot-based diet containing ergopeptines and various other alkaloids commonly associated with the perennial ryegrass endophyte, *N. lolii*. In this study, the ergopeptines, including ergotamine and ergovaline, were most effective among the various alkaloids at deterring beetle feeding. Surprisingly, peramine, a compound demonstrated to deter feeding of other insects, including the economically important argentine stem weevil (64,68), was ineffective against the adult beetles. Similarly, lolitrem B, an indole-diterpene alkaloid with great mammalian toxicity produced by *N. lolii*, did not deter beetle feeding under the conditions used. Thus, ergopeptines may deter feeding of certain insect species at concentrations observed in plants in the field, and certain other species when supplied at high enough concentrations. Their ultimate contribution to insect feeding deterrence in light of the other related and unrelated alkaloids produced in infected grasses is difficult to assess. The development of near-isogenic strains of endophytic fungi differing only in production of specific alkaloids should facilitate studies to assess the impact of these alkaloids on insect feeding deterrence and other traits.

Greater than 90% of the tall fescue pastures in the United States contain endophyte-infected plants, and the mean percentage of infected plants in these pastures is greater than 50% (66). With over 34 million acres of tall fescue pastures in the United States (73), toxicity of tall fescue is a very serious concern. The economic cost to U.S. agriculture attributed to *N. coenophialum* is staggering. In 1990, the estimated loss to the beef cattle industry alone was $600 million (74). Considering inflation and the losses to other grazing animals (e.g., dairy cattle, horses, and sheep) it is not unreasonable to suggest that this is now a $1 billion per year problem. Applications of biotechnology that may help ameliorate this problem are discussed in section 4.1.

3.2. Significance of fungal peptide toxins in plant pathogenesis

Whereas the impact of ergopeptines is on the consumer of the plant rather than the plant itself, several fungi in the genera *Cochliobolus, Alternaria* and *Fusarium* produce peptides that are highly phytotoxic (refer to section 1.2).

Our awareness of HC-toxin and victorin came from similar circumstances, since in both cases sensitivity to the toxin and thus susceptibility to the pathogen were inadvertently bred into host plant varieties. *Cochliobolus carbonum* race 1 was unknown prior to the planting of new varieties of maize in the late 1930's. After a few years of leaf spot disease outbreaks, the pathogen was characterized and a source of resistance was sought (75). Analysis of breeding lines showed that most were resistant to the pathogen and conscious inclusion of this resistance has suppressed the disease caused by *C. carbonum* race 1 for the decades since its discovery.

The significance of the peptide HC-toxin to disease development has been supported by genetic, physiological, and molecular experiments. Scheffer et al. (10) showed that the ability to produce HC-toxin and to cause disease on susceptible varieties of maize cosegregated in crosses of toxin-producing (race 1) and toxin-nonproducing (race 2) isolates of the fungus. Similarly, the addition of HC-toxin to the infection site allowed the typically non-pathogenic race 2 of the fungus to infect varieties of maize susceptible to race 1 of *C. carbonum* (76). Additional proof of the singular importance of the toxin to disease was provided by cloning of genes involved in synthesizing HC-toxin and their inactivation by gene knock out experiments, resulting in concomitant loss of toxin production and loss of pathogenicity (14,15).

Resistance in maize to the pathogen has been shown to be due to a dominant nuclear gene named *Hm1*. In the absence of *Hm1*, the dominant allele at another locus, named *Hm2*, provided intermediate resistance to the fungus (77). *Hm1* encodes a carbonyl reductase, named HC-toxin reductase, that detoxifies HC-toxin through a simple chemical reduction of an essential carbonyl group (16,17).

From a strictly agricultural perspective HC-toxin can be viewed as a problem created through breeding that has been successfully solved by inclusion of the *Hm1* allele in breeding lines of maize. However, the impact of HC-toxin and peptides with similar properties (such as chlamydocin and cyl-2, see ref. 78) on our crop plants may be considerable. HC-toxin reductase, or its functional equivalent, appears to be common to nearly all genotypes of maize, as well as to several other monocot crop species including wheat, sorghum, oats, and barley (17). Because this enzyme is found among several grasses and grain crops, Meeley et al. (17) hypothesized that potential pathogens producing HC-toxin or related peptides have exerted enough selection pressure on grasses that either an ancestral species, or each species independently, evolved an enzymatic means of detoxification. The alternative hypothesis would be that HC-toxin reductase is widespread because it has a common or housekeeping role in the plant and that its detoxification activity is strictly fortuitous. This latter hypothesis appears less tenable because mutants at *Hm1*, either naturally occurring or constructed by transposon-tagging (16), have no apparent phenotype other than toxin sensitivity. Naturally occurring mutants that have lost HC-toxin reductase activity could not retain any alternate housekeeping activity because the recessive allele, *hm1*, is not transcribed in susceptible plants (16).

The pathogenic fungus *C. victoriae*, which produces the peptide toxin victorin, was first discovered in the 1940's when oats derived from the variety 'Victoria' became widely planted due to the inclusion of the *Pc-2* gene which controlled resistance to certain races of the crown rust fungus, *Puccinia coronata*. Under environmental conditions conducive to the pathogen, these oats were devastated by *C. victoriae*. Experiments with culture filtrates of the pathogen showed shortly thereafter that the exceptional, cultivar-specific virulence of *C. victoriae* was due to its production of a small toxic metabolite (79). The structure of that metabolite eluded scientists for years but was finally elucidated as the unusual pentapeptide victorin (19) described in section 1.2.

The significance of victorin to the pathogenicity of *C. victoriae* was indicated by the pioneering work of Meehan and Murphy (18,79) and confirmed through genetic crosses of Scheffer et al. (10), who demonstrated cosegregation of victorin production and pathogenicity to oats. Recently, Wolpert and colleagues have studied the effects of victorin on susceptible and resistant oats and have found that victorin appears to bind to components of the glycine decarboxylase complex of sensitive mitochondria. Consistent with this proposed site of action, competitive inhibition of victorin binding by pyrodoxal phosphate reduces victorin effects, as does incubating plant tissue under conditions that eliminate photorespiration (80).

Similar to the case of HC-toxin, natural resistance to this previously unknown pathogen had been bred out of oats in the quest to obtain resistance to crown rust. As breeders became aware of this fact, the pathogen has been controlled by maintenance of the homozygous recessive condition at the *vb* or *pc-2* locus. Interestingly, the *Vb* allele for sensitivity to victorin has never been separated from the *Pc-2* allele for resistance to certain races of crown rust. Thus, the impact of victorin on oat production was direct and severe for a few years in the 1940's. Since that time victorin's impact has been indirect, through demands it still places on the genetic constitution of oats planted in areas where the threat of infection by this pathogen persists.

Another interesting and important peptide from several *Cochliobolus* species has not yet been purified and characterized, though it role as a general virulence factor for two *Cochliobolus* species has been shown by gene knock out experiments with the gene thought to encode it (81). This cryptic peptide is assembled by the product of the *CPS1* gene originally isolated from *Cochliobolus heterostrophus* and shown by gene knock out to be necessary for full virulence of *C. heterostrophus* and *C. victoriae* on their respective host plants (81).

Homologues of this gene occur in several other species of *Cochliobolus* and some related fungi.

Among the peptide phytotoxins that are not host-specific, the enniatins have been shown most clearly to contribute to the virulence of plant pathogenic *Fusarium* species to their hosts. Enniatins cause chlorosis and necrosis in leaves of many plant species, necrosis in potato tubers, reductions in wheat seed germination, and reductions in root and shoot growth of wheat seedlings (29,30,82). Potato tubers infected with certain species of *Fusarium* may contain 4- to 50-fold the concentration of enniatin necessary to cause necrosis in potato tissue (30). The gene for enniatin synthetase in *F. avenaceum* was inactivated by gene knock out, producing enniatin-non-producing mutants that had significantly reduced virulence to potato tubers when compared to the wild type and three transformants that did produce enniatins (31). The non-producing mutants could still infect potato tubers, indicating that enniatins are not necessary for pathogenicity but are important for virulence. Enniatins may be involved in many plant/pathogen interactions between *Fusarium* species and their plant hosts considering the wide variety of *Fusarium* species that produce these compounds (30).

3.3. Role of fungal peptides in fungus/insect interactions

Beauveria bassiana is the most common fungus isolated from moribund insects in nature (35) and has one of the widest host ranges with many hosts among the Lepidoptera, Coleoptera, and Hemiptera, and some additional hosts in the Diptera, Hymenoptera, and Orthoptera (35,83). *Beauveria bassiana* is currently commercially available (Mycotrol™ES, Mycotech Corp., Butte, MT) as a biocontrol agent for selected agricultural insect pests and is being evaluated as a biocontrol agent for Colorado potato beetle, blueberry maggot, and other agricultural pests. There is wide variation in the pathogenicity and virulence of different strains of the fungus to different species of insects (35).

The role of beauvericin in the pathogenicity of *B. bassiana* has been difficult to assess because the fungus has a wide host range and the toxicity of beauvericin varies depending upon the insect species and stage tested. *In vitro* production of beauvericin in different strains of *B. bassiana* has been correlated to virulence to larvae of *Galleria mellonella* (84). Beauvericin is highly toxic to mosquito larvae (*Aedes aegypti*), but less toxic to adult blowflies (*Calliphora erythrocephala)* and corn earworm larvae (*Heliothis zea*) (32,39). Ninety percent of corn earworm larvae died when infected by *B. bassiana*, but beauvericin was not detected in the hemolymph of the larvae (32). Unfortunately the beauvericin assay used in this study had a high detection limit of 0.5 μg and was not specific because it detected the toxin by measuring its antibiotic activity to a species of *Bacillus*. Nevertheless, injections of up to 6 μg of beauvericin into the hemolymph of adult blowflies and corn earworm larvae were not lethal (32,39). Ingestion of beauvericin-impregnated silica beads by mosquito larvae (*Culex pipiens autogenicus)* causes extensive damage to the epithelium of the midgut with intensive vacuolization of cells and visible damage to mitochondria, membranes and nuclei (38). The toxin acts as an ionophore, disrupting membrane permeability (85), and may have an effect on the immune system of insects rather than a direct lethal effect. Beauvericin and some of its analogs also are toxic to Gram-positive bacteria and fungi (37) and may have an antibiotic effect by protecting insect cadavers from colonization by other microorganisms (35).

Other cyclic peptides produced by entomopathogenic fungi may be involved in pathogenicity to insects. Bassianolide may be involved in pathogenicity of *B. bassiana* since it is toxic to silkworm larvae at 13 ppm and similar concentrations of toxin can be detected in *B. bassiana* infected silkworm larvae (44). Cornworm larvae were paralyzed shortly after injections of bassianolide into their hemolymph, but all of the larvae recovered after 3 days (32). The enniatin complex (a mix of four enniatin species) may also be involved in pathogenicity of *Fusarium lateritium*, a pathogen of the scale insect *Hemiberlesia rapax* (39). These peptides are toxic to adult blowflies (39) and can be produced in large quantities by *Fusarium* species (30). The beauverolides are produced in low quantities by *B. bassiana*, *B. tenella* and *Paecilomyces fumoso-roseus* and are not insecticidal but may affect the immune system of insects (43).

Metarhizium anisopliae has a wide host range and at least ten of the destruxins it produces affect numerous insects (86). Lepidoptera and Diptera species are highly sensitive to destruxins but the level of toxicity and the effect depend upon the species of insect (46,87). Destruxins appear to be metabolized in the tissue of some insects, and this may affect their toxicity to different insects (88). Destruxins can be detected in the hemolymph of fungus-infected mosquito and silkworm larvae in amounts similar to those that cause paralysis and death with injections of pure destruxin (87). Symptoms, including paralysis, caused by injection with pure destruxins were similar to those caused by natural infection with *M. anisopliae* (46,47). Destruxins affect calcium membrane fluxes and protein phosphorylation in lepidopteran insect cell lines and specifically inhibit vacuolar-type ATPase (89,90) which may explain some of the effects caused by destruxins in insect tissue. Destruxins have specific inhibitory effects on malpighian tubule fluid secretion in desert locusts (*Schistocerca gregaria*) and blowfies (*Calliphora* spp.) and are cytotoxic to cells in these organs (91). In some insects, destruxins also inhibit plasmatocyte cells of the host's immune system affecting the morphology and activity of these cells (92,93). Purified destruxins have been found to have similar cytotoxic effects on insect cells *in vitro* compared to those seen during infection of insects by *M. anisopliae* (92).

The peptide toxins produced by *B. bassiana* and *M. anisopliae* may be involved in pathogenicity of these fungi by suppressing the immune system of insects and affecting localized areas of insect tissue to allow penetration by the fungi. Other factors including the many forms of proteases and chitinases produced by *M. anisopliae* and *B. bassiana* may be also involved in the virulence of these fungi (94). The production of an extracellular protease has been related to virulence of isolates of *B. bassiana* (95). The recent cloning of a portion of a cyclic peptide synthetase gene from *M. anisopliae* (96) and the cloning of a gene for enniatin synthetase (62), which is similar in function to beauvericin synthetase (40,42), may facilitate the production of toxin non-producing mutants and direct testing of the involvement of these toxins in pathogenicity.

4. POTENTIAL FOR BIOTECHNOLOGICAL MANIPULATION OF FUNGAL PEPTIDES IN THE AGRI-FOOD INDUSTRY

4.1. Manipulation of ergopeptine accumulation in forage grasses

It would be highly desirable to develop isolates of *Neotyphodium coenophialum* and, to a lesser extent, *N. lolii* that do not synthesize or accumulate ergovaline but still confer many or all of the beneficial endophyte-associated agronomic traits. Grasses infected with such an endophyte could be established in new pastures, overseeded in existing pastures, and planted in place of currently available varieties on disturbed lands (a purpose for which these grasses are commonly used).

Tall fescue can be cured of its endophytic fungus and endophyte-free seed is sold commercially. However, plants that are free of endophyte are much less vigorous and durable (65). Rather than planting endophyte-free tall fescue, management strategies have been employed to reduce ergopeptine intake in grazing animals. For example, whenever possible, animals are moved from tall fescue pastures before they graze the grass to the point where they are grazing near the crown of the plant where the fungi and their ergopeptines are concentrated (65,68).

One biotechnological strategy that may reduce accumulation of ergopeptines (in particular ergovaline) in endophyte-infected tall fescue is to create a null mutation in a gene in the ergopeptine biosynthetic pathway. Two target genes have been cloned. The *dmaW* (or *cpd*) gene encoding dimethylallyltryptophan (DMAT) synthase, which catalyzes the first committed step in the biosynthesis of all ergot alkaloids, was originally cloned from what is likely an isolate of *Claviceps fusiformis* (97) and, more recently, from *C. purpurea* (63,98). The function of this gene has been demonstrated by expression in *Saccharomyces cerevisiae* (97). Genes encoding a peptide synthetase that is very likely involved in the assembly of the highly toxic ergopeptines have recently been cloned from *C. purpurea* (63,98,99) and *N. lolii* (100). In *C. purpurea*, this gene is very tightly linked to the DMAT synthase gene (63,98,99). This type of clustering is typical of genes involved in the synthesis of fungal secondary metabolites (101). Sequence analysis (63) indicates that this gene encodes lysergyl peptide synthetase 1 (LPS1) previously isolated from *C. purpurea* (102).

Fragments of the genes encoding DMAT synthase or the peptide synthetase may be used in gene knock out or replacement strategies (see, e.g., ref. 14,15,31), in which a copy of the cloned gene is rendered nonfunctional *in vitro* and then re-introduced into the fungus by transformation, replacing the native copy of the target gene. Strains of the *Neotyphodium* endophytes in which the DMAT synthase gene or the peptide synthetase gene have been knocked out should produce, respectively, no ergot alkaloids at all, or clavine alkaloids and simple derivatives of lysergic acid but no ergopeptines. Such engineered strains should be very useful for testing the contribution of these different classes of alkaloids to mammalian toxicity and to the fitness enhancement of the host grass. If these altered strains are shown through careful studies to be less toxic to grazing animals and to confer the desirable agronomic traits of wild type isolates of *Neotyphodium* spp., they may prove useful for development of improved forage grass varieties.

Various aspects of the *N. coenophialum* life cycle make it highly unlikely that an engineered, toxin-deficient strain would become toxigenic by recombination with wild, toxin-producing isolates. Vegetative hyphae of *Neotyphodium* spp. grow intercellularly in the leaf sheaths, crowns, and stems of grass plants and, most importantly, grow into developing seeds. The fungus survives in the seed and is transmitted vertically to seedlings of the next generation. Seed transmission is the only known means of dispersal of *N. coenophialum* (64,66). *Neotyphodium coenophialum* is excluded from pollen and thus is passed on as a maternally inherited trait (64,66). Although the related fungal endophytes found in *Poa rigidifolia* and *Agrostis hiemalis* have been observed to produce conidia sparsely on their hosts (103), *N. coenophialum* is not known to sporulate in nature and it is not known to infect plants exogenously. Finally, *N. coenophialum* has no known sexual state (64,66). For all these reasons, there is a low likelihood of recombination between an ergovaline-deficient endophyte and wild ergovaline-producing endophytes. Thus, ergopeptine-deficient endophyte strains should be durable in agriculture, provided that they confer upon their hosts the fitness enhancements of wild isolates that are found in the "competing" varieties of grass.

4.2. Manipulation of insect toxins

Beauveria bassiana is currently commercially available as a biocontrol agent (Mycotrol™ES) for grasshoppers, crickets, locusts and whiteflies in a broad range of crops (45). Commercial preparations of *Metarhizium anisopliae* and *Paecilomyces fumoso-roseus* also are being used for biocontrol of insects. There is a wide variation in the pathogenicity and virulence of different strains of *B. bassiana* and a wide range in the sensitivity of different insect species and their stages to different fungal toxins (32,35,39,87). Strains of *B. bassiana* with selective production of particular toxins could be developed to target specific stages of an insect. The wide variety of analogs of beauvericin and other insecticidal peptides, such as enniatins and bassianolides, provide a pool of compounds that can be tested for selective toxicity. Beauvericin synthetase has been purified from *B. bassiana* and found to operate by a mechanism very similar to the enzyme that synthesizes insecticidal enniatins, which are very similar in structure to beauvericin (40,42). Enniatin synthetases purified from different species of *Fusarium* all have the necessary activities to manufacture enniatins but have different specificities for amino acids substrates (34,104). The gene encoding enniatin synthetase has been cloned and characterized from *F. scirpi* (62,105). Cloning of the gene encoding beauvericin synthetase from *B. bassiana* or enniatin synthetase genes from other *Fusarium* species may be possible by hybridization-based strategies with the cloned gene for enniatin synthetase. *Beauveria bassiana* has successfully been transformed by addition of a benomyl resistant β-tubulin gene (106), and the transfer of enniatin synthetase genes into *B. bassiana* or modification of the beauvericin synthetase gene in *B. bassiana* may allow the development of more virulent strains or tailoring of strains designed for biocontrol of specific insect pests.

Any applications that would increase the amount of beauvericin in or on crops should be approached with caution. Concern about the mycotoxic properties of beauvericin is growing, since it is lipophilic and has the potential to bioaccumulate (107). Beauvericin can alter membrane transport functions and can induce cell death similar to apoptosis in cultured cells (85,108). Beauvericin is produced by plant pathogenic fungi including many toxigenic *Fusarium* species that are common contaminants of grains (109). Four strains *of F. subglutinans* that were toxic to animals were found to produce beauvericin and other toxic metabolites (110). The toxicity of beauvericin to organisms other than insects needs to be studied further.

Strategies similar to those described for *B. bassiana* could be applied for the improvement of *Metarhizium anisopliae* as a biocontrol agent. The wide variety of destruxins and the variation of susceptibility of *M. anisopliae* hosts to them suggest there is room for designing destruxins to target specific insect species. A strain selected for increased virulence against mosquito larvae also had increased mycotoxin production as well as increased *in vitro* spore germination (111). Destruxin analogs with altered amino acid residues have been designed and synthesized *in vitro* allowing testing of the effects of altered amino acids on their structures and activities (112). A portion of a peptide synthetase gene has been cloned from *M. anisopliae* (96). Future cloning of the destruxin synthetase gene may facilitate engineering of new destruxins with improved toxicity or specificity to insect pests.

4.3. Pharmaceuticals and Farmaceuticals

The modular nature of peptide synthetases provides an opportunity for producing new peptides by switching or modifying modules. Collectively, the known nonribosomally synthesized peptides contain over 300 different amino acid (or hydroxy acid) residues (113). Individual modules have specificity for particular amino acids or their derivatives (42). These modules, each recognizing, adenylating, and thioesterifying their substrates, have been switched between different peptide synthetases through genetic manipulation (114,115). The middle module of the three-module peptide synthetase SrfA-A of *Bacillus subtilis* was replaced by modules from other bacterial and fungal peptide synthetases (115). The inserted modules maintained their specificity for their original amino acid and did not affect the specificity of the flanking modules. These data demonstrate that new peptides can be designed through module switching. The length of recombinant peptides also may be determined by moving the carboxyl-terminal thioesterase domain to the module for the last desired amino acid (116). The ability to engineer new peptides may allow the design of new peptides with increased toxicity or altered specificity. Engineering of recombinant peptide synthetase genes also will allow their expression in systems better suited for large scale production of compounds. Expression systems may include traditional microbial fermentation cultures, transgenic plants, or transgenic fungi (e.g., endophytes) growing within plants.

Presently, fungal peptides have crucial uses as pharmaceuticals and the prospect of future uses of these compounds is even greater. The antibiotics penicillin and cephalosporin and their derivatives are widely used in the treatment of bacterial infections. The immunosuppressant cyclosporin is extremely important in the treatment of auto-immune diseases and of organ-transplant patients. New derivatives of these drugs with increased specificity or activity, or that elude bacterial resistance mechanisms, are constantly being sought. The genes encoding ACV synthetase have been cloned from several fungi (see section 2.2). The modules of the ACV synthetase gene of *P. chrysogenum* that activate cysteine and valine retained their specificity when they were used to replace the middle module in the *srf*A-A gene and the single module of the *srf*A-C gene for surfactin synthesis of *Bacillus subtilis* (114,115). These experiments indicate that development of new antibiotics by manipulation of peptide synthetase genes will be possible in the near future.

Other fungal peptides have current and possible future uses as pharmaceuticals. Uses for ergopeptines include the exploitation of their vasoconstrictive activity in the treatment of migraines (117), and exploitation of their serotonin and dopamine agonist activities to treat Parkinson's disease, depression, and hypertension (117,118,119,120). Amatoxins, particularly α-amanitin, which is a specific inhibitor of eukaryotic RNA polymerase II, are being explored for use as targeted cancer treatments (53). The insecticidal destruxin B suppresses expression of hepatitis B viral surface antigen on human cells and may have use as a specific anti-viral drug (121). Beauvericin and enniatins are used in studies of membrane transport and function (85,108) and also may have use in disease treatment. Once mechanisms of these compounds are determined, new more specific or active compounds may be designed and manufactured by manipulating the sequence and types of modules of recombinant peptide synthetase genes.

Plants are being examined as living 'factories' for the production of chemicals, pharmaceuticals and vaccines (122). Plants have advantages for use as factories for complex bioactive proteins compared to animal cell lines or transgenic animals (123). Transgenic plants can be produced on a large scale for relatively low cost, and they do not serve as hosts for infectious agents of animals. Presently, plants are being evaluated for large-scale production of oils and specialty polymers, such as plastics (122). Transgenic plants have been developed that produce complex proteins including vaccines, human antibodies, and enzymes (123,124,125). The cost of large-scale purification of these complex compounds will depend upon the compound and plant species used.

Although plants present the advantage of being able to grow a large amount of biomass quickly and easily for production of expressed compounds, transgenic plants do have disadvantages (126). Production of transgenic plants is time and labor intensive and transgene expression is often variable. Conversely, the production and evaluation of recombinant fungi is very common and requires less time and space than that required for transgenic plants. Recombinant fungi can be grown in fermentation culture for the production of recombinant proteins or their products. The possibility exists for taking advantage of both the easy production of large amounts of biomass with plant growth and the quick production of recombinant fungal strains by using fungal endophytes as factories for the production of recombinant enzymes or secondary products. Effective transformation systems for *N. lolii* and *N. coenophialum* have been developed (127,128). Expression of marker genes has been detected *in vitro* and in infected ryegrass plants (127). Because the foreign genes are expressed in the fungus while it grows throughout the grass plant, transformation of the fungus serves as a surrogate for transformation of the plant. Expression of the fungal peptide synthetases in a fungus such as *N. lolii* should be simpler and more efficient than creating constructs for plant expression and expressing those constructs in a suitable crop plant.

REFERENCES

1. Rutschman, J. and Stadler, P.A. 1978. Chemical background, in *Ergot Alkaloids and Related Compounds*, Vol. 49, Berde, B., and Schild, H.O. (eds.). Springer-Verlag, New York, pp. 29-85.
2. Waller, G.R. and Dermer, O.C. 1981. Enzymology of alkaloid metabolism in plants and microorganisms, in *The Biochemistry of Plants*, Vol. 7, Conn, E.E. (ed.). Academic Press, New York, pp. 317-402.
3. Hudler, G. 1998. Magical Mushrooms, Mischievous Molds. Princeton University Press.
4. Matossian, M.K. 1989. *Poisons of the Past: Molds, Epidemics, and History*. Yale University Press, New Haven, CT.
5. Berde, B., and Stürmer, E. 1978. Introduction to the pharmacology of ergot alkaloids and related compounds as a basis of their therapeutic application, in *Ergot Alkaloids and Related Compounds*, Vol. 49, Berde, B., and Schild, H.O. (eds.). Springer-Verlag, New York, pp. 1-28.
6. Clark, B.J., Chu, D., and Aellig, W.H. 1978. Actions on the heart and circulation, in *Ergot Alkaloids and Related Compounds*, Vol. 49, Berde, B., and Schild, H.O. (eds.). Springer-Verlag, New York, pp. 321-420.

7. Schumann, G.L. 1991. *Plant Diseases: Their Biology and Social Impact*, APS Press, St. Paul, Minnesota, pp. 209-216.
8. Kavaler, L. 1965. *Mushrooms, Molds, and Miracles: The Strange Realm of Fungi*. The John Day Company, Ltd., New York.
9. Pringle, R.B., and Scheffer, R.P. 1967. Isolation of the host-specific toxin and a related substance with nonspecific toxicity from *Helminthosporium carbonum*. *Phytopathology* 57:169-1172.
10. Scheffer, R.P., Nelson, R.R., Ullstrup, A.J. 1967. Inheritance of toxin production and pathogenicity in *Cochliobolus carbonum* and *Cochliobolus victoriae*. *Phytopathology* 57:1288-1291.
11. Walton, J.D., Earle, E.D., and Gibson, B.W. 1982. Purification and structure of the host-specific toxin from *Helminthosporium carbonum* race 1. *Biochem. Biophys. Res. Comm.* 107:785-794.
12. Pope, M.R., Ciuffetti, L.M., Knoche, H.W., McCrery, D., Daly, J.M., Dunkle, L.D. 1983. Structure of the host-specific toxin produced by *Helminthosporium carbonum*. *Biochemistry* 22:3502-3506.
13. Rasmussen, J.B., and Scheffer, R.P. 1988. Isolation and biological activity of four selective toxins from *Helminthosporium carbonum*. *Plant Physiol.* 86:187-191.
14. Panaccione, D.G., Scott-Craig, J.S., Pocard, J.-A., and Walton, J.D. 1992. A cyclic peptide synthetase gene required for pathogenicity of the fungus *Cochliobolus carbonum* on maize. *Proc. Natl. Acad. Sci., USA* 89:6590-6594.
15. Ahn, J-H., and Walton, J.D. 1997. A fatty acid synthase gene in *Cochliobolus carbonum* required for production of HC-toxin, cyclo(D-prolyl-L-alanyl-D-alanyl-L-2-amino-9,10-epoxy-8-oxodecanoic acid). *Mol. Plant-Microbe Interact.* 10:207-214.
16. Johal, G.S., and Briggs, S.P. 1992. Reductase activity encoded by the *HM1* disease resistance gene in maize. *Science* 258:985-987.
17. Meeley, R.B, Johal, G.S., Briggs, S.P., and Walton, J.D. 1992. A biochemical phenotype for a disease resistance gene of maize. *Plant Cell* 4:71-77.
18. Meehan, F., and Murphy, H.C. 1946. A new *Helminthosporium* blight of oats. *Science* 104:413.
19. Wolpert, T.J., Macko, V., Acklin, W., Juan, B., Seibl, J., Meili, J., and Arigoni, D. 1985. Structure of victorin C, the major host-selective toxin from *Cochliobolus victoriae*. *Experientia* 41:1524-1529
20. Wolpert, T.J., Macko, V., Acklin, W., Juan, B., and Arigoni, D. 1986. Structure of minor host-selective toxins from *Cochliobolus victoriae*. *Experientia* 42:1296-1301.
21. Okuno, T., Ishita, Y., Sawai, K., and Matsumoto, T. 1974. Characterization of alternariolide, a host-specific toxin produced by *Alternaria mali* Roberts. *Chem. Lett.* 1974:635-638.

22. Kohmoto, K., Otani, H., and Tsuge, T. 1995. *Alternaria alternata* pathogens, in *Pathogenesis and Host Specificity in Plant Diseases*, vol. II, Kohmoto, K., Singh, U.S., and Singh, R.P. (eds.). Pergamon, New York, pp. 51-63.
23. Bains, P.S., and Tewari, J.P. 1987. Purification, chemical characterization and host-specificity of the toxin produced by *Alternaria brassicae. Physiol. Mol. Plant Path.* 30:259-271.
24. Ayer, W.A., and Pena-Rodriguez, L.M. 1987. Metabolites produced by *Alternaria brassicae*, the black spot pathogen of canola. Part1, the phytotoxic components. *J. Nat. Prod.* 50:400-407.
25. Buchwaldt, L., and Jensen, J.S. 1991. HPLC purification of destruxins produced by *Alternaria brassicae* in culture and leaves of *Brassica napus. Phytochemistry* 30:2311-2316.
26. Buchwaldt, L., and Green, H. 1992. Phytotoxicity of destruxin B and its possible role in the pathogenesis of *Alternaria brassicae. Plant Path.* 41:55-63.
27. Shivanna, K.R. and Sawhney, V.K. 1993. Pollen selection for *Alternaria* resistance in oilseed brassicas: responses of pollen grains and leaves to a toxin of *A. brassicae. Theor. Appl. Genet.* 86:339-344.
28. Sharma, T.R., and Tewari, J.P. 1996. Flow cytometric analysis of *Brassica juncea* cell and pollen cultures treated with destruxin B, a toxin produced by *Alternaria brassicae. Physiol. Mol. Plant Path.* 48:379-387.
29. Burmeister, H.R., and Plattner, R.D. 1987. Enniatin production by *Fusarium tricinctum* and its effect on germinating wheat seeds. *Phytopathology* 77:1483-1487.
30. Hermann, M., Zocher, R., and Haese, A. 1996. Enniatin production by *Fusarium* strains and its effect on potato tuber tissue. *Appl. Environ. Microbiol.* 62:393-398.
31. Hermann, M., Zocher, R., and Haese, A. 1996. Effect of disruption of the enniatin synthetase gene on the virulence of *Fusarium avenaceum. Mol. Plant-Microbe Interact.* 9:226-232.
32. Champlin, F.R., and Grula, E.A. 1979. Noninvolvement of beauvericin in entomopathogenicity of *Beauveria bassiana. Appl. Environ. Microbiol.* 37:1122-1125.
33. Billich, A., and Zocher, R. 1988. Constitutive expression of enniatin synthetase during fermentative growth of *Fusarium scirpi. Appl. Environ. Microbiol.* 54:2504-2509.
34. Zocher, R., Keller, U., and Kleinkauf, H. 1982. Enniatin synthetase, a novel type of multifunctional enzyme catalyzing depsipeptide synthesis in *Fusarium oxysporum. Biochemistry* 21:43-48.
35. Tanada, Y., and Kaya, H.K. 1993. *Insect Pathology.* Academic Press, Inc., San Diego.
36. Ainsworth, G.C. 1976. *Introduction to the history of mycology.* Cambridge University Press, New York.

37. Hamill, R.L., Higgens, C.E., Boaz, H.E., and Gorman, M. 1969. The structure of beauvericin, a new depsipeptide antibiotic toxic *to Artemia salina. Tetra. Letters* 49:4255-4258.
38. Zizka, J., and Weiser, J. 1993. Effect of beauvericin, a toxic metabolite of *Beauveria bassiana*, on the ultrastructure of *Culex pipiens autogenicus* larvae. *Cytobios* 75:13-19.
39. Grove, J.F., and Pople, M. 1980. The insecticidal activity of beauvericin and the enniatin complex. *Mycopathologia* 70:103-105.
40. Peeters, H., Zocher, R., Madry, N., Kleinkauf, H. 1983. Incorporation of radioactive precursors into beauvericin produced by *Paecilomyces fumoso-roseus. Phytochemistry* 22:1719-1720.
41. Gupta, S., Montllor, C., and Hwang, Y.-S. 1995. Isolation of novel beauvericin analogues from the fungus *Beauveria bassiana. J. Nat. Prod.* 58:733-738.
42. Kleinkauf, H., and von Döhren, H. 1990. Nonribosomal biosynthesis of peptide antibiotics. Eur. J. Biochem. 192:1-15.
43. Jegorov, A., Sedmera, P., Matha, V., Òimek, P., Zahradnícková, H., Landa, Z., and Eyal, J. 1994. Beauverolides L and La *from Beauveria tenella* and *Paecilomyces fumoso-roseus*. Phytochemistry 37:1301-1303.
44. Suzuki, A., Kanaoka, M., Isogai, A., Murakoshi, S., Ichinoe, M., and Tamura, S. 1977. Bassianolide, a new insecticidal cyclodepsipeptide from *Beauveria bassiana* and *Verticillium lecanii. Tetrahedron Lett.* 25:2167-2170.
45. Carlton, B.C. 1996. Development and commercialization of new and improved biopesticides. Ann. New York Acad. Sci. 792:154-163.
46. Roberts, D.W. 1966. Toxins from the entomogenous fungus *Metarhizium anisopliae* II. Symptoms and detection in moribund hosts. *J. Invert. Pathol.* 8:222-227.
47. Roberts, D.W. 1969. Toxins from the entomogenous fungus *Metarhizium anisopliae:* Isolation of destruxins from submerged cultures. *J. Invert. Pathol.* 14:82-88.
48. Jegorov, A., Sedmera, P., and Matha, V. 1993. Biosynthesis of destruxins. *Phytochemistry* 33:1403-1405.
49. Martin, J.F., and Liras, P. 1989. Organization and expression of genes involved in the biosynthesis of antibiotics and other secondary metabolites. *Annu. Rev. Microbiol.* 43:173-206.
50. Weber, G., Schörgendorfer, K., Schneider-Scherzer, E., and Leitner, E. 1994. The peptide synthetase catalyzing cyclosporine production in *Tolypocladium niveum* is encoded by a giant 45.8-kilobase open reading frame. *Curr. Genet.* 26:120-125.
51. Weiser, J., and Matha, V. 1988. The insecticidal activity of cyclosporines on mosquito larvae. *J. Invert. Path.* 51:92-93.

52. Weiland, T., and Faulstich, H. 1978. Amatoxins, phallotoxins, phallolysin, and antamanide: the biologically active components of poisonous *Amanita* mushrooms. *CRC Crit. Rev. Biochem.* 5:185-260.
53. Weiland, T., and Faulstich, H. 1991. Fifty years of amanitin. *Experientia* 47:1186-1193.
54. Lipmann, F. 1971. Attempts to map a process evolution of peptide biosynthesis. *Science* 173:875-884.
55. Lee, S.G., and Lipmann, F. 1975. Tyrocidine synthetase system. *Meth. Enzymol.* 43:585-602.
56. Marahiel, M.A., Stachelhaus, T., and Mootz, H.D. 1997. Modular peptide synthetases involved in nonribosomal peptide synthesis. *Chem. Rev.* 97:2651-2673.
57. Diez, B., Gutiérrez, S., Barredo, J.L., van Solingen, P., van der Voort, L.H.M., and Martin, J.F. 1990. The cluster of penicillin biosynthetic genes. Identification and characterization of the *pcbAB* gene encoding the α aminoadipyl-cysteinyl-valine synthetase and linkage to the *pcbC* and *penDE* genes. *J. Biol. Chem.* 265:16358-16365.
58. Smith, D.J., Earl, A.J., and Turner, G. 1990. The multifunctional peptide synthetase performing the first step of penicillin biosynthesis in *Penicillium chrysogenum* is a 421 073 protein similar to *Bacillus brevis* peptide antibiotic synthetases. *EMBO J.* 9:2743-2750.
59. Gutiérrez, S., Diez, B., Montenegro, E., and Martin, J.F. 1991. Characterization of the *Cephalosporium acremonium pcbAB* gene encoding α -aminoadipyl-cysteinyl-valine synthetase, a large multidomain peptide synthetase: linkage to the *pcbC* gene as a cluster of early cephalosporin biosynthetic genes and evidence of multiple functional domains. *J. Bacteriol.* 173:2354-2365.
60. MacCabe, A.P., van Liempt, H., Palissa, H., Unkles, S.E., Riach, M.B.R., Pfeifer, E., von Döhren, H., and Kinghorn, J.R. 1991. δ-(L-α-aminoadipyl)-L-cysteinyl-D-valine synthetase from *Aspergillus nidulans*. *J. Biol. Chem.* 266:12646-12654.
61. Scott-Craig, J.S., Panaccione, D.G., Pocard, J.-A., and Walton, J.D. 1992. The cyclic peptide synthetase catalyzing HC-toxin production in the filamentous fungus *Cochliobolus carbonum* is encoded by a 15.7-kb open reading frame. *J. Biol. Chem.* 267:26044-26049.
62. Haese, A., Schubert, M., Hermann, M., and Zocher, R. 1993. Molecular characterization of the enniatin synthetase gene encoding a multifunctional enzyme catalysing N-methyldepsipeptide formation in *Fusarium scirpi*. *Mol. Microbiol.* 7:905-914.
63. Tudzynski, P., Holter, K., Correia, T., Arntz, C., Grammel, N., and Keller, U. 1999. Evidence for an ergot alkaloid gene cluster in *Claviceps purpurea*. *Mol. Gen. Genet.* 261:133-141.
64. Schardl, C.L. 1996. *Epichloë* species: fungal symbionts of grasses. *Annu. Rev. Phytopathol.* 34:109-130.

65. Schardl, C.L., and Phillips, T.D. 1997. Protective grass endophytes: Where are they from and where are they going? *Plant Dis.* 81:430-437.
66. Siegel, M.R., Latch, G.C.M., and Johnson, M.C. 1987. Fungal endophytes of grasses. *Annu. Rev. Phytopathol.* 25:293-315.
67. Lyons, P.C., Plattner, R.D., and Bacon, C.W. 1986. Occurrence of peptide and clavine ergot alkaloids in tall fescue grass. *Science* 232:487-489.
68. Siegel, M.R., and Bush, L.P. 1997. Toxin production in grass/endophyte associations, in *The Mycota V, Part A*, Carroll, G., and Tudzynski, P. (eds.). Springer-Verlag, Berlin-Heidelberg, pp. 185-204.
69. Rowan, D.D. 1993. Lolitrems, peramine, and paxilline – mycotoxins of the ryegrass-endophyte interaction. *Agric. Ecosyst. Environ.* 44:103-122.
70. Porter, J.K., 1994. Chemical constituents of grass endophytes, in *Biotechnology of Endophytic Fungi of Grasses*, Bacon, C.W., and White, J.F. (eds.). CRC Press, Boca Raton, FL, pp. 103-124.
71. Clay, K., and Cheplick, G.P. 1989. Effect of ergot alkaloids from fungal endophyte-infected grasses on fall armyworm (*Spodoptera frugiperda*). *J. Chem. Ecol.* 15:169-182.
72. Ball, O.J.-P., Miles, C.O., and Prestidge, R.A. 1997. Ergopeptine alkaloids and *Neotyphodium lolii*-mediated resistance in perennial ryegrass against adult *Heteronychus arator* (Coleoptera, Scarabaeidae). *J. Econ. Entomol.* 90:1382-1391.
73. Putnam, M.R., Bransby, I., Schumacher, J., Boosinger, T.R., Bush, L., Shelby, R.A., Vaughan, J.T., Ball, D., Brendemuhl, J.P. 1991. Effects of the fungal endophyte *Acremonium coenophialum* in fescue on pregnant mares and foal viability. *Am. J. Vet. Res.* 52:2071-2074.
74. Hoveland, C. 1993. Importance and economic significance of the *Acremonium* endophytes to performance of animals and grass plants. *Agr. Ecosyst. Environ.* 44:3-12.
75. Ullstrup, A.J. 1941. Two physiologic races of *Helminthosporium maydis* in the corn belt. *Phytopathology* 31:508-521.
76. Comstock, J.C., and Scheffer, R.P. 1973. Role of host-selective toxin in colonization of corn leaves by *Helminthosporium carbonum*. *Phytopathology* 63:24-29.
77. Nelson, O.E., and Ullstrup, A.J. 1964. Resistance to leaf spot in maize; genetic control of resistance to race 1 of *Helminthosporium carbonum* Ull. J. Heredity 55:195-199.
78. Walton, J.D., Earle, E.D., Stähelin, H., Grieder, A., Hirota, A., and Suzuki, A. 1985. Reciprocal biological activities of the cyclic tetrapeptides chlamydocin and HC-toxin. *Experientia* 41:348-350.
79. Meehan, F., and Murphy, H.C. 1947. Differential toxicity of metabolic by-products of *Helminthosporium victoriae*. *Science* 106:270.
80. Wolpert, T.J., Navarre, D.A., and Lorang, J.M. 1998. Victorin-induced oat cell death, in *Molecular Genetics of Host-Specific Toxins in Plant Disease*, Kohmoto, K., and Yoder, O.C. (eds.). Kluwer, The Netherlands, pp. 105-114.

81. Yoder, O.C. 1998. A mechanistic view of the fungal/plant interaction based on host-specific toxin studies, in *Molecular Genetics of Host-Specific Toxins in Plant Disease*, Kohmoto, K., and Yoder, O.C. (eds.). Kluwer, The Netherlands, pp. 3-15.
82. Hershenhorn, J., Park, S.H., Stierle, A., and Strobel, G.A. 1992. *Fusarium avenaceum* as a novel pathogen of spotted knapweed and its phytotoxins, acetamido-butenolide and enniatin B. *Plant Sci.* 86:155-160.
83. Jeffs, L.B., Feng, M.G., Falkowsky, J.E., and Khachatourians, G.G. 1997. Infection of the migratory grasshopper (Orthoptera, Acrididae) by ingestion of the entomopathogenic fungus *Beauveria bassiana*. *J. Econ. Entomol.* 90:383-390.
84. Ferron, P. 1981. Pest control by the fungi *Beauveria* and *Metarhizium*, in *Microbial control of pests and plant diseases 1970-1980*, Burges, H.D. (ed.). Academic Press, New York, pp. 465-482.
85. Benz, R. 1978. Alkali ion transport through lipid bilayer membranes mediated by enniatin A and B and beauvericin. *J. Membrane Biol.* 43:367-394.
86. Gupta, S., Roberts, D.W., and Renwick, J.A.A. 1989. Insecticidal cyclodepsipeptides from *Metarhizium anisopliae*. *J. Chem. Soc. Perkin Trans. I* 2347-2357.
87. Roberts, D.W. 1981. Toxins of entomopathogenic fungi, in *Microbial control of pests and plant diseases 1970-1980*, Burges, H.D. (ed.). Academic Press, New York, pp. 441-464.
88. Jegorov, A., Matha, V., and Hradec, H. 1992. Detoxification of destruxins in *Galleria mellonella* L. larvae. *Comp. Biochem. Physiol.* 103C:227-229.
89. Dumas, C., Matha, V., Quiot, J-M., and Vey, A. 1996. Effects of destruxins, cyclic depsipeptide mycotoxins, on calcium balance and phosphorylation of intracellular proteins in lepidopteran cell lines. *Comp. Biochem. Physiol.* 114C:213-219.
90. Muroi, M., Shiragami, N., and Takatsuki, A. 1994. Destruxin B, a specific and readily reversible inhibitor of vacuolar-type H^+-translocating ATPase. *Biochem. Biophys. Res. Comm.* 205:1358-1365.
91. James, P.J., Kershaw, M.J., Reynolds, S.E., and Charnley, A.K. 1993. Inhibition of desert locust (*Schistocerca gregaria*) malpighian tubule fluid secretion by destruxins, cyclic peptide toxins from the insect pathogenic fungus *Metarhizium anisopliae*. *J. Insect Physiol.* 39:797-804.
92. Vilcinskas, A., Matha, V., and Götz, P. 1997. Effects of the entomopathogenic fungus *Metarhizium anisopliae* and its secondary metabolites on morphology and cytoskeleton of plasmatocytes isolated from the greater wax moth, *Galleria mellonella*. *J. Insect Physiol.* 43:1149-1159.
93. Vilcinskas, A., Matha, V., and Götz, P. 1997. Inhibition of phagocytic activity of plasmatocytes isolated from *Galleria mellonella* by entomogenous fungi and their secondary metabolites. *J. Insect Physiol.* 43:475-483.

94. St. Leger, R.J., Charnley, A.K., and Cooper, R.M. 1986. Cuticle-degrading enzymes of entomopathogenic fungi: synthesis in culture on cuticle. *J. Invert. Pathol.* 48:85-95.
95. Bidochka, M.J. and Khachatourians, G.G. 1990. Identification of *Beauveria bassiana* extracellular protease as a virulence factor in the pathogenicity toward the migratory grasshopper, *Melanoplus sanguinipes*. *J. Invert. Pathol.* 56:362-370.
96. Bailey, A.M., Kershaw, M.J., Hunt, B.A., Paterson, I.C., Charnley, A.K., Reynolds, S.E., and Clarkson, J.M. 1996. Cloning and sequence analysis of an intron-containing domain from a peptide synthetase-encoding gene of the entomopathogenic fungus *Metarhizium anisopliae*. *Gene* 173:195-197.
97. Tsai, H.-F., Wang, H., Gebler, J.C., Poulter, C.D., and Schardl, C.L. 1995. The *Claviceps purpurea* gene encoding dimethylallyltryptophan synthase, the committed step for ergot alkaloid biosynthesis. *Biochem. Biophys. Res. Comm.* 216:119-125.
98. Wang, J., Panaccione, D.G., and Schardl, C.L. 1998. Ergot alkaloid biosynthesis genes in *Claviceps* and *Neotyphodium* species. *Phytopathology* 88:S133 (abstract).
99. Panaccione, D.G., Wang, J., Schardl, C.L., Annis, S.L., and Damrongkool, P. 1998. Association of a *Claviceps purpurea* peptide synthetase gene with ergopeptine biosynthesis. *Phytopathology* 88:S131 (abstract).
100. Panaccione, D.G., Wang, J., Young, C.A., Scott, D.B., Schardl, C.L., and Damrongkool, P. 1999. Cloning of an ergopeptine-associated peptide synthetase gene from Clavicipitaceous fungi. Phytopathology 89:(in press)(abstract).
101. Keller, N.P., and Hohn, T.M. 1997. Metabolic pathway gene clusters in filamentous fungi. *Fungal Genet. Biol.* 21:17-29.
102. Riederer, B., Han, M., and Keller, U. 1996. D-Lysergyl peptide synthetase from the ergot fungus *Claviceps purpurea*. *J. Biol. Chem.* 271:27524-27530.
103. White, J.F. Jr., Martin, T.I., and Cabral, D. 1996. Endophyte-host associations in forage grasses. XXII. Conidia formation by *Acremonium* endophytes on the phylloplanes of *Agrostis hiemalis* and *Poa rigidifolia*. *Mycologia* 88:174-178.
104. Pieper, R., Kleinkauf, H., and Zocher, R. 1992. Enniatin synthetases from different Fusaria exhibiting distinct amino acid specificities. *J. Antibiot.* 45:1273-1277.
105. Haese, A., Pieper, R., von Ostrowski, T., and Zocher, R. 1994. Bacterial expression of catalytically active fragments of the multifunctional enzyme enniatin synthetase. *J. Mol. Biol.* 243:116-122.
106. Pfeifer, T.A., and Khachatourians, G.G. 1992. *Beauveria bassiana* protoplast regeneration and transformation using electroporation. *Appl. Microbiol. Biotechnol.* 38:376-381.
107. Thakur, R.A., and Smith, J.S. 1997. Liquid chromatography/thermospray/ mass spectrometry analysis of beauvericin. J. Agric. Food Chem. 45:1234-1239.

108. Ojcius, D.M., Zychlinsky, A., Zheng, L.M., and Young, J. D-E. 1991. Ionophore-induced apoptosis: role of DNA fragmentation and calcium fluxes. *Exp. Cell Res.* 197:43-49.
109. Logrieco, A., Moretti, A., Castella, G., Kostecki, M., Golinski, P., Ritieni, A., and Chelkowski, J. 1998. Beauvericin production by *Fusarium* species. *Appl. Environ. Microbiol.* 64:3084-3088.
110. Moretti, A., Logrieco, A., Bottalico, A., Ritieni, A., Randazzo, G., and Corda, P. 1995. Beauvericin production by *Fusarium subglutinans* from different geographical areas. *Mycol. Res.* 99:282-286.
111. Al-Aidroos, K., and Roberts, D.W. 1978. Mutants of *Metarhizium anisopliae* with increased virulence towards mosquito larvae. *Can. J. Genet. Cyto.* 20:211-219.
112. Cavelier, F., Verducci, J., Andre, F., Haraux, F., Sigalat, C., Traris, M., and Vey, A. 1998. Natural cyclopeptides as leads for novel pesticides: tentoxin and destruxin. *Pest. Sci.* 52:81-89.
113. Kleinkauf, H., and von Döhren, H. 1996. A nonribosomal system of peptide biosynthesis. *Eur. J. Biochem.* 236:335-351.
114. Stachelhaus, T., Schneider, A., and Marahiel, M.A. 1995. Rational design of peptide antibiotics by targeted replacement of bacterial and fungal domains. *Science* 269:69-72.
115. Schneider, A., Stachelhaus, T., and Marahiel, M.A. 1998. Targeted alteration of the substrate specificity of peptide synthetases by rational module swapping. *Mol. Gen. Genet.* 257:308-318.
116. de Ferra, F., Rodriguez, F., Tortora, O., Tosi, C., and Grandi, G. 1997. Engineering of peptide synthetases. *J. Biol. Chem.* 272:25304-25309.
117. Ohno, S., Koumori, M., Adachi, Y., Mizukoshi, K., Nagasaka, M., and Ichihara, K. 1994. Synthesis and structure-activity relationships of new (5R,8S,10R)-ergoline derivatives with antihypertensive or dopaminergic activity. *Chem. Pharmaceut. Bull.* 42:2042-2048.
118. Brown, A.M., Patch, T.L., and Kaumann, A.J. 1991. The antimigraine drugs ergotamine and dihydroergotamine are potent 5-HT1C receptor agonists in piglet choroid plexus. *Brit. J. Pharmacol.* 104:45-48.
119. Markstein, R., Seiler, M.P., Jaton, A., and Briner, U. 1992. Structure activity relationship and therapeutic uses of dopaminergic ergots. *Neurochemistry International* 20:S211-S214.
120. Muck-Seler, D., and Pericic, D. 1993. Possible antidepressant dihydroergosine preferentially binds to 5-HT1B receptor sites in the rat hippocampus. *J. Neural Transmission* 92:1-9.
121. Chen, H.C., Chou, C.K., Sun, C.M., and Yeh, S.F. 1997. Suppressive effects of destruxin B on hepatitis B virus surface antigen gene expression in human hepatoma cells. *Antiviral Res.* 34:137-144.
122. Moffat, A.S. 1995. Plants as chemical factories. *Science* 268:659.

123. Cramer, C.L., Weissenborn, D.L., Oishi, K.K., Grabau, E.A., Bennett, S., Ponce, E., Grabowski, G.A., and Radin, D.N. 1996. *Ann. New York Acad. Sci.* 792:62-71.
124. Ma, J. K-C., and Hein, M.B. 1996. Antibody production and engineering in plants. *Ann. New York Acad. Sci.* 792:72-81.
125. Beachy, R.N., Fithcen, J.H., and Hein, M.B. 1996. Use of plant viruses for delivery of vaccine epitopes. *Ann. New York Acad. Sci.* 792:43-49.
126. Smith, G., Walmsley, A., and Polkinghorne, I. 1997. Plant-derived immunocontraceptive vaccines. *Reprod. Fertil. Dev.* 9:85-89.
127. Murray, F.R., Latch, G.C.M., and Scott, D.B. 1992. Surrogate transformation of perennial ryegrass, *Lolium perenne*, using genetically modified *Acremonium* endophyte. *Mol. Gen. Genet.* 233:1-9.
128. Tsai, H.-F., Siegel, M.R., and Schardl, C.L. 1992. Transformation of *Acremonium coenophialum*, an endophyte of tall fescue. *Curr. Genet.* 22:399-406.

Applied Mycology and Biotechnology
Volume 1. Agriculture and Food Production
G.G. Khachatourians and D.K. Arora (editors)

Plant antifungal peptides and their use in transgenic food crops

Adrienne E. Woytowich and George G. Khachatourians

Department of Applied Microbiology and Food Science, University of Saskatchewan, 51 Campus Drive, Saskatoon, SK, Canada. S7N 5A8.

1. INTRODUCTION

Fungal, bacterial and viral diseases of plants contribute to significant losses in crop yields and value worldwide. To confront the bacterial and fungal pathogens, a major goal of plant biotechnology is to introduce defense genes into food crops. As is the case with animals, plants are continuously exposed to herbivorous vertebrate and invertebrate animals and pathogenic microorganisms throughout their life span. To defend against attack by microbes and herbivores, plants have evolved very effective mechanisms that restrict the growth and spread of microorganisms within and outside their population of cells and tissues. Whether it is through the presence of microbial pathogens or wounding a plant or animal responds through a similar signaling pathway where plant systemin and animal cytokines are the first phase of the response to such trauma (Ryan and Pearce, 1998). Later, the synthesis of linolenic acid, phytodienoic acid and jasmonic acid in plants and arachidonic acid and prostaglandins in animals signal the necessary responses in the form of plant defense and inflammation in plants and animals respectively. These reactions bring about the production of about 20 wound response proteins in plants and about the same numbers of acute phase response proteins in animals to direct the overall strategies for defense against the pathogen or healing of the wound (Ryan and Pearce, 1998). Several hundreds of peptide antibiotics are produced as defense molecules. In addition to the immune response, phagocytosis, complement activation many small antimicrobial proteins and peptides (AMPs) such as thrombin-induced platelet microbial protein-1 and human neutrophil defensin-1 operate to ward off infection in animals (Elsbach, 1990; Yeaman, 1997; Xiong *et al.*, 1999). Entomopathogenic fungi also produce a series of linear or cyclic peptide cause toxemia, injury and death in insects (Khachatourians, 1991; 1996). Bacteriocins and zymocins, are killer peptides which are synthesized by producer cells against members of the same and/or different species as killer substances. Several cationic and neutral bacterial peptides (bacteriocins) are secreted from Gram-positive and -negative bacteria and act as AMPs (Nissen Meyer and Ness, 1997; Hancock and Chapple, 1999).

2. PLANT ANTIMICROBIAL PEPTIDES

Plants do not have an immunoglobulin-based system for defense against trauma. Instead, several anatomical features and chemical factors constitutively present to defend against microbial infections and wounding (Ryan and Pearce, 1998). There are three groups of plant genes which code for defense proteins acting as: 1) antifungal peptides (AFPs); 2) pathogenicity factors against microbial toxins (e.g., tablotoxin of *Pseudomanas syringae* pv. *tabaci*, causative agent of the wildfire disease); and 3) signals determining the need for synthesis of other natural defense products. The genetic expression of antimicrobial plant products can be divided into (a) adaptive and (b) constitutive. The constitutive ones are found in herbs and spices and include for example phenolics, essential oils (allicin) and various thiocyanates. In the presence of microbial plants pathogens, there is the convenience of signals for the production of lectins, enzymes, hydrogen peroxides, AMPs and peptides. In recent years many potent AMPs have been isolated and thoroughly characterized from several plant species. Several detailed reports on characterization of various AMPs have been published (Broekaert *et al.,* 1997; Caaveiro *et al.,* 1997; Whipps, 1997; Shewry and Lucas, 1997).

The term antimicrobial can be used to define any compound which exhibits an inhibition of fungal or bacterial growth. There are many plant peptides that possess strong activity *in vitro* against plant-pathogenic fungi, of which the antifungal peptides (AFPs) belong to a group of AMPs referred to as disulfide-linked peptides. These peptides are approximately 50 amino acids in length and contain cysteine residues that allow for the formation of disulfide bonds. These internal disulfide bridges dictate the conformation of the molecules for biological activity. For example, all plant AMPs that have been isolated and characterized to date have even (4, 6 or 8) numbers of cysteines connected by disulfide bridges contributing to their high stability (Cammue *et al.,* 1992; Broekaert *et al.,* 1997). Table 1 shows a compilation of these proteins and their cys-cys residue spacing. Froy and Gravitz (1998) compared the structure and function of a group of membrane potential modulators including plant γ-thionins, scorpion toxins, insect and scorpion defensins, bee venom apapmin, snake sarafotoxis, and human endothelins. In spite of phylogenetic origins and structural differences, comparative analysis reveals commonalty of the effects on membrane potential, common cystein stabilized α-helical (CSH) motif and gene organization. Except for insect defensins which are intronless, others have an intron that splits a codon toward the end of the leader sequence and orientation of CSH motif thus suggesting an evolutionary linkage.

Based on homologies at the primary structure level, there are several families of plant AMP defensins (Hoffman and Heru, 1992: Lehrer *et al.,* 1993; Evans and Harmon, 1995), thionins (Bohlmann and Apel, 1991), brevenin (Morikawa *et al.,* 1992), lipid transfer proteins (Matsuzaki *et al.,* 1995), thaumatin-like AFPs (Richardson *et al.,* 1987) and hevein- and knottin-type AFPs (Broekaert *et al.* 1992).

3. ISOLATION AND CHARACTERIZATION OF PLANT AFPs

Several taxonomically distinct plant species have been searched for novel plant AMPs. Although it has been found that several different parts of a plant, such as leaves, tubers, roots and flower organs, produce AMPs, the most abundant source appears to be the seeds. Potent AMPs have been isolated and purified from the

Table 1. Comparison of the size (number of amino acids) and cysteine spacing/connectivities, where known, to the adjacent amino acid residues.

Protein Name	Example	Polypeptide Size (kDa)	Spacing of cysteine residues
Defensins[1]	Rs-AFP2	51	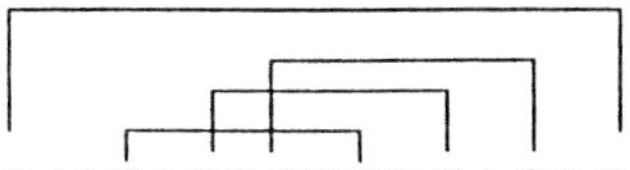 3 C 10 C 5 C 3 C 9 C 8 C 1 C 3 C
Knottin -type[2]	MjAMP1	36	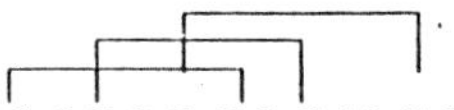 1 C 6 C 8 C C 3 C 10 C 3
Hevein-type[3]	Ac-AMP2	30	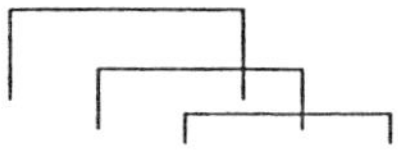 3 C 4 C 4 C C 5 C 6 C 2
Macadamia[4]	Mi-AMP1	76	10 C 9 C 1 C 25 C 14 C 11 C
Thionin[5]	α-purothionin	45	 2 C C 7 C 3 C 8 C 3 C 1 C 8 C 6
Impatiens[6]	Ib-AMP1	20	5 C C 8 C 3 C

References
[1] Terras *et al.*, 1995; [2] Cammue *et al.*, 1992; [3] Broekaert *et al.*, 1992; [4] Marcus *et al.*, 1997; [5] Ohtani *et al.*, 1977; [6] Tailor *et al.*, 1997.

seeds, kernels and roots of a number of plant species such as Mj-AMP1 in *Mirabilis jalapa* (Cammue *et al.*, 1992), *Raphanus sativus* L., (Terrras *et al.*, 1992a,b), *Amaranthus caudatus* (DeBolle *et al.*, 1996), *Picea abies* (Sharma and Lonneberg, 1996), *Macadamia integrifolia* (Marcus *et al.*, 1997), *Impatiens balsamina* (Tailor *et al.*, 1997), *Atriplex nummularia* (Last *et al.*, 1997), and zein in *Zea mays* (Roberts and Selitrennikoff, 1991), *Avena sativa, Hordeum vulgare, Linum ustiatissimum, Sorghum bicolor,* and *Triticum aestivum* (Vigers *et al.*, 1991). Table 2 provides the common and Latin names of these plant. The antimicrobial potency of these proteins have been assessed *in vitro* using large numbers of plant pathogens.

Table 2
Common and Latin names of plants with AMPs

Common name	Latin name
Alum root	*Heuchera sanguinea*
Amarantus	*Amarathus caudatus*
Barley	*Hordeum vulgare*
Common balsam	*Impatiens balsamina*
Corn	*Zea mays*
Dahlia	*Dahlia merckii*
Flax	*Linum ustiatissimum*
Four o'clock	*Mirabilis jalapa*
Giant saltbrush	*Atriptex nummularia*
Macadamia	*Macadamia integrifolia*
Mistletoe	*Viscum album*
Norway spruce	*Picea abies*
Oat	*Avena sativa*
Pokeweed	*Phytoalacca americana*
Radish	*Raphanus sativus*
Sorghum	*Sorghum bicolor*
Spinach	*Spinacia oleracea*
Wheat	*Triticum aestivum*

Although for bioassays of various AMPs, researchers have used different microorganisms and plating media, a set of common inhibitory values can be derived. Table 3 shows the IC_{50} values. For example, the potency of AMPs Mj-AMP1 and Mj-AMP2 isolated from *M. jalapa* seeds was examined using 15 different phytopathogenic fungi and four bacteria (Cammue *et al.*, 1992). The antimicrobial potential of the peptide isolated from *M. integrifolia* was tested against 11 fungal phytopathogens, two oomycete phytopathogens, two pathogenic bacteria and three mycopathogens of humans (Marcus *et al.*, 1997). In addition 13 fungal and nine bacterial strains were employed by Tailor *et al.* (1997) to determine the antimicrobial activity of the cysteine-rich peptides purified from *I. balsamina.*

Most of the research undertaken in this area requires assaying various plant tissues against the phytopathogenic fungi. Once inhibition of fungal growth has been observed the proteins can be isolated, purified and their inhibitory effect confirmed. To study the release and potency of two 5kDa defensins produced by *R. sativus, in vitro*, Terras *et al.* (1995) developed a plate bioassay, where radish seeds

Table 3. Known *in vitro* antimicrobial activity of different plant antimicrobial peptides. IC_{50} equals the concentration required for 50% growth inhibition of the test organism.

Test Organism	IC_{50} (μg/ml) AMP2	Ib-AMP4	MiAMP1
Fungi			
Alternaria brassicola	6	-	-
A. helianthi	-	-	2-5
A. longipes	-	3	-
Ascochyta pisi	6	-	-
Botrytis cinerea	2	6	5-10
Ceratocystis parodoxa	-	-	20
Cercospora beticola	2	-	-
Colletotrichum gloeosporioides	-	-	2-5
C. llindemuthianum	1	-	-
Fusarium culmorum	3	1	-
F. oxysporum f.sp. *pisi*	5	-	2-5
Leptosphaeria maculans	-	-	5
Macrophomina phaseolina	-	-	<25
Microsporum gypseum	-	-	>100
Nectria haematococca	0.5	-	-
Penicillium digitatum	-	3	-
Phoma betae	6	-	-
Phytophomins cryptogea	-	-	5-10
Pythium graminicola	-	-	5
Pyrenophora tritici-repentis	20	-	-
Pyricularia oryzae	0.5	-	-
Rhizoctonia solani	15	-	-
Saccharomyces cerevisiae	-	-	2-5
Sclerotium rolfsii	-	-	>100
Sclerotinia sclerotiorum	-	-	5
Trichoderma viride	-	6	-
Venturia inaequalis	1	-	-
Verticillium alboatrum	-	6	-
V. dahliae	0.5	-	2
Bacteria			
Bacillus megaterium	2	-	-
B. subtilis	-	20	-
Clavibacter michiganensis	-	<10	-
Erwinia amylovora	-	>100	-
E. carotovora	>500	-	-
Escherichia coli	>500	>500	-
Micrococcus luteus	-	5	-
Proteus vulgaris	-	>500	-
Pseudomonas solanancearum	-	>100	-
Sarcina lutea	50	-	-
Staphylococcus aureus	-	20	-
Streptomyces faecalis	-	5	-
Xanthomonas campestris	-	6	-
X. oryzae	-	15	-

were germinated on agar medium supporting the growth of a fungal colony. They examined the edges of the expanding fungal colony as it approached the germinating seed to test for growth inhibition. As well, they tested the ability of germinating seeds to tolerate 1μg of the purified AMP. The positive controls for the bioassay included an endoprotease, to show abolition of the inhibition zones, autoclaved seeds and proteins, and abscissic acid which inhibited seed germination.

Whole plant assays for assessing the response of specific transgenic plants to fungal pathogens are simple. Further growth of the seedlings and extent of disease symptoms can be scored over a period of a week or two (Yang and Verma, 1992; Woytowich, 1997). Specifically, in these assays the transgenic plants are allowed to grow for usually three to four weeks at which time their leaves are inoculated with the spore suspension of the phytopathogenic fungus such as *A. longipes* or *R. solani*. The infected plants are further incubated and disease symptoms and severity are scored for seven days after inoculation. Conversely seedlings grown in sterile potting soil can also be assayed by their transfer to potting soil that has been inoculated with the fungal phytopathogen. Other assays using molecular tools are also used e.g., immunoblot analysis of proteins from several transgenic lines and *in vitro* tests for antifungal activity by quantitative microplate assay (Broekaert *et al.,* 1990). Further tests are employed to determine if antifungal gene products display the same levels of activity *in vivo* as they do *in vitro* bioassays. There is a need for standardization of protocols in this area of research.

Due to the small size of these peptides (3-6 kDa) either the peptides or the DNA can easily be sequenced, once purified. The nucleotide sequence can be deduced from a peptide sequence and deoxyoligunucleotide probes can be generated for isolating the corresponding gene(s). The final step in the construction of transgenic food crops containing AFP gene(s). Figure 1 shows one scheme that was employed for AMP-2 by Woytowich (1997).

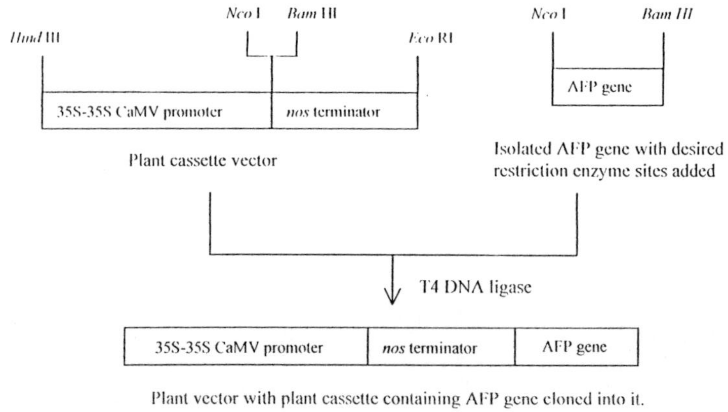

Figure 1. A scheme showing steps in the construction of a plant vector containing 35S CaMV promoter, nopaline synthase (nos) terminator and an AFP gene.

4. DEFENSINS

Plant defensins are peptides which are inhibitory to phytopathogenic filamentous fungi such as *A. brassicola, A. longipes, C. beticola, F. culmorum, L. maculans, N. crassa, P. tritici-repentis, P. oryzae,* and several bacterial pathogens of plants.

4.1. Genes and peptide structure

Some of the first plant AMPs to be isolated were found to be structurally related to insect and mammal defensins and therefore were named "plant defensins". Some of these have a high antifungal activity. These AFPs inhibit fungal growth by one of two mode of action: 1) those that act through morphological distortions of the fungal hyphae; and 2) those that inhibit fungal growth without morphological distortion (Broekaert *et al.*, 1997).

Defensins are highly cationic peptides, consist of 29 to 54 amino acids and nonglycosylated isolated from a number of different plants (Lehrer *et al.*, 1993; Evans and Harmon, 1995). Generally, these peptides contain six invariant cysteines and are arginine rich. In the case of insect defensins which contain three disulfide bridges, they are produced by the insect fat body and are secreted in to the hemolymph. Their secondary structure consists of three distinct domains: an amino-terminal loop, an amphipathic α-helix; and α-carboxyl-terminal, double-stranded, antiparallel ß-sheet (Bonmatin *et al.,* 1992). The three dimensional structure of plant defensins has also been determined. They contain a predominant set of disulfide bonds and b-sheet structures. Plant defensins are dominated by a triple-stranded, antiparallel ß-sheet and a single α-helix that lies in parallel with the ß-sheet (Bruix *et al.,* 1993). Mammalian defensins are produced by a number of specialized cells in the body and their structure consists of three antiparallel ß-strands stabilized by three disulfide bridges (Hill *et al.,* 1991). A highly basic peptide, Ib-AMP3 produced by *I. balsamina* which has four cysteine residues and two intramolecular disulfide bridges (Tailor *et al.*, 1997) has been reported to reduce the viability of germinated conidia of *Aspergillus flavus* or *Fusarium moniliforme.* It was effective on the non-germinated conidia *F. moniliforme* but not those of *A. flavus* (De Lucca and Walsh, 1999). Plant defensins are structurally related to insect and mammalian defensins except that mammalian defensins lack the cysteine-stabilized α-helix motif and that insect defensins lack the domain, which corresponds to the amino-terminal ß- strand of plant defensins.

The first plant defensins, originally termed γ -thionins, were isolated from rabbit wood, wheat and the barley (Vernon *et al.*, 1985; Bohlman *et al.*, 1988; Collila *et al.*, 1990; Mendez *et al.*, 1990). Since 1985 plant defensins have been isolated from more than a dozen plant species (Broekaert *et al.*, 1995b, 1997) and have been found in leaves, tubers, seeds, pods and flower organs. Defensins preferentially accumulate in the peripheral cell layer and in the case of healthy sugar beet leaves, in the xylem, stomatal cells and the cell walls lining the substomatal cavities (Kragh *et al.*, 1995; Terras *et al.*, 1995). The presence of defensins in these locations is consistent with their role in the prevention of invading microorganisms. Two defensins isoforms, Rs-AFP1 and Rs-AFP2, isolated from the seeds of radishes were the first plant defensins found to possess antifungal activity against the following fungi; *P. tritici-repentis, F. culmorum, P. oryzae* and *A. brassicola* (Terras *et al.,* 1992). Anomalous fungal growth has been reported for plant defensins *in vitro* by a purified radish thionin on *F. culmorum* (Terras *et al.,* 1995; 1996). Segura and co-workers (1998) found novel defensin peptides (So-D1-7) from crude cell wall preparations of spinach (*Spinacia oleracea* cv. Matador) leaves. Judging from their amino acid sequences, six (So-D2-7) represented a novel structural subfamily of plant defensins (group IV) which were preferentially distributed in the epidermal cell layer of leaves and in the sub epidermal region of stems. Group-IV defensins were also functionally distinct from those of groups I-III. These AMPs were active at concentrations < 20 μM against Gram-positive (*Clavibacter michiganensis*) and Gram-negative (*Ralstonia solanacearum*) bacterial pathogens, and fungi, such as *F. culmorum, F. solani, Bipolaris maydis,* and *Colletotrichum lagenarium*. Fungal inhibition occurred without hyphal branching.

The cDNAs encoding the pathogen-inducible Rs-AFP isoforms from radish leaves accumulate at high levels at or immediately around the infection sites in leaves inoculated with *A. brassicola*, as well as in healthy tissue further away from the infection sites and in non-infected leaves from injected plants (Terras *et al.,* 1998). The expression of Rs-AFP genes is systemically triggered upon fungal infection and is activated upon treatment with methyl jasmonate, ethylene, and paraquat. In comparison, in these infected plants the pathogenesis related, PR-1, gene expression is only activated locally and is not induced by external application of salicylic acid (Terras *et al.,* 1998).

From a morphological perspective, antifungal assays distinguish at least two groups of plant defensins: 1) the morphogenic plant defensins, isolated from the genus Brassicaceae and Saxifragaceae, which cause reduced hyphal elongation with a concomitant increase in branching; and 2) the non-morphogenic defensins, such as Dm-AMP from dahlia seeds, Ah-AMP1 from horse chestnut seeds and Ct-Amp1 from *Clitoria tematea* seeds which allow for slow hyphal extension, but do not cause marked morphological distortions (Terras *et al.,* 1992; Broekaert *et al.,* 1995b, Osborn *et al.,* 1995). Treatment of hyphae of *N. crassa* with antifungal plant defensins, Rs-AFP2 and Dm-AMP1 isolated from radish and dahlia seeds, respectively, induced a rapid K^+ efflux, Ca^{++} uptake, and alkalinization of the incubation medium (Thevissen *et al.,* 1996). Further studies on a defensin from the seeds of *Heuchera sanguinea* have shown that specific, high-affinity binding sites are present on *N. crassa* hyphae and microsomal membranes (Thevissen *et al.,* 1997). Binding is competitive, reversible, and saturable. A similarity in binding affinity was found between hyphae and microsomal membrane interactions indicating that binding sites reside on the plasma membrane. Thevissen *et al.,* (1997) suggested that the binding of *H. sanguinea* defensins to their receptor sites is linked to their antifungal effects. The most recent publication of this group indeed demonstrates that binding-site-mediated insertion of the defensin into fungal plasma membrane is the primary cause of growth inhibition induced by a number of plant defensins (Thevissen *et al.*, 1999).

The antifungal activity of both groups of defensins is affected by the presence of cations (Ca^{++}, Ba^{++} and Mg^{++}), especially divalent cations, in the fungal growth assay medium (Terras *et al.,* 1992; Terras *et al.,* 1993). In this regard, the activity of the mammalian "defensin", lactoferricin, an 18 amino acid long residue of an enzymatic product of lactoferrin, against *Candida albicans* is reduced when Ca^{++}, Fe^{++} and Mg^{++} are present (Yamauchi *et al.,* 1993). The antagonistic effect of these ions on the antifungal activity of defensins strongly depends on the type of defensin and the specific fungus (Osborn *et al.,* 1995). For example, it has been demonstrated that the defensin Hs-AFP1, isolated from *H. sanguinea,* inhibits *Fusarium culmorum* with an IC_{50} value of 3 μg/ml in the presence of 1mM $CaCl_2$ and 50mM KCl. De-Samblanx *et al.* (1996) synthesized a series of overlapping 15-mer peptides based on the amino acid sequence of Rs-AFP2. Peptides 6, 7, 8 and 9, comprising the region from cysteine 27 to cysteine 47 of Rs-AFP2 showed substantial antifungal activity against several fungal species (minimal inhibitory concentrations of 30-60 micrograms/mL), but no activity towards bacteria (except peptide 6 at 100 micrograms/mL). The active peptides were shown to be sensitive to the presence of cations in the medium and to the composition and pH of the medium. When present at a subinhibitory concentration (20 μg/mL), peptides 1, 7, 8 and 10 potentiated the activity of Rs-AFP2 from 2.3-fold to 2.8-fold. By mapping the characteristics of the active peptide by nuclear magnetic resonance, the active region of the antifungal protein appears to involve ß-strands 2 and 3 in combination with the loop connecting those strands. A cyclized synthetic mimic of the loop, cysteine 36 to cysteine 45, was shown to have antifungal activity. Substitution of tyrosine 38 by alanine in the cyclic peptide substantially reduced the antifungal activity, indicating the importance of this residue for the activity of Rs-AFP2 as demonstrated by mutational analysis. Later, De-Samblanx *et al.* (1997) used radish (*Raphanus sativus* L.) Rs-AFP2 gene to perform site directed mutagenesis by polymerase chain reaction. Several mutations generated peptide variants with either reduced or enhanced antifungal activity against *F. culmorum..* Substituting single amino acids by arginine yielded mutants with more activity than wild-type Rs-AFP2 in media with high ionic strength. When added to *F. culmorum* in a high ionic strength medium, Rs-AFP2 stimulated Ca^{++} uptake by up to 20-fold. An arginine substitution variant with enhanced antifungal activity caused increased Ca^{++} uptake by up to 50-fold, whereas a variant that was virtually devoid of antifungal activity did not stimulate Ca^{++} uptake. However, it was found not to be active against *Leptosphaeria maculans.* On the other hand, the horse chestnut defensin, Ah-AMP1 is active against *L. maculans* with an IC_{50} value of 6 μg/ml, but is inactive against *F. culmorum.* Depending on the source plant some defensins are also active against *Bacillus subtilis, Clavibacter michigansis* and *Pseudomonas solanacearum* (Osborn *et al.,* 1995; Moreno *et al.,* 1994).

4.2. Transgenic plants containing defensin-like peptides

The ability of plant defensins to control fungal pathogens in planta has been studied by generating transgenic plants carrying chimeric gene constructs. A chimeric Rs-AFP2 gene isolated from the seeds of radish (*Raphanus sativus*) was put under the control of the constitutive cauliflower mosaic virus 35S-promoter and transferred to tobacco plants (Terras *et al.,* 1995). Based on immunoblot analysis, it was found that Rs-AFP2 was expressed at levels ranging from 0.6-2.4 μg per mg of total leaf protein. The leaves of the Rs-AFP2-expressing plants were also found to display a decreased susceptibility to *A. longipes* infection when compared to untransformed tobacco plants. However, it was also determined that Rs-ARP2 expression in transgenic plants must reach a threshold level before displaying an increased disease tolerance phenotype. This result was consistent with the fact that when Terras *et al.* (1995) examined the purified seed Rs-AFP2 they also observed a sharp dose-dependent growth inhibition effect in their *in vitro* assays.

Recently a pea defensin gene has been isolated and transferred to *Brassica napus* to determine if its constitutive expression confers resistance to blackleg (*L.*

maculans) (Wang *et al.,* 1999). Transgenic lines containing a single-copy T-DNA insert of the defensin gene were screened for both the cotyledon- and adult plant-resistance and it was found that these lines showed slight enhancement of resistance when compared with non transgenic lines. As well, antifungal activity of extracts from defensin transgenic plants were also examined and it was demonstrated that they inhibit fungal spore germination *in vitro*.

Research has also been undertaken to study the promoter regions of defensin genes to determine if they are induced by the presence of plant pathogens. The promoter of the plant defensin, PDF1.2, which was previously shown to accumulate systemically via a salicylic acid-independent pathway in leaves of *Arabidopsis* plants upon challenge by fungal pathogens has been introduced into *Arabidopsis* C-24 plants (Manners *et al.,* 1998). By using the reporter gene ß-glucuronidase (GUS) Manners and co-workers (1998) found that when the transgenic plants were challenged with the pathogens *A. brassicicola* and *B. cinerea* local and systemic induction of the reporter gene occurred. As well the gene was induced upon treatment of the plants with jasmonates or the active oxygen generating compound paraquat. However, GUS, activity was absent upon wounding of the plants or treatment with salicylate and its functional analogues.

Finally, recent research on defensins and their possible resistance role in transgenic plants has not only focused on plant defensins, but also mammalian defensins. Fu *et al.* (1998) have isolated and transferred the rabbit defensin NP-1 gene to tobacco plants to determine if increased resistance could be obtained. Northern hybridization analysis of the transgenic plants revealed that the expression cassette was normally transcribed and when challenged with the bacterial wilt pathogen, *R. solanacearum,* partial resistance was observed. Further research on the antifungal properties of this protein still remains to be performed.

5. THIONINS

Thionins are inhibitory to single cell- and filamentous- fungi such as *S. cerevisiae* and *A. brassicola, F. culmorum, N. crassa,* and *Plasmodiophora brassicae,* respectively, and several bacterial pathogens of plants.

5.1. Genes and peptide structure

Thionins are a group of small, cysteine-rich, basic plant peptides of 45-47 amino acids in length and molecular weights of approximately 5000 Da (Bohlmann, 1994). Thionins are synthesized as preproteins and consist of a signal sequence, the thionin portion and an acidic domain. Thionins are separated into three distinct subgroups; (a) peptides possessing eight cysteines that are interconnected by four disulfide bridges, (b) peptides containing six cysteines which form three disulfide bridges, and (c) thionins that are truncated and lack a C-terminal non peptide portion (Castagnero *et al.,* 1995). The amino acid sequences of various thionins are highly divergent, although conserved residues include six cysteines (at positions 3, 4, 16, 27, 33, and 41), the aromatic residue (at position 13) and the arginine (at position 10) (Broekaert *et al.,* 1997). The three dimensional structures of subgroups (a) and (b) have been elucidated as a compact L- shaped molecule where the long arm is formed by two antiparallel α-helices and the short arm by a ß-sheet (Rao *et al.,* 1995; Stec *et al.,* 1995). As well, thionins contain clustered hydrophobic- and hydrophilic-amino acid residues at particular loci of the L- shaped molecule. Certain thionins contain a phospholipid binding site (Broeakert *et al.*, 1997).

Thionins are found in many mono- and di-cotyledonous plants and in their seed endosperm, roots, stems and in pathogen-stressed leaves (Bohlmann and Apel, 1991; Bohlmann, 1994). In barley two groups of thionins are found, those which are leaf specific and those which are endosperm-specific α- and ß-hordothionins (Bohlmann and Apel, 1987; Gausing, 1987). The thionins expressed in these different organs are encoded by different genes displaying organ-specific expression (Garcia-Olmedo *et al.*, 1992; Holtorf *et al.*, 1995). It has been estimated that the number of α-hordothionins genes does not exceed more than 2-4 per haploid genome (Rodriguez-Palenzuela *et al.*, 1988). However, it has been found that leaf thionins are encoded by a complex multigene family with approximately 50 genes per haploid genome (Bohlmann *et al.*, 1988; Holtorf *et al.*, 1995). Two possible explanations are offered for the large number of leaf thionin genes present in barley. First, it is possible that different genes may be specialized for expression in different tissues or for induction by specific external stimuli. Second, the large number of different genes could have arisen due to co-evolution with a genetically variable pathogen that would drive the variability of the protein coding region of the thionin gene (Bohlmann 1994). Doughty *et al.* (1998) reports that PCP-A1, a defensin-like *Brassica* pollen coat protein is also characterized by the presence of a structurally important motif consisting of eight cysteine residues shared by the plant defensins which binds the S locus glycoprotein (SLG), the product of gametophytic gene expression.

Plant thionin genes are regulated by many agents. Physical agents, e. g. light, strongly represses the accumulation of thionin transcripts in barley leaves, but not in meristemic cells of the leaves protected by the sheath of the leaf underneath (Reimann-Philipp *et al.*, 1989a). Among chemical agents, heavy metals (Fischer *et al.*, 1989), and jasmonic acid (Andresen *et al.*, 1992; Epple *et al.*, 1995) produce accumulation of thionin transcripts. Methyl jasmonate, a hormone-like compound accumulates in plants upon wounding, in the presence of microbial elicitors or in the presence of fungal pathogens (Bohlmann *et al.*, 1988, Creelman *et al.*, 1992; Mueller *et al.*, 1993; Boyd *et al.*, 1994, Penninckx *et al.*, 1996). It is uncertain, however, if these stimuli regulate the whole gene family in concert or only specific genes.

All thionins are synthesized as preproteins and the mature thionin domain is located centrally between the N-terminus signal sequence and the C-terminus. There are a number of special features of the prodomain that define specificity of thionins interactions (Broekaert, *et al.*, 1997). It is thought that the prodomain of the polypeptide may have one of two possible roles. Firstly, it may act as an intramolecular chaperone and assist the mature domain to form its desired structure during the folding process. Secondly, it may act to prevent the occurrence of certain harmful interactions between the domain and the components of intracellular trafficking machinery (Broekaert *et al.*, 1997). Kushmerick *et al.* (1998) examined the functional and structural features of γ--zeathionins, γ-1- and γ-2-zeathionins purified from maize seeds and found remarkable structural similarity to scorpion neurotoxins and insect defensins. Caldwell and co-worker's (1998) recently reported that the thermostable sweet tasting protein brazzein from the fruit of *Pentadiplara brazzeana* Baillon, resembles that of γ-thionins, and contains amino acid sequences similar to defensine serine protease inhibitors.

Immunocytochemical studies have located thionin's presence in vacuolar compartments 40-fold higher than in cell walls of leaf although similar isoform mixtures are present in each compartment (Reimann-Philipp *et al.,* 1989b). Those found in the cell walls appeared to be tightly bound by ionic interactions. Therefore, thionins present in the cell walls are a result of an overflow from the main trafficking route that leads to the vacuoles. Due to a plant's hypersensitive response or the action of necrosis-inducing toxins from the phytopathogen, vacuole's rupture leads to the release of thionins.

Using the whole-cell patch clamp technique Kushmerick *et al.* (1998) found that γ-1- and γ-2-zeathionins purified from maize seeds at 50μM concentration caused an inhibition of the sodium current in the GH3 cells (a cell line obtained from a rat pituitary tumor). Inhibitory effect was rapid and reversible without change in the kinetics or voltage dependent activation or inactivation phenomena. This is the first report of an AFP that inhibits the sodium channel. Finally, the antifungal activity of thionins can be enhanced when combined with rape 2S albumin or with barley trypsin inhibitor (Terras *et al.*, 1993a, 1993b).

5.2. Thionin containing transgenic plants

Barley thionin gene-containing transgenic tobacco plants have had mixed results. Enhanced resistance to specific bacterial pathogens *Pseudomonas syringae* pv. *tabaci* or *Pseudomonas syringae* pv. *syringae* was reported by Carmona *et al.* (1993). Florack and co-workers (1994) could not show differences in the resistance of transformed and untransformed tobacco plants confronted with *P. syringae* pv. *tabaci*. Florack *et al.* (1994) used two different gene constructs for the construction of transgenic tobacco plants. The first construct contained the barley thionin gene with the prodomain and the second construct lacked the prodomain. Upon examining the sub cellular targeting of the thionins in transgenic plants containing either construct, no differences were found, indicating that the prodomain is probably not involved in targeting the protein. When the expression levels of these different plants were assayed it was determined that those plants transformed with the construct lacking the prodomain had a tenfold lower level of thionin expression. This observation suggest a role for the prodomain in maturation of the protein.

Current attention is focused on a gene isolated from *A. thaliana* (Epple *et al.,* 1997a; Epple *et al.,* 1998). Epple and co-workers have found that the *Thi2.1* gene of *A. thaliana*, which is inducible by phytopathogenic fungi, enhances the resistance of the susceptible ecotype Columbia to *F. oxysporum* f. sp. *matthiolae* when constitutively overexpressed. The resulting transgenic lines supported significantly less growth for the pathogen on the cotyledons and lesser amounts of chlorophyll loss. Further, enhanced resistance is due to direct effect of the over expressed thionin gene comes from the observation that growth anomalies of the fungal hyphae on the transgenic plants. Epple *et al.* (1997b) employed random sequencing of expressed sequence tags (ESTs) in *A. thaliana* to show several defensin genes can be grouped into two subfamilies. One EST, Pdf2.3, is constitutively expressed in seedlings, rosettes, flowers, and siliques and is not inducible in seedlings either by methyl jasmonate, salicylate, ethephon, and silver nitrate or by several different phytopathogenic fungi. The expression of a second gene, Pdf1.2 is in untreated plants is only detectable in rosettes and in seedlings and is inducible by methyl jasmonate, silver nitrate, and different phytopathogenic fungi, notably *F. oxysporum*

f. sp. *matthiolae*. The regulation of Pdf1.2 resembles that of the pathogen-inducible thionin gene Thi2.1 (Epple *et al.*, 1997b).

More recently the viscotoxin A3 gene has been isolated from *Viscum album*, put under the control of the constitutive CaMV-omega promoter and transferred to *A. thaliana* C24 (Holtorf *et al.*, 1998). When transgenic lines displaying high viscotoxin A3 levels in all parts of the plant were tested for resistance to *Plasmodiophora brassicae,* the fungus responsible for club root, they were determined to be more resistance than the parental lines.

6. KNOTTIN-LIKE PEPTIDES

The name knottin was coined because of the knotted structure of a group of bioactive peptides which were reactive with serine proteases (Le Nguyen *et al.*, 1990). Knottine-like peptides have inhibitory effects against phytopathogenic filamentous fungus *A. longipes,* and the opportunistic fungus *T. hamatum*. These peptides also have antimicrobial activity against certain Gram-positive bacteria.

6.1. Genes and peptide structure

Some of the most extensively studied antifungal peptides *in vivo* and *in vitro* were isolated from the seeds of *M. jalapa* plants in the early 1990's (Cammue *et al.*, 1992). Two peptides, Mj-AMP1and Mj-AMP2, are knottin-like peptides that show strong antimicrobial activity against a wide range of phytopathogens and Gram-positive bacteria. Knottin proteins are characterized by a compact triple-stranded ß-sheet and a long loop connecting the first and second strands thus forming a knot-like fold. Both Mj-AMP1 and Mj-AMP2 are very small peptides, 37 and 36 amino acids, respectively that are highly basic. They each contain six cysteine residues and only differ from each other at four positions in their sequence. The antimicrobial activity of these two peptides *in vitro* indicates that their effectiveness are reversed in the presence of cations, especially divalent cations, above a concentration of 1mM (Cammue *et al.*, 1995). For example, it was observed that when the media contained $CaCl_2$ and KCl at concentrations of 1mM and 50mM, respectively, these peptides did not affect fungal growth at concentrations below 100 ul/ml. However, they are effective at concentrations below 10 μg/ml in low ionic strength media. Shao *et al.* (1999) have found a highly basic APF, termed PAFP from the seeds of the pokeweed. This 38 amino acid long peptide, with three disulfide bridges and a molecular mass of 3929 has significant similarity and the same cysteine motif with Mj-AMP. It has a broad spectrum of antifungal activity.

6.2. Knottin-like transgenic plants

Transgenic tobacco plants carrying the Mj-AMP2 gene show increased resistance to plant pathogens (DeBolle *et al.*, 1996). To study the processing, sorting and biological activity of this peptide, two different gene constructs, a wild-type and a vacuolar targeting signal sequence mutant Mj-AMP2 in tobacco plants were constructed. Analysis of the transgenic plants for processing and localization of these peptides indicated that plants transformed with the wild-type and the mutant constructs respectively accumulate Mj-AMP2 extracellularly and intracellularly. These results were consistent with the results of Reimann-Philipp *et al.* (1989) who had shown that thionins which are vacuole bound show higher levels of resistance as compared to those which are extracellular.

DeBolle and co-workers (1996) demonstrated that the antifungal activity of the different precursor AMP proteins purified from the transgenic tobacco plants was very similar to that of the authentic proteins isolated from *M. jalapa* seeds. When the transgenic plants were challenged *in vivo* with the plant pathogens *Botrytis cinerea* and *Alternaria longipes*, it was found that none of the plants showed enhanced resistance. Since knottin-like antimicrobial peptides of *M. jalapa* are extremely sensitive to the ionic strength of the bioassay medium, the presence of the same ions within plant cells could affect antifungal activity and limit their use in engineering resistant plants. Further research to increase levels of expression or functional properties of the AMPs is needed for the application of AMPs in this case.

7. HEVEIN-LIKE PEPTIDES

Hevein-like peptides have inhibitory effects against phytopathogenic filamentous fungi such as *A. longipes* and *B. cinerea*.

7.1. Genes and peptide structure

Hevein is a 43-residue peptide that can bind chitin and is one of the most abundant proteins in rubber tree latex (Walujono *et al.*, 1975). The three-dimensional structure of hevein consists of a triple-stranded b-sheet and a short single turn α-helix connecting the second ß-strand to the third (Andersen *et al.*, 1993). Although hevein has weak antifungal properties, other proteins homologous to hevein, are much more potent in their inhibitory properties. Hevein-like AMPs affecting growth of several plant pathogens have been isolated from the seeds of amaranth and sweet pepper (DeBolle *et al.*, 1993; Broekaert *et al.*, 1994). The peptide isolated from sweet pepper contains eight cysteines residues that are linked by disulfide bonds while the peptides from amaranth contain six. Examination of one of the amaranth cDNA clones revealed that the protein has three distinct domains: 1) an amino-terminal putative signal peptide; 2) a domain that corresponds to the mature peptide; and 3) a carboxyl-terminal propeptide containing a putative N-glycosylation site (DeBolle *et al.*, 1993). Unlike hevein and the sweet pepper AMP the carboxyl-terminal of the amaranth protein lacks the portion of amino acid sequence that contains the seventh and eighth cysteine residues. More recently, two hevein-like peptides, Pn-AMP1 and Pn-AMP2, with 41 and 40 amino acids in length, respectively, were purified from the seeds of *Pharbitis nil* (Koo *et al.*, 1998). These highly basic peptides (pI 12.02) with their characteristics of cysteine/glycine rich chitin-binding domains have strong antifungal activities. Assays indicated that these peptides possess potent antifungal activity. Pn-AMPs penetrate very rapidly into the fungal hyphae and localize at septa and hyphal tips bringing about fungal bursting, resulting in disruption of the fungal membrane and leakage of the cytoplasmic materials. Although Pn-AMP1 and Pn-AMP2 are hevein-like, Koo *et al.* (1998) show their fungicidal effects to be similar to thionins.

7.2. Transgenic plants with hevein- like peptides

In the case of hevein-like proteins, the *A. caudatus* cDNA clone Ac-AMP2 has been transferred to tobacco and its antimicrobial activity *in vitro* and *in vivo* has been studied (DeBolle *et al.*, 1996). As with the case of the cDNA clone from *M. jalapa,* when the Ac-AMP2 gene was transferred to tobacco, the resulting transgenic plants did not show enhanced resistance to the pathogenic fungi, *A. longipes* or *B. cinerea*. Once again the purified transgenic AMP protein showed antifungal activity *in vitro* only after passing through a reversed-phase column. It has been suggested that as in the case of knottin-like proteins the antimicrobial activity is affected by the presence of cations in the bioassay medium or the plant cell.

8. CONCLUSIONS

In general, AMPs and AFPs pose interesting challenges and opportunities for biotechnology. These peptides while only recent discoveries, evolved to a limited extent so as to conserve their defensive role for various species against pathogenic and predatory organisms. Charlet *et al.* (1996) discovered several cysteine-rich antimicrobial peptides and a novel 6.2-kDa antifungal peptide containing 12 cysteines from the blood of a mollusk, *Mytilus edulis*. Charlet *et al.* (1996) suggest that molluscan and arthropod defensins have a common ancestry which predated the separation between molluscs and arthropods at the root of the Cambrian, about 545 million years ago. The study of Landon *et al.* (1997) for three-dimensional structure study of 1H 2D NMR, in aqueous solution for drosomycin, an inducible antifungal protein of *Drosophila melanogaster* t, shows that α-helix and a twisted three-stranded ß-sheet, is stabilized by three disulfide bridges. From comparative study of the corresponding Cysteine Stabilized α-ß motif, also found in insect defensin A, short- and long-chain scorpion toxins, plant thionins and potent plant AFP defensins, Landon *et al.* (1997) commented that these genes appear remarkably persistent along the evolutionary time frame.

The last 15 years of research on antifungal peptides has made it very clear that small AMPs and AFPs are useful materials for developing plants with superior resistance towards a wide variety of plant pathogens (Salmeron and Vernooi, 1998). The necessary prerequisite, however, is ease of transformation of particular crop plants with a given antifungal gene(s) and demonstration of the highest levels of resistance in field conditions. As illustrated in this chapter, all plant AMPs characterized to date contains multiple disulfide-bridges, yet various levels of resistance observed in transgenic plants depend on the specific peptide tested. In some cases peptides that showed high levels of antimicrobial activity *in vitro* did not necessarily show any resistance upon transfer of the AMP-AFP gene(s) to the genome of plants. Therefore, some aspects of the acquisition and expression of resistance to specific plant pathogens is not understood. Likewise, research in the area of transferring mammalian and insect antimicrobial genes to plants is very limited. Some of these peptides are disulfide-linked while others, such as those belonging to the magainin family, are classified as linear peptides. Since improper folding and disulfide bond formation of disulfide-linked proteins in planta may play a role in the levels of resistance obtained, it may be equally useful to pursue the use of linear AMPs. There is no doubt that solving the mysteries of inconsistent results observed by several research groups using different antimicrobial genes for varying levels of resistance peptides in plants remains a challenge.

One hopeful avenue of pursuit comes from an approach taken by Smith and co-workers (1998). During research on production of a filamentous phage-fd display libraries for cellulose binding domain of cellohydrolase I of *Trichoderma reesei*, Smith *et al.* (1998) found a number of small knottin like peptides. These peptides were selected for binding to cellulose or to one of three enzymes, α- amylase, alkaline phosphatase and β-glucouronidase. Because this system could be used as a combinatorial repertoire of knottins displayed on a bacteriophage, it should have great potential for *in vitro* mutational improvement of AFPs structural and functional diversity and plant specificity.

A number of additional questions such as; 1) should the expression of the peptides be constitutive or tissue-specific, and 2) should the peptides be secreted or located intracellularly, must be posed and hopefully answered. The stability and the ability of the peptides to undergo correct folding in planta must also be examined. The clarity of our understanding of the mechanisms of inter-action of AMPs and their target fungal pathogens, as a recent commentary indicates, needs further research before one could determine their potential as practical phytoprotectants (Osbourn, 1999). Finally, because of the rise in yeasts and fungi resistant to conventional antimycotic drugs, new research on AMPs and AFPs and their development for animal therapy should produce clinically useful and important alternatives to current antifungal drugs (De-Lucca and Walsh, 1999; Hancock and Lehrer, 1998; Hancock and Chapple, 1999). By using the tools of biotechnology we can understand the mechanisms and applications of these molecules in applied mycology. In due course it is hoped that natural AMPs and AFPs will play a key role in the production of agricultural plants destined for foods.

REFERENCES

Andresen, I., W. Becker, K. Schluter, J. Burges, B. Parthier and K. Apel, Plant Mol. Biol. 19 (1992) 193.

Andresen, NH., B. Cao, A. Rodriguez-Romero and B. Arreguin, Biochemistry 32 (1993)1407.

Bohl, S. and K. Apel, Plant J. 3 (1993) 887.

Bohlmann, H. Crit. Rev. Plant. Sci. 13(1994) 1.

Bohlmann, H. and K. Apel, Mol. Gen. Genet. 207 (1987) 446.

Bohlmann, H. and K. Apel, Annu. Rev. Plant Physiol. Plant Mol. Biol. 42(1991) 227.

Bohlmann, H., S. Clausen, S. Behnke, H. Giese, C. Hiller, G. Schrader, V. Barkholt and K. Apel, EMBO J. 7 (1988) 1559.

Bonmatin, J.M., J.L. Bonnat, X. Gallet, F. Vovelle, M. Ptak, J.M. Reichhart, J.A. Hoffman, E. Keppi, M. Legrain and T. Achstetter, J. Biomol. NMR 2 (1992) 235.

Boyd, L.A., P.H. Smith, R.M. Green and J.K.M. Brown, Mol. Plant-Microbiol. Interact. 7 (1994) 401.

Broekaert, W.F., W. Marien, F.R.G. Terras, M.F.C. De Bolle, P. Proost, J. Van Damme, L. Dillen, M. Claeys, S.B. Rees, J. Vanderleyden and B.P.A. Cammue, Biochemistry, 31 (1992) 4308.

Broekaert, W.F., F.R.G. Terras, B.P.A. Cammue and J. Vanderleyden, Lett. FEMS 69 (1992) 55.

Broekaert, W.F., B.P.A. Cammue, M.F.C. De Bolle, K. Thevissen, G.W. De Samblanx and R.W. Osborn, Crit. Rev. Plant Sci. 16 (1995a) 297.

Broekaert, W.F., F.R.G. Terras, B.P.A. Cammue and R.W. Osborn, Plant Physiol. 108 (1995b) 1353.

Broglie, K., I. Chet, M. Holliday, R. Cressman, P. Biddle, S. Knowlton, C.J. Mauvais and R. Broglie, Science 254 (1991) 1194.
Bruix, M., M.A. Jimenez, J. Santoror, C. Gonzales, F.J. Colilla, E. Mendez, and M. Rico, Biochemistry 32 (1993) 715.
Caldwell, J.E., F. Abildgaard, Z. Dzakula, D. Ming, G. Hellekant and J.L. Markley, Nat. Struct. Biol. 5 (1998) 427.
Cammue, B.P.A, M.F.C. De Bolle, F.R.G. Terras, P. Proost, J. Van Damme, S.B. Rees, J.C. Vanderleyden and W.F. Broekaert, J. Biol. Chem. 267 (1992) 2228.
Carmona, M.J.M., A. Molina, J.A. Fernandez, J.J. Lopez-Fando and F. Garcia-Olmendo, Plant J. 3 (1993) 457.
Castagnero, A., A. Segura and F. Garcia-Olmedo, Plant Physiol. 107 (1995) 1475.
Caaveiro, J.M.M., A. Molina, M.J.M. Gonzalez, P.P. Rodriguez and F. Garcia-Olmedo, FEBS Lett. 410 (1997) 338.
Charlet, M., S. Chernysh, H. Philippe, C. Hetru, J.A. Hoffmann, and P. Bulet, J. Biol. Chem. 271 (1996) 21808.
Clore, G.M., M. Nilges, D.K. Sukumaran, A.T. Brunger, M. Karplus and A.M. Gronenborn, EMBO J. 5 (1986) 2729.
Colilla, F.J., A. Rocher and E. Mendez., FEBS Lett. 270 (1990) 191.
Creelman, R.A., M.L. Tierney, and J.E. Mullet, Proc. Natl. Acad. Sci. USA 89 (1992) 4938.
DeBolle, M.F.C., K. Eggermont, R.E. Duncan, R.W. Osborn, F.R.G. Terras and W.F. Broekaert, Plant Mol. Biol. 28 (1995) 713.
DeBolle, M.F.C., R.W. Osborn, I.J. Goderis, L. Noe, D. Acland, C.A. Hart, S. Torrekens, F. Van Leuven and W.F. Broekaert, Plant Mol. Biol. 31 (1996) 993.
De-Lucca, A.J. and T.J. Walsh, Antimicrob. Agents. Chemother. 43 (1999) 1.
De-Samblanx, G.W., A. Fernandez-del-Carmen, L. Sijtsma, H.H. Plasman, W.M. Schaaper, G.A., G.A. Posthuma, F. Fant, R.H. Meloen, W.F. Broekaert and A. van-Amerongen, Pept. Res. 9 (1996) 262.
De-Samblanx, G.W., I.J. Goderis, K. Thevissen, R. Raemaekers, F. Fant, F. Borremans, D.P. Acland, R.W. Osborn, S Patel and W.F. Broekaert, J. Biol. Chem. 272 (1997) 1171.
Doughty, J., S. Dixon, S.J. Hiscock, A.C. Willis, I.A. Parkin and H.G. Dickinson, Plant-Cell. 10 (1998) 1333.
Elsbach, P. Trends Biotechnol. 8 (1990) 26.
Epple, P., K. Apel and H. Bohlmann, Plant Physiol. 109 (1995) 813.
Epple, P., K. Apel and H. Bohlmann, Plant Cell 9 (1997a) 509.
Epple, P., K. Apel and H. Bohlmann (1997b) FEBS-Lett. 400, 168.
Epple, P., A. Vignutelli, K. Apel and H. Bohlmann, Mol. Plant Microb. Interact. 11 (1998) 523.
Evans E.W. and B.G. Harmon, Vet. Clin. Pathol. 24 (1995) 109.
Fink, J., A. Boman, H.G. Boman and R.B. Merrifield, Int. J. Peptide Protein Res. 33 (1989) 412.
Florack, D.E.A., W.G. Dirkse, B. Visser, F. Heidekamp and W.J. Stiekema, Plant Mol. Biol. 24 (1994) 83.
Frank, R., R. Gennaro, K. Schneider, M. Przybylski and D. Romeo, J. Biol. Chem. 265 (1990) 18871.
Froy D. and M. Gravitz, FASEB J. 12 (1998) 1793.
Fu, R.Z., Y.F. Peng, G.C.C. Cao, J.S. Ma, C.X. Chen, L. M. Zhang, W.B. Li and Y.R. Sun, Chinese Sci. Bull. 43 (1998) 1544.

Garcia-Olmedo, F., G. Salcedo, R. Sanchez-Monge, C. Hernandez-Lucas, M.J. Carmona, J.J. Lopez-Fando, J.A. Fernandez, L. Gomez, J. Royo, F. Garcia-Maroto, A. Castagnaro and P. Carbonero (1992) in P.R. Shewry (ed) Barley, Genetics, Biochemistry, Molecular Biology and Biotechnology CAB International, Walllingford, p. 335-350.

Gausing, K. Planta 171 (1987) 241.

Gincel, E., J.P. Simorre, A. Caille, D. Marion, M. Ptak and F. Vovelle, Eur. J. Biochem. 226 (1994) 413.

Hancock, R.E.W. and R. Lehrer, Trends Biotechnol. 16 (1998) 82.

Hancock, R.E.W. and D.S. Chapple, Antimicrob. Ag. Chemother. 43 (1999) 1317.

Hill et al., Science 251 (1991) 1481.

Hoffmann, J.A. and C. Hetru, Immunol. Today 13 (1992) 411.

Holtorf, S., K. Apel and H. Bohlmann, Plant Mol. Biol. 29 (1995) 673.

Holtorf, S., J. Ludwig-Muller, K. Apel and H. Bohlmann, Plant Mol. Biol. 36 (1998) 673.

Jach, G., S. Logemann, G. Wolf, A. Oppenheim, I. Chet, J. Schell and J. Logemann Biopractice, 1 (1992) 1.

Kader, J.C. M. Julienne and C. Vergnolle, Eur. J. Biochem. 139 (1984) 411-.

Khachatourians, G.G. (1991) In, Handbook of applied mycology, Arora, D.K., L. Ajello, and K.G. Mukerji, Marcel Decker Inc, New York, vol 2, 613-661.

Khachatourians, G.G. (1996) In The Mycota, , Ed. D.H. Howard and J.D. Miller. Springer-Verlag, Berlin. vol 6, 331-363.

Koltunow, A.M., J. Truettner, K.H. Cox, M. Wallroth and R.B. Goldberg, Plant Cell 2 (1990) 1201.

Koo, J.C., S.Y. Lee, H.J. Chun, Y.H. Cheong, J.S. Choi, S. Kawabata, M. Miyagi, S. Tsunasawa, K.S. Ha and D.W. Bae, Biochim. Biophys. Acta. 1382 (1998) 80.

Kragh, K.M., J.E. Nielsen, K.K. Nielsen, S. Dreboldt and J.D. Mikkelsen, Mol. Plant Microbe Interact. 8 (1995) 424.

Kushmerick, C., M. de-Souza-Castro, J. Santos-Cruz, C. Bloch Jr, P. S. Beirao, FEBS-Lett. 440 (1998) 302.

Landon. C., P. Sodano, C. Hetru, J. Hoffmann and M. Ptak, Protein-Sci. 6 (1997) 1878.

Last, D.I. and D.J. Llewellyn, New Zealand J. Bot. 35 (1997) 385.

Lehrer, R.I., A.K. Lichtenstein and T. Ganz, Annu. Rev. Immunol. 11 (1993) 105.

Le Nguyen, A. Heitz, L. Chiche, B. Castro, R.A. Biogegrain, R.A. Favel, and M.A. Coletti-Previero, Biochimie 72 (1990) 431.

Logemann, J., G. Jach, H. Tommerup, J. Mundy and J. Schell, Biotechnology, 10. (1992) 305.

Manners, J.M., I.A.M.A. Penninckx, K. Vermaere, K. Kazan, R.L. Brown, A. Morgan, D.J. Maclean, M.D. Curtis, B.P.A. Cammue and W.F. Broekaert, Plant Mol. Biol. 38 (1998) 1071.

Marcus, J.P., K.C. Goulter, J.L. Green, S.J. Harrison and J.M. Manners, Eur. J. Biochem. 244 (1997) 743.

Matsuzaki, K., O. Murase, N. Fujii and K. Miyajima, Biochemistry 34 (1995) 6521.

Mendez, E., A. Moreno, F. Colilla, F. Pelaez, G.G. Limas, R. Mendez, F. Soriano, M. Salinas and C. DeHaro, Eur. J. Biochem. 194 (1990) 533.

Molina, A. and F. Garcia-Olmedo, Plant J. 4 (1993) 983.

Moreno, M., A. Segura and F. Garcia-Olmedo, Eur. J. Biochem. 223 (1994) 135.

Morikawa, N., K. Hagiwara and T. Nakajima, Biochem. Biophys. Res. Comm. 189 (1992) 184.
Mueller M.J., W. Brodschelm, E. Spannagl and M.H. Zenk, Proc . Natl. Sci. USA 90 (1993) 7490-.
Nissen-Meyer, J. and I.F. Ness , Arch. Microbiol. 167(1997) 67.
Osbourn, A.E. Fung. Genet. Biol. 26(1999) 163.
Osborn, R.W., G.W. De Samblanx, K. Thevissen, I. Goderis, S. Torrekens, F. Van Leuven, S. Attenborough, S.B. Rees and W.F. Broekaert, FEBS Lett. 368 (1995) 257.
Penninckx, I.A.M.A., K. Eggermont, F.R.G. Terras, B.P.H.J. Thomma, G.W. De Samblanx, A. Buchala, J.P. Metraux, J.M. Manners and W.F. Broekkaert, Plant Cell 8 (1996) 2309.
Ponz, F., J. Paz-Ares, C. Hernandez-Lucas, F. Garcia-Olmedo and P. Carbonero, Eur. J. Biochem. 156 (1986) 131.
Rao, U., B. Stec and M.M. Teeter, Acta. Crystallogr. D. 51 (1995) 904.
Reimann-Philipp, U., S. Behnke, A. Batschauer, E. Schafer and K. Apel, Eur. J. Biochem. 182 (1989a) 283.
Reimann-Philipp, U., G. Schrader, E. Martinoia, V. Barkholt and K. Apel, J. Biol. Chem. 264 (1989b) 8978.
Richardson, M., S. Valdes-Rodriguez and A. Blanco-Labra, Nature 327 (1987) 432.
Roberts, W.K. and C.P. Selitrennikoff, J. Gen. Microbiol. 136 (1991) 1771.
Rodriguez-Palenzuela, P., J-A. Pintor-Toro, P. Carbonero and F. Garcia-Olmedo, Gene 70 (1988) 271.
Ryan, C.A. and G. Pearce, Annu. Rev. Cell Dev. Biol. 14 (1998) 1.
Salmeron, J.M. and J.B. Vernooi, Curr. Opin. Plant Biol. 1 (1998) 347.
Schrader-Fischer, G. and K. Apel, Plant Mol. Biol. 23 (1993) 1233.
Schrader-Fischer, G. and K. Apel, Mol. Gen. Genet. 245 (1994) 380.
Segura, A., M. Morina and F. Gracia-Olmendo, FEBS Lett. 332 (1993) 243.
Segura, A., M. Morina, A. Molina and F. Gracia-Olmendo, FEBS-Lett. 435 (1998) 159.
Shao, F., Z. Hu, Y.M. Xiong, Q.Z. Huang, C.G. Wang, R.H. Zhu, and D.C. Wang, Biochim. Biophys. Acta 1430 (1999) 262.
Sharma, P. and A. Lonnenberg, Plant Mol. Biol. 31 (1996) 702.
Shewry, P. R. and J. A. Lucas, Adv. Bot. Res. 26 (1997) 135.
Stec B., U. Rao, M.M. Teeter, Acta Crystallogr. 51 (1995) 914.
Tailor, R.H., D.P. Acland, S. Attenborough, B.P.S. Cammue, I.J. Evans, R.W. Osborn, J.A. Ray, S.B. Rees, and W.F. Broekaert, J. Biol. Chem. 272 (1997) 24480.
Teeter M. M., X-Q. Ma, U. Rao and M. Whitolow, Proteins, Structure Funct. Genet. 8 (1990) 118.
Terras, F.R.G., I.J. Goderis, F. Van Leuven, J. Vanderleyden, B.P.A. Cammue and W.F. Broekaert, Plant Physiol. 100 (1992a) 1055.
Terras, F.R.G., H. Schoofs, M.F.C. De Bolle, F. Van Leuven, S.B. Rrees, J. Vanderleyden, B.P.A. Cammue and W.F. Broekaert, J. Biol. Chem. 267 (1992b) 15301.
Terras, F.R.G., S. Torrekens, F. Van Leuven, R.W. Osborn, J. Vanderleyden, B.P.A. Cammue and W.F. Broekaert, Plant Physiol. 103 (1993a) 1311.
Terras, F.R.G., S. Torrekens, F. Van Leuven, R.W. Osborn, J. Vanderleyden, B.P.A. Cammue and W.F. Broekaert, FEBS Lett. 316 (1993b) 233.

Terras, F.R.G., K. Eggermont, V. Kovaleva, N.V. Raikhel, R.W. Osborn, A. Kester, S.B. Rees, J. Vanderleyden, B.P.A. Cammue and W.F. Broekaert, Plant Cell 7 (1995) 573.
Terras, F.R.G., S. Torrekens, F. Van Leuven and W.F. Broekaert, Plant Physiol. Biochem. 34 (1996) 599.
Terras, F.R.G., I.A. Penninckx, I.J. Goderis, and W.F. Broekaert, Planta 206. (1998) 117.
Thevissen, K., A. Ghazi, G.W. De-Samblanx, C. Brownlee, R.W. Osborn, and W.F. Broekaert, J. Biol. Chem. 271 (1996) 15018.
Thevissen, K., Terras, F.R.G. and W.F. Broekaert, Appl. Environ. Microbiol. 65 (1999) 5451.
Thevissen, K., A. Ghazi, G.W. De-Samblanx, C. Brownlee, R.W. Osborn, and W.F. Thoma, S., Y. Kaneko and C. Sommerville, Plant J. 3 (1993) 427.
Vernon L.P., G.E. Evett, R.D. Zeikus and W.R. Gray, Arch. Biochem. Biopys. 238 (1985) 18.
Vigers, A.J., Roberts, W.K. and C.P. Selitrennikoff, Mol. Plant-Microbe Interact.4 (1991) 315.
Walujono K., R.A. Scholma, J.J. Beintema, A.M. Arono and A.M. Hahn, Proc. Int. Rubber Conf. Kuala Lumpur 2 (1975) 518.
Wang, Y.P., G. Nowak, D. Culley, L.A. Hadwiger and B. Fristensky, Mol. Plant-Microbiol. Interact. 12 (1999) 410.
Whipps, J. M. Adv. Bot. Res. 26 (1997) 1.
Woytowich, A.E. (1997) Ph.D. Thesis, University of Saskatchewan, Saskatoon, Canada.
Yamauchi, K., M. Tomita, T. J. Gierhl and R.T. Ellison, Infect. Immun. 61 (1993) 719.
Yang, J. and P.R. Verma, Crop Protect. 11 (1992) 443.
Yeaman, M.R.Clin. Infect. Dis. 25 (1997) 951.
Xiong, Y-Q., M.R. Yeaman, and A.S. Bayer, Antimicrob. Ag. Chemother. 43 (1999) 1111.

Applied Mycology and Biotechnology
Volume 1. Agriculture and Food Production
G.G. Khachatourians and D.K. Arora (editors)
2001 Elsevier Science B.V.

Clustered Metabolic Pathway Genes in Filamentous Fungi

J. W. Cary, P.-K. Chang and D. Bhatnagar*

Food and Feed Safety Research Unit, U.S. Department of Agriculture, Agricultural Research Service, Southern Regional Research Center, New Orleans, Louisiana 70124 USA

1. INTRODUCTION

Clustering of genes can be defined as a group of two or more genes demonstrating a close linkage in the genome that contributes to a common metabolic or developmental pathway. This phenomenon is more commonly observed in prokaryotes. But, relatively recently (since 1989) clusters of genes governing specific metabolic pathways have been reported in eukaryotes, particularly from filamentous fungi with the discovery of a cluster of genes governing proline catabolism (1, 2). Preliminary evidence of clustering of genes for some metabolic pathways in *A. nidulans* was subsequently reported (3), without complete characterization of specific pathways.

Research in the last decade on fungal gene clusters has centered around defining the genetics of primary and secondary metabolism. Keller and Hohn (1) in their review of metabolic pathway gene clusters in filamentous fungi discussed the presence of numerous pathways in these organisms, which carried out "dispensable" metabolic functions. They used the term "dispensable metabolic pathways" for those pathways "that either are not required for growth or are only required for growth under a limited range of conditions." Keller and Hohn (1) divided their review on the basis of two major types of "dispensable metabolic pathways," (1) catabolic pathways describing the utilization of low-molecular-weight nutrients (nutrient utilization pathways) and biosynthetic pathways for other low-molecular-weight compounds (natural product pathways). In this review, we have expanded their discussion of the literature to include an update of the information on the earlier described "dispensable" pathways, but have also included several additional fungal metabolic pathways. We have, consequently, divided these pathways into three categories (Table 1): (a) nutrient utilization pathways such as nitrogen and carbon utilization or utilization of proline or quinate; (b) secondary metabolite pathways including antibiotics, mycotoxins, etc.; and (c) host-specific toxin biosynthetic pathways. All

*Corresponding author (email: dbhatnagar@nola.srrc.usda.gov)

Table 1
Metabolic pathway gene clusters in filamentous fungi[a]

Pathways	Approx. Cluster size (kb)	No. of ORFs[c]	Genetic composition of the cluster[b] Regulatory genes	Structural genes	Transport genes
1. Nutrient utilization					
Proline	13	5	1	2	1
Ethanol	15	7	1	1	0
Nitrate	12	7	1	2	1
Sugar	4	3-4	1	2	1
Quinate	17	7	2	3	1
2. Secondary metabolite synthesis					
Aflatoxins	75	23	1	21	1
Sterigmatocystin	60	25	1	14	0
Trichothecene	25	10	1	6	1
Penicillin	15	3	0	3	0
Cephalosporin					
Early	15	2	0	2	0
Late	5	2	0	2	0
Gibberellins	18	6	0	6	0
Melanin	30	3	0	3	0
Fumonisins	20	9	ND[d]	2	ND
Paxilline	60	ND	ND	1	ND
3. Host-specific toxin synthesis					
HC-toxin	550	7	1	5	1
AK-Toxin	8	2	0	2	0
CPS1	35	13	1	3	ND

[a]Modified from Keller and Hohn (1)
[b]Gene function determined or based on identity to other known genes
[c]ORFs (open reading frames)
[d]ND, not determined

of these pathways consist of a group of genes that are found to be clustered in one or several fungal species. Even though several of these pathway clusters are not completely elucidated, a few common features are emerging. For example, in addition to structural genes, each of the pathway clusters seems to have its own pathway-specific, positive-acting or negative-acting regulatory gene(s) (4, 5, 6), as well as a gene for transport of the final product of the pathway or some other critical metabolite of the process (Table 1).

In the gene clusters, where these features have not been reported, it is unclear if these do not exist on the cluster and are not closely linked, or if they have not been discovered so far. There is an increasing body of evidence suggesting that regulation of secondary metabolite pathways may be tied into the regulation of other pathways, for example, in some cases metabolite synthesis pathways have been shown to be controlled by regulatory elements governing fungal development (7, 8, 9).

2. NUTRIENT UTILIZATION PATHWAYS

2.1. Proline utilization

The proline utilization pathways, which convert proline to glutamate, of *Aspergillus nidulans* and *Saccharomyces* sp. are biochemically identical (10). However *A. nidulans* can use proline as a source of nitrogen and carbon, but *S. cerevisiae* can use proline only as a nitrogen source. Arst and MacDonald (11) first identified the proline utilization gene cluster in *A. nidulans*. All the genes responsible for converting proline to glutamate are clustered in a 13-kb region on chromosome VII (2, 4). The proline gene cluster has the gene order of *prnA-prnX-prnD-prnB-prnC* (11); *prnB, prnD,* and *prnC* encode a proline-specific permease, a proline oxidase, and a Δ^1-pyrroline-5-carboxylate dehydrogenase, respectively.

In fungal mycelia, the expression of the *prn* structural genes is co-regulated in response to proline induction as well as to nitrogen metabolite repression and carbon catabolite repression (12). Thus, the expression of *prn* genes is fully repressed by the simultaneous presence of ammonium and glucose. To use proline, cells must activate the expression of the *prn* genes that encode the proline permease and the two catabolic enzymes. The *prnA* gene encodes a transcription factor that mediates the expression of other *prn* genes (13), including *prnX* (4), a gene located between *prnA* and *prnD* but whose function is not yet known. However, unlike other pathway-specific regulatory genes, the transcription of *prnA* appears to be neither self-regulated nor affected by nitrogen metabolite repression or carbon catabolite repression (14). In *A. nidulans*, ammonium and glucose repression of the *prn* genes are through the action of AreA, a positive-acting Cys2-type transcription factor belonging to the GATA family (5), and CreA, a negative-acting Cys2/His2-type zinc-finger protein (15). The prime target of nitrogen metabolite repression and carbon catabolite repression is *prnB*, but *prnD* and *prnA* are also repressed. There are seven CreA-binding sites in the *prnD-prnB* intergenic promoter region; two are necessary for CreA-mediated repression of the *prn* genes (16).

Two models with regard to how CreA and AreA regulate *prn* gene expression have been proposed. One is the AreA/CreA competition model, which assumes that the function of AreA is to prevent the action of CreA. However, available genetic and physiological evidence argues against such a model (2). Another model proposed by Scazzocchio (4) involves an enhancer element which can by-pass the need for AreA in the absence of CreA. When this element is silenced by CreA, then AreA becomes absolutely necessary for the full expression of *prn* genes. Gonzalez *et al.* (17) has identified a 290 bp region in the *prnD-prnB* intergenic region that is essential for the integration of nitrogen metabolite repression and carbon catabolite repression. They called it the ADA element (for absolute dependence on AreA). ADA most probably contains a binding site for an unidentified positive-acting transcription factor.

2.2. Ethanol utilization

Ethanol utilization in *A. nidulans* requires two enzymes: alcohol dehydrogenase I (ADH1) and aldehyde dehydrogenase (AldDH) which are encoded by *alcA* and *aldA*, respectively (18). Ethanol is oxidized to acetate via acetaldehyde. Although ADHI and AldDH are co-induced by acetaldehyde, *aldA* which is located on chromosome VIII (19) is not linked to *alcA*. But, the ethanol utilization regulon which consists of *alcA*, the regulatory *alcR* gene and closely linked genes of unknown function designated *alcM*, *alcS*, *alcO*, *alcP*, and *alcU*, is on chromosome VII (20).

The AlcR protein, a Cys6-type transcription factor, induces the expression of *alcA* but not the unlinked *aldA* (see below). And *alcM*, *alcS*, *alcO*, *alcP*, and *alcU*, although clustered, are differentially regulated by AlcR (20). AlcR binds *in vitro* to a single consensus half-site, 5'-CCGCN-3', with high affinity (21). However, for AlcR sites to be functional *in vivo*, they have to be organized as inverted or direct repeats. In addition, a strong synergistic activation of *alcA* transcription by AlcR requires that all AlcR-binding sites, the three clustered repeats, are present (22).

The *alcR* gene itself is inducible, and the inducibility depends on the expression of the *alcR* gene itself (23). The *aldA* gene, although with a AlcR half-site in its promoter (21), is subjected only to the control of the CreA repressor (24). On the other hand, *alcM*, *alcS*, *alcO*, *alcP*, and *alcU* are not all under the direct control by CreA (20).

Ethanol strongly induces the transcription of the *alc* genes, whereas glucose totally represses the transcription of *alcR*. As a consequence, *alcA* is not expressed. The CreA repressor is the sole determinant responsible for carbon catabolite repression of the *alc* genes. CreA acts both on the regulatory gene *alcR* and directly on *alcA* and *aldA* (25). Among the seven putative CreA sites identified in the *alcA* promoter region, two different divergent CreA sites, of which one overlaps a functional AlcR inverted repeat site, are largely responsible for repression of *alcR*/*alcA* expression. The removal of one of the CreA targets results in a 50% derepression of the *alcR* gene (26). Totally derepressed *alcA* expression is achieved when both CreA sites are disrupted in addition to another single site, which overlaps the functional palindromic AlcR inverted repeat site (27). The repression of *alcR* expression by the competition of CreA for the region containing AlcR/CreA binding sites is sufficient to explain the carbon catabolite repression of ADH1 and AldDH.

2.3. Nitrate assimilation

Nitrate assimilation in filamentous fungi has been well studied in *A. nidulans* and *Neurospora crassa* (28, 29 and references therein). Nitrate is taken up by a permease, and reduced to nitrite by nitrate reductase. Nitrite is further reduced to utilizable ammonium by nitrite reductase. Genes encoding the above products have been identified from several fungi and yeast, and they are organized in a diverse fashion. For example, the nitrate reductase (*niaD*) and nitrite reductase (*niiA*) genes in *Aspergillus* species such as *A. nidulans*, *A. fumigatus*, *A. parasiticus*, *A. oryzae* and *A. niger* as well as in *Penicillium chrysogenum* are transcribed divergently from an intergenic promoter region (30, 31, 32, 33, 34). Whereas, the *niaD* and *niiA* genes in *Leptosphaeria maculans* are closely linked but transcribed in the same direction (35). The *Ustilago maydis* nitrate reductase (*nar1*) and nitrite reductase (*nir1*) genes are also closely linked, but the directions of transcription have not been determined yet (36). In contrast, these two genes in *N. crassa* and *Gibberella fujikuroi* are located on separate chromosomes (37, 38, 39). The nitrate transporter (*YNT1*), nitrite reductase (*YNI1*) and nitrate reductase (*YNR1*) genes are clustered in the yeast *Hansenula polymorpha*, and they are transcribed in the same direction (40). Most recently, the *YNA1* gene which encodes a Cys6-type transcription factor with similarity to the pathway-specific transcription factors of *A. nidulans* NirA and *N. crassa* NIT4 has been identified from this gene cluster, and the gene order is *YNT1*-*YNI1*-*YNA1*-*YNR1*(41).

The assimilation of nitrate is a tightly regulated process and involves the interplay of positive-acting and negative-acting factors. The expression of nitrate and nitrite reductase genes requires both the lifting of nitrogen metabolite repression and specific induction by nitrate. Two positive-acting transcription factors, one, a Cys2-type GATA-binding wide-domain nitrogen regulatory protein (NIT2 in *N. crassa* and AreA in *A. nidulans* and other *Aspergillus* species) (5, 42, 43), and another, a Cys6-type pathway-specific nitrogen regulatory protein (NIT4 in *N. crassa* and NirA in *A. nidulans*) (44, 45) , are necessary for transcriptional activation of the *nit* genes. The negative-acting NMR protein encoded by *N. crassa nmr*-1 (for nitrogen metabolite repression) interacts with two distinct regions of the NIT2 protein, a short alpha-helical motif within the NIT2 DNA-binding domain and a second motif at its carboxyl terminus (46). Deletion of the gene *nmrA* in *A. nidulans*, a homolog of *nmr*-1, results in partial derepression of activities subject to nitrogen metabolite repression (47). Thus, protein-protein interaction(s) likely exists for AreA and NmrA in *A. nidulans*. Haas *et al.* (48) identified the *nreB* gene from *P. chrysogenum* that encodes a Cys2-type GATA-binding factor with homology to Dal80p/Uga43p and Gzf3p/Nil2p, both repressors of nitrogen metabolism in *Saccharomyces cerevisiae* (49.). Overexpression of *nreB* leads to repression of the nitrate assimilatory gene cluster. Their finding cast doubts on the simple view that nitrate assimilation is regulated by only one negative-acting regulator in filamentous fungi.

In the *A. nidulans niaD-niiA* intergenic promoter region, four NirA binding sites, 5'-CTCCGHGG-3', are present. It appears that site 1 is necessary for the inducibility of *niiA* alone, while sites 2, 3, and 4 act bidirectionally (50). Of the ten AreA sites in the *niaD-niiA* intergenic region, the central clustered sites 5-8 are responsible for 80% of the transcriptional activity of both genes (51). The NIT2 and NIT4 proteins both bind to specific sites in the nitrate reductase (*nit-3*) gene promoter. In addition, a direct interaction between NIT2 and NIT4 is essential for optimal *nit-3* expression (52). A similar interaction was reported for AreA and NirA in *A. nidulans*. Muro-paster *et al.* (51) showed that AreA is essential for chromatin remodeling of the *niaD-niiA* intergenic region, but this process is independent from NirA-mediated transcription activation. Chromatin remodeling probably is necessary for the recruitment of NirA and other transcription factors during gene activation (53).

2.4. Sugar Utilization

Recently, our lab (Yu *et al.*, unpublished results) has discovered four clustered genes with the order *nadA-hxtA-glcA-sugR* that are related to sugar utilization and located next to the established aflatoxin biosynthetic gene cluster in *A. parasiticus*. This sugar utilization gene cluster was separated from the aflatoxin gene cluster by a 5 kb DNA spacer region, in which no known open-reading-frame was identified. According to sequence homology, *sugR* encodes a Cys6-type transcription factor; *hxtA*, encodes a membrane-bound transporter protein for hexoses; *glcA* encodes a α-glucosidase, and *nadA* encodes a NADH oxidase. Aflatoxin biosynthesis has been known to be stimulated by simple sugars (ribose, glucose, xylose, sucrose, etc) as opposed to complex sugars (starch); and the expression of *hxtA* was concurrent with the aflatoxin pathway cluster genes (Yu *et al.*, unpublished results). Therefore, the possibility of co-regulation of gene clusters involved in primary (sugar utilization) and secondary metabolism (aflatoxin synthesis), and the genetic and biological effects of this sugar utilization cluster on aflatoxin production are being investigated. In the closely related non-aflatoxigenic *Aspergillus* strain *A. oryzae*, a cluster of two genes, *agd4* (GenBank accession #D45179), encoding an α-glucosidase and, *amyR* (GenBank accession #AB012945), encoding the *agdA* transcription activator has been cloned.

The observation of this sugar cluster from *Aspergilli* mimics those observed in non-filamentous fungi. In *S. cerevisiae*, following the cloning of the glucose transport-related gene, *SNF3* (54), and a galactose permease gene, *GAL2* (55), over a dozen high and low affinity hexose transporter genes have been identified (56, 57, 58, 59). These gene products contain 12 hydrophobic regions associated with the membrane and are necessary for the uptake of 6-carbon sugars such as glucose and galactose. At the *MAL6* locus in yeast, a cluster of three genes required for maltose uptake and utilization are referred to as *MAL61* (maltose permease), *MAL62* (maltase), and *MAL63* (MAL-activator) (60, 61). The *CASUCI* gene in *Candida albicans*, encoding a zinc-finger containing protein with 28% identity to that of *MAL63* of *S. cerevisiae*, was identified as a possible regulator of sucrose utilization and α-glucosidase activity (62).

2.5. Quinate utilization

In *N. crassa* and *A. nidulans*, genes involved in quinate utilization are clustered. The quinate utilization gene cluster comprises genes that control quinate-shikmate utilization. In *N. crassa*, this 17.3 kb *qa* gene cluster contains five structural genes and two regulatory genes tightly linked to the centromere of linkage group VII (63). The *qa* gene order has been established as *qa-x, qa-2, qa-3, qa-4, qa-y, qa-1S, qa-1F*. However, this gene order is not conserved in *A. nidulans qut* gene cluster which is located in a 11.9 kb region with the order of *qutC-D-B-E-A-R* on chromosome VIII (64). *qa-y(=qutD)* encodes quinate permease; *qa-2*(=*qutE*), *qa-3*(=*qutB*) and *qa-4*(*qutC*) encode 3-dehydroquinase, quinate dehydrogenase and dehydratase, respectively. The function of the *qa-x* is still unclear. The gene *qa-1F* (=*qutA*) encodes a Cys6-type transcription factor that binds to 5'-GGRTAARYRYTTATCC-3' sites found in the promoters of the *qa* genes, including *qa-1F* (65). *qa-1S*(=*qutR*) encodes a repressor protein (63, 66).

The expression of *qa* genes is induced by quinate as well as coordinately regulated by *qa-1F* and *qa-1S*. QA-1F (QUTA) controls the production of *qa-1F* (*qutA*) mRNA as well as that of *qa-1S* (*qutR*) (67). Lamb *et al.* (68) found that overexpression of *qutA* and *qutR* was correlated with derepressed and super-repressed phenotypes in *A. nidulans*, respectively. They proposed that a direct interaction between QUTA and QUTR mediates *qut* gene expression.

Transcription of *qut* genes not only is induced by quinate but also by the quinate pathway intermediates, 3-dehydroquinate (DHQ) and dehydro-shikimate (DHS). These two compounds are also intermediates of the shikimate pathway that leads to the synthesis of aromatic amino acids (69, 70). DHQ and DHA are metabolized by AROM, a protein that consists of pentafunctional domains and is encoded by *aromA* (71, 72). Comparison of amino acid sequences of QUTA, QUTR and AROM suggests that *qutA* and *qutR* are derived from the splitting of a duplicated copy of *aromA*, with the two amino-terminal coding regions giving rise to *qutA* and the three carboxy-terminal coding regions to *qutR* (73). Hawkins *et al.* (74) extended their findings to the genesis of wide-domain and pathway-specific regulatory proteins, such as NIT2/AreA and AlcR, and proposed that evolution of regulatory proteins by enzyme recruitment is a general route. However, this controversial hypothesis was reaccessed by Arst *et al.* (75) and found that the conclusion based on the observations of Hawkins *et al.* (74) was incorrect.

3. SECONDARY METABOLIC PATHWAYS

3.1. Aflatoxins and Sterigmatocystin

Aflatoxins are toxic and carcinogenic polyketide-derived secondary metabolites that have been shown to be produced by five species of *Aspergillus*. These include numerous isolates of *A. flavus*, *A. parasiticus*, and *A. nomius*, while only one isolate of the previously believed nonaflatoxigenic species *A. tamarii* (76) and *A. ochraceoroseus* (77) have been found to produce the toxin (78). A significant body of literature has been generated on the chemistry, enzymology, and genetics of aflatoxins mainly through the study of both *A. flavus* and *A. parasiticus*. These two fungi are most commonly associated with preharvest aflatoxin contamination of food and feed crops (reviewed in 79).

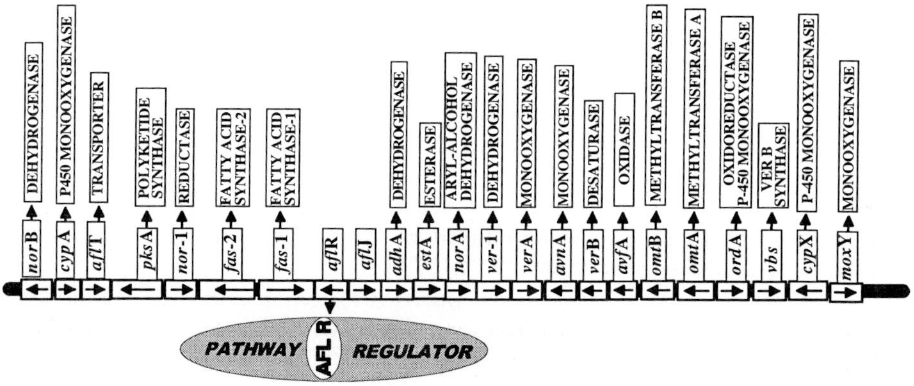

Figure 1. Organization and orientation of genes in the aflatoxin biosynthetic pathway gene cluster from *Aspergillus parasiticus* (modified from 79).

Sterigmatocystin, the penultimate precursor to aflatoxin is produced by a number of fungi including *Aspergillus nidulans*, which has served as the model system for the study of sterigmatocystin biosynthesis. Molecular genetic studies have shown that the genes for both the aflatoxin and sterigmatocystin biosynthetic pathways are clustered (80, 81, 82). A high degree of conservation is seen between the aflatoxin clusters in *A. flavus* and *A. parasiticus*. The order of the genes on the pathway and their direction of transcription are identical. In addition, there is a high degree of sequence conservation (>95%) at both the nucleotide and amino acid level (81). Comparison of the aflatoxin cluster to the *A. nidulans* sterigmatocystin gene cluster shows a high degree of similarity with respect to gene function and structure though sequence homology, but the order of the genes on the clusters between these *Aspergillus* species is not as highly conserved (Figure 2).

DNA sequence analysis and transcript mapping of mRNAs produced under conditions conducive to aflatoxin formation has identified as many as 23 genes covering a span of approximately 75 kb of the *A. parasiticus* genome that have been proven or are believed to be involved in aflatoxin biosynthesis (Fig. 1). These techniques were also used to identify 25 co-regulated transcripts covering approximately 60 kb of the *A. nidulans* genome that are responsible for the biosynthesis of sterigmatocystin (82; reviewed in 1). For many of these genes, the function of their products in the biosynthesis of aflatoxin and sterigmatocystin has been proven using techniques such as reverse genetics, recombinational inactivation with disrupted versions of the genes and genetic complementation of pathway mutants (reviewed in 79).

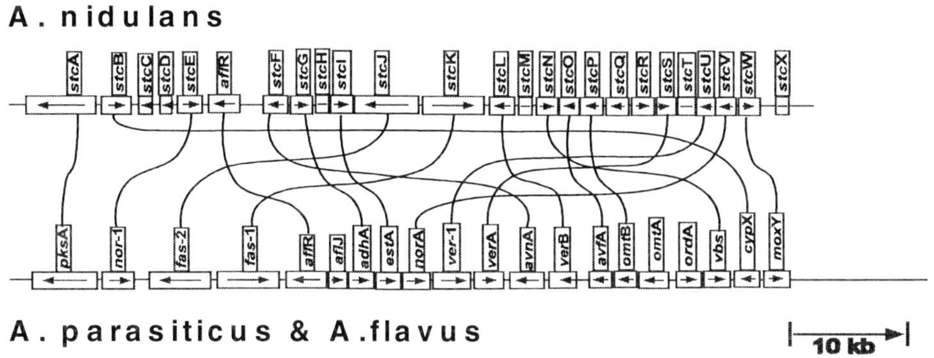

Figure 2. Comparison of the organization of genes in the aflatoxin biosynthetic pathway gene cluster from *Aspergillus flavus* and *A. parasiticus* and in the sterigmatocystin pathway cluster from *A. nidulans* (adapted from information in 79, 82).

However, a number of the genes present on both the aflatoxin and sterigmatocystin gene clusters have yet to have their exact functions elucidated. The putative function of these genes is often based on homology to other genes identified from examination of sequence databases in conjunction with enzyme activities known to be required for biosynthesis but whose genes have not been accounted for as of yet. With respect to aflatoxin biosynthesis, examples of these include; *afl*T, believed to encode a transporter for excretion of aflatoxin from the fungal mycelia (P.-K. Chang, unpublished data); *est*A (a homolog of *A. nidulans stc*I), proposed to encode an esterase involved in the conversion of versiconal hemiacetal acetate to versiconal (J. Cary, unpublished data); two dehydrogenases, *nor*A and *nor*B, believed to be involved in the conversion of norsolorinic acid to averantin and the production of norsolorinic acid from noranthrone respectively (83; J. Cary and D. Bhatnagar *et al.*,unpublished results). Additionally, three *A. parasiticus* genes, *cyp*A, *cyp*X, and *mox*Y, present at each end of the gene cluster (Fig. 1) demonstrate homology to monooxygenases (84). Expression of all of the above mentioned genes is regulated in a manner similar to that of known aflatoxin biosynthetic genes and determination of their functions is being pursued.

Common to both the aflatoxin and sterigmatocystin gene clusters is the presence of a gene, designated *afl*R, that encodes a zinc binuclear cluster-type, sequence specific DNA-binding protein that has been shown to be necessary for expression of the genes in both clusters (85, 86, 87). In *A. nidulans*, binding studies using a recombinant form of the AflR protein and either a region of the *stc*U promoter or oligonucleotides based on sequence from within the promoter fragment demonstrated binding of AflR to the palindromic sequence 5'-TCGN_5CGA-3' (88). The validity of this binding site was confirmed in a survey of 11 aflatoxin pathway genes from *A. parasiticus* using the recombinant AflR with oligonucleotides designed to the upstream regions of the 11 genes (89). Based on these studies the consensus AflR binding sequence was determined to be 5'-TCGSWNNSCGR-3'. Another gene present in both the gene clusters whose function remains to be established is *afl*J. In the aflatoxin pathway, *afl*J is found adjacent to *afl*R and they are divergently transcribed. Disruption of *afl*J resulted in "reduced" aflatoxin biosynthesis and no conversion of pathway intermediates (90). The disrupted strain did, however, accumulate pathway gene transcripts under conditions conducive to aflatoxin formation. Another series of experiments looking at the effects of *afl*J on the expression both early and late aflatoxin pathway genes showed that in the presence of *afl*R expression, *afl*J increased aflatoxin biosynthesis by increasing the expression of only the early pathway genes. No increase in expression of the late pathway genes *ver*-1 and *omt*A was observed (Payne, unpublished results). The exact mechanism by which *afl*J modulates transcription of these pathway genes in concert with *afl*R remains to be determined. There also appear to be genes that are not linked to either the aflatoxin or sterigmatocystin gene clusters whose products regulate *afl*R expression. Mutagenesis of a norsolorinic acid-accumulating strain of *A. nidulans* identified 23 mutations that were not linked to the sterigmatocystin cluster, three of which resulted in loss of expression of *afl*R (91). The unlinked dominant *afl*-1 mutation of *A. flavus* results in loss of toxin synthesis in diploids (92). Mutants did not produce transcripts to known pathway genes but *afl*R transcripts were detected, suggesting that *afl*R is necessary but not sufficient for aflatoxin gene expression.

A number of nutritional and environmental factors have been shown to influence the production of both aflatoxin and sterigmatocystin (93; reviewed in 79). These include temperature, nitrogen and carbon source, and pH. Interestingly, there appears to be a significant difference in the regulatory mechanisms controlling toxin synthesis in *A. nidulans* and *A. parasiticus*. It was shown that while high temperature and nitrate support sterigmatocystin production in *A. nidulans* the opposite was seen with *A. parasiticus* as these culture conditions repress synthesis of aflatoxin (93). pH has also been shown to regulate biosynthesis of sterigmatocystin and aflatoxin. A mutant of *A. nidulans* that constitutively produced the pH regulatory factor, PacC, synthesized 10-fold less sterigmatocystin than the wild type strain (94). This result is not suprising, as PacC is involved in the positive regulation of alkali expressed genes such as *ipn*A of the *A. nidulans* penicillin biosynthetic pathway, and sterigmatocystin and aflatoxin production are highest under acidic growth conditions. Interestingly, a PacC binding site has been identified in the promoter of the *afl*R gene though no direct evidence for the interaction of PacC and *afl*R has been established (95).

3.2. Trichothecenes

Trichothecenes consist of a family of over 60 sesquiterpenoid compounds produced by a number of genera of filamentous fungi. Perhaps the best studied of the trichothecene producing fungi is *Fusarium* which is responsible for trichothecene contamination of grains such as maize, wheat, barley, and rye (reviewed in 96). Examples of trichothecenes produced by *Fusarium* species include diacetoxyscirpenol (DAS), deoxynivalenol (DON), and T-2 toxin, while the more structurally complex macrocyclic trichothecenes are produced by fungal genera such as *Myrothecium*, *Stachybotrys*, and *Trichothecium*. Ingestion of trichothecene contaminated foods and feeds is associated with several human and animal diseases including acute toxicoses that can lead to death (97). Though the exact reason for the toxic effects attributable to trichothecene poisoning is not well understood it is believed to be the result of inhibition of protein synthesis that induces the disease symptoms.

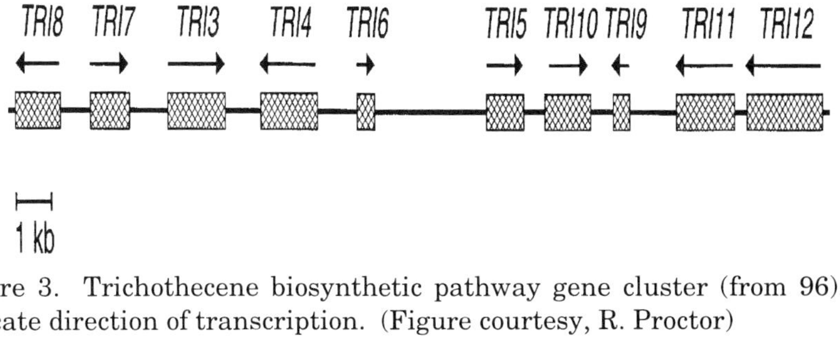

Figure 3. Trichothecene biosynthetic pathway gene cluster (from 96). Arrows indicate direction of transcription. (Figure courtesy, R. Proctor)

The biosynthesis of trichothecenes begins with the cyclization of farnesyl pyrophosphate by trichodiene synthase to produce trichodiene. Trichodiene then undergoes an ordered series of oxygenation, isomerization, and esterification reactions leading to the formation of the various members of the trichothecenes. Complementation analysis using two overlapping cosmid clones harboring TRI5, encoding trichodiene synthase, has led to the identification in *F. sportrichioides* of 10 clustered genes (Figure 3) that appear to be involved in trichothecene biosynthesis (reviewed in 96). To date, the function of six of the 10 genes has been determined. The predicted amino acid sequences of the TRI4 and TRI11 gene products were similar to those of cytochrome P450 monooxygenases. Their exact function in biosynthesis was determined using inactivation of the wild type genes via homologous recombination with their respective disrupted versions (98, 99). TRI4 was found to catalyze the oxygenation of trichodiene at the C-2 position to yield 2-hydroxytrichodiene while TRI11 oxygenates the C-15 position of isotrichodermin to yield 15-decalonectrin. Although no identity could be established for the TRI3 gene product based on similarity to other known protein sequences, gene disruption showed that it encodes an O-acetyltransferase catalyzing the conversion of 15-decalonectrin to calonectrin via acetylation at the C-15 position (100). TRI6 encodes a Cys_2His_2 type zinc-finger protein required for the expression of trichothecene biosynthetic pathway genes. TRI6 was shown to bind in the promoter regions of nine pathway genes at the minimum consensus site 5'-YNAGGCC-3' (101). Additionally, a site-directed mutation, C187A, in the predicted Cys_2His_2 zinc-finger motif of the carboxy-terminus of TRI6 abolished binding of the protein to promoter DNA fragments, supporting the role of TRI6 and specifically the C-terminus in DNA binding. TRI12 encodes a trichothecene efflux pump with sequence similarities to known members of the major facilitator superfamily of transporters. Disruption of TRI12 resulted in reduced growth on complex media and reduced levels of trichothecene biosynthesis (102). Expression of TRI12 and TRI3 in *pdr5* yeast mutants (sensitive to trichothecene toxicity) fed 15-decalonectrin accumulated more of the expected product, calonectrin, than a strain only expressing TRI3. This result was mimiced by yeast strains expressing the PDR5 transporter (known to mediate trichothecene resistance in yeast) and TRI12/TRI3 but not expressing TRI3 alone, supporting the role of TRI12 in *F. sporotrichioides* self-protection against trichothecenes. In addition, another gene designated TRI101 from *F. graminearum*, produces a trichothecene 3-O-acetyltransferase that also appears to play a role in self-protection against trichothecenes. The TRI101 homolog in *F. sporotrichioides* (FsTRI101) apparently does not reside within the trichothecene gene cluster but its disruption leads to reduced growth on trichothecene-containing media compared to the wild-type strain suggesting a similar role for FsTRI101 in self-protection against the toxic effects of trichothecenes (103). Disruption of the TRI7 gene results in accumulation of HT-2 toxin which differs from T-2 toxin by an acetyl group at C-4 indicating that TRI7 is an acetylase. TRI8 disruptants accumulate 4,15-diacetoxyscirpenol. Feeding of isovalerate to TRI8 disruptants resulted in the conversion of HT-toxin to T-2 toxin indicating that TRI8 is involved in the esterification of isovalerate at C8 (Susan McCormick, personal communication).

Production of the structurally more complex macrocyclic trichothecenes such as roridin E, verrucrin, and baccharinoid B7 is most commonly associated with members of the genus *Myrothecium*. The macrocyclic trichothecenes exhibit about 10-fold more toxicity than the *Fusarium* trichothecenes. Screening of a *M. roridum* cosmid library with a *F. sporotrichioides* TRI5 gene probe identified three overlapping cosmid clones that were used for gene isolation and cosmid mapping studies (104). This work led to the identification in *M. roridum* of homologs for the *F. sporotichioides* TRI5 (MrTRI5), TRI4 (MrTRI4), and TRI6 (MrTRI6) genes. Interestingly, these genes spanned a region of about 40 kb in *M. roridum* while the *F. sporotrichioides* genes only span about 8 kb. In addition, the MrTRI4 and MrTRI6 genes are convergently transcribed in *M. roridum* while their *F. sporotrichioides* counterparts are divergently transcribed. The deduced amino acid sequence of the MrTRI5 product demonstrated between 72% and 75% with trichodiene synthases from four *Fusarium* species. The predicted MrTRI6 product was found to be almost twice as large as the FsTRI6 product and the overall amino acid identity was quite low. However, a C-terminal region of MrTRI6 demonstrated 78% identity with the C-terminal region of FsTRI6 corresponding to the zinc-finger motifs. All of the Cys and His residues believed to play a role in zinc-finger binding were conserved between these two proteins. The MrTRI4 product was shown to share significant identity (64%) with FsTRI4 and also functionality as the MrTRI4 gene was able to restore T-2 toxin biosynthesis in $TRI4^-$ mutant of *F. sporotrichioides*.

3.3. Penicillins and Cephalosporins

Penicillins are β-lactam antibiotics that are produced exclusively by a few filamentous fungi, most notably *Aspergillus nidulans* and *Penicillium chrysogenum*. The β-lactam antibiotics, cephalosporins, are produced by the fungus *Acremonium chrysogenum* (syn. *Cephalosporium acremonium*) as well as a number of bacteria. The biosynthesis of penicillins and cephalosporins have the first two steps in common. The first step involves the condensation of amino acid precursors to form the tripeptide δ-(L-α-aminoadipyl)-L-cysteinyl-D-valine (ACV). This reaction is carried out by a single, multifunctional enzyme, ACV synthase (ACVS), that is encoded by the gene *acv*A. The second step involves production of the bicyclic ring structure (the β-lactam ring fused to the thiozolidine ring) and is catalyzed by isopenicillin N synthase (IPN) which is encoded by *ipn*A. At this point the penicillin and cephalosporin biosynthetic pathways diverge. In the penicillin pathway, an acyl CoA:IPN acyltransferase encoded by *aat*A, produces penicillin while the products of *cef*D, *cef*EF, and *cef*G result in the production of cephalosporin (reviewed in 105).

The genes for both penicillin and cephalosporin biosynthesis are clustered. In both *A. nidulans* and *P. chrysogenum*, the *acv*A, *ipn*A, and *aat* genes are localized within a region of approximately 15 kb, while in *A. chrysogenum* the *acv*A and *ipn*A genes (early cluster) also reside within about a 15 kb region of the genome (106, 107). In all three fungi the *acv*A gene is divergently transcribed from the *ipn*A gene while in *A. nidulans* and *P. chrysogenum* the *aat*A gene is transcribed in the same direction as *ipn*A. The cephalosporin pathway-specific genes, *cef*EF and *cef*G (late cluster), are located on a separate chromosome (chromosome II) in *A. chrysogenum* and are separated by a 938 bp intergenic region from which both genes are divergently transcribed (108). The location of the *cef*D gene has yet to be determined in *A. chrysogenum*.

A number of different nutritional and environmental factors have been identified that regulate the biosynthesis of penicillins and cephalosporins. Due to the availability of mutants and ease of genetic manipulation, the majority of regulatory studies have focused on the penicillin biosynthetic pathway of *A. nidulans* (reviewed in 105). Carbon source appears to affect the production of both penicillin and cephalosporin. Those carbon sources that result in the most rapid growth of the fungus (i.e. glucose and glycerol) have a negative effect on β-lactam production. The involvement of cAMP in penicillin production is controversial and little if any involvement of regulatory factors such as CreA, CreB, or CreC has been noted (109, 110). Penicillin biosynthesis has been shown to be under the control of pH, being greatest at alkaline pH (111). PacC is a wide-domain regulatory protein which has been shown to mediate pH regulation of several genes in *A. nidulans*, including *ipn*A (112). PacC activates transcription of alkaline-expressed genes and PacC mutants of *A. nidulans* that mimic growth at alkaline conditions produced about twice as much penicillin as wild-type strains. Four PacC binding sites have been identified in the *acv*A and *ipn*A intergenic region and shown to bind PacC at the core consensus sequence GCCARG (112, 113). CCAAT- boxes have been identified in the promoter regions of a number of eukaryotes including filamentous fungi and have been shown to bind protein complexes (termed HAP-complexes in *S. cerevisiae*) that regulate transcription of genes (114). Recently, CCAAT-containing DNA binding motifs were identified in the *acv*A-*ipn*A intergenic region and also upstream of the *aat*A gene (115, 116). Both of these regions were found to bind a HAP-like complex of proteins termed PENR1, for penicillin regulator 1. Mutatagenesis of these CCAAT box regions demonstrated that PENR1 is of major importance as a positive acting factor for *ipn*A and *aat*A expression while it appears to negatively regulate expression of *acv*A.

3.4. Gibberellins

Gibberellins are a large family of diterpenoid hormones that are produced by green plants, fungi, and bacteria. The rice pathogen *Gibberella fujikuroi* (mating population C) produces large amounts of gibberellic acid and it is a main commercial source of the bioactive gibberellins, particularly GA_3. Gibberellin biosynthesis follows the early steps in the general isoprenoid biosynthetic pathway until it branches from synthesis of carotenoids at geranylgeranyl-diphosphate (GGDP) (reviewed in 117). The various gibberellins are then produced via a complex series of oxidations and hydroxylations. Genes known to be specifically involved in gibberellin biosynthesis have been found to be clustered (118). Differential screening of a *G. fujikuroi* expression library identified a gene sequence that encoded a cytochrome P450 monooxygenase, designated P450-1, believed to be involved in gibberellin biosynthesis. Use of the P450-1 cDNA as a probe to screen a *G. fujikuroi* resulted in the identification of another P450-like gene, P450-2, that shared an intergenic region of about 1.5 kb with P450-1. In order to identify additional genes believed to be involved in gibberellin biosynthesis, genome walking experiments were conducted using lambda genomic clones that harbored P450-1 and P450-2. These studies resulted in the identification of the genes *ggs2*, encoding the geranylgeranyl diphosphate synthase gene; *cps/ks*, encoding *ent*-copalyl diphosphate synthase; and two more cytochrome P450 genes, also believed to play a role in gibberellin biosynthesis (117, 118). These six genes spanned a region of about 18 kb of the *G. fujikuroi* genome and all were expressed at high levels under conditions conducive to gibberellin production. The *cps/ks* gene had previously been cloned and characterized from *G. fujikuroi* and disruption of this gene resulted in loss of gibberellin production (119). Disruption of both P450-1 and P450-2 also resulted in loss of gibberellin biosynthesis (118). Interestingly, earlier studies had identified another geranylgeranyl diphosphate synthase gene, *ggs1*, that was constitutively expressed and thought to be involved in synthesis of other isoprenoids, while the GGS2 enzyme is now believed to be responsible for supplying the CPS/KS enzyme with the substrate GGDP for initiation of gibberellin biosynthesis (117).

Very little is known about the regulation of gibberellin production at the molecular genetic level. To date a positive-acting, gibberellin pathway-specific regulator protein has not been identified in *G. fujikuroi*. However, studies have shown that production of gibberellin is regulated by nitrogen (120). AreA-GF of *G. fujikuroi* is a homolog of the global nitrogen regulator AreA of *A. nidulans*. Disruption of *are*A-GF was shown to significantly reduce gibberellin production and complemention of the *are*A-GF mutant with a wild-type *are*A-GF. Northern blot analysis performed on five of the gibberellin pathway genes showed that transcript levels are drastically increased under conditions of ammonium limitation (gibberellin inducing) which would be expected from the action of a positive-acting regulatory factor like AreA-GF (118). Potential GATA binding motifs for AreA-GF have been identified upstream of gibberellin biosynthetic genes.

3.5. Melanins

A number of fungi produce melanins which are the black or near-black pigments formed by oxidative polymerization of phenolic compounds produced by the dihydroxynaphthalene (DHN)-melanin pathway. Melanin has been shown to be a virulence factor in plant, animal, and human pathogenic fungi and it also functions in survival and longevity in nature of fungal propagules (reviewed in 121). In addition to its function in UV-tolerance, studies of melanin mutants of *Alternaria alternata* also showed that melanins play a role in conidial development (122). Melanin biosynthesis initiates through the action of a polyketide synthase which converts acetate to the intermediate compound 1,3,6,8-tetrahydroxy-naphthalene (T4HN) which is subsequently reduced by scytalone reductase to scytalone. Dehydration of scytalone forms 1,3,8-trihydroxynaphthalene (T3HN) which is then converted to 1,8-dihydroxynaphthalene (DHN) after an additional reduction and dehydration step. DHN is then polymerized and oxidized to yield melanin.

The genes involved in melanin biosynthesis have been found to be clustered in some fungi while in others they are not. Complementation of *A. alternata* melanin-deficient mutants with cosmid DNA allowed the identification of an approximately 30 kb region of the *A. alternata* genome that harbored three genes encoding the polyketide synthase (*ALM*), scytalone dehydratase (*BRM1*) and T3HN reductase (*BRM2*) involved in melanin biosynthesis. Subsequent gene disruption experiments confirmed the function of these genes in melanin biosynthesis (123). Expression of these three genes was synchronous with the onset of mycelial melanization. A developmentally regulated gene cluster involved in pigment biosynthesis has been characterized in the human opportunistic pathogen, *Aspergillus fumigatus*, which produces a blue-green conidial pigment (124). Sequence and gene disruption analysis of DNA from a cosmid clone complementing an *A. fumigatus* conidial color mutant identified six genes within a 19 kb region that were involved in pigment biosynthesis. All six genes were developmentally regulated, being expressed during conidiation. The gene products of *alb*1, *arp*1, and *arp*2 demonstrated significant identity at the amino acid level with polyketide synthases, scytalone dehyrdratases, and T3HN reductases of brown and black fungi, respectively. Disruption of the *alb*1 gene of *A. fumigatus* significantly reduced virulence in a murine model just as disruption of melanin genes in plant pathogenic fungi severely reduces virulence (125). Similarity searches of sequence databases showed that of the other three genes, *abr*1 encoded a putative multicopper oxidase and *abr*2 a putative laccase, while no homologies were found for *ayg*1. The identification of the *abr*2 gene product as a laccase would be in keeping with the proposed requirement for an oxidase for polymerization of the phenolic precursor to the pigment (DHN to melanin in the case of brown and black fungi) though this gene has not been identified in the melanin pathway of the brown and black fungi (126). Genes homologous to the *A. alternata* DHN-melanin biosynthetic genes have been identified in plant pathogenic fungi such as *Colletotrichum lagenarium* and *Magnaporthe grisea* (127, 128). However, none of the genes have been shown to be closely linked, though the expression of the *C. lagenarium* genes appear to be developmentally controlled as in the *A. alternata* melanin gene cluster. A melanin pathway-specific regulatory protein has yet to be identified in any of the fungi investigated.

3.6. Fumonisins

Fumonisins are polyketide mycotoxins that are commonly produced in maize kernels by the fungus, *Gibberella fujikuroi* mating population A (MP-A). These toxins have been shown to cause or be associated with a number of diseases in animals and humans including cancer following consumption of contaminated grain (129). Fumonisins consist of a linear 19- or 20-carbon backbone that is substituted at various positions with hydroxyl, methyl, and tricarballylic acid moieties and an amino group at C-2 (reviewed in 96). Until recently it was not known if fumonisins were the product of either polyketide or fatty acid biosynthesis. However, disruption and complementation analysis of a cloned *G. fujikuroi* gene, FUM5, encoding a polyketide synthase demonstrated that fumonisins are a product of polyketide biosynthesis (130).

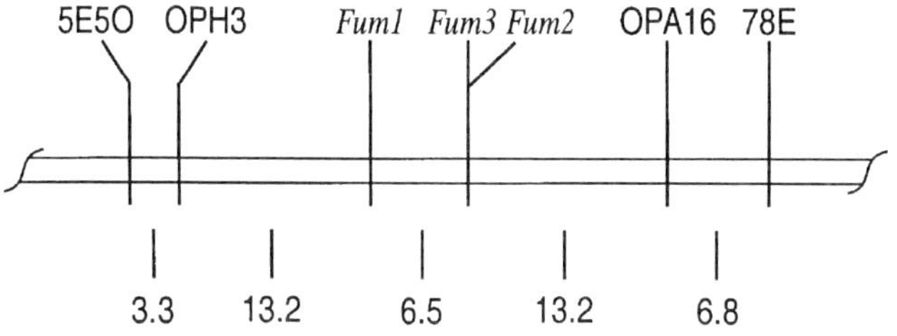

Figure 4. Linkage map of *Fum* loci, RAPD markers OPA16 and OPH3, and RFLP markers 5E50 and 78E on linkage group 1 of *G. fujikuroi* MP-A. Xu and Leslie (133) identified the RFLP markers on linkage group 1 and determined their positions relative to *Fum1*; Desjardins *et al.* (131) determined the recombination frequency between *Fum1*, *Fum2* and *Fum3*, and Proctor *et al.* (96) determined the recombination freqeuncies between 5350, OPH3, 78E, and OPA16. (Figure courtesy of R. Proctor.)

Classical genetic analyses of fumonisin (*Fum*) mutants of *G. fujikuroi* MP-A indicated that the genes for fumonisin biosynthesis are closely linked on chromosome 1 (131). Complementation analysis of *Fum* mutants with cosmid DNA harboring the Fum5 gene supported the meiotic analyses indicating that the genes involved in fumonisin biosythesis are clustered (130). Complementation of *Fum*2 and *Fum*3 mutants that produced only FB_2 and FB_3 respectively, did not result in production of FB_1 by transformants but did lead to their overproduction of FB2 and FB3. This increase in fumonisin production may have resulted from the expression of other fumonisin biosynthetic genes present on the transformed cosmid DNA. Introduction of the *Fum*5-harboring cosmid DNA into a *Fum*1 mutant that produced no fumonisins did restore toxin production. In order to identify other genes involved in fumonisin production that may reside near *Fum*5, the DNA regions surrounding this gene have been sequenced (96, 132; Proctor, personal communication). Sequencing approximately 7 kb downstream and 15 kb upstream of *Fum*5 resulted in the identification of 5 ORFs in the upstream and 3 ORFs in the downstream regions. Based on sequence homologies to known genes and proteins, four of the ORFs appear to encode proteins that would be expected to be involved in fumonisin production. One of these ORFs encodes a putative cytochrome P450 monooxygenase.

3.7. Paxilline

Paxilline is a tremorgenic, indole-diterpenoid, mycotoxin produced by the filamentous fungus, *Penicillium paxilli* (134). Though little is known about the biosynthesis of these mycotoxins, geranylgeranyl pyrophosphate (GGPP) and indole are presumed to be precursors. Utilizing plasmid-tagged mutagenesis and REMI-mutagenesis techniques, a number of paxilline negative, deletion mutants of *P. paxilli* have been isolated and characterized (135). Analysis of *P. paxilli* genomic DNA flanking and including the deleted regions of the mutants identified a cluster of ORFs with similarities to prenyltransferases and monooxygenases (136). Targeted disruption of a GGPP synthase confirmed that this gene was required for toxin synthesis but not for primary metabolism. The boundaries of the cluster have been defined to a size of about 60 kb based on deletion analysis. Sequence analysis of this region has identified other putative paxilline biosynthetic genes, including a potential zinc binuclear cluster type DNA binding protein (136).

4. HOST-SPECIFIC TOXINS

4.1. HC-toxin

HC-toxin is an epoxide-containing cyclic tetrapeptide that is a critical virulence determinant in the pathogenic interaction between *Cochliobolus carbonum* and maize (137). Classical genetic analysis of HC-toxin producing (Tox 2^{+}) isolates of *C. carbonum* indicated that production was under control of a single locus, designated TOX2. However further molecular analysis showed that TOX2 is highly complex consisting of duplicated copies of a number of genes extending over 550 kb of a single chromosome (138). These genes include, HTS1, encoding a tetrapartite cyclic peptide synthetase; TOXA, encoding an HC-toxin membrane transporter of the major facilitator superfamily; TOXC, encoding a fatty acid synthase beta subunit believed to be responsible in part for the synthesis of the decanoic acid backbone of 2-amino-9,10-epoxi-8-oxodecanoic acid (Aeo) moiety of the cyclic tetrapeptide (reviewed in 139). TOXE, encodes a novel regulatory protein involved in HC-toxin biosynthesis. The TOXE protein has a bZIP basic DNA binding domain but no discernable leucine zipper characteristic of other bZIP transcription factors. Interestingly, at its carboxy terminus, TOXE has four ankyrin repeats found in other regulatory proteins but to date these repeats have never been associated with a bZIP-type DNA binding domain. TOXE expression was shown to be required for expression of all the other toxin genes except HTS1 (140). TOXD, encodes a protein of unknown function and TOXD mutants still produce toxin and are pathogenic. However, because TOXD is only present in $Tox2^{+}$ isolates, is linked to HTS1 and TOXC, and is positively regulated by TOXE, it is believed to be involved in toxin synthesis. It has been theorized that TOXD shares an overlapping activity with another enzyme involved in production of one of the other cyclic peptides produced by *C. carbonum* and this enzyme is capable of compensating for the loss of TOXD (reviewed in 139). None of the above mentioned genes are present in HC-toxin non-producing ($Tox2^{-}$) strains. All natural $Tox2^{+}$ isolates of *C. carbonum* have two copies of HTS1 and TOXA which are located adjacent to one another being separated by only 386 bp and divergently transcribed. The two copies of HTS1 are about 270 kb apart from one another. Three copies of TOXC span about 540 kb of the chromosome and are interspersed around the copies of HTS1: two copies flank the two copies of HTS1 while the third copy is located between the two. Most $Tox2^{+}$ isolates have two copies of TOXE, however unlike the other genes involved in toxin production which are present on either the 3.5 Mb (type 1 strains) or 2.2 Mb chromosome (type 2 strains), in strain SB111 (type 1) one copy of TOXE is on the 3.5 Mb chromosome while the other is found on a 0.7 Mb chromosome (140). This finding was explained by a reciprocal translocation that had occurred during the evolution of $Tox2^{+}$ isolates in which the 3.5 Mb chromosome is broken into fragments of 1.5 and 2.0 Mb with the 0.7 Mb chromosome joining with the 1.5 Mb chromosome to produce chromosomes of 2.2 Mb and 2.0 Mb found in many type 2 strains of *C. carbonum*. Recently, two additional genes, designated TOXF and TOXG, have been shown to be involved in HC-toxin production and linked to the TOX2 locus (141, 142). Both genes are unique to $Tox2^{+}$ isolates of *C. carbonum* and targeted disruption of either gene abolished toxin production and pathogenicity. The deduced amino acid sequence of TOXF has moderate homology to many known branched-chained, amino acid transaminases suggesting that TOXF may function to aminate a precursor of the Aeo moiety of HC-toxin (141). TOXG is predicted to encode an alanine racemase (142).

4.2. AK-toxin

The Japanese pear pathotype of *Alternaria alternata* causes black spot of Japanese pear by producing the host-specific toxin, AK-toxin (143). Utilizing restriction enzyme-mediated integration (REMI) mutagenesis of *A. alternata* pear pathotype strain 15A resulted in the isolation of fungal transformants that produced no AK-toxin (144). Two genes, AKT1 and AKT2, were identified within an 8 kb region of the genome which when disrupted resulted in loss of AK-toxin production and pathogenicity. Disruption of either gene resulted in isolation of transformants that failed to produce the AK-toxin precursor 9,10-epoxy-8-hydroxy-9-methyl-decatrienoic acid. AKT1 demonstrated homology with carboxyl-activating enzymes and it was theorized that its product activates an earlier precursor of 9,10-epoxy-8-hydroxy-9-methyl-decatrienoic acid for further modification by other enzymes involved in AK-toxin production. AKT2 demonstrated no similarity to known proteins. Interestingly, homologs of both AKT1 and AKT2 were detected in Southern blots of DNA from *A. alternata* tangerine (ACT-toxin) and strawberry (AF-toxin) pathotypes but not in other pathotypes or non-pathogenic strains of *A. alternata*. All three of these pathotypes share the common precursor moiety 9,10-epoxy-8-hydroxy-9-methyl-decatrienoic acid suggesting the possibility of horizontal transfer of these genes between these pathotypes (144). No information has been gleaned as to the molecular mechanisms controlling AK-toxin biosynthesis.

4.3. CPS1

Recently, a gene cluster has been identified from the corn pathogen *Cochliobolus heterostrophus* that is required for nonribosomal peptide biosynthesis and general virulence (145, 146). *C. heterostrophus* is best known for its production of the polyketide,T-toxin, which is required for strains of race T to be highly virulent to T-cytoplasm corn. The genes for T-toxin synthesis have been identified and were found to be unlinked (145). Though T-toxin and other host-specific toxins have been implicated in the host range of the pathogen, it has long been hypothesized that additional molecules are necessary for pathogenesis (145). Using REMI mutagenesis of *C. heterostrophus* race O, a single mutant was identified that caused lesions on corn leaves about half the normal size of wild-type lesions. Sequence analysis of DNA flanking the marker DNA identified a gene, designated CPS1, that encodes an as of yet unidentified multifunctional peptide synthetase. Rescue and retransformation of the linearized plasmid DNA into wild-type *C. heterostrophus* resulted in disruptants of CPS1 that produced lesions similar to those of the REMI mutant. Southern blot analysis has shown the presence of CPS1 homologs in many fungal pathogens. Disruption of the CPS1 homolog of the oat pathogen *C. victoriae* (produces the toxin, victorin) resulted in the mutant being drastically reduced in virulence on V*b* oats even though it continued to produce wild-type levels of victorin (145). This indicated that a general virulence factor, presumably the hypothetical CPS1-encoded peptide, may be required to facilitate the action of other virulence factors such as host-specific toxins. Sequence analysis of *C. heterostrophus* cosmid DNA representing about 33 kb of DNA surrounding CPS1 identified an additional 13 open reading frames (ORFs)(146). Only four of these ORFs demonstrated similarity to known proteins in databases examined. These include a thioesterase (TES1); a DNA-binding protein (DBZ1); a coenzyme transferase (COT1); and a decarboxylase (DEC2). As observed for CPS1, disruption of DBZ1 also caused a drastic reduction in the virulence of three different pathogens, *C. victoriae* and race O and T of *C. heterostrophus* (146).

5. SIGNIFICANCE OF GENE CLUSTERS

The body of knowledge obtained on fungal gene clusters continues to grow but the reasons why genes are organized in clusters and the benefits they impart on the fungus are still far from clear. The fact that these gene clusters appear to be "dispensable" in fungi would argue that they provide little if any benefit to the fungus. Though gene clusters may appear to be dispensable under the conditions and over the narrow "window" of time that these organisms have been studied; it is quite possible that over the course of "evolutionary time" these clusters have played an important role in the growth, survival, and evolution of these organisms. Information gained from the study of gene clusters with respect to the regulation of pathway gene expression, structure of genes and their organization, and evolutionary origins should continue to shed some light as to their significance to the host organism.

5.1 Gene structure and organization

The structure of the coding regions of genes within fungal gene clusters varies little with respect to that of other non-clustered fungal genes. Codon usage bias and the presence of introns appears to be quite similar. An exception to this would be the *acvA* and *ipnA* genes involved in penicillin biosynthesis in *A. nidulans* and *P. chrysogenum* and cephalosporin biosynthesis in *A. chrysogenum*. These genes lack introns and demonstrate high G+C content similar to their prokaryotic homologs which suggests that these fungi acquired these genes by horizontal transfer from bacterial β-lactam producers (147, 148). However, unlike their bacterial counterparts, these genes are transcribed from different promoters and not as polycistronic mRNAs. Comparison of the organization of genes within functionally similar pathway gene clusters of distantly related fungi show that gene arrangement on the cluster typically does not follow any particular bias. This can be seen from comparison of the quinate utilization pathway of *A. nidulans* and *N. crassa* (see section 2.5) as well as the aflatoxin pathway cluster of *A. flavus/parasiticus* and the sterigmatocystin gene cluster of *A. nidulans* (see Fig. 2, section 3.1). Additionally, in the case of fairly well characterized gene clusters, there does not seem to be any requirement for the arrangement of the genes on the pathway cluster to coincide with the biosynthetic steps of the pathway. The only known exception to this is the penicillin gene cluster of *A. nidulans* and *P. chrysogenum* and the melanin biosynthetic gene cluster of *A. alternata*.

5.2 Regulation of pathway gene expression

A primary advantage of gene clustering may be for the purpose of coordinated gene expression. Clustering of genes allows regulatory elements to be shared, as is seen with the promoter regions of the divergently transcribed *acvA-ipnA* genes in the penicillin gene cluster (see section 3.3) and the *niiA-niaD* genes of *Aspergillus sp.* nitrate assimilation gene cluster (see section 2.3). There is also evidence that gene clustering may influence gene regulation through modulation of localized chromatin structure. Transformation of *A. parasiticus* with the promoter region of the aflatoxin biosynthetic gene *ver-1A* fused to a GUS reporter gene showed that the site of integration within the fungal genome affects gene expression (149). Homologous integration of the *ver-1A*::GUS reporter within the aflatoxin gene cluster did not significantly affect GUS expression while integration outside the cluster at the *niaD* locus resulted in 500-fold lower GUS activity. However, close association of pathway biosynthetic genes does not appear to be a prerequisite for coordinated gene expression as demonstrated by the melanin biosynthetic pathway of *C. lagenarium* (150) and *M. grisea* (127), and the nitrogen assimilation pathway of *N. crassa* (38) and *Gibberella fujikuroi* (39), in which the genes are unlinked yet regulated in a coordinate fashion. The clustered organization of fungal pathway genes may have some intrinsic significance in gene regulation in that expression of one cluster may affect the expression of other genes adjacent to the gene cluster. An example of this is the juxtaposition of the aflatoxin biosynthetic gene cluster and the sugar utilization cluster in *A. parasiticus* (Yu *et al.* unpublished results), with aflatoxin production being dependent on the type of carbon source metabolized.

As additional details on the regulation of metabolic pathways in fungi emerge, aspects of cluster chromatin structure may be better understood and the significance of clustering of genes will become clearer. The significance or advantage to the organism of a gene cluster in reduced expenditure of precious energy resources for turning on or off of a pathway, may also become obvious. Keller and Hohn (1) have appropriately noted that, "Expression of the appropriate catabolic pathways can be critically important for survival (of fungi) under limiting nutrient growth conditions." They go on to say, "To ensure that the required catabolic pathways are expressed appropriately in response to changing nutritional conditions while simultaneously limiting the loss of cellular resources due to unnecessary pathway, gene expression, fungi have developed complex regulatory systems."

5.3 Evolutionary significance

In addition to regulatory advantages of gene clusters, the most plausible explanation for these clusters is the horizontal genetic transfer of metabolic pathways between prokaryotes and fungi, and subsequently between fungi. This hypothesis most likely emerged not only from the observation that prokaryotic pathway genes are generally clustered, but also from the early discoveries of gene clusters of the penicillin and cephalosporin pathways in fungi. These pathways are also found in prokaryotes, and the fungal genes in the clusters for the synthesis of penicillin and cephalosporin also contain features of the prokaryotic pathway genes, such as the absence of introns and high G+C content (151). Granted that several individual genes discovered recently in various metabolic pathways of fungi have prokaryotic homologs , but several of the metabolic pathways and the corresponding gene clusters seem to be unique to fungi, e.g., the mycotoxin and host-specific toxin biosynthetic gene clusters. And within gene clusters with similarity to prokaryotic systems, several genes seem to be of fungal origin e.g., the *cef*G gene in cephalosporin C pathway of *A. chrysogenum* and the *aat* gene of the penicillin pathway of *A. nidulans* and *P. chrysogenum* (152). This indicates that these genes were "recruited" from the fungal genome for the purpose of β- lactam biosynthesis, but why were they positioned so as to form part of a gene cluster? Again, this would suggest that clustering of genes is functionally important, perhaps reflecting that coordinate regulation of gene expression is imparted by localized chromatin structure. Therefore, the possibility does exist that due to evolutionary pressures, individual genes (or in some cases significant parts of a cluster) were acquired by fungi from prokaryotes. But aside from the penicillin and cephalosporin genes there is very little evidence of horizontal transfer of pathway gene clusters from prokaryotes to fungi in nature.

Yet another plausible hypothesis put forth for evolutionary significance of gene clusters in fungi is that having acquired individual genes for specific enzyme activities from either prokaryotes (e.g., penicillin pathway genes; 151) or other fungi (reductase in melanin and aflatoxin pathways; 153) by horizontal gene transfers, fungi organized these genes into clusters for specific metabolic functions. Coupled with this was the successful transformation and expression of the entire penicillin pathway from *P. chrysogenum* into *N. crassa* and *A. niger*, neither of which possess a penicillin pathway (154). This fueled the possibility of horizontal gene transfer of pathway clusters between fungi. However, while the possibility of horizontal transfer of individual or small groups of genes is quite plausible, what are the chances for the transfer of an entire gene cluster, even between related fungal species? To date there is little evidence for horizontal transfer of a gene cluster between fungi. However, the frequency of gene cluster transfer between related or unrelated fungal species may actually be much higher than predicted, but detection of these events has been limited by the number of fungal samples obtained from nature for analysis. Indeed, recent research has identified one isolate of *A. tamarii* (section *Flavi*) and one isolate of *A. ochraceoroseus* (section *Circumdati*) that are capable of producing aflatoxin (78). In addition, the aflatoxin biosynthetic pathway cluster in *A. flavus* and *A. parasiticus* is very similar (Figure 2), but the sterigmatocystin pathway cluster in *A. nidulans*, performing essentially the same function as the aflatoxin pathway (sterigmatocystin is a precursor of aflatoxins), and regulated in a similar manner, is organized very differently. This observation argues against transfer of metabolic pathways as clusters. However, *Aspergillus flavus* and *A. parasiticus*, like several fungi, lack a sexual cycle, and can cross by parasexual recombination, whereas *A. nidulans* undergoes sexual reproduction. Therefore, it could be postulated that after horizontal transfer of gene clusters in the fungi, the *A. nidulans* gene cluster could have been rearranged through meiotic crossover in sexual recombination in this species, an event that was not possible in parasexual recombination in other *Aspergilli*. But what would be the evolutionary or physiological advantage to *A. nidulans* for this rearrangement? *A. flavus*, and to some extent *A. parasiticus*, predominantly produce aflatoxins in nature, whereas *A. nidulans* makes sterigmatocystin under very specific conditions. Maybe, this suggests that *A. nidulans* is in the evolutionary process of dissolution of the original toxin pathway gene cluster. It may have already rid itself of the last two steps in aflatoxin synthesis (sterigmatocystin → O-methylsterigmatocystin → aflatoxin B_1). Keller and Hohn (1) describe this process, "Gene cluster dissolution may in turn signal the first step in the eventual loss of the pathway or may simply reflect the lack of selection pressure required to maintain the cluster." The absence of clustering of the genes for melanin biosynthesis in some fungi may also suggest dissolution of gene clusters. The melanin biosynthetic pathway gene cluster is supposedly in the process of "break-up" to a differing extent in several fungi (150); with no apparent reason for this process. A consequence of this dissolution could be the utilization of one or more genes of the cluster towards related functions in other pathways. These arguments will be substantiated as more metabolic pathways are characterized, and as one or more of these pathway clusters are detected in other related or unrelated fungi. For example, with the discovery of aflatoxin pathway gene clusters in the non-aflatoxigenic *Aspergilli*, *A. oryzae* and *A. sojae* (Section *Flavi*) (155, 156) and the presence of interrupted regulatory elements in an atoxigenic *Aspergillus* sp. (157), more clues as to the reasons for the "dissolution of pathway gene clusters" may be uncovered ,as well as to what may be the evolutionary significance of several of the metabolic pathway gene clusters.

Research in filamentous fungi is currently underway to establish (a) the cascade of regulatory elements that govern processes expressed by the identified gene clusters; (b) the relationship between genes or gene clusters involved in primary and secondary metabolism; (c) ecological or evolutionary significance of gene clusters; (d) the relationship between genes or gene clusters in prokaryotes to those found in fungi; and finally (e) the identification and characterization of gene clusters that may be responsible for several other metabolic pathways in fungi. Research tools such as sequencing of fungal genomes, phylogenetic studies, and proteomics will enable mycologists to rapidly decipher the biological need or significance of metabolic pathway gene clusters in filamentous fungi.

REFERENCES

1. E. P. Hull, P. M. Green, H. N. Arst, Jr., and C. Scazzocchio, Mol. Microbiol. 3 (1989) 553.

2. N. P. Keller and T. M. Hohn, Fungal Genetics and Biology. 21 (1997) 17.

3. A. J. Clutterbuck, Sexual and parasexual genetics of *Aspergillis* species. In: J. W. Bennett and M. A. Klich (Eds.), *Aspergillus*: Biology and Industrial Applications. Butterworth-Heinemann, Boston, 1992, pp. 3-18.

4. C. Scazzocchio, Control of gene expression in the catabolic pathways of *Aspergillus nidulans*: A personal and biased account. In: J. W. Bennett and M. A. Klich (Eds.), *Aspergillus*: Biology and Industrial Applications. Butterworth-Heinemann, Boston, 1992, pp. 43-68.

5. B. Kudla, M. X. Caddick, T. Langdon, N. M. Marinez-Rossi, C. F. Bennet, S. Sibley, R. W. Davis and H. N. Arst, Jr., EMBO J. 9 (1990) 1355.

6. C. E. Dowzer and J. M. Kelly, Mol. Cell. Biol. 11 (1991) 5701.

7. S. P. Kale, J. W. Cary, D. Bhatnagar and J. W. Bennett, Appl. Environ. Microbiol. 62 (1996) 3399.

8. J. K. Hicks, J.-H. Yu, N. P. Keller, and T. H. Adams, EMBO J. 16 (1997) 4916.

9. D. Guzmán-de-Peña and J. Ruiz-Herrera, Fung. Gen. and Biol. 21 (1997) 198.

10. M.C. Brandriss and B. Magasanik, J. Bacteriol. 140 (1979) 498.

11. H. N. Arst Jr. and D. W. MacDonald, Nature. 254 (1975) 26.

12. C. Scazzocchio, Prog. Ind. Microbiol. 29 (1994) 221.

13. K. K. Sharma and H. N. Arst Jr., Curr. Genet. 9 (1985) 299.

14. B. Cazelle, A. Pokorska, E. Hull, P. M. Green, G. Stanway and C. Scazzocchio, Mol. Microbiol. 28 (1998) 355.

15. C. E. Dowzer and J. M. Kelly, Mol. Cell. Biol. 11 (1991) 5701.

16. B. Cubero and C. Scazzocchio, EMBO J. 13 (1994) 407.

17. R. Gonzalez, V. Gavrias, D. Gomez, C. Scazzocchio, and B. Cubero, EMBO J. 16 (1997) 2937.

18. B. Felenbok, H. M. Sealy-Lewis, Prog. Ind. Microbiol., 29 (1994) 141.

19. M. Pickett, D. I. Gwynne, F. P. Buxton, R. Elliott, R. W. Davies, R. A. Lockington, C. Scazzocchio and H. M. Sealy-Lewis, Gene. 51 (1987) 217.

20. S. Fillinger and B. Felenbok, Mol. Microbiol. 20 (1996) 475.

21. F. Lenouvel, I. Nikolaev, and B. Felenbok, J. Biol. Chem. 272 (1997) 15521.

22. C. Panozzo, V. Capuano, S. Fillinger and B. Felenbok, J. Biol. Chem. 272 (1997) 22859.

23. R. Lockington, C. Scazzocchio, D. Sequeval, M. Mathieu and B. Felenbok, Mol. Microbiol. 1 (1987) 275.

24. S. Fillinger, C. Panozzo, M. Mathieu and B. Felenbok, FEBS Lett. 368 (1995) 547.

25. P. Kulmburg, M. Mathieu, C. Dowzer, J. Kelly and B. Felenbok, Mol. Microbiol. 7 (1993) 847.

26. M. Mathieu and B. Felenbok, EMBO J. 13 (1994) 4022.

27. C. Panozzo, E. Cornillot and B. Felenbok, J. Biol. Chem. 273 (1998) 6367.

28. M. X. Caddick, D. Peters, A. N. Platt, and Antonie Van Leeuwenhoek. 65 (1994) 169.

29. G. A. Marzluf, Microbiol. Mol. Biol. Rev. 61 (1997) 17.

30. I. L. Johnstone, P. C. McCabe, P. Greaves, S. J. Gurr, G. E. Cole, M. A. Brow, S. E. Unkles, A. J. Clutterbuck, J.R. Kinghorn, and M. A. Innis, Gene. 15 (1990) 181.

31. H. Haas and G. A. Marzluf, Curr. Genet. 28 (1995) 177.

32. P.-K. Chang, K. C. Ehrlich, J.E. Linz, D. Bhatnagar, T. E. Cleveland, J. W. Bennett, Curr. Genet. 30 (1996) 68.

33. N. Kitamoto, T. Kimura, Y. Kito, K. Ohmiya, N. Tsukagoshi, Biosci. Biotechnol. Biochem. 59 (1995) 1795.

34. Y. G. Amaar and M. M. Moore, Curr. Genet. 33 (1998) 206.

35. R.S. Williams, M. A. Davis, and B. J. Howlett, Gene. 158 (1995) 153.

36. G. R. Banks, P. A. Shelton, N. Kanuga, D. W. Holden, and A. Spanos, Gene. 131 (1993) 69.

37. P. M. Okamoto, Y.H. Fu, and G.A. Marzluf, Mol. Gen. Genet. 227 (1991) 213.

38. G. E. Exley, J.D. Colandene, and R.H. Garrett, J. Bacteriol. 175 (1993) 2379.

39. B. Tudzynski, K. Mende, K. M. Weltring, J. R. Kinghorn, and S. E. Unkles, Microbiology. 142 (1996) 533.

40. M. D. Perez, C. Gonzalez, J. Avila, N. Brito, and J. M. Siverio, Biochem. J. 321 (1997) 397.

41. J. Avila, C. Gonzalez, N. Brito, and J. M. Siverio, Biochem. J. 335 (1998) 647.

42. Y. H. Fu and G. A. Marzluf, Proc. Natl. Acad. Sci. 87 (1990) 5331.

43. P.-K. Chang, J. Yu, D. Bhatnagar, and T. E. Cleveland, Biochim. Biophy. Acta. (2000) In press.

44. G. F. Yuan, Y. H. Fu, and G. A. Marzluf, Mol. Cell. Biol. 11 (1991) 5735.

45. G. Burger, J. Strauss, C. Scazzocchio, and B. F. Lang, Mol. Cell. Biol. 11 (1991) 5746.

46. H. Pan, B. Feng, and G. A. Marzluf, Mol. Microbiol. 26 (1997) 721.

47. A. Andrianopoulos, S. Kourambas, J. A. Sharp, M. A. Davis, and M. J. Hynes, J. Bacteriol. 180 (1998) 1973.

48. H. Haas, K. Angermayr, I. Zadra, and G. Stoffler, J. Biol. Chem. 272 (1997) 22576.

49. J. Hofman-Bang, Mol. Biotechnol. 12 (1999) 35.

50. P. J. Punt, J. Strauss, R. Smit, J. R. Kinghorn, C. A. van den Hondel, and C. Scazzocchio, Mol. Cell. Biol. 15 (1995) 5688.

51. M. I. Muro-Pastor, R. Gonzalez, J. Strauss, F. Narendja, and C. Scazzocchio, EMBO J. 18 (1999) 1584.

52. B. Feng and G. A. Marzluf, Mol. Cell. Biol. 18 (1998) 3983.

53. A. J. Small, M. J. Hynes, and M. A. Davis, Genetics. 153 (1999) 95.

54. J. L. Celenza, L. Marshall-Carlson, and M. Carson, Proc. Natl. Acad. Sci. 85 (1988) 2130.

55. J. O. Nehlin, M. Carlberg, and H. Ronne, Gene. 85 (1989) 313.

56. A. Kruckeberg and L. F. Bisson, Mol. Cell. Biol. 10 (1990) 5903.

57. D. A. Lewis and L. F. Bisson, Mol. Cell. Biol. 11 (1991) 3804.

58. C. H. Ko, H. Liang, and R. F. Gaber, Mol. Cell. Biol. 13 (1993) 638.

59. M. Johnston, Trends Genet. 15 (1999) 29.

60. R. B. Needleman, Mol. Microbiol. 9 (1991) 2079.

61. A. W. Gibson, L. A. Wojciechowicz, S. E. Danzi, B. Zhang, J. H. Kim, Z. Hu, and C. A. Michels, Genetics. 146 (1997) 1287.

62. R. Kelly and K. J. Kwon-Chung, J. Bacteriol. 174 (1992) 222.

63. H. Giles, M.E. Case, J. Baum, R. Geever, L. Huiet, V. Patel, and B. Tyler, Microbiol. Rev. 49 (1985) 338.

64. A. R. Hawkins, H. K. Lamb, M. Smith, J. W. Keyte, and C. F. Roberts, Mol. Gen. Genet. 214 (1988) 224.

65. J. A. Baum, R. Geever, and N. H. Giles, Mol. Cell. Biol. 7 (1987) 1256.

66. A. R. Hawkins, H. K. Lamb, and C. F. Roberts, Gene. 110 (1992) 109.

67. I. Levesley, G. H. Newton, H. K. Lamb, E. van Schothorst, R. W. Dalgleish, A. C. Samson, C. F. Roberts, and A. R. Hawkins, Microbiology. 142 (1996) 87.

68. H. K. Lamb, G. H. Newton, L. J. Levett, E. Cairns, C.F. Roberts, and A.R. Hawkins, Microbiology. 142 (1996) 1477.

69. A.R. Hawkins, H.K. Lamb, J.D. Moore, I.G. Charles, and C.F. Roberts, J. Gen. Microbiol. 139 (1993) 2891.

70. D. H. Griffin, Fungal Physiology. (1994) 215.

71. I.G. Charles, J. W. Keyte, W. J. Brammar, M. Smith, and A.R. Hawkins, Nucleic Acids Res. 14 (1986) 2201.

72. A. R. Hawkins, Curr. Genet. 11 (1987) 491.

73. A. R. Hawkins, H. K. Lamb, J.D. Moore, and C.F. Roberts, Gene. 136 (1993) 49.

74. A. R. Hawkins, H. K. Lamb, A. Radford, and J. D. Moore, Gene. 146 (1994) 145.

75. H. N. Arst, Jr., D. W. Holden, and M. X. Caddick, Gene. 173 (1997) Suppl:S1.

76. T. Goto, D. T. Wicklow, and Y, Ito, Appl. Environ. Microbiol. 62 (1996) 4036.

77. J.C. Frisvad and R.A. Samson, New producers of aflatoxin. In: Samson, R.A. and Pitt, J.I. (Eds.), Integration of molecular and morphological approaches to Aspergillus and Penicillium taxonomy. Plenum, Reading, UK, 1999, pp.

78. M. A. Klich, E. J. Mullaney, C. B. Daly, and J. W. Cary, Appl. Microbiol. Biotechnol. (2000) In press.

79. J. W. Cary, D. Bhatnagar and J. E. Linz, Aflatoxins; Biological significance and regulation of biosynthesis. In: J. W. Cary, J. E. Linz, and D. Bhatnagar, (Eds.), Microbial foodborne diseases: Mechanisms of pathogenesis and toxin synthesis. Technomic, Lancaster, PA, 2000, pp. 317-361.

80. F. Trail, N. Mahanti, M. Rarick, R. Mehigh, S. H. Liang, R. Zhou, and J. E. Linz, Appl. Environ. Microbiol. 61 (1995) 2665.

81. J. Yu, P.-K. Chang, J. Cary, M. Wright, D. Bhatnagar, T. E. Cleveland, G. A. Payne and J. E. Linz, Appl. Environ. Microbiol. 61 (1995) 2365.

82. D. W. Brown, J.-H. Yu, H. S. Kelkar, M. Fernandes, T. C. Nesbitt, N. P. Keller, T. H. Adams, and T. J. Leonard, Proc. Natl. Acad. Sci. 93 (1996b) 1418.

83. J. W. Cary, M. Wright, D. Bhatnagar, R. Lee, and F. S. Chu, Appl. Environ. Microbiol. 62 (1996) 360.

84. J. Yu, P.-K. Chang, D. Bhatnagar and T. E. Cleveland, Appl. Microbiol. Biotechnol. (2000) in press.

85. P.-K. Chang, K. C. Ehrlich, J. Yu, D. Bhatnagar, and T. E. Cleveland, Appl. Environ. Microbiol. 61 (1995) 2372.

86. J.-H. Yu, R. A. E. Butchko, M. Fernandes, N. P. Keller, T. J. Leonard, and T. H. Adams, Curr. Genet. 29 (1996) 549.

87. K.C. Ehrlich, B.G. Montalbano, D. Bhatnagar, and T. E. Cleveland, Fungal Genet. Biol. 23 (1998) 279.

88. M. Fernandes, N.P. Keller, and T. H. Adams, Mol. Microbiol. 28 (1998) 1355.

89. K.C. Ehrlich, B.G. Montalbano, and J. W. Cary, Gene. 230 (1999) 249.

90. D. M. Meyers, G. O'Brian, W. L. Du, D. Bhatnagar, and G.A. Payne, Appl. Environ. Microbiol. 64 (1998) 3713.

91. R. A. E. Butchko, T. H. Adams, and N. P. Keller, Genetics. 153 (1999) 715.

92. C. P. Woloshuk, G. L. Yousibova, J. A. Rollins, D. Bhatnagar, and G.A. Payne, Appl. Environ. Microbiol. 61 (1995) 3019.

93. G. H. Feng and T. J. Leonard, Appl. Environ. Microbiol. 64 (1998) 2275.

94. N. P. Keller, C. Nesbitt, B. Sarr, T. D. Phillips, and G. D. Burow, Phytopathology. 87 (1997) 643.

95. K. C. Ehrlich, J. W. Cary, and B. G. Montalbano, Biochim. Biophys. Acta. 1444 (1999) 412.

96. R. H. Proctor, Fusarium toxins: Trichothecenes and Fumonisins. In: J. W. Cary, J. E. Linz, and D. Bhatnagar (Eds.), Microbial foodborne diseases: Mechanisms of pathogenesis and toxin synthesis. Technomic, Lancaster, PA, 2000, pp. 363-381.

97. R. P. Sharma and Y. W.Kim, Trichothecenes. In: R. P. Sharma, and D. K. Salunkhe (Eds.), Mycotoxins and Phytoalexins. CRC Press, Boca Raton, FL, 1991, pp. 339-359.

98. S. P. McCormick and T.M. Hohn, Appl. Environ. Microbiol. 63 (1997) 1685.

99. N. J. Alexander, T.M. Hohn, and S.P. McCormick, Appl. Environ. Microbiol. 64 (1998) 221.

100. S. P. McCormick, T. M. Hohn, and A. E. Desjardins, Appl. Environ. Microbiol. 62 (1996) 353.

101. T. M. Hohn, R. Krishna, and R.H. Proctor, Fungal Genet. Biol. 26 (1999) 224.

102. N. J. Alexander, S. P. McCormick, and T.M. Hohn, Mol. Gen. Genet. 261 (1999) 977.

103. S. P. McCormick, N. J. Alexander, S. E. Trapp, and T.M. Hohn, Appl. Environ. Microbiol. 65 (1999) 5252.

104. S. C. Trapp, T. M. Hohn, S. P. McCormick, and B. B. Jarvis, Mol. Gen. Genet. 257 (1998) 421.

105. A.A. Brakhage, Microbiol. Mol. Biol. Rev. 62 (1998) 547.

106. A. P. MacCabe, M. B. R. Riach, S. E. Unkles, J. R. Kinghorn, EMBO J. 9 (1990) 279.

107. S. Gutierrez, B. Diez, E. Montenegro, and J. F. Martin, J. Bacteriol. 173 (1991) 2354.

108. S. Gutierrez, J. Velasco, F. J. Fernandez, and J. F. Martin, J. Bacteriol. 174 (1992) 3056.

109. E. A. Espeso, J. M. Fernandez-Canon, and M. A. Penalva, FEMS Microbiology Lett. 126 (1995) 63.

110. E. A. Espero, J. Tilburn, H. N. Arst, and M. A. Penalva, EMBO J. 12 (1993) 3947.

111. A. J. Shah, J. Tilburn, M. W. Adlard, and H.N. Arnst, Jr., FEMS Microbiol. Lett. 77 (1991) 209.

112. J. Tilburn, S. Sarkar, D. A. Widdick, E. A. Espeso, M. Orejas, J. Mungroo, M. A. Penalva, and H. N. Arst, Jr., EMBO J. 14 (1995) 779.

113. K. Then Bergh and A. A. Brakhage, Appl. Environ. Microbiol. 64 (1998) 843.

114. A. A. Brakhage, A. Andrianopoulos, M. Kato, S. Steidl, M.A. Davis, N. Tsukagoshi, and M.J. Hynes, Fungal Genet. Biol. 27 (1999) 243.

115. K. Then Bergh, O. Litzka, and A.A. Brakhage, J. Bacteriol. 178 (1996) 3908.

116. O. Litzka, P. Papagiannopolous, M. A. Davis, M. J. Hynes, and A. A. Brakhage, Eur. J. Biochem. 251 (1998) 758.

117. B. Tudzynski, Appl. Microbiol. Biotechnol. 52 (1999) 298.

118. B. Tudzynski and K. Hölter, Fungal Genet. Biol. 25 (1998) 157.

119. B. Tudzynski, H. Kawaide, and Y. Kamiya, Curr. Genet. 34 (1998) 234.

120. B. Tudzynski, V. Homann, B. Feng, and G.A. Marzluf, Mol. Gen. Genet. 261 (1999) 106.

121. M.J. Butler and A. W. Day, Can. J. Microbiol. 44 (1998) 1115.

122. C. Kawamura, T. Tsujimoto, and T. Tsuge, Mol. Plant-Microbe Interact. 12 (1999) 59.

123. N. Kimura and T. Tsuge, J. Bacteriol. 175 (1993) 4425.

124. H-F. Tsai, M. H. Wheeler, Y. C. Chang, and K.J. Kwon-Chung, J. Bacteriol. 181 (1999) 6469.

125. H-F. Tsai, Y. C. Chang, R. G. Washburn, M. H. Wheeler, and K. J. Kwon-Chung, J. Bacteriol. 180 (1998) 3031.

126. M. H. Wheeler and A. A. Bell, Melanins and their importance in pathogenic fungi. In: M. R. McGinnis (Ed.), Current topics in medical mycology, vol. 2. Springer-Verlag, New York, 1988, pp. 338-387.

127. F. G. Chumley and B. Valent, Mol. Plant-Microbe Interact. 3 (1990) 135.

128. Y. Kubo, Y. Takano, N. Endo, N. Yasuda, S. Tajima, and I. Furusawa, Appl. Environ. Microbiol. 62 (1996) 4340.

129. W. F. O. Marasas, Fumonisins: history, world-wide occurrence and impact. In: L. S. Jackson, J. W. DeVries, and L. B. Bullerman, (Eds.), Fumonisins in food, Plenum Press, New York, 1996, pp. 1-17.

130. R. H. Proctor, A. E. Desjardins, R. D. Plattner, and T. M. Hohn, Fungal Genet. Biol. 27 (1999) 100.

131. A. E. Desjardins, R. D. Plattner, and R.H. Proctor, Appl. Environ. Microbiol. 62 (1996) 2571.

132. R. H. Proctor, A. E. Desjardins, R. D. Plattner, and J-A. Seo, Mycotoxins. (2000) In press.

133. J.-R. Xu and J. F. Leslie, Genetics. 143 (1996) 175.

134. J. Cole, J. W Kirksey, and J. M. Wells, Can. J. Microbiol. 20 (1974) 1159.

135. C. Young, Y. Itoh, R. Johnson, I. Garthwaite, C. O. Miles, S. C. Munday-Finch, and B. Scott, Curr. Genet. 33 (1998) 368.

136. C. Young, L. McMillan, and B. Scott, 20th Fungal Genetics Conference, Pacific Grove, CA, March 23-28, 1999, abstract #284, p. 121.

137. J. D. Walton, Plant Cell. 8 (1996) 1723.

138. J-H. Ahn and J. D. Walton, Plant Cell. 8 (1996) 887.

139. J. D. Walton, J-H. Ahn, J. W. Pitkin, Y. Q. Cheng, A. N. Nikolskaya, R. Ransom, and S. Wegener, Enzymology, molecular genetics, and regulation of biosynthesis of the host-selective toxin HC-toxin. In: K. Kohmoto and O. C. Yoder (Eds.), Molecular genetics of host-specific toxins in plant disease, Kluwer, Dordrecht, Netherlands, 1998, pp. 25-34.

140. J-H. Ahn and J.D. Walton, Mol. Gen. Genet. 260 (1998) 462.

141. Y. Q. Cheng, J-H. Ahn, and J.D. Walton, Microbiol. 145 (1999) 3539.

142. Y. Q. Cheng and J-H. Ahn, 20th Fungal Genetics Conference, Pacific Grove, CA, March 23-28, 1999, abstract #214, p. 100.

143. H. Otani, K. Kohmoto, S. Nishimura, T. Nakashima, T. Ueno, and H. Fukami, Ann. Phytopathol. Soc. Jpn. 51 (1985) 285.

144. A. Tanaka, H. Shiotani, M. Yamamoto, and T. Tsuge, Mol. Plant-Microbe Interact. 12 (1999) 691.

145. O.C. Yoder, A mechanistic view of the fungal/plant interaction based on host-specific toxin studies. In: K. Kohmoto and O. C. Yoder (Eds.), Molecular genetics of host-specific toxins in plant disease, Kluwer, Dordrecht, Netherlands, 1998, pp. 3-15.

146. S. W. Lu, B.G. Turgeon, and O.C. Yoder, 20th Fungal Genetics Conference, Pacific Grove, CA, March 23-28, 1999, abstract #245, p. 109.

147. S. Gutiérrez, F. Fierro, J. Casqueiro, and J. F. Martin, Antoine van Leevwenhoek. 75 (1999) 1058.

148. L. Mathison, C. Soliday, T. Stephan, T. Aldrich, and J. Rambosek, Curr. Genet. 23 (1993) 33.

149. S.-H. Liang, T.-S. Wu, R. Lee, F. S. Chu and J. F. Linz, Appl. Environ. Microbiol. 63 (1997) 1058.

150. T. Kubo, Y. Takano and I. Furusawa, Molecular genetic analysis of melanin biosynthetic genes essential for appressorium function in *Colletorichum lagenarium*. In: D.Mills, H. Kunoh, N. Keen and S. Mayama (Eds.), Molecular Aspects of Pathogenicity: Requirements for Signal Transduction. APS Press, St. Paul, 1996, pp. 76-82.

151. B. J. Weigel, S. G. Burgett, V. J. Chen, P. L. Skatrud, C. A. Frolik, S. W. Queener, and T. D. Ingolia, J. Bacteriol. 170 (1988) 3817.

152. L. Mathison, C. Soliday, T. Stepan, T. Aldrich and J. Rambosek, Curr. Genet. 23 (1993) 33.

153. A. Vidal-Cros, F. Viviani, G. Labesse, M. Boccara, and M. Gaudry, Eur. J. Biochem. 219 (1994) 985.

154. M. W. Smith, D. Feng, and R. F. Doolittle, Trends Biochem. Sci. 17 (1992) 489.

155. P.-K. Chang, D. Bhatnagar, T. E. Cleveland and J. W. Bennett, Appl. and Environ. Microbiol. 61 (1995) 40.

156. M. A. Klich, J. Yu, P.-K. Chang, E. J. Mullaney, D. Bhatnagar, and T. E. Cleveland, Appl. Microbiol. Biotechnol. 44 (1995) 439.

157. D. M. Geiser, J. I. Pitt, and J. W. Taylor, Proc. Natl. Acad. Sci. 95 (1998) 388.

Applied Mycology and Biotechnology
Volume 1. Agriculture and Food Production
G.G. Khachatourians and D.K. Arora (editors)

Molecular transformation, gene cloning, and gene expression systems for filamentous fungi

Scott E. Gold[1], John W. Duick[1], Regina S. Redman[2,3], and Rusty J. Rodriguez[2,3,*]

[1]Plant Pathology, University of Georgia, Athens, GA, 30602-7274; [2]Botany Department, University of Washington, Seattle, WA, 98195; [3]Western Fisheries Research Center, USGS/BRD, Seattle, WA, 98115

1. INTRODUCTION

The development of molecular biological techniques has allowed researchers to begin elucidating the genetic and biochemical mechanisms responsible for complex phenotypes in a variety of organisms. One of the most important techniques, molecular transformation, has revolutionized our understanding of fungal biology and resulted in numerous biotechnological developments. Molecular transformation involves the movement of discrete amounts of DNA into cells, the expression of genes on the transported DNA, and the sustainable replication of the transforming DNA. The ability to move DNA and study the expression of specific genes in fungi has significantly increased scientific understanding of developmental processes, symbiosis, response to environmental stress, primary and secondary metabolism, taxonomy, genetics, and meiosis.

Over the last decade, there have been a limited number of excellent reviews on fungal transformation (Fincham, 1989). Several reviews covering expression systems for specific groups of fungi have also been published (Ballance, 1986; Ballance, 1991; Bodie, 1994; Frommer and Ninnemann, 1995; MacKenzie et al., 1993; Radzio and Kuck, 1997; Turner, 1994; Verdoes, 1995). Rather than reiterate most of the details described in those reviews, this chapter will focus primarily on more recent developments. Specifically, we focus on transformation strategies and vectors for fungi, mechanisms of transformation, exploiting transformation to identify and clone genes, and the expression of gene sequences in heterologous and homologous cells. Most of the discussion is limited to the higher fungal taxa of the Ascomycetes and the Basidiomycetes. In addition, citation of the literature is intended to be illustrative of the discussion and is not exhaustive.

* Corresponding Author

2. TRANSFORMATION SYSTEMS AND GENE CLONING

2.1. Historical development of transformation

Molecular transformation of fungi resulted from research beginning as early as the 1920's and is still ongoing. To encompass important early achievements, we will begin with a historical perspective on the development of molecular biology and the transformation process.

The ability to transform fungi is dependent on the stable replication and expression of genes located on the transforming DNA. Three phenomena observed in bacteria (competence, plasmids, and restriction enzymes to facilitate cloning) were responsible for the development of molecular transformation in fungi. Observations involving these three aspects of bacterial physiology began with studies on Streptococcus pneumoniae in the 1920's and ended in 1972 with the first sucessful DNA-mediated transformation of *Escherichia coli* (Cohen et al., 1972).

Competence involves the ability of an organism to take-up DNA and was first reported by F. Griffith (1928) who discovered that *S. pneumoniae* isolated from patients with pneumonia produced colonies with a smooth (S) appearance (Griffith, 1928). When the S colonies were subcultured in vitro, colonies with a rough (R) appearance appeared that were no longer virulent. It was also observed that if R cells were grown in the presence of heat-killed S cells, the R cells would convert to S cells. It was later determined that cell free extracts of S colonies contained a substance capable of transforming R cells back to S cells with a concomitant reversion back to virulence. It was not until 1944 that the transforming factor was isolated and identified as DNA (Avery et al., 1944). A series of experiments followed that involved the conversion of nutritional mutants of bacteria (auxotrophs) to wild type (wt) isolates by exposing the auxotrophs to sheared DNA from wt bacteria. The competence of the bacteria is dependent on the physiological state of the cells and has been intensively studied (Dubnau, 1991; Lorenz and Wackernagel, 1994).

The discovery of bacterial plasmids began in 1946 when Lederberg and Tatum pursued transformation experiments with auxotrophic mutants of *E. coli* (Lederberg and Tatum, 1946). When viable cells, having different auxotrophic markers were mixed together, wt cells could be produced by a process termed conjugation (cell contact without fusion). Eventually, it was determined that there were two mating types in *E. coli* designated a genetic donor (male) and a genetic recipient (female) [see review, (Clark and Adelberg, 1962)]. Transformation involved the transfer of a genetic element, through a conjugation tube, that converted females to males and was designated the fertility or F plasmid (extrachromosomal hereditary determinant). Auxotrophic complementation occurred as a result of F plasmid recombination with genomic DNA such that functional genes could be transferred to the F plasmids and subsequently transferred to recipient cells via conjugation.

The characterization of F plasmids paved the way to the discovery of a new class of plasmids that could move between cells without conjugation. During the 1940ís bacterial dysentery was a serious problem in Japan and was treated with sulfonamide. By 1950, strains of the causative agent, *Shigella*, were isolated that were resistant to sulfonamide. In response, three additional antibacterial drugs (tetracycline, streptomycin, and chloramphenicol) were used to control the bacterium. By 1960 isolates of *Shigella* were found that possessed resistance to all of the drugs used to combat the disease. This form of resistance could be transmitted between cells independent of the conjugation process and the genetic elements designated resistance or R plasmids [see historical review, (Anderson, 1968)]. The R plasmids contained genes that conferred resistance to chemicals and were later exploited to develop transformation vectors that allowed scientists to identify transformants based on the expression of drug resistance.

The final discovery that led to the development of transformation protocols began with observastions in the 1920ís that resulted in the discovery of bacteriophage [see historical review, (Lwoff, 1953)]. In 1950, Lwoff et al. found that specific strains of Bacillus megaterium exposed to ultraviolet light would lyse 45 minutes after exposure releasing virus particles into the medium (Lwoff, 1953). In addition, several studies indicated that bacteriophage could exist in a dormant (lysogenic) or in an active (lytic) form in cells and that bacteria lysogenized with lambda were resistant to infection by other phages (incompatibility) [see review, (Lwoff, 1953)]. Research from several laboratories demonstrated that host cells controlled phage incompatibility (Arber, 1965) and in 1962 Dussoix and Arber found that incompatible phages were degraded soon after infection (Dussoix and Arber, 1962). In 1968, Meselson and Yuan purified and characterized the host enzyme restriction system which was defined as the first type 1 restriction enzyme [see review, (Boyer, 1971)]. In 1970, the first type II restriction enzyme was isolated and shown to have specific sequence recognition sites (Old and Primrose, 1989).

By combining the information and tools generated from 50 years of research on cell competence, plasmid biology, and restriction enzymes, Cohen et al. transformed *E. coli* with three plasmids that collectively conferred resistance to tetracyline, streptomycin, kanamycin, neomycin, chloramphenicol, and sulphoamide, (Cohen et al., 1972). Based on work by Mandel and Higa (Mandel and Higa, 1970), it was observed that the addition of high concentrations of Ca^{++} ions resulted in a significant increase in transformation efficiency (Cohen et al., 1972). This was the first demonstration that DNA manipulated in vitro could transform the cells by the expression of foreign genes. Between 1960 and 1973, several labs began experiments to determine if auxotrophic fungal mutants could be transformed in the same manner as bacteria. Auxotrophic mutants of *Neurospora crassa* (Mishra, 1977; Mishra, 1979; Mishra and Tatum, 1973) were transformed at very low rates to prototrophy by exposing the mutants to sheared wtDNA. The frequency of fungal transformation was dependent on the level of cell competence and transformation occurred with intra- or inter-species DNA. A major breakthrough in fungal transformation came with the realization that when the cell walls were removed, the resulting protoplasts showed increased competence levels by several orders of magnitude (Hinnen et al., 1978) In 1978, *Saccharomyces cerevisiae* was the first eukaryotic organism to be transformed (Beggs, 1978; Hinnen et al., 1978) followed by *Neurospora crassa* in 1979 (Case, 1982). These events were followed by two decades of research to better understand the transformation process and develop strategies to transform numerous prokaryotic and eukaryotic taxa with high efficiency.

2.2. Transformation strategies for fungi

2.2.1. Protoplasts

Initial transformation success with filamentous fungi, involving complementation of auxotrophic mutants by exposure to sheared genomic DNA or RNA from wt isolates, occurred with low transformation efficiencies. In addition, it was difficult to retrieve complementing DNA fragments and isolate genes of interest. This prompted the development of transformation vectors and methods to increase efficiencies. During the 1960ís, physiological studies performed with fungi indicated that the cell wall could be removed to generate protoplasts see review, (Davis, 1985). In 1973, it was realized that protoplasts could be transformed with significantly greater efficiencies than walled-cells (Hinnen et al., 1978). Between 1973 and 1984 methods for protoplast formation and transformation vectors were developed for both single celled- and filamentous-fungi. (Davis, 1985). These protocols involved the enzymatic degradation of cell walls with a variety of enzymes derived from fungi (Peberdy, 1985). Preparation of cell wall digesting enzymes was laborious and commonly resulted in end products that were inconsistent with regard to the efficiency of cell wall removal. In addition, cell wall structure varies between fungal species so effectiveness of cell wall degrading enzyme preparations may be limited to specific species or isolates (Farkas, 1985). Eventually, commercial preparations of cell wall degrading enzymes such as Novozyme 234 from *Trichoderma viride* became available. Although the commercial preparations eliminated the laborious tasks of enzyme preparation, inconsistencies between preparations has continued to be a problem. In fact, the laborious nature of protoplast production together with the inconsistent activity and availability of cell wall degrading enzyme preparations present the greatest difficulty in fungal transformation for many laboratories. As a result, several strategies have been developed in an attempt to transform hyphal cells, spores, and single celled fungi so protoplasting difficulties can be avoided (see below). Hopefully, future protocols will circumvent the need for protoplasting cells to achieve high levels of competence.

2.2.2. Transformation strategies

DNA has been introduced into the nuclei of fungal protoplasts, spores, germlings, and hyphal fragments by chemical, electrical, or physical treatments. The most common protocol used for transformation involves exposing protoplasts to DNA in the presence of polyethylene glycol (PEG) and calcium chloride ($CaCl_2$). Typically, the PEG used is approximately 4000MW and $CaCl_2$ can be replaced with other alkali cations such as LiCl (Ito et al., 1983). Over the last decade, species representing numerous fungal taxa have been transformed using the PEG/$CaCl_2$ method. Although the PEG/$CaCl_2$ method is used most frequently, it is a time consuming task due to the need to generate competent protoplasts followed by the regeneration of protoplasts after DNA uptake. More importantly, as mentioned above, variation in commercial enzyme preparations makes this strategy unreliable and limits the number of species that may be manipulated.

Over the last 15 years, several protocols have been developed in an attempt to overcome difficulties associated with protoplast generation, to increase efficiency of DNA incorporation, and to symplify the transformation process. These methods include vortexing cells in the presence of DNA and glass beads, exposing cells to lithium acetate, electroporation, biolistic bombardment, and agrobacterium mediated transformation. The first alternatives to the protoplast-PEG/$CaCl_2$ method involved exposing fungal cells to 0.1M lithium acetate and transforming DNA (Dhawale et al., 1984) or vigourous mixing of cells in a vortex in the presence of transforming DNA and glass beads (Costanzo and Fox, 1988). Although these methods are simple and rapid, transformation efficiencies are very low compared to the protoplast methods (Allison et al., 1992; Dhawale et al., 1984; Dickman, 1988; Soni et al., 1993).

Another relatively simple physical transformation method involves the electrical permeabilization of biomembranes by a process designated electroporation (Potter et al., 1984; Riggs and Bates, 1986). This process is based on exposing cells or protoplasts to a short-duration, high amplitude electric field resulting in a reversible permeabilization of cell membranes. The membrane permeabilization allows for the uptake of DNA and can result in the molecular transformation of cells (if the vector DNA is transported and expressed in the nucleus). While this method has not been used widely, species of *Colletotrichum, Neurospora, Aspergillus, Beauveria, Penicillium, Leptosphaeria, Fusarium,* and *Ustilago* have been transformed by electroporation of protoplasts, conidia, germinated conidia, and hyphal fragments (Bakkeren and Kronstad, 1993; Chakraborty et al., 1991; Pfeifer and Khachatourians 1992, 1993; Redman and Rodriguez, 1994; Richey et al., 1989).

Biolistic transformation is based on the super-sonic delivery of microprojectiles to intact cells (Klein et al., 1987; Sanford et al., 1993). The microprojectiles consist of small diameter particles (0.1-30μm) made from either tungsten, gold, glass, lambda phage, or dried *E. coli* cells and are either coated with or contain transforming DNA. Cells are placed on a substrate (often agar media) and placed in a biolistic chamber where a vacuum is generated followed by microprojectile bombardment. Sanford et al. (Sanford et al., 1993) describes various aspects of the mechanisms and protocols for the biolistic process in detail. Although this transformation strategy requires expensive equipment, it avoids difficulties in generating protoplasts and has been used to successfully transform a number of fungi including *S. cerevisiae, Aspergillus nidulans, Paecilomyces fumosoroseus, Trichoderma harzianum, Gliocladium virens,* and the Zygomycete, *Mucor circinelloides* (Armaleo et al., 1990; Barreto et al., 1997; Gonzalez-Hernandez et al., 1997; Herzog et al., 1996; Klein et al., 1992; Lorito et al., 1993). In some fungi, such as *Neocallimastix frontalis,* this technique has not resulted in stable transformation but has allowed researchers to study the transient expression of a selectable gene (Durand et al., 1997).

An intriguing recent transformation strategy involves the use of *Agrobacterium tumefaciens* to deliver plasmid DNA into fungal cells. *Agrobacterium tumefaciens* possesses large plasmids (designated Ti) and causes crown galls on a large variety of dicotyledenous plants and some monocots (Lippincott and Lippincott, 1975; Van Veen et al., 1988). Gall formation is due to the transfer of a portion of the Ti plasmid (designated T-DNA) to the plant nuclear genome (Chilton et al., 1977). The T-DNA contains several genes that modify plant physiology and development resulting in tumorogenesis and gall formation (Joos et al., 1983; Nester and Gordon, 1988). Transfer of the T-DNA is dependent on two 25 bp repeats and several vir genes located on the Ti plasmid. Upon activation of the *vir* genes by the wound-induced plant biochemical acetosyringone, a linear single stranded (ss) DNA molecule is generated that comprises the sequence between the two 25 bp repeats but does not include the *vir* genes (Bundock et al., 1995). The linear ssDNA is covalently linked with the *vir*D2 gene protein that is responsible for transporting the molecule to the plant nucleus (Bundock et al., 1995).

In 1984, Bevan reported the construction of a binary plant transformation system based on the *Agrobacterium* Ti plasmid system (Bevan, 1984). The binary system consisted of two plasmids that worked in concert to transfer foreign DNA into plants. One plasmid (Bin19) contained the T-DNA, the left and right boarder regions from the Ti plasmid, and a prokaryotic kanamycin resistance gene. The second plasmid (Ti) contained all of the vir genes required for the transfer of T-DNA to plants.

Both plasmids were tansformed into *A. tumefaciens* and subsequent transfer of the T-DNA from Bin19 to plants required the trans acting vir gene products from the Ti plasmid.

Recently, Bin19 was modified to contain genes that allow for selection of fungal transformation events (Bundock et al., 1995; Bundock and Hooykaas, 1996; de Groot et al., 1998). When agrobacterial strains harboring the modified Ti plasmids were mixed with cells, protoplasts, and conidia of different fungal species, transformation was observed. The transformation process was dependent on the presence of an inducer of the *vir* genes such as acetosyringone. Interestingly, the modified Bin19 vectors transformed both Ascomycetes and Basidiomycetes. The simplicity and efficiency of this strategy may result in the generation of a series of modified agrobacterial plasmids to transform a wide taxonomic array of fungi.

2.3. Transformation vectors and selection of transformants

2.3.1. Selection of transformants

There are two types of transformation vectors available for fungi, autonomously replicating and genomic integrating vectors. Both vectors contain genes that allow transformants to be differentiated from non-transformed cells. These genes confer resistance to antibiotics, complement auxotrophic mutations, confer novel metabolic abilities, or produce pigments. The resistance genes most commonly used in fungi confer resistance either to hygromycin B (Punt and van den Hondel, 1992), phleomycin (Punt and van den Hondel, 1992), benomyl (Orbach et al., 1986), or kanamycin/G418 (Jimenez and Davies, 1980).

The most widely used fungal transformation vectors, involving resistance genes, are based on either hygromycin B or phleomycin selection. Hygromycin B is an aminoglycoside antibiotic from *Streptomyces hygroscopicus* (Pittenger et al., 1953) that inhibits peptide chain elongation in both prokaryotes and eukaryotes (Gonzalez et al., 1978). A gene (*hph*) from *E. coli* was isolated that confers resistance to hygromycin by expressing an enzyme (hygromycin B phosphotransferase) that detoxifies the inhibitor (Gritz and Davies, 1983). Phleomycin is a metalloglycopeptide antibiotic that causes DNA strand breakage (Kross et al., 1982). Genes (*ble)* have been isolated from *E. coli* and *Streptoalloteichus hindustanus* that confer resistance by encoding proteins that bind and inactivate phleomycin (Punt and van den Hondel, 1992). The *hph* and *ble* genes have been fused to a variety of fungal promoter sequences to ensure expression in different species.

A series of metabolic genes have also been used in transformation vectors that allow transformants to be identified. These genes have been used to complement auxotrophic mutants or confer new metabolic activity. Auxotrophic mutants deficient in either amino acid (leucine, arginine, and tryptophan), uracil, nitrogen metabolisms, or the catabolism of quinic acid have been used to develop complementing transformation vectors. The vectors have contained the *leu, arg,* and *trp* genes, or *ura* and *pyr* genes to complement amino acid and pyrimidine auxotrophs, respectively (Baek and Kenerley, 1998; Banks and Taylor, 1988; van Hartingsveldt et al., 1987; Wada et al., 1996; Weidner et al., 1998). The qa2 gene has been used to complement mutants in quinic acid catabloisms (Case et al., 1979) and mutants defective in nitrogen metabolism have been used to develop complementation vectors containing the *nia*D gene which encodes the enzyme nitrate reductase (Levis et al., 1997; Tudzynski et al., 1996; Unkles et al., 1989). The *amdS* gene from *Apsergillus nidulans* has allowed the development of vectors that confer new metabolic activity allowing transformants to grow on acetamide as sole nitrogen and/or carbon sources (Hynes et al., 1983).

Another class of genes used to develop transformation vectors allows transformants to be identified visually. The ß-lactamase (*bla*), ß-galactosidase (*lacZ*), β-glucoronidase (GUS), and green fluorescent protein (*gfp*) genes have been used to allow transformants to produce pigments for visualization (Chalfie et al., 1994; Du et al., 1999; Herd et al., 1997; Jefferson et al., 1986; Judelson, 1993; Kolar et al., 1988; Leger et al., 1995; Penttila et al., 1987; Redkar et al., 1998; Rohe et al., 1996; Spellig et al., 1996). For the detection of β-lactamase, β-galactosidase and β-glucoronidase expressing transformants to be visualized, they must be cultured on media containing potassium iodide, X-Gal (5-bromo-4-chloro-3-indolyl-β-D-galacto-pyranoside), and a β-glucoronide, respectively. Since *gfp* produces an auto-fluorescing protein, no substrates are required and transformants may be visualized by epifluorescence (Heim et al., 1994; Prasher et al., 1992).

2.3.2. Transformation vectors

Transformation vectors/plasmids are designed so that they can replicate and be selected for both in *E. coli* and the fungi of interest. Bacterial replication of vectors allows researchers to obtain large quantities of the plasmid DNA for transformation studies and in vitro manipulations. The basic structure of a fungal transformation vector includes an *E. coli* origin of DNA replication, a gene that confers antibiotic resistance to *E. coli*, a number of restriction enzyme sites that occur only once in the vector, and a gene that allows fungal transformants to be identified. Additional modifications allow vectors to incorporate and transform differing amounts of DNA. The basic plasmid design described above may be used to transfer less that 15kb of DNA efficiently. The addition of "cos" sequences from phage lambda allows up to 45kb of DNA to be transformed (An et al., 1996; Diaz-Perez et al., 1996; Orbach, 1994). Other vectors such as YAC (yeast artificial chromosome) and BAC (bacterial artificail chromosome) vectors behave as small chromosomes in yeast and *E. coli*, respectively, and are able to transfer several hundred kilobases of DNA to the fungi of interest during the transformation process (Burke et al., 1987; Centola et al., 1994; Diaz-Perez et al., 1996). Although vectors may vary in size depending on the amount of genomic DNA, there is little documented evidence regarding the relationship between vector size and transformation efficiencies in fungi.

Identification of fungal transformants is dependent on adequate expression of selectable genes. Therefore, the regulatory region used to drive selectable genes are often derived from the fungal species of interest and fused, in frame, with selectable gene sequences. Promoters used to regulate the expression of selectable genes in fungal transformation studies have been obtained from primary and secondary metabolic genes that are expressed in a constitutive or induced manner. Fortunately, there appears to be significant crossover between related fungal species with regard to promoter recognition. For example, the *trp, gdp, nia,* β–tubulin, and *amd*S regulatory regions from *A. nidulans* and *N. crassa* have been used to express selectable markers in a variety of fungal species encompassing several fungal genera (Ballance and Turner, 1985; Barrett et al., 1990; Beri and Turner, 1987; Bogo et al., 1996; Churchill et al., 1990; Daboussi et al., 1989; Kistler and Benny, 1988; Leung et al., 1990; Malardier et al., 1989; Mort-Bontemps and Fevre, 1997; Oliver et al., 1987; Pfeifer and Khachatourians, 1992; Seip et al., 1990; van de Rhee et al., 1996; Wnendt et al., 1990) However, several fungi have required the incorporation of homologous promoters into transformation vectors for biochemical recognition and gene (Tudzynski et al., 1996; Turgeon et al., 1987; Upchurch et al., 1991; van Hartingsveldt et al., 1987) .

There are two basic types of fungal transformation vectors, autonomously replicating and those that integrate into, and replicate as part of the genome. Autonomously replicating vectors contain sequences that are recognized by fungi that allow replication to occur without integration into a chromosome. The first sequence used to design autonomously replicating vectors in fungi was the autonomously replicating sequence, ARS, (*ars*1) from *S. cerevisiae* (Stinchcomb et al., 1979). In filamentous fungi, a number of different sequences have been identified that confer autonomously replicating abilities to transformation vectors. In *Aspergillus* two sequences (ANS1 and AMA1) have been identified that allow for the autonomous replication of transformation vectors (Aleksenko et al., 1996; Aleksenko and Clutterbuck, 1995; Gems et al., 1991). In *Fusarium, Colletotrichum,* and *Cryptococcus* species, telomeric sequences appear to allow for autonomous replication (Edman, 1992; Garcia-Pedrajas and Roncero, 1996; Kistler and Benny, 1992; Redman and Rodriguez, 1994). A unique sequence from *Ustilago maydis* functions as an ARS in transformation vectors for basidiomycetes but not ascomycetes (Tsukuda et al., 1988). Autonomously replicating vectors have often resulted in significantly greater transformation efficiencies than integrative vectors (see below).

2.4. Transformation efficiencies and mechanisms

2.4.1. Transformation efficiencies

The efficiency of transformation is defined as the number of transformants generated from one microgram of transforming vector DNA. Factors that affect transformation efficiency include the number of recipient cells/protoplasts, the competence of cells/protoplasts, vector conformation, vector composition, vector amplification, and chemical/physical aspects of the transformation procedure. Although all of these factors are known to be important in the transformation process, few studies have been completed in sufficient detail to define the importance of each factor among fungi. The only generality that may be summarized for fungi is that there is tremendous variation between species with regard to which of the above factors are most important for achieving high transformation efficiencies. This is best represented by the fact that fungal transformation efficiencies have been shown to vary between <1 to >100,000 transformants/μg vector.

The majority of protocols reported for fungal transformation involve the generation of protoplasts. There is significant variation between protoplast preparations due to cell wall differences between species and isolates, variation in commercial enzyme preparations, and a dearth of knowledge defining the basis of competence. As a result, variation in transformation efficiencies ranging from 2-20 fold between experiments with the same isolates is not uncommon (Barrett et al., 1990; Bogo et al., 1996; Churchill et al., 1990; Kistler and Benny, 1988; Mort-Bontemps and Fevre, 1997; Oliver et al., 1987; Tsuge et al., 1990; Tudzynski et al., 1996; Unkles et al., 1989).

In addition to variation in protoplasts, optimization of chemical and physical aspects of transformation protocols greatly impact transformation efficiencies and may be species or isolate specific. For example, the transformation efficiency of *Cryphonectria parasitica* was increased from 57 to 100,000μg DNA by protocol modification (Churchill et al., 1990). Optimization of chemical and physical aspects of the transformation protocol is most likely required for each species and possibly individual isolates in order to achieve maximal efficiencies. Specifically, the concentrations and quality of chemicals, temperature, use, and duration of heat shock, and post transformation incubation, and the order of addition of transformation constituents to protoplasts/cells all appear to affect transformation efficiency (Churchill et al., 1990).

Transformation strategies other than the PEG/$CaCl_2$ method (Lithium acetate, biolistic, and electroporation) also require optimization to achieve maximum transformation efficiencies (Chakraborty et al., 1991; Dhawale et al., 1984; Dickman, 1988; Redman and Rodriguez, 1994; Richey et al., 1989; Riggs and Bates, 1986; Sanford et al., 1993; Soni et al., 1993). As described for the PEG/$CaCl_2$ method, optimization of these other transformation protocols may be required for each fungal species and perhaps specific isolates. There have been several studies reported comparing the efficiencies of the PEG/$CaCl_2$ method to biolistic and/or electroporation methods (Barreto et al., 1997; Bogo et al., 1996; Herzog et al., 1996; Lorito et al., 1993; Redman and Rodriguez, 1994). In these comparisons, transformation efficiencies with electroporation and biolistic protocols were significantly higher than those observed with the PEG/$CaCl_2$ method. Although it has not been tested extensively, the agrobacterial plasmid transformation method may be more efficient that all other methods (de Groot et al., 1998). However, the variation in transformation efficiencies observed with the agrobacterial method suggests that this protocol will also require species-specific optimization (de Groot et al., 1998).

Several studies have indicated that the conformation and composition of vector DNA can greatly influence transformation efficiency. The conformation of vectors can be circular or linear, and in many fungi, linear molecules transform with higher efficiency (Fotheringham and Holloman, 1990; Garnand and Nelson, 1995; Unkles et al., 1989; Upchurch et al., 1991; Weidner et al., 1998). For example, in *U. maydis* integrative transformation efficiencies of certain vectors can be increased 100 fold if the vectors are linearized prior to transformation (Fotheringham and Holloman, 1989; Fotheringham and Holloman, 1990). In *N. crassa* transformation efficiency is less affected by vector conformation. However, the stability of transformants increases if the vector is linearized (approximately 10% unstable) versus circular vector (approximately 50% unstable) (Garnand and Nelson, 1995). In the Zygomycete, *Rhizomucor pusillus*, transformation efficiency is lower with linear versus circular vector molecules (Wada et al., 1996) while in the Stremenopile, *Phytophthora infestans* efficiency was not effected by vector conformation (Judelson, 1993).

Vector composition has also been shown to affect transformation efficiencies. The inclusion of both functional and nonfunctional sequences in vectors have increased transformation efficiencies in several fungi. Specifically, these sequences have including selectable marker genes, homologous genomic DNA, and the promoter sequences that regulate expression of marker genes. Several reports indicate that the use of homologous marker genes (from the same organism being transformed) can greatly increase transformation efficiency (Levis et al., 1997; Tudzynski et al., 1996; van Hartingsveldt et al., 1987; Weidner et al., 1998). For example, the transformation efficiency of *Giberella fujikuroi* was increased from 1-10 to 200/μg DNA when the heterologous gene *nia*D from *A. nidulans* was replaced with *nia*D from *G. fujikuroi* (Tudzynski et al., 1996). The inclusion of non-selectable homologous sequences has also increased transformation efficiencies in some fungi. The incorporation of ribosomal DNA into the hygromycin vector pDH25 (128) increased transformation efficiency in *Alternaria alternata* from 4.5 to 89/μg vector DNA (Tsuge et al., 1990). In fact, the increase in transformation efficiency was directly proportional to the amount of the ribosomal DNA repeat incorporated into the vector.

The least studied component of fungal transformation efficiencies involves the vector amplification process in bacteria. When vectors are amplified in *E. coli*, it is assumed that the amplified product is the same regardless of what strain is used. However, one report indicates that there may be significant variation in the ability of *E. coli* strains to produce vectors that have the same transformation efficiencies in fungi (Redman and Rodriguez, 1994). Although the same quantitative amounts of vector were produced by different *E. coli* strains, the ability of those vectors to transform *Colletotrichum* species varied from 0 to 20,000/μg DNA. One *E. coli* strain, DH1, never produced vector that was functional in *Colletotrichum*. All other strains produced vector with varying ability to transform *Colletotrichum*. For example, 30 single colony derived cultures of strain DH5α produced equivalent quantities of the vector pHA1.3. However, the ability of the 30 plasmid preparations to transform Colletotrichum varied from 0-20,000/μg DNA (Redman and Rodriguez, 1994). The plasmids used in that study could not be differentiated by restriction enzyme analysis. Although the basis of the bacterial variation remains an enigma, it may represent a major impediment to increasing transformation efficiencies and transforming additional fungal species.

2.4.2. Mechanisms of transformation

Fungal transformation has been shown to occur by integration of vectors into either homologous or non-homologous (ectopic) sequences in the genome, autonomous replication and expression of vectors, gene conversion without vector replication, and co-transformation. The molecular mechanisms for the first three transformation processes have been described in detail elsewhere (Fincham, 1989) , thus we will focus this discussion on co-transformation and on the frequency of homologous integration events.

Co-transformation involves the pooling of non-selectable DNA with a selectable vector during the transformation process. The non-selectable DNA often interacts with selectable vector DNA resulting in intermolecular recombination prior to integration into the fungal genome or autonomous replication of the vector (Gems and Clutterbuck, 1993; Judelson, 1993). This has allowed researchers to transform fungi with unselectable genes such as nutritional genes (Judelson, 1993; Unkles et al., 1989) or β-glucoronidase and β-galactosidase that produce visible pigments (Cooley et al., 1990; Judelson, 1993; Kolar et al., 1988; St. Leger et al., 1995; Penttila et al., 1987; Rohe et al., 1996). Although the frequency of co-transformation has varied between <10-100% of the selectable transformants screened, the majority of fungi co-transform at frequencies greater than 50% (Cooley et al., 1990; Glumoff et al., 1989; Kolar et al., 1988; Nicolaisen and Geisen, 1996; Penttila et al., 1987; Punt et al., 1987; Rohe et al., 1996). In some species, the frequency of co-transformation events is dependent on the conformation of both the selectable and non-selectable DNA. For example, in Phytophthora infestans co-transformation occurred in 10% of the transformants when circular DNA was used and near 100% of the transformants when the DNA was linearized (Judelson, 1993).

One potential problem with studying genes that have been introduced by integration into chromosomes involves position effects with regard to where the vector recombines in the genome and how that affects gene expression (discussed later in this chapter). In addition, ectopic integration may cause the disruption or modification of resident genes resulting in aberrant phenotypes. One strategy for avoiding position effects and genome modifications is to incorporate homologous DNA into transformation vectors either as the selectable genes (Baek and Kenerley, 1998; Banks and Taylor, 1988; Case et al., 1979; Levis et al., 1997; Tudzynski et al., 1996; Unkles et al., 1989; Upchurch et al., 1991; van Hartingsveldt et al., 1987; Wada et al., 1996; Weidner et al., 1998) or non-selectable sequences (Tsuge et al., 1990). This allows vector DNA to recombine at homologous sites in the genome. The frequency of homologous integration appears to be dependent on the amount of homology in the vector (Tsuge et al., 1990) although no minimal amount has been identified in filamentous fungi. In S cerevisiae, between 35bp and 250bp of homology are required for homologous integration depending on the sequences utilized (Rothstein, 1991; Wach et al., 1994). Most of the homologous vectors have greater than 1000bp of homology and transformation occurs by gene conversion, homologous integration, or ectopic integration (Baek and Kenerley, 1998; Banks and Taylor, 1988; Case et al., 1979; Levis et al., 1997; Tsuge et al., 1990; Tudzynski et al., 1996; Unkles et al., 1989; Upchurch et al., 1991; van Hartingsveldt et al., 1987; Wada et al., 1996; Weidner et al., 1998). When homologous selectable genes are used, gene conversion occurs most frequently (40-100% of the transformants) followed by single integration events at the homologous site. Homologous integration also allows researchers to test and/or confirm the functionality of isolated genes by performing gene disruption experiments and replacement of nonfunctional genes with wt copies. As indicated above, the amount of homologous DNA in the transformation vector may influence the frequency of homologous integration. However, there appear to be undefined factors that result in significant variation between species with regard to the frequency of homologous integration.

2.5. Gene cloning strategies

The development of molecular transformation has allowed researchers to begin isolating and studying the effects of single genes on complex physiological processes. The genome size of most filamentous fungi is approximately 2-5 x 10^7 bp and the average gene size is 1000-5000 bp (Aramayo et al., 1989; Fotheringham and Holloman, 1989; Vollmer and Yanofsky, 1986). Depending on the amount of repeptitive and non-coding DNA in the genome, the number of genes/nucleous may vary from 1000 to 25,000. Therefore, cloning specific genes requires strategies that increase the probability of identifying and isolating a small subset of genomic sequences. There are three basic strategies that are commonly used for isolating gene sequences: complementation of mutants or physiological variants, disruption tagging of wild type genes, and screening for differential gene expression.

2.5.1. Complementation

This cloning strategy requires the construction of genomic libraries consisting of sections of genomic DNA ligated into transformation vectors. This involves the digestion of genomic DNA with restriction enzymes to generate sub-genomic fragments appropriate for either plasmid (<20 kb), cosmid (35-45 kb), or YAC (100-1000 kb) libraries. Although the difficulty of in vitro manipulations increases with the size of DNA, the number of transformants required to ensure genome coverage is inversely proportional to the amount of DNA in transformation vectors. For example, a cosmid library with 35-45 kb genomic DNA inserts requires 4-6000 transformants to be screened to have a 99% chance of identifying any given specific gene in a genome of 4 X 10^7 bp (Clarke and Carbon, 1976). A plasmid library with 15 kb inserts requires 10-15,000 transformants to be screened to cover the same genome size. Once the library is constructed, the vectors are used to generate transformants that can be screened to identify genes of interest. The success of this strategy is dependent on the transformation efficiency and the selection required to identify genes of interest. For example, when libraries are used to isolate genes that complement nutritional mutants, the selection simply requires transformants to be screened on media devoid of the required biochemical. This strategy has been used to isolate genes involved in many aspects of cell metabolism including the biosynthesis of amino acids (Baek and Kenerley, 1998; Case et al., 1979; Fotheringham and Holloman, 1989; Vollmer and Yanofsky, 1986) , pyrimidines (Banks and Taylor, 1988) , nitrogen assimilation (Malardier et al., 1989; Unkles et al., 1989), and fungicide resistance (Gold et al., 1991; Keon et al., 1991; Orbach et al., 1986).

When this strategy is used to complement non-nutritional mutants or physiological variants, the selection process may be time consuming and labor intensive. For example, to clone genes involved in host specificity, pathogenicity, toxin production, or other symbiotic interactions between fungi and plants, transformants usually need to be screened individually on plants. In addition, if transformation efficiencies are low it may be difficult to generate adequate numbers of transformants. As a result, various strategies have been developed to identify genes responsible for complex physiology such as symbiosis. An example of this is represented by the isolation of a gene (ToxA) encoding a plant-host specific protein toxin from *Pyrenophera tritici-repentis*, a pathogen of wheat (Ciuffetti et al., 1997). The production of the toxin correlated with virulence but its role in pathogenicity was unknown. To clone the ToxA gene, a phage cDNA expression library was first constructed and screened with antibodies against the toxin. Positive clones were sequenced and the deduced protein sequence compared to sequence of the protein toxin to confirm the identity of the clones. A positive clone was then used as a probe to isolate a genomic copy of ToxA flanked with regulatory sequences. The genomic clone was then transformed into an isolate of the fungus that did not produce toxin or cause plant disease. Non-pathogenic/non-toxin producing isolates of *P. tritici-repentis* transformed with the genomic copy of ToxA produced protein toxin and were virulent on wheat plants indicating that the toxin was the primary determinant of pathogenicity.

It has become routine to use previously cloned genes from heterologous fungi to generate probes or oligonucleotide primers for polymerase chain reaction (PCR) amplification of genomic copies from species of interest. This allows researchers to identify potential inserts in DNA libraries to reduce the number of transformants to be screened. This strategy has been used in several systems to isolate gene in various aspects of cell metabolism including the biosynthesis of amino acids (Wada et al., 1996), pyrimidines (Weidner et al., 1998) , cytoskeletal proteins (Upchurch et al., 1991), extracellular enzymes (Templeton et al., 1994) , kinases (Bencina et al., 1997; Clarke et al., 1997) and nitrogen assimilation (Levis et al., 1997; Tudzynski et al., 1996; Unkles et al., 1989).

2.5.2. Gene disruption

To avoid some of the time and labor demands required for screening DNA libraries, cloning strategies have been developed that involve either the disruption or deletion (VanEtten et al., 1998) of wild type genes sequences. However, gene disruption is the most commonly used alternative to complementation cloning. This arose from observations that during the transformation process a certain percentage of homologous vector systems integrated at ectopic sites. Occasionally, ectopic integrations resulted in a low frequency (0.3%) of gene disruptions resulting in the generation of mutants (Diallinas and Scazzocchio, 1989). Studies with *S. cerevisiae* indicated that if vectors were linearized with a restriction enzyme and the same enzyme was included in the transformation process the linearized vector would integrate at genomic sites compatible with the restriction enzyme (Schiestl and Petes, 1991). This process was designated restriction enzyme-mediated *i*ntegration (REMI) and was hypothesized to be based on the uptake of the restriction enzyme and linearized vector by cells resulting in limited digestion of genomic DNA such that entry points for the linearized vector were generated (Schiestl and Petes, 1991). This process was later used to disrupt, tag, and clone developmental genes in *Dictostelium discoideum* (Kuspa and Loomis, 1992). Transforming *D. discoideum* with EcoR1 linearized vector and in the presence of abundant EcoR1 enzyme resulted in approximately 0.25% of the REMI transformants expressing aberrant developmental phenotypes (Kuspa and Loomis, 1992). In addition, the transformation efficiencies increased up to 62-fold as a result of restriction enzyme addition. The tagged genes were re-isolated by digesting genomic DNA with a restriction enzyme that did not cut within the vector, ligating the linearized genomic DNA, and transforming the ligated DNA into *E. coli*. Since the original transformation vector contained a prokaryotic resistance gene such as ampicillin, *E. coli* transformants containing the re-isolated vector could be selected on media containing ampicillin.

REMI process has been used to transform a variety of fungi including Ascomycetes, Basidiomycetes, and imperfect species. In general, REMI transformation results in increased transformation efficiencies of up to 100-fold (Brown et al., 1996; Brown et al., 1998; Redman and Rodriguez, 1994; Sanchez et al., 1998; Shi et al., 1995). The increase in transformation efficiency is dependent of the concentration and type of restriction enzyme used and may be different with each fungal species.

Recently, REMI transformation has been used to tag genes involved in developmental biology (Epstein et al., 1998) , toxin production (Akamatsu et al., 1997; Lu et al., 1994), and pathogenicity (Bolker et al., 1995) in a variety of fungal species. The REMI mutation rates in these fungal systems have varied from 0.15-2.0% making this form of mutagenesis comparable to chemical and physical methods but with the tremendous advantage of generating tagged mutations.

3. GENE EXPRESSION SYSTEMS

Fungal gene expression systems have important applications in research and in industry. These systems are dependent upon the fusinion of genes of interest to regulatory sequences such as promoters, enhancers, and terminators to obtain desired levels of expression and biologically active gene products. As a result, fungal research employs a wide variety of regulatory sequences for various levels and types of expression. For industrial purposes high level gene expression is often a major goal so that high yields of valuable products may be realized. Generally, strong continuous expression is desired for industrial applications but under certain circumstances specific timing of expression is necessary. Fungi are used for commercial production of valuable compounds and as the subjects of diverse studies in cell biology. Specific areas of study on the genetic control of cell function in fungal systems include primary and secondary metabolism, protein secretion, signal transduction, self non-self recognition and mating systems, cell cycle control, cytoskeletal and cell wall structure, morphogenesis and pathogenicity. For some of these studies constitutive expression systems are perferred whereas for others the ability to induce expression on command is important for unraveling the role of specific genes in various processes. A few specific examples of expression systems and their uses in research are provided for illustration later in this chapter.

3.1. Applications of gene expression systems

3.1.1. Industrial protein and compound synthesis

One of the most common applications of fungal gene expression systems is in the production of industrially important compounds of various chemistries and useful proteins. Many of these production systems have relied on naturally occurring fungal strains which have been minimally modified. However, the use of molecular tools will undoubtedly have an increasing impact to enhance the yield and quality of industrially important fungal products. Important industrial compounds produced by fungi include metabolic products such as citric acid and ethanol as well as molecules that would otherwise be prohibitively expensive or impossible to synthesize artificially such as the β-lactam antibiotics including the penicillins produced commercially by *Penicillium chrysogenum* and the cephalosporins produced by *Acremonium chrysogenum* (*Cephalosporium acremonium*) (Bellgardt, 1998; Diez et al., 1996; Fernandez-Canon and Penalva, 1995; Litzka et al., 1998; Penalva et al., 1998; Smith et al., 1991; Timberlake and Marshall, 1989; Usher et al., 1992). Citric acid production is an example of a low molecular weight fungal metabolite that is widely used in the food, beverage, and pharmaceutical industries. Over 80% of the worldís supply of citric acid is produced by microbial fermentation (Bodie, 1994). Citric acid is used as a buffer and anti-oxidant in detergents and other industrial applications.

Gene expression systems are important in the synthesis of specific fungal proteins that are of economic value. These include enzymes used as additives in detergents, paper pulping or for food processing. The most significant enzymes produced by fungi comercially are from *Aspergillus* and *Trichoderma* species. These enzymes include: glucoamylase used for production of glucose syrups from starch (for example, high fructose corn syrup commonly used in sodas); α-amylase also used for glucose syrups and for production of bread, maltose syrups and fermentation products from certain Asian countries; plant cell wall degrading enzymes such as pectinolytic enzymes (including pectin methylesterase, polygalacturonase and pectate lyase) and cellulases (including cellulase and hemicellulase) are used in wine and fruit juice production; lactase used to convert lactose to glucose in milk and whey, and sold in over-the-counter digestive aids for lactose intolerant individuals; proteases used in bread and beer manufacture as well as fermentation of Asian foods and beverages; catalase used for the breakdown of hydrogen peroxide in various processes; glucose oxidase used in enzyme kits to determine glucose concentration in bodily fluids and in the food and beverage industries to remove glucose or oxygen; lipase used for flavor enhancement, particularly in cheese ripening (in the mold cheeses, e.g. Roquefort) in the dairy industry; and xylanases used in brewing, baking, animal feed and paper pulp bleaching industries (Bodie, 1994).

3.1.2. Novel gene isolation

The identification of genes for commercial or experimental use can be accomplished by several methods including: i) homology based strategies such as gene library screening (Gold et al., 1991; Minetoki et al., 1995; van Nistelrooy et al., 1996; Wu et al., 1996) or production of primers for PCR amplification based on highly conserved regions of known genes encoding enzymes of interest (Nikolskaya et al., 1995); ii) mutant complementation (Gold et al., 1994; Mayorga and Gold, 1998); iii) screening of expression libraries for cDNA clones with antibodies produced against purified enzyme (Dean and Timberlake, 1989); iv) utilizing purified peptide sequence data to design degenerate oligonucleotides to be used as probes or PCR primers (Scott-Craig et al., 1998; Scott-Craig et al., 1990); and; v) functional cloning, a modification of mutant complementation in which a different expression host organism is transformed and assayed for the acquired ability to express the gene of interest (Agnan et al., 1997; Christgau et al., 1996; Dalboge and Heldt-Hansen, 1994; Pauly et al., 1999). With the advent of complete genome sequencing programs with fungal species (Agnan et al., 1997; Bennett, 1997), many potentially useful genes will be identified by sequence homology and/or expression pattern studies.

The functional cloning approach has great potential to identify novel enzymes for commercial application because it allows the efficient screening of an organismís genome for any assayable enzyme encoding gene without the necessity for available sequence data. For this approach very few tools pertaining to the manipulation of the donor organism are necessary since screening occurs in a highly tractable recipient organism. Generally, *S. cerevisiae* has been used as the expression host for a transcriptionally fused full-length cDNA library from the organism of interest (Christgau et al., 1996). For secreted enzymes, for which this process has primarily been employed, the translational start and functional secretion signal sequence must be present in the clone. This system may also be modified for automated high throughput screening (Dalboge, 1997). Thus the potential exists to develop systems to rapidly identify useful novel enzymes. Together with the ability to introduce the genes encoding these enzymes in high expression fungal systems their discovery opens the door for a continuous supply of new industrially important products.

3.1.3. Research applications of gene expression systems

The use of gene disruption and expression sytsems in fungal research has revolutionized the rate at which knowledge can be gained on various cellular processes. There are hundreds of examples in the literature where gene expression systems in fungi have contributed to our understanding of these organisms. One illustrative example of the use of a fungal gene expression system in fungal research is described here. In work by Yang et al. (1994) a regulated heterologous promoter was used to successfully perform fungal-toxin research that had previously been thwarted by technical difficulties. The *pel*A promoter of *A. nidulans* (Dean and Timberlake, 1989) was employed to inducibly drive the expression of a gene (*T-urf13*) in *Cochliobolus heterostrophus*. The *T-urf13* gene confers sensitivity to the polyketide toxin called T-toxin. In 1970 T-toxin producing strains (race T) of *C. heterostrophus* caused a devastating disease on the Texas male sterile cytoplasm maize cultivars that were nearly uniformly planted at the time. *Cochliobolus heterostrophus* is insensitive to T-toxin. The gene conferring sensitivity to toxin was cloned from maize and shown to encode an aberrant protein that integrates into the mitochondrial membrane and binds toxin (reviewed in (Levings and Siedow, 1992). To analyze the mechanism of T-toxin synthesis, mutants incapable of producing T-toxin were desired. However, earlier attempts to identify such mutants had been relatively unsuccessful. A clever scheme was employed in which *C. heterostrophus* transformants were engineered to experience inducible growth inhibition by becoming sensitive to their own toxin when *T-urf13* was expressed (Yang et al., 1994). These strains grew normally when provided with glucose but growth was inhibited when polygalacturonic acid (PGA), the inducer of the pelA promoter, replaced glucose. Next a *pelA-urf13* containing transformant was exposed to chemical mutagenesis and plated on PGA medium to induce expression of *T-urf13* and thus sensitivity to T-toxin. The efficiency of this procedure is indicated by the fact that 9 of 362 mutagenesis survivors no longer produced T-toxin. Thus the use of a heterologous promoter that is under the same regulation as in the homologous organism was employed successfully to generate important research tools

3.3. Determinants of fungal gene expression
3.3.1. Molecular strain improvement

A potent way to generate improved industrial strains and research tools is through the use of molecular biological techniques to identify (discussed above) and effectively deploy new genes. Early strain improvement efforts for industrial strains were primarily via muatgenesis of natural or engineered producers followed by screening for hyper producers of the desired product (Archer et al., 1994; Diez et al., 1996; Radzio and Kuck, 1997). This approach has been successfully employed many times. For example, selected deoxyglucose resistant mutants of *Aspergillus awamori* produce higher levels of *glaA* driven chymosin (Dunn-Coleman et al., 1991) and various *Aspergillus* and *Penicillium* strains improved for production of the compound verbenol were obtained following exposure to a number of mutagens (Agrawal et al., 1999). These improvements tend to be relatively small, generally on the order of 2 to 3-fold (Verdoes, 1995) but occasionally are higher (Agrawal et al., 1999). With the advent of molecular tools more precise methods for strain improvement by increased expression have become possible (see below). An additional area not discussed further in this chapter, that is likely to have major impact in the future, is protein engineering using rational design to produce strains with optimal protein production and quality (Aikawa et al., 1992; Pitts et al., 1991; Pitts et al., 1993; Schmidt-Dannert et al., 1998).

3.3. Components of fungal expression systems
3.3.1. The fungal promoter

Plasmid expression vectors have been designed to drive expression in the species of interest. The basic arrangement of expression vectors is 5' promoter - gene of interest - terminator 3'. Depending on the purpose, translational fusions may be useful. Additional components such as secretion signal sequences or even the entire coding region of a strongly expressed homologous protein encoding gene together with signal peptidase and/or KEX2-like protease cleavage recognition sites may also be incorporated (Contreras et al., 1991; Ward et al., 1990). These vectors also invariably carry a selectable marker gene (unless they are used for cotransformation in combination with another selectable plasmid) to allow the isolation of transformants. Promoters chosen for fungal expression may produce constitutive expression or they may be regulated depending on their intended purpose.

The fungal promoter is a rather ill-defined and highly variable entity (Ballance, 1986; Ballance, 1991). Deletion analysis or site directed mutagenesis is often the only way to identify functional elements within promoters and these often remain somewhat incompletely defined (Bell-Pedersen et al., 1996; Bibbins et al., 1998; Hamer and Timberlake, 1987; Ilmen et al., 1998; Punt et al., 1990; Steiner and Philippsen, 1994). Abundant examples exist in the literature in which conserved upstream elements are noted but in general functional analysis was not performed. Some, but not all, fungal promoters have recognizable TATA box elements that have, in some cases, been shown to function in the localization of transcriptional initiation (Bell-Pedersen et al., 1996). Generally, the start of transcription in fungal genes possessing TATA box elements is 30-60 base pairs 3í of the TATA box although several different transcriptional start sites may be found in this region (Ballance, 1991). There seems to be a general correlation between the presence and absence of the TATA box and the inducibility versus solely constitutive nature of the promoter, respectively (Ozer et al., 1998). In *S. cerevisiae* this has clearly been demonstrated for the GCY1/RIO1 divergently transcribed gene pair (Angermayr and Bandlow, 1997). The GCY1 gene has a consensus (TATAAA) TATA box while RIO1 does not. GCY1 is galactose inducible while RIO1 is constitutive and not induced by galactose. RIO1 has a sequence (TATAGA) resembling the TATA box but it is not within the consensus sequence for a functional element. By site directed mutagenesis the TATAAA of GCY1 was converted to TATAGA and the TATAGA sequence of RIO1 was converted to TATAAA resulting in the modification of expression for both genes. RIO1 became inducible and the basal and regulated expression of GCY1 were both significantly reduced. The level of GCY1 induction by galactose over the basal rate was reduced from about 24 to 4-fold. Thus TATA box like elements alone play important roles in determining the inducibility of yeast genes. In the filamentous fungi good consensus TATA box elements are often found in inducible promoters.

A constant constitutive promoter that does not vary at all with development is probably very rare (Bell-Pedersen et al., 1996). This fact appears to be supported by the recent results in *S. cerevisiae* in which genome wide expression studies have been carried out (Bowtell, 1999; Chu et al., 1998; DeRisi et al., 1997; Eisen et al., 1998; Lashkari et al., 1997; Shalon et al., 1996). Some promoters are very complex and have multiple elements involved in gene activation or repression. For example the eas (ccg-2) promoter of *N. crassa* has three conditions which induce its transcription. These include distinguishable *cis*-elements involved in circadian rhythms, blue light induction and development (Bell-Pedersen et al., 1996) .

Box elements, CAAT (CCAAT) are also found in many 5' regions of fungal genes, however, except in a relatively few cases (Chen et al., 1998; Davis et al., 1993; Kato et al., 1997; Nagata et al., 1993; Tsukagoshi, 1998; van Heeswijck and Hynes, 1991), their function is not well documented. CAAT box function has clearly been shown in *S. cerevisiae* where a genetic screen has identified genes regulated by a CCAAT-box binding factor (Dang et al., 1994). Subunits of the CAAT binding factors have been identified in several fungi (Chen et al., 1998; Kato et al., 1997; McNabb et al., 1995). However, the function of these CAAT binding factors in the filamentous fungi is currently not known.

3.4. Genomic approaches to fungal promoter analysis

With the advent of full scale genomic sequencing new opportunities have arrived for understanding various determinants of gene expression. Sequencing of the entire genome of *S. cerevisiae* has allowed researchers to identify all of the organsism's open reading frames (ORFs), but understanding the function of each individual gene remains a major challenge. One powerful feature of the yeast genome database is that the upstream promoter sequences of all the genes are now accessible. This allows a search of promoter sequences in the genomic database to identify all the genes possessing similar potential *cis*-regulating elements.

Microarray technology has the potential of leading to an understanding of genome wide gene expression. Microarrays (as reviewed in DeRisi et al., 1997) may be constructed by several methods. The most common microarray method involves the adherance of PCR amplification products from yeast ORFs, onto glass slides in a specific order. Gene expression can then be studied by isolating mRNA from cells and synthesizing fluorescently tagged cDNA *in vitro.* Fluorescently labeled cDNA is prepared by reverse transcription in the presence of Cy3 (green) or Cy5 (red)-labeled deoxyuridine triphosphate (dUTP), and subsequently hybridized to microarrays. For example, to determine the changes in gene expression levels due to stress, cDNA prepared from cells under stressed conditions (labeled with Cy5 red) is mixed with cDNA from reference (unstressed) cells (labeled with Cy3 green), and the difference in fluorescence intensity is measured. For more information on these experimental procedures see the Brown lab web site at:

http://cmgm.stanford.edu/pbrown/mguide/index.html

Microarrays can be used to determine the level of expression of a set of genes simultaneously by comparing a geneís transcript level to other genes, under a given set of experimental conditions. Clustered gene expression patterns have been explored in yeast under conditions such as diauxic shift (metabolic shift from fermentation to respiration) and sporulation (Chu et al., 1998; DeRisi et al., 1997).

Defining genes for which no mutant phenotype has been identified is difficult using standard technologies. However, microarray technology may provide clues as to the physiological roles of functionally undefined genes based on their expression patterns. For example during the diauxic shift approximately 1000 genes that had a variance in their expression pattern were of unknown function (DeRisi et al., 1997). An uncharacterized geneís expression pattern in a metabolic or developmental pathway may be an instrumental first step toward understanding its function.

Microarray analysis used to determine gene expression patterns in S. cerevisiae during sporulation, identified genes involved at various stages of this developmental process. Genes involved in sporulation have been found to contain upstream promoter sequences with specific transcription factor consensus binding sites (Chu et al., 1998). The gene expression program of sporulation has been previously characterized as a transcriptional cascade involving four temporal classes of transcripts, early, middle, mid-late, and late (as discussed in Chu et al. 1998). Known transcription factors that control the expression of genes in the early and middle phase are IME1 and Ndt80, respectively. In the early induction period of sporulation, 43% of the genes differentially expressed were found to have the core Upstream Repression Sequence (URS1) consensus site that is the *cis* -element recognized by the regulatory protein IME1p. In the middle (70% of the genes differentially expressed) and mid-late phases (36% of the genes differentially expressed), genes contain the 5í mid sporulation element (MSE) site recognized by *Ndt80* encoded transcription factor. The mid-late genes are probably controlled by an additional unknown negative regulatory site that delays their induction (Chu et al., 1998). These results illustrate the potential of using upstream promoter sequences to search a genomic database to identify genes that are potentially induced under specific conditions. The *cis*-regulatory elements will then become available for manipulation for industrial or research purposes and this will aid in the discovery of new promoter sequences useful for specific expression vector construction. Now that full scale genome sequencing projects are underway in *Magnaporthe grisea, A. nidulans, U. maydis* and other filamentous fungi, searching databases with promoter sequences will aid in the identification of genes regulated under similar conditions such as dimorphism, pathogenecity, and sporulation.

3.4.1. Strong constitutive or inducible promoters

For most applications a strong constitutive promoter is sufficient. Various fungal promoters have been shown to have a wide range of strength (Vanwert and Yoder, 1994). In general, industrial applications for fungal products employ strong constitutive promoters such as the *A. nidulans gpdA* (Punt et al., 1990) promoter to drive the expression of genes encoding products of interest. Generally, for standard transformation vectors, constitutive promoters driving selectable markers are employed because drug resistance or prototrophy should be constantly expressed to promote the growth of transformants under all selection schemes. The common promoters and terminators for transformation vectors are the native promoters of various metabolic genes used to select for prototrophy (reviewed for *Aspergillus* in, Turner, 1994) or homologous host derived drug resistance (Giles et al., 1985; Gold et al., 1994; Keon et al., 1991; May et al., 1985) and 5' and 3' UTRs of various housekeeping genes transcriptionally fused to drug resistance markers (Gold et al., 1994; Turner, 1994; Wang et al., 1988). The most important application for inducible promoters is to allow for temporal triggering of gene expression. For industrial purposes this may be necessary if the gene product is toxic to the cell or may be used as a method to avoid protein degradation by reducing the time the product is exposed to proteases. Inducible promoters, as exemplified by the T-toxin sensitive *C. heterostrophus* described above, are invaluable for regulated expression for morphogenetic or physiological research.

Some typical examples of constitutive promoters used in various organisms include: for use in *Aspergillus* species (Davies, 1994), the *A. nidulans gpdA* and the modified *aldA* promoter (Devchand and Gwynne, 1991) , *adhA* in *A. niger*, and *glaA* in *A. awamori* (Berka et al., 1991) ; for *N. crassa* the ß-tubulin promoter provides constitutive expression (Nakano et al., 1993); and for use in the relatively new commercial production host *Fusarium graminearum* (the Quorn mycoprotein fungus) the *F. oxysporum* trypsin promoter and terminator (Royer et al., 1995). The strong constitutive promoter derived from the *A. nidulans gpdA* gene has been analyzed in some detail. A 50 base pair region termed the gpd box was shown to increase the level of reporter gene product 30-fold when inserted upstream of the *amdS* gene (Punt et al., 1990). For use in Basidiomycetes, promoters that give useful continuous expression include: for *U. maydis* the *hsp70* promoter and terminator (Wang et al., 1988), the gap promoter (Gold et al., 1994; Kinal et al., 1991) and the *tef* promoter (Spellig et al., 1996); the GPD promoter in *Schizophyllum commune* (Harmsen et al., 1992) and; in *Coprinus cinereus* the ß-tubulin promoter (M. Zolin, personal communication).

The important features of an inducible promoter are low basal activity and a high transcription rate under induction. The inducer should be low cost, readily available and convenient to use (Davies, 1994). An excellent example of a promoter that meets these criteria is the A. nidulans alcohol dehydrogenase I (alcA) promoter (Felenbok, 1991). Although the alcA promoter is not completely silent under non-induced catabolite repressed conditions (Waring et al., 1989) transcript is undetectable in northern blots (Davies, 1994). Catabolite repression is mediated by CREA and indirectly by the limited availability of the *alcA* positive regulator encoded by *alcR* (discussed further below) which itself is under catabolite repression (Davies, 1994; Felenbok, 1991). Catabolite repression is avoidable by use of non-repressing carbon sources such as oils or fatty acids in place of glucose. A 320 base pair version of the *alcA* promoter lacking the *creA* binding site is still under indirect catabolite repression due the fact that *alcR* is repressed by *creA* (Davies, 1994). Specific mutation of the *alcA* CREA binding site also relieves the direct catabolite repression by CREA (Hintz and Lagosky, 1993). The *alcA* promoter is thus ideal for the expression of toxic peptides because protein encoding genes under its control should yield little product until the fungus is placed under inducing conditions. Within one hour under glucose depletion and full induction with ethanol or threonine, (Turner, 1994) or a variety of inexpensive industrial ketones about 1% of newly synthesized cellular protein is derived from this promoter (Davies, 1994). The *alcA* promoter has been very useful in research on the developmental biology of *Aspergillus* (Adams and Timberlake, 1990; Felenbok, 1991; Marhoul and Adams, 1995; McGoldrick et al., 1995; Osherov et al., 1998).

The *glaA* promoter of *A. niger* is also inducible and catabolite repressible. This promoter however has a higher basal level of activity than that of *alcA* and thus would be a poor choice for the induced expression of toxic products. However, this is a popular promoter for industrial purposes because it has a very high level of activity when induced with inexpensive starch (Davies, 1994). The *glaA* signal sequence is also convenient for directing protein secretion. Similar to the *alcA* promoter discussed above, the A. nidulans acetate inducible amdS gene (Davis et al., 1993) is activated by the positive acting *amdA* gene (Lints et al., 1995). Recently a new expression system for the industrial fungus *A. awamori* has been developed based on the D-xylose inducible *exlA* gene which is reported to be three-fold more efficient than the *A. niger glaA* promoter (Gouka et al., 1996). For *Trichoderma* the cbh1 promoter and terminator allow highly inducible expression (Cheng and Udaka, 1991; Harkki et al., 1991). The inducing compound for induction of cbh1 is cellulose but for this to be effective the fungus must secrete extracellular ß-glucosidase (Fowler and Brown, 1992). Additionally, although the induction is on the order of 1100-fold, a low basal level of cellulase activity is required for its own induction (Carle-Urioste et al., 1997).

Several promoters have been shown to have value for controlled gene induction in Basidiomycetes. For example the *U. maydis crg1* promoter (Bottin et al., 1996) is inducible by arabinose; the *Cryptococcus neoformans* GAL7 promoter is inducible by galactose (Chang et al., 1995; Del Poeta et al., 1999; del Poeta et al., 1999; Wickes and Edman, 1995) and; the *Coprinus cinereus acu-7* promoter is inducible by acetate.

In some plant genes introns can function as enhancers of transcription by binding specific transcription factors (Gidekel et al., 1996). The importance of sequences within introns for the rate of transcription in fungi is not well documented. However, Corrochano et al. found that deletion of the first intron of the *con-10* gene reduced the level of the ß-galactosidase reporter activity in translational fusions

(Corrochano et al., 1995). In *Podospora* the S12 ribosomal protein gene requires an intron for full expression (Dequard-Chablat and Rotig, 1997).

3.5. Heterologous gene expression in fungi

Heterologous expression systems rely on the premise that the transcription and translation machinery in all organisms share many features (reviewed in, Frommer and Ninnemann, 1995). In some cases animal and plant promoters function in yeast and there is considerable conservation between fungi and these groups in terms of gene regulatory mechanisms. For example the maize transcription factor *Opaque-2* can substitute for *GCN4* in yeast (Mauri et al., 1993). This is however far from a universal scenario. The recognition and proper regulation of transcription in closely related species, especially within the same genus is the norm. However, the more distant the taxa tested are the lower the likelihood of properly regulated expression. For example, between Ascomycetes and Basidiomycetes the expression of genes from their native promoter in an organism from the other class seems to be possible in about half of the cases studied (Banks and Taylor, 1988; Caddick et al., 1998; Casselton and Herce, 1989; Keon et al., 1994; Straffon et al., 1996) . Even when driven by proven fungal promoters much lower levels of protein synthesis is achieved for human proteins in fungal cells than homologous proteins or proteins encoded by the introduced genes of other fungi (Fowler and Berka, 1991; Gouka et al., 1996; Ward et al., 1992). There are a number of reasons that lower yield may occur from heterologous genes expressed from homologous promoters including inefficient translation (see codon usage below) and incorrect protein processing. Proper protein folding is required for transport from the endoplasmic reticulum and if auxiliary factors are missing that are required for folding to occur the protein may be mistargetted and/or degraded (Frommer and Ninnemann, 1995). Because gene expression systems generally only involve the introduction of a single gene of interest, any required auxiliary factors are not included.

3.6. Multi-copy gene introduction for increased expression levels

To boost gene expression levels, a common approach has been to introduce multiple copies of the gene of interest by transformation. In fungi mutiple copies of a gene can be introduced by one of two methods. The first and most common is by integration of multiple copies of a sequence into one or more chromosomal locations of the transformant. An alternative available for a few fungi is the introduction of autonomously replicating episomal plasmids that are in multiple copies per cell.

3.6.1. Multicopy integration

Many fungi show ectopic integration of heterologous DNA which can occur at different loci and be variable in copy number. In some cases there is a somewhat imprecise positive correlation between copy number and level of expression (Graessle et al., 1997; Verdoes, 1995). Generally this linear increase in expression exists up to 7 to 10 copies of a gene integrated into the genome in *Aspergillus* species (Davies, 1994). However, this is far from universal and in some species there is no observable dependence on copy number (Radzio and Kuck, 1997; van den Hondel, 1991). In fact, in some fungi multiple copies of a transgene yields gene silencing by various mechanisms (see below). The integration position can have a large effect on expression (Verdoes et al., 1994). This problem has generally been dealt with by screening a sufficient number of transformants and choosing those with high levels of expression. In this way the fact that some locations of integration enhance and others suppress expression has generally been ignored. However, for careful comparative expression studies as in the study of *cis*-acting elements by deletion analysis it is very helpful that position effects be eliminated. This can be accomplished by employing a homologous integration scheme in which all constructs to be studied are inserted in the same locus (Bell-Pedersen et al., 1996; Hamer and Timberlake, 1987). The potential to employ insertions in the regions of the genome with chromatin structure particularly conducive to high level expression is an area that may yield improved expression levels (Verdoes, 1995). The identification of the mediator of transcription complex in yeast (reviewed in (Bjorklund and Kim, 1996) and the involvement of SIN4 a chromatin structure modifying component of this complex (Jiang and Stillman, 1995; Macatee et al., 1997) may assist in the long term for rational design of optimal integration vectors for high expression.

3.6.2. Multicopy episomes

High level expression is apparent from high copy number autonomously replicating episomal plasmids in *S. cerevisiae* and has repeatedly been used as the basis for multicopy suppressor genetics to identify additional genes that interact with the mutated gene under study (Blachly-Dyson et al., 1997; Formosa and Nittis, 1998; Schwer and Shuman, 1996; Sivadon et al., 1997; Van Dyck et al., 1995; Yang and Bisson, 1996). As discussed earlier in this chapter autonomously replicating plasmids are employed in a growing number of fungi because they increase transformation rates. There are few studies, however, quantifying the effect of multicopy plasmids on gene expression in the filamentous fungi. Although multicopy plasmids are now available in several filamentous fungi, until fairly recently they were rarely used. In *U. maydis*, however, multicopy plasmids have been commonly employed for more than a decade. The level of gene expression from these plasmids, which are present at about 25 copies per cell (Tsukuda et al., 1988), is high and can yield interesting information. The expression level of a *U. maydis* gene from an ARS vector is likely proportional to the copy number. For example, we found that introduction of a wild type copy of the *ubc4* gene (encoding a MAPKK kinase) on an ARS vector yielded partial remediation of a *ubc2* mutation even though there was already a wt copy of *ubc4* present in the genome of this mutant (Mayorga and Gold, 1998). Thus, as with multicopy suppression studies in S. cerevisiae, these results indicate that there is a genetic interaction between *ubc2* and *ubc4*.

Aleksenko et al. has shown that in *A. nidulans* transcription of the strongly expressed gene *bgaS* stimulates the expression of the *argB* gene located on the same AMA1 autonomously replicating plasmid irrespective of orientation (Aleksenko et al., 1996). This effect also occurs with chromosomally integrated vectors, but to a lesser extent. These autonomously replicating vectors are present in about 10 copies per nucleus. The expression level of the *argB* gene is approximately proportional to vector copy number. However, the amount of mRNA transcribed from highly expressed *bgaS* on the AMA1 plasmid does not exceed that from single copy integrants (Aleksenko, 1994). These data suggest that a single copy of *bgaS* gene is sufficient for a saturation level of expression, while further copies are limited by the amount of transcriptional or RNA-stabilizing factors, so that the total amount of *bgaS* mRNA in the cell is not increased. The *argB* gene, however, exhibits clear dependence of level of expression upon copy number, in both integrative and replicative transformants (Aleksenko et al., 1996).

3.7. Factors influencing transgene transcription

As with other organisms, the regulatory sequences 5í of the start of transcription play the major role in the transcriptional level of fungal genes. However, a number of other factors may strongly influence the expression of genes in particular organisms. Some of these factors are discussed below.

3.7.1. Availability of regulatory factors

Multiple copies of a particular promoter in a cell can titrate the supply of trans-acting factors thus altering the transgene's regulation. This may, at least in part, be the cause of less than a linear increase in expression in multicopy transformants (Davies, 1994; Verdoes, 1995). In some cases it is possible to overcome this problem by increasing the gene dosage of *trans*-acting factors. For example, the expression from the *alcA* promoter in multicopy strains can be improved by adding additional copies of *alcR* (Davies, 1994). Although this could also be an effect of increased successful competition for cis-elements by ALCR over CREA (Mathieu and Felenbok, 1994). Thus, it would be wise to consider introduction of additional copies of positive regulators to enhance yield of desired products.

3.7.2. Zinc regulatory factors

A note on the transcriptional regulatory proteins in fungi seems in order here. This pertains to the ability of heterologous genes to be expressed and/or regulated properly in fungal cells. As noted in Todd and Andrianopoulos (1997) transcription factors have DNA binding motifs which allow them to be grouped into various classes. Fungi have members from a number of these transcription regulatory factor classes that are also found in plants and/or animals. These include common motifs such as the homeodomain (Hobert and Ruvkun, 1998; Purugganan, 1998), MADS box (Molkentin and Olson, 1996; Riechmann and Meyerowitz, 1997), HMG (Bianchi and Beltrame, 1998) , and various classes of zinc finger proteins (GATA, C2H2, C6). Interestingly, the C6 (Zn(II)2Cys6) class of zinc finger regulatory proteins has only been found in the fungi (Todd and Andrianopoulos, 1997). Approximately 80 genes potentially encoding such C6 proteins have been identified in the fungi with the *S. cerevisiae* genome encoding at least 56 (of which 34 have as yet unknown functions) (Todd and Andrianopoulos, 1997). Several of these proteins regulate the promoters used in common expression vectors. A few examples (reviewed in (Todd and Andrianopoulos, 1997)) include: GAL4 in *S. cerevisiae* that regulates GAL1 through GAL10; the *A. nidulans AlcR* and *AmdR* proteins that regulate *alcA* and *amdS*, respectively; QA1F that regulates the qa (quinic acid) cluster genes; AflR, the positive regulator of aflatoxin biosynthetic genes in various aflatoxin or sterigmatocystin producing *Aspergillus* species and including the non-toxin producing common Asian fermentation fungi *A. oryzae* and *A. sojae* (Watson et al., 1999). Only one C6 type regulator encoding gene has been identified from a Basidiomycete. The PRIB protein is involved in the regulation of fruiting in *Lentinus edodes* (Endo et al., 1994). Thus, this class of regulators is likely found throughout at least the higher fungi. Some C6 genes can function in other fungi, particularly if they are close relatives (Whiteway et al., 1992). However, as stated above, the C6 zinc finger proteins are fungal specific which is consistent with the idea that fungal specific gene regulation mechanisms occur and points out once again the importance of employing fungal control signals for efficient transgene expression in the fungi.

3.7.3. Transcriptional gene silencing

Transcriptional gene silencing resulting from disparate mechanisms has been described for fungi in both the sexual (reviewed in; Cogoni and Macino, 1999 and Singer and Selker, 1995) and vegetative phases (reviewed in; Cogoni and Macino, 1999 and Sachs, 1998) of growth. The phenomenon of repeat induced point mutations (RIP) in *N. crassa* is a permanent gene silencing which occurs premiotically after mating (Selker, 1990). Another sexual gene silencing phenomenon called methylation induced premiotically (MIP) is found in *Ascobolus immersus* and results in a reversible transcriptional gene silencing (Barry et al., 1993).

3.7.4. Transcriptional terminators

Transcriptional termination in eukaryotes is an ill-defined process. Work with the human immunodeficiency virus suggests that transcriptional termination may be a process that requires inhibition to avoid premature termination rather than a signal marking a specific site of termination. RNA polymerase II can terminate at various locations (0.5-2 kb) downstream of the polyadenylation site in most mammalian transcriptional units (Lodish et al., 1995). Most expression systems in fungi employ the 3' end of a characterized gene to control transcriptional termination. However even in circumstances where no 3í sequence is employed mRNA of a consistent length can be produced (Davies, 1994) .

3.8. Posttranscriptional regulation of fungal gene expression

3.8.1. Posttranscriptional gene silencing

Quelling (reviewed in; Cogoni and Macino, 1997, and Sachs, 1998) is a silencing phenomenon found in *N. crassa* and is similar to cosuppression in plants. Quelling occurs when multiple copies of a gene are introduced into a nucleus. The phenomenon is posttranscriptional and the same amount of primary transcript is synthesized from quelled genes. The quelling process is not well understood but with analysis of recently isolated quelling deficient (*qde*) mutants (Cogoni and Macino, 1997) more information on the phenomenon may be soon forthcoming. The first *qde* gene has been identified as an RNA polymerase which is interesting since quelling involves posttranscriptional silencing (Cogoni and Macino, 1999) . Quelling is not limited to *N. crassa* , a similar phenomenon termed cosuppression occurs and has been employed for genetic analysis in the chestnut blight fungus *C. parasitica* (Chen et al., 1996; Choi et al., 1995; Wang and Nuss, 1995). Quelling obviously has important implications in choosing a method for high level gene expression and suggests that multicopy insertions in certain fungal species is counter productive.

3.8.2. RNA stability

According to Sachs (1998) only one example of the effect of 3' sequences on the mRNA stability in the fungi is currently known (Platt et al., 1996). This is with the A. nidulans nitrogen metabolism regulatory factor gene areA. The *are*A mRNA is significantly more stable in cells under low versus high nitrogen conditions. When the 3' untranslated regions (UTR) is deleted from *are*A, however, this instability is eleviated and the transcript levels are equal to those found under low nitrogen conditions.

Polyadenylation is commonly si AATAAA in the 3' UTR of transcribed transcripts and the level of polyade stability (Ross, 1996; Wickens et al possessing recognizable polyadenylat polyadenylated and thus other signals control of polyadenylation in fungi is *nidulans* (Marhoul and Adams, 199 developmental switch from vegetative a putative poly(A)-binding protein.

3.9. Translational control

3.9.1. Translational initiation signa

The translation alignment region conforms to the sequence generally defined the CA C/A C/A ATG(G)C con artificial genes should employ this con

3.9.2. Codon usage

The effects of natural selection a organisms. Early studies on *E. coli* abundant tRNA species correlated wit Highly expressed genes were found t tRNAs whereas in genes expressed (Sharp and Matassi, 1994). These res highly expressed mRNA sequences w greater abundance (Sharp and Matas important role in the expression of Designing human genes for optimal ex optimization of codon usage to facilit yield a commercially viable protein pr etc. The importance of considering cc been demonstrated in the developmen Since the plant's translational machi plant's codon bias allowed efficient tra resistance (Iannacone et al., 1997; S yeast *Candida maltosa* has been stud of hydrophobic compounds. Heterolo closely related *Candida* species was hi a universal leucine codon, being read 1995). Recently it was further deter studied used codon CUG to encode leu encode serine (Sugita and Nakase, 19

What is known about the co filamentous fungi is very limited. Or and *S. cerevisiae* had the same subs divergent in genomic G+C content (Sharp, 1992).

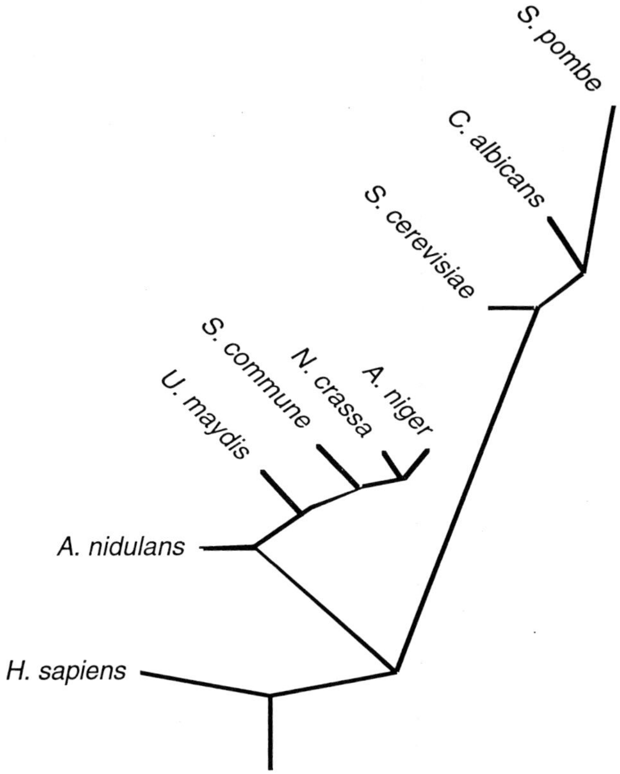

Figure 1. Rooted phylogenetic tree based on codon usage bias.

Data points were for analysis were based on the 64 different values of codon usage from each species (Table 1.). Twenty one characteristics (amino acids and terminator) were compared for each organism and possible alleles were determined by the number of codons per amino acid (i.e. Phe 2, ter 3, Leu 4, etc.). The web based PHYLIP (http://evolution.genetics.washington.edu.phylip.html) program was used to determine allelic frequencies using the CONTML program. The out group was *H. sapiens* and two distinct clades are present. One clade consisted of the Ascomycete yeasts, *S. cerevisiae* and *S. pombe* and the dimorphic species *C. albicans*. The second clade consisted of the filamentous fungi including members of both the Ascomycetes and the Basidiomycetes.

3.10. Posttranslational modifications
3.10.1. Glycosolation

Correct glycosylation patterns may occasionally be important for the function of a given protein but they are often important for in vivo protein stability (Aikawa et al., 1992) , (Davies, 1994). Additionally, glycosylation may be important for enhanced fungal secretion (Ward, 1989). Fungi lack the ability to add certain moieties found in mammalian glycosylated proteins (Davies, 1994). The imprecise glycosylation patterns of mammalian proteins produced in fungi limits their use in drug therapy (Davies, 1994) due to the importance of these moieties for function (Kukuruzinska and Lennon, 1998; Reuter and Gabius, 1999). Post synthesis treatments or expression of cloned genes involved in mammalian glycosylation that are absent in fungi may allow this problem to be eliminated (Jeenes et al., 1991; Maras et al., 1997).

3.10.2. Subcellular transport

For extracellular transport a 13 to 50 amino acid secretion signal sequence, consisting of an N-terminal series of mainly positively charged amino acids followed by a hydrophobic stretch and a signal peptidase cleavage site, must be present at the amino terminus of the protein (Radzio and Kuck, 1997). One advantage to using the filamentous fungi for mammalian gene expression is that the mammalian signal sequences are often recognized by the signal peptidases of these organisms as opposed to common failure of mammalian signal peptides in yeast (Davies, 1994). Nonetheless, a fungal signal may be used to replace the native signal sequence for more efficient secretion or to direct the secretion of normally non-secreted proteins.

3.10.3. Protein stability

Polypeptide folding, posttranslational modification and sorting are important in protein stability. Protein stability has important implications for industrial production of enzymes. If proteins are produced at sufficient levels but are unstable or toxic to host cells the use of an inducible promoter is advised (Radzio and Kuck, 1997). The process of secretion protects proteins from exposure to degradation (Radzio and Kuck, 1997). Additionally, mutant production strains deleted for specific proteases may be useful (Mattern et al., 1992). Intracellular protein degradation is often the result of ubiquitination, thereby targeting the protein for degradation (Hershko, 1997; Hilt and Wolf, 1995; Hochstrasser et al., 1995; Tanaka, 1998). Control of the cell cycle in yeast and likely the other fungi involves ubiquitination of key cell-cycle regulating proteins (reviewed in (Sachs, 1998)). Another example is the temporal degradation of the *frq* gene encoded polypeptide in *N. crassa* which is important in circadian rhythms (Ruoff et al., 1999). As mentioned above, proper folding and transport of proteins is important for avoiding degradation by proteases.

Industrially produced proteins are often secreted which has several advantages including easier purification, reduced exposure to intracellular host proteases and reduced affect on host cellular processes (Radzio and Kuck, 1997). Translational fusion of heterologous genes to highly expressed stable fungal proteins has proved useful for boosting yield (reviewed in; MacKenzie et al., 1993; and Radzio and Kuck, 1997). Fusion of the sequence for mature chicken lysozyme to the first 498 amino acids of the *A. niger* glucoamylase gene resulted in a 20-fold increase in secreted lysozyme (Jeenes et al., 1993; MacKenzie et al., 1993) . The sequence encoding the KEX2 endoproteolytic cleavage site was placed at the fusion junction of the glucoamylase and the lysozyme encoding regions and it was found that all the produced lysozyme was correctly processed by the *A. niger* KEX2-like protease. A similar approach was recently applied to taumatin production in A. awamori (Moralejo et al., 1999). Fusion of the entire coding sequence of the fungal glucoamylase to two different mammalian protein encoding genes improved their (Contreras et al., 1991; Ward et al., 1990). In *T. reesei* a 150-fold increase in Fab antibody fragment production occurred in a translational fusion of the CBHI gene N-terminal to the Fab sequence (Nyyssonen et al., 1993).

4. CONCLUSIONS

Through systematic study to further understand fungal gene regulation and cellular events such as protein processing and secretion, the ability to produce important products more efficiently and at lower cost through genetic engineering appears inevitable. Additional discoveries of relevance from allied fields may also generate unexpected benefits for improved expression. An example of such a potentially useful result comes from the study of plant viruses. In a study by Ishikawa et al. the introduction of plasmids capable of generating galactose inducible transcription of the brome mosaic virus (BMV) RNA3 molecule in *S. cerevisiae* was carried out (Ishikawa et al., 1997). Upon galactose induction transcripts were generated that encoded BMV viral replication proteins. These proteins carried out the functions within yeast cells that they normally do within plant cells and generated negative strand copies of the BMV RNA3 molecule. In the cell these negative RNA3 strands were used as templates to generate 45-fold higher levels of positive RNA3 molecules than DNA derived transcripts. Thus an effective amplification process occurred leading to a 500,000-fold increase in induced over uninduced levels of reporter gene activity. Once induction was initiated the process could not be reversed in this system because of the continued availability and synthesis of the BMV replication components.

A final cautionary note however, must be made. When attempting to optimize gene expression in fungi it is worth recalling that in *S. cerevisiae* an association has been found for high rates of spontaneous mutation in genes with high transcription levels (Datta and Jinks-Robertson, 1995). This should serve as a note of caution for continuously used strains with optimized expression. Stocks of the original strain should be stored and used as inoculum to avoid the generation of mutants over time.

Acknowledgements

We would like to extend our graditude to Dr. Russell Malmberg, Botany Department, UGA for assistance with the phylogenetic analysis, and the Dr. Richard Hurts and Dr. Harry Merkin for contructive reviews.

REFERENCES

Adams, T.H. and W.E.Timberlake, Proc. Natl. Acad. Sci. USA 87, (1990) 5405.
Agnan, J., Korch, C. and C. Selitrennikoff, Fungal Genet. Biol. 21, (1997) 292.
Agrawal, R., Deepika, N. U. and R.Joseph, Biotechnol. Bioeng. 63, (1999) 249.
Aikawa, J., Nishiyama, M. and T. Beppu, Scand. J. Clin. Lab. Invest. Suppl. 210, (1992) 51.

Akamatsu, H., Itoh, Y., Kodama, M., Otani, H. and K. Kohmoto, Phytopathology 87, (1997) 967.
Aleksenko, A., Gems, D. and J. Clutterbuck, Mol Microbiol 20, (1996) 427.
Aleksenko, A., Nikolaev, I., Vinetski, Y. and Clutterbuck, A.J. Mol. Gen. Genet. 253, (1996) 242.
Aleksenko, A.Y. Curr. Genet. 26, (1994) 352.
Aleksenko, A.Y. and Clutterbuck, A.J. Curr. Genet. 28, (1995) 87.
Allison, D.S., Rey, M.W., Berka, R.M., Armstrong, G. and Dunn-Coleman, N.S. Curr. Genet. 21, (1992) 225.
An, Z., Farman, M. L., Budde, A., Taura, S. and Leong, S. A. Gene 176, (1996) 93.
Anderson, E.S. Annu. Rev. Microbiol. 22, (1968) 131.
Angermayr, M. and Bandlow, W. J. Biol. Chem. 272, (1997) 31630.
Aramayo, R., Adams, T.H. and Timberlake, W.E. Genetics 122, (1989) 65.
Arber, W. Annu. Rev. Microbiol. 19, (1965) 365.
Archer, D.B., Jeenes, D.J. and Mackenzie, D.A. Ant. Van Leeuw. 65, (1994) 245.
Armaleo, D., Ye, G.N., Klein, T.M., Shark, K.B., Sanford, J.C. and Johnston, S.A. Curr. Genet. 17, (1990) 97.
Avery, O.T., MacLeod, C.M. and McCarty, M. J. Exp. Med. 79, (1944) 137.
Baek, J.M. and Kenerley, C.M. Fungal Genet. Biol. 23, (1998) 34.
Bakkeren, G. and Kronstad, J.W. Plant Cell 5, (1993) 123.
Ballance, D.J. Yeast 2, (1986) 229.
Ballance, D.J. and Turner, G. Gene 36, (1985) 321.
Ballance, D.J. Transformation systems for filamentous fungi *In* Molecular Industrial Mycology: Systems and Applications for Filamentous Fungi (S.A. Leong and R.M. Berka, eds.), pp.1-29. Marcel Dekker Inc. New York, 1991.
Banks, G.R. and Taylor, S.Y. Mol Cell Biol 8, (1988) 5417.
Barreto, C.C., Alves, L.C., Aragao, F.J., Rech, E., Schrank, A. and Vainstein, M.H. FEMS Microbiol Lett. 156, (1997) 95.
Barrett, V., Dixon, R.K. and Lemke, P.A. Appl. Microbiol. Biotechnol. 33, (1990) 313.
Barry, C., Faugeron, G. and Rossignol, J.L. Proc. Natl. Acad. Sci. USA 90, (1993) 4557.
Beggs, J.D. Nature 275, (1978) 104.
Bell-Pedersen, D., Dunlap, J.C. and Loros, J.J. Mol. Cell. Biol. 16, (1996) 513.
Bellgardt, K.H. Adv Biochem. Eng. Biotechnol. 60, (1998) 153.
Bencina, M., Panneman, H., Ruijter, G.J., Legisa, M. and Visser, J. Microbiol. 143, (1997) 1211.
Bennett, J.W. Fungal Genet Biol 21, (1997) 3.
Beri, R.K. and Turner, G. (1987) Curr. Genet. 11, 639.
Berka, R.M., Kodama, K.H., Rey, M.W., Wilson, L.J. and Ward, M. (1991) Biochem. Soc. Trans. 19, 681.
Bevan, M. (1984) Nucleic Acids Res 12, 8711.
Bianchi, M.E. and Beltrame, M. (1998) Am. J. Hum. Genet. 63, 1573.
Bibbins, M., Sheffield, P.J., Gainey, L.D., Mizote, T. and Connerton, I.F. (1998) Biochim Biophys Acta 1442, 320.
Bjorklund, S. and Kim, Y.J. (1996) Trends Biochem. Sci. 21, 335.
Blachly-Dyson, E., Song, J., Wolfgang, W.J., Colombini, M. and Forte, M. (1997) Mol. Cell. Biol. 17, 5727.

Bodie, E.A., Bower, B., Berka, R.M. and Dunn-Coleman N.S. In: Martinelli, S. D.; Kinghorn, J.R., eds. *Aspergillus*: 50 Years On, Vol. 29 (New York: Elsevier Science).1994.
Bogo, M.R., Vainstein, M.H., Aragao, F.J., Rech, E. and Schrank, A. FEMS Microbiol. Lett. 142, (1996) 123.
Bolker, M., Bohnert, H.U., Braun, K.H., Gorl, J. and Kahmann, R. Mol. Gen. Genet. 248, (1995) 547.
Bottin, A., Kamper, J. and Kahmann, R. Mol. Gen. Genet.253, (1996) 342.
Bowtell, D.D. Nat Genet 21, (1999) 25.
Boyer, H.W. Annu. Rev. Microbiol. 25, (1971) 153.
Brown, D.H., Slobodkin, I.V. and Kumamoto, C.A. Mol. Gen. Genet. 251, (1996) 75.
Brown, J. S., Aufauvre-Brown, A. and Holden, D.W. Mol. Gen. Genet. 259, (1998) 327.
Bundock, P., den Dulk-Ras, A., Beijersbergen, A. and Hooykaas, P.J. EMBO J. 14, (1995) 3206.
Bundock, P. and Hooykaas, P.J. Proc. Natl. Acad. Sci. USA. 93, (1996) 15272.
Burke, D.T., Carle, G.F. and Olson, M.V. Science 236, (1987) 806.
Caddick, M.X., Greenland, A.J., Jepson, I., Krause, K.P., Qu, N., Riddell, K.V., Salter, M.G., Schuch, W., Sonnewald, U. and Tomsett, A. B. Nature Biotechnol. 16, (1998) 177.
Carle-Urioste, J.C., Escobar-Vera, J., El-Gogary, S., Henrique-Silva, F., Torigoi, E., Crivellaro, O., Herrera-Estrella, A. and El-Dorry, H. J. Biol. Chem. 272, (1997) 10169.
Case, M.E. Basic Life Sci. 19, (1982) 87.
Case, M.E., Schweizer, M., Kushner, S.R. and Giles, N.H. Proc. Natl. Acad. Sci. USA.76, (1979) 5259.
Casselton, L.A. and Herce, A.D. Curr. Genet.16, (1989) 35.
Centola, M.B., Yamashiro, C.T., Martel, L.S., Royer, J. C. and Schmidhauser, T.J. Fungal Genetics Newsletter 41, (1994) 23.
Chakraborty, B.N., Patterson, N.A. and Kapoor, M. Can. J. Microbiol. 37, (1991) 858.
Chalfie, M., Tu, Y., Euskirchen, G., Ward, W.W. and Prasher, D.C. Science 263, (1994) 802.
Chang, Y.C., Wickes, B.L. and Kwon-Chung, K.J. Gene 167, (1995) 179.
Chen, B., Gao, S., Choi, G.H. and Nuss, D.L. Proc. Natl. Acad. Sci USA 93, (1996) 7996.
Chen, H., Crabb, J.W. and Kinsey, J.A. Genetics 148, (1998) 123.
Cheng, C. and Udaka, S. Agric. Biolo. Chem. 55, (1991) 1817.
Chilton, M.D., Drummond, M.H., Merio, D.J., Sciaky, D., Montoya, A.L., Gordon, M.P. and Nester, E.W. Cell 11, (1977) 263.
Choi, G.H., Chen, B. and Nuss, D.L. Proc. Natl. Acad. Sci. USA 92, (1995) 305.
Christgau, S., Kofod, L.V., Halkier, T., Andersen, L.N., Hockauf, M., Dorreich, K., Dalboge, H. and Kauppinen, S. Biochem. J. 319, (1996) 705.
Chu, S., DeRisi, J., Eisen, M., Mulholland, J., Botstein, D., Brown, P.O. and Herskowitz, I. Science,282, (1998) 699.
Churchill, A.C.L., Ciuffetti, L.M., Hansen, D.R., Van Etten, H.D. and Van Alfen, N. K. Curr. Genet. 17, (1990) 25.
Ciuffetti, L.M., Tuori, R.P. and Gaventa, J.M. Plant Cell 9, (1997) 135.
Clark, A.J. and Adelberg, E.A. Annu. Rev. Microbiol. 16, (1962) 289.
Clarke, D.L., Newbert, R.W. and Turner, G. Curr. Genet. 32 (1997) 408.
Clarke, L. and Carbon, J. Cell 9, 91.

Cogoni, C. and Macino, G. (1999) Nature 399, (1976) 166.
Cogoni, C. and Macino, G. Proc. Natl. Acad. Sci. USA 94, (1997) 10233.
Cohen, S.N., Chang, A.C.Y. and Hsu, L. Proc. Natl. Acad. Sci. USA. 69, (1972) 2110.
Contreras, R., Carrez, D., Kinghorn, J.R., van den Hondel, C.A. and Fiers, W. Biotechnol. 9, (1991) 378.
Cooley, R.N., Franklin, F.C.H. and Caten, C.E. Mycol. Res. 94, (1990) 145.
Corrochano, L.M., Lauter, F.R., Ebbole, D.J. and Yanofsky, C. Dev Biol 167, 190-(1995) 200.
Costanzo, M.C. and Fox, T.D. Genetics 120, (1988) 667.
Daboussi, M.J., Djeballi, A., Gerlinger, C., Blaiseau, P.L., Bouvier, I., Cassan, M., Lebrun, M.H., Parisot, D. and Brygoo, Y. Curr. Genet. 15, (1989) 453.
Dalboge, H. FEMS Microbiol. Rev. 21, (1997) 29.
Dalboge, H. and Heldt-Hansen, H.P. Mol. Gen. Genet. 243, (1994) 253.
Dang, V.D., Valens, M., Bolotin-Fukuhara, M. and Daignan-Fornier, B. Yeast 10, (1994) 1273.
Datta, A. and Jinks-Robertson, S. Science 268, (1995) 1616.
Davies, R.W. Prog. Ind. Microbiol. 29, (1994) 527.
Davis, B. (1985) Factors influencing protoplast isolation. In Fungal Protoplasts: applications in biochemicstry and genetics, J. F. Peberdy and L. Ferenczy, eds. (New York: Marcel Dekker, Inc.), pp. 45-71.
Davis, M.A., Kelly, J.M. and Hynes, M.J. Genetica 90, (1993) 133.
de Groot, M.J., Bundock, P., Hooykaas, P.J. and Beijersbergen, A.G. Nature Biotechnol. 16, (1998) 839.
Dean, R.A. and Timberlake, W.E. Plant Cell 1, (1989) 275.
Del Poeta, M., Toffaletti, D.L., Rude, T.H., Dykstra, C.C., Heitman, J. and Perfect, J.R. Genetics 152 (1999) 167.
del Poeta, M., Toffaletti, D.L., Rude, T.H., Sparks, S.D., Heitman, J. and Perfect, J.R. Infect. Immun. 67 (1999) 1812.
Dequard-Chablat, M. and Rotig, A. Mol. Gen. Genet. 253 (1997) 546.
DeRisi, J.L., Iyer, V.R. and Brown, P.O. Science 278, (1997) 680.
Devchand, M. and Gwynne, D.I. (1991) J. Biotechnol. 17 3.
Dhawale, S.S., Paietta, J.V. and Marzluf, G.A. Cur. Genet. 8 (1984) 77.
Diallinas, G. and Scazzocchio, C. Genetics 122 (1989) 341.
Diaz-Perez, S.V., Crouch, V.W. and Orbach, M.J. Fungal Genet. Biol. 20 (1996) 280.
Dickman, M.B. Cur. Genet. 14 (1988) 241.
Diez, B., Mellado, E., Fouces, R., Rodriguez, M. and Barredo, J.L. Microbiologia 12 (1996) 359.
Du, W., Huang, Z., Flaherty, J.E., Wells, K. and Payne, G.A. Appl Environ. Microbiol. 65 (1999) 834.
Dubnau, D. Microbiol. Rev. 55 (1991) 395.
Dunn-Coleman, N.S., Bloebaum, P., Berka, R.M., Bodie, E., Robinson, N., Armstrong, G., Ward, M., Przetak, M., Carter, G.L. and LaCost, R. Biotechnol. 9 (1991) 976.
Durand, R., Rascle, C., Fischer, M. and Fevre, M. Curr. Genet. 31 (1997) 158.
Dussoix, D. and Arber, W. J. Molec. Biol. 5 (1962) 37.
Edman, J.C. Mol. Cell. Biol. 12 (1992) 2777.
Eisen, M.B., Spellman, P.T., Brown, P.O. and Botstein, D. Proc. Natl. Acad. Sci. USA. 95 (1998) 14863.
Endo, H., Kajiwara, S., Tsunoka, O. and Shishido, K. Gene 139 (1994) 117.
Epstein, L., Lusnak, K. and Kaur, S. Fungal Genet. Biol. 23 (1998) 189.

Farkas, V. (1985) The fungal cell wall. In Fungal Protoplasts: applications in biochemicstry and genetics, J. F. Peberdy and L. Ferenczy, eds. (New York: Marcel Dekker, Inc.), pp. 3-30.
Felenbok, B. J. Biotechnol. 17 (1991) 11.
Fernandez-Canon, J.M. and Penalva, M.A. Mol. Gen. Genet. 246 (1995) 110.
Fincham, J.R. [published erratum appears in Microbiol. Rev. 1991 Jun;55(2):334]. Microbiol. Rev. 53 (1989) 148.
Formosa, T. and Nittis, T. Mol. Gen. Genet. 257 (1998) 461.
Fotheringham, S. and Holloman, W.K. Mol. Cell Biol. 9 (1989) 4052.
Fotheringham, S. and Holloman, W.K. Genetics 124 (1990) 833.
Fowler, T. and Berka, R.M. Curr. Opin. Biotechnol. 2 (1991) 691.
Fowler, T. and Brown, R.D., Jr. Mol. Microbiol. 6 (1992) 3225.
Frommer, W.B. and Ninnemann, O. Annu. Rev. Plant Physiol. Plant Molec. Biol. 46 (1995) 419.
Garcia-Pedrajas, M.D. and Roncero, M.I. Curr. Genet. 29 (1996) 191.
Garnand, K. and Nelson, M.A. Fungal Genet. Newsletter 42 (1995) 29.
Gems, D., Johnstone, I.L. and Clutterbuck, A.J. Gene 98 (1991) 61.
Gems, D.H. and Clutterbuck, A.J. Curr. Genet. 24 (1993) 520.
Gidekel, M., Jimenez, B. and Herrera-Estrella, L. Gene 170 (1996) 201.
Giles, N.H., Case, M.E., Baum, J., Geever, R., Huiet, L., Patel, V. and Tyler, B. Microbiol. Rev. 49 (1985) 338.
Glumoff, V., Kappeli, O., Fiechter, A. and Reiser, J. Gene 84 (1989) 311.
Gold, S., Duncan, G., Barrett, K. and Kronstad, J. Genes Dev. 8 (1994) 2805.
Gold, S.E., Bakkeren, G., Davies, J.E. and Kronstad, J.W. Gene 142 (1994) 225.
Gold, S. E., Casale, W. L. and Keen, N. T. Mol. Gen. Genet. 230 (1991) 104.
Gonzalez, A., Jimenez, A., Vazquez, D., Davies, J.E. and Schindler, D. Biochim. Biophys. Acta 521 (1978) 459.
Gonzalez-Hernandez, G.A., Herrera-Estrella, L., Rocha-Ramirez, V., Roncero, M.I.G. and Gutierrez-Corona, J.F. Mycol. Res. 101(1997) 953.
Gouka, R.J., Hessing, J.G., Punt, P.J., Stam, H., Musters, W. and Van den Hondel, C.A. Appl. Microbiol. Biotechnol. 46 (1996) 28.
Gouka, R.J., Punt, P.J., Hessing, J.G. and van den Hondel, C.A. Appl. Environ. Microbiol. 62 (1996) 1951.
Graessle, S., Haas, H., Friedlin, E., Kurnsteiner, H., Stoffler, G. and Redl, B. Appl. Environ. Microbiol. 63 (1997) 753.
Griffith, F. J. Hygene 27 (1928) 113.
Gritz, L. and Davies, J. Gene 25 (1983) 179.
Hamer, J.E. and Timberlake, W.E. Molec. Cell. Biol. 7(1987) 2352.
Harkki, A., Mantyla, A., Penttila, M., Muttilainen, S., Buhler, R., Suominen, P., Knowles, J. and Nevalainen, H. Enzyme Microb. Technol. 13 (1991) 227.
Harmsen, M.C., Schuren, F.H., Moukha, S., van Zuilen, C.M., Punt, P.J. and Wessels, J.G. Curr. Genet. 22 (1992) 447.
Heim, R., Prasher, D.C. and Tsien, R.Y. Proc. Natl. Acad. Sci. USA. 91 (1994) 12501.
Herd, S., Christensen, M.J., Saunders, K., Scott, D.B. and Schmid, J. Microbiol. 143 (1997) 267.
Hershko, A. Curr. Opin. Cell Biol. 9 (1997) 788.
Herzog, R.W., Daniell, H., Singh, N.K. and Lemke, P.A. Appl. Microbiol. Biotechnol. 45 (1996) 333.
Hilt, W. and Wolf, D.H. Mol. Biol. Rep. 21 (1995) 3.

Hinnen, A., Hicks, J. B. and Fink, G. R. Proc. Natl. Acad. Sci. USA. 75 (1978) 1929.
Hintz, W.E. and Lagosky, P. A. Biotechnology 11 (1993) 815.
Hobert, O. and Ruvkun, G. Biol. Bull. 195 (1998) 377.
Hochstrasser, M., Papa, F.R., Chen, P., Swaminathan, S., Johnson, P., Stillman, L., Amerik, A.Y. and Li, S.J. Cold Spring Harb. Symp. Quant. Biol. 60 (1995) 503.
Hynes, M.J., Corrick, C.M. and King, J.A. Mol. Cell Biol. 3 (1983) 1430.
Iannacone, R., Grieco, P.D. and Cellini, F. Plant Mol. Biol. 34 (1997) 485.
Ilmen, M., Onnela, M.L., Klemsdal, S., Keranen, S. and Pentila, M. Mol. Gen. Genet. 257 (1998) 386.
Ishikawa, M., Janda, M., Krol, M.A., and Ahlquist, P. J. Virol. 71 (1997) 7781.
Ito, H., Fukuda, Y., Murata, K. and Kimura, A. J. Bacteriol. 153 (1983) 163.
Jeenes, D.J., Mackenzie, D.A., Roberts, I.N. and Archer, D.B. Biotechnol. Genet. Eng. Rev. 9 (1991) 327.
Jeenes, D.J., Marczinke, B., MacKenzie, D.A. and Archer, D.B. FEMS Microbiol Lett. 107 (1993) 267.
Jefferson, R.A., Burgess, S.M. and Hirsh, D. Proc. Natl. Acad. Sci. USA. 83 (1986) 8447.
Jiang, Y.W. and Stillman, D.J. Genetics 140 (1995) 103.
Jimenez, A. and Davies, J. Nature 287 (1980) 869.
Joos, H., Inze, D., Caplan, A., Sormann, M., Van Montagu, M. and Schell, J. Cell 32 (1983) 1057.
Judelson, H.S. Mol. Gen. Genet. 239 (1993) 241.
Kato, M., Aoyama, A., Naruse, F., Kobayashi, T. and Tsukagoshi, N. Mol. Gen. Genet. 254 (1997) 119.
Keon, J.P., James, C.S., Court, S., Baden-Daintree, C., Bailey, A.M., Burden, R.S., Bard, M. and Hargreaves, J.A. Curr. Genet. 25 (1994) 531.
Keon, J.P., White, G.A. and Hargreaves, J. A. Curr. Genet. 19 (1991) 475.
Kinal, H., Tao, J. and Bruenn, J.A. Gene 98 (1991) 129.
Kistler, H.C. and Benny, U. Gene 117 (1992) 81.
Kistler, H.C. and Benny, U.K. Cur. Genet. 13 (1988) 145.
Klein, T.M., Arentzen, R., Lewis, P.A. and Fitzpatrick-McElligott, S. Biotechnology 10 (1992) 286.
Klein, T.M., Wolf, E.D., Wu, R. and Sanford, J.C. Nature 327 (1987) 70.
Kolar, M., Punt, P.J., van den Hondel, C.A. and Schwab, H. Gene 62 (1988) 127.
Kozak, M. J. Mol. Biol. 196 (1987) 947.
Kross, J., Henner, W.D., Hecht, S.M. and Haseltine, W.A. Biochemistry 21 (1982) 4310.
Kukuruzinska, M.A. and Lennon, K. Crit. Rev. Oral Biol. Med. 9 (1998) 415.
Kuspa, A. and Loomis, W.F. Proc. Natl. Acad. Sci. USA. 89 (1992) 8803.
Lashkari, D.A., DeRisi, J.L., McCusker, J.H., Namath, A.F., Gentile, C., Hwang, S.Y., Brown, P.O. and Davis, R. W. Proc. Natl. Acad. Sci. USA. 94 (1997) 13057.
Lederberg, J. and Tatum, E.L. Nature 158 (1946) 558.
Leung, H., Lehtinen, U., Karjalainen, R., Skinner, D., Tooley, P., Leong, S. and Ellingboe, A. Curr. Genet. 17 (1990) 409.
Levings, C.S.D. and Siedow, J.N. Plant Mol Biol 19 (1992) 135.
Levis, C., Fortini, D. and Brygoo, Y. Curr. Genet. 32 (1997) 157.
Lints, R., Davis, M.A. and Hynes, M.J. Mol. Microbiol. 15 (1995) 965.
Lippincott, J.A. and Lippincott, B.B. Annu. Rev. Microbiol. 29 (1975) 377.

Litzka, O., Papagiannopolous, P., Davis, M.A., Hynes, M.J. and Brakhage, A.A. Eur. J. Biochem. 251 (1998) 758.
Lloyd, A.T. and Sharp, P.M. Nucleic Acids Res. 20 (1992) 5289.
Lodish, H., Baltimore, D., Berk, A., Zipinsky, S.L., Matsudaira, P. and Darnell, J. Molecular Cell Biology, 3rd Edition: Scientific American Books. 1995.
Lorenz, M.G. and Wackernagel, W. Microbiol. Rev. 58 (1994) 563.
Lorito, M., Hayes, C.K., Di Pietro, A. and Harman, G.E. Curr. Genet. 24 (1993) 349.
Lu, S., Lyngholm, L., Yang, G., Bronson, C., Yoder, O.C. and Turgeon, B. G. Proc. Natl. Acad. Sci. USA. 91 (1994) 12649.
Lwoff, A. Bacteriol. Rev. 17 (1953) 269.
Macatee, T., Jiang, Y.W., Stillman, D.J. and S.Y. Roth, Nucl. Acids Res. 25 (1997) 1240.
MacKenzie, D. A., Jeenes, D.J., Belshaw, N.J. and Archer, D.B. J. Gen. Microbiol. 139 (1993) 2295.
Malardier, L., Daboussi, M.J., Julien, J., Roussel, F., Scazzocchio, C. and Brygoo, Y. Gene 78 (1989) 147.
Mandel, M. and Higa, A. J. Molec. Biol. 53 (1970) 159.
Maras, M., Saelens, X., Laroy, W., Piens, K., Claeyssens, M., Fiers, W. and Contreras, R. Eur. J. Biochem. 249 (1997) 701.
Marhoul, J.F. and Adams, T.H. Genetics 144 (1996) 1463.
Marhoul, J.F. and Adams, T.H. Genetics 139 (1995) 537.
Mathieu, M. and Felenbok, B. EMBO J. 13 (1994) 4022.
Mattern, I.E., van Noort, J.M., van den Bend: American Society of Plant Physiologists), pp. 1-10.
Mauri, I., Maddaloni, M., Lohmer, S., Motto, M., Salamini, F., Thompson, R. and Martegani, E. Mol. Gen. Genet. 241 (1993) 319.
May, G.S., Gambino, J., Weatherbee, J.A. and Morris, N.R. J. Cell Biol. 101 (1985) 712.
Mayorga, M. E. and Gold, S. E. Fungal Genet. Biol. 24 (1998) 364.
McGoldrick, C.A., Gruver, C. and May, G.S. J. Cell Biol. 128 (1995) 577.
McNabb, D.S., Xing, Y. and Guarente, L. Genes Dev. (1995) 47.
Minetoki, T., Gomi, K., Kitamoto, K., Kumagai, C. and Tamura, G. Biosci. Biotechnol. Biochem. 59 (1995) 1516.
Mishra, N.C. Genet. Res. 29 (1977) 9.
Mishra, N.C. J. Gen. Microbiol. 113 (1979) 255.
Mishra, N.C. and Tatum, E.L. (1973). Proc. Natl. Acad. Sci. USA. 70, 3875-9.
Molkentin, J.D. and Olson, E.N. (1996). Proc. Natl. Acad. Sci. USA. 93, 9366-73.
Moralejo, F.J., Cardoza, R.E., Gutierrez, S. and Martin, J.F. (1999). Appl. Environ. Microbiol. 65, 1168-74.
Mort-Bontemps, M. and Fevre, M. Curr. Genet. 31 (1997) 272.
Nagata, O., Takashima, T., Tanaka, M. and Tsukagoshi, N. Mol. Gen. Genet. 237 (1993) 251.
Nakano, E.T., Fox, R.D., Clements, D.E., Koo, K., Stuart, W.D. and Ivy, J.M. Fungal Genet. Newslett. 40 (1993) 54.
Nester, E.W. and Gordon, M.P. (1988) Early events in the transformation of higher plants by *Agrobacteriaum*. In Physiology and biochemistry of palnt-microbe interactions, N.T. Keen, T. Kosuge and L.L. Walling, eds. (Rockville, Maryland: American Society of Plant Physiologists), pp. 1-10.
Nicolaisen, M. and Geisen, R. Microbiol. Res. 151 (1996) 281.

Nikolskaya, A.N., Panaccione, D.G. and Walton, J.D. Gene 165 (1995) 207.
Nyyssonen, E., Penttila, M., Harkki, A., Saloheimo, A., Knowles, J.K.C. and Keranen, S. Bio/Technology 11 (1993) 591.
Old, R.W., and Primrose, S.B. (1989) Cutting and joining DNA molecules. In Principles of Gene Manipulation, an introduction to genetic engineering, N. G. Carr, J.L. Ingraham and S.C. Rittenberg, eds. (London: Blackwell Scientific Pub.), pp. 14-35.
Oliver, R.P., Roberts, I.N., Harling, R., Kenyon, L., Punt, P.J., Dingemanse, M.A. and van den Hondel, C. A. M. J. J. Cur. Genet. 12 (1987) 231.
Orbach, M.J. Gene 150 (1994) 159.
Orbach, M.J., Porro, E.B. and Yanofsky, C. Mol. Cell Biol. 6 (1986) 2452.
Osherov, N., Yamashita, R.A., Chung, Y.S. and May, G.S. (1986) J. Biol. Chem. 273 (1986) 27017.
Ozer, J., Lezina, L.E., Ewing, J., Audi, S. and Lieberman, P.M. Mol. Cell Biol. 18 (1998) 2559.
Pauly, M., Andersen, L.N., Kauppinen, S., Kofod, L.V., York, W.S., Albersheim, P. and Darvill, A. Glycobiology 9 (1999) 93.
Peberdy, J.F. (1985) Mycolytic enzymes. In Fungal Protoplasts: applications in biochemicstry and genetics, J.F. Peberdy and L. Ferenczy, eds. (New York: Marcel Dekker, Inc.), pp. 31-44.
Penalva, M.A., Rowlands, R.T. and Turner, G. Trends Biotechnol. 16 (1998) 483.
Penttila, M., Nevalainen, H., Ratto, M., Salminen, E. and Knowles, J. Gene 61 (1987) 155.
Pfeifer, T.A. and Khachatourians, G.G. Appl. Microbiol. Biotechnol. 38 (1992) 376.
Pfeifer, T.A. and Khachatourians, G.G. J. Invertebr. Pathol. 62 (1993) 231.
Pittenger, R. ., Wolfe, R.N., Hoehn, M.M., Marks, P.N., Daily, W.A. and McGuire, J.M. Antibiot. Chemother. 111 (1953) 1269.
Pitts, J.E., Quinn, D., Uusitalo, J. and Penttila, M. Biochem. Soc. Trans. 19 (1991) 663.
Pitts, J.E., Uusitalo, J.M., Mantafounis, D., Nugent, P.G., Quinn, D.D., Orprayoon, P. and Penttila, M.E. J. Biotechnol. 28 (1993) 69.
Platt, A., Langdon, T., Arst, H., Kirk, D., Tollervey, D., Sanchez, J.M. and Caddick, M.X. EMBO J. 15 (1996) 2791.
Potter, H., Weir, L. and Leder, P. Proc. Natl. Acad. Sci. USA. 81(1984) 7161.
Prasher, D.C., Eckenrode, V.K., Ward, W.W., Prendergast, F.G. and Cormier, M.J. Gene 111 (1992) 229.
Punt, P.J., Dingemanse, M.A., Kuyvenhoven, A., Soede, R.D., Pouwels, P.H. and van den Hondel, C.A. Gene 93 (1990) 10.
Punt, P.J., Oliver, RP., Dingemanse, M.A., Pouwels, P.H. and van den Hondel, C.A. Gene 56 (1987) 117.
Punt, P.J. and van den Hondel, C.A. Methods Enzymol. 216 (1992) 447.
Purugganan, M.D. Bioessays 20 (1998) 700.
Radzio, R. and Kuck, U. Appl. Microbiol. Biotechnol. 48 (1997) 58.
Redkar, R.J., Herzog, R.W. and Singh, N.K. Appl. Environ. Microbiol. 64 (1998) 222.
Redman, R.S. and Rodriguez, R.J. Exp. Mycol. 18 (1994) 230.
Reuter, G. and Gabius, H.J. Cell Mol. Life Sci. 55 (1999) 368.
Richey, M.G., Marek, E.T., Schardl, C.L. and Smith, D.A. Phytopathology 79 (1989) 844.
Riechmann, J.L. and Meyerowitz, E.M. Biol. Chem. 378 (1997) 1079.
Riggs, C.D. and Bates, G.W. Proc. Natl. Acad. Sci. USA. 83 (1986) 5602.

Rohe, M., Searle, J., Newton, A.C. and Knogge, W. Curr. Genet. 29 (1996) 587.
Ross, J. Trends Genet. 12 (1996) 171.
Rothstein, R. (1991) Targeting, disruption, replacement, and allele rescue: integrative DNA transformation in Yeast. In Methods in Enzymology, C. Guthrie and G.R. Fink, eds. (San Diego: Academic Press INC.), pp. 281-301.
Royer, J.C., Moyer, D.L., Reiwitch, S.G., Madden, M.S., Jensen, E.B., Brown, S.H., Yonker, C.C., Johnston, J.A., Golightly, E.J. and Yoder, W.T. Biotechnology 13 (1995)1479-83.
Ruoff, P., Vinsjevik, M., Mohsenzadeh, S. and Rensing, L. J. Theoret. Biol. 196 (1999) 483.
Sachs, M.S. Fungal Genet. Biol. 23 (1998) 117.
Sanchez, O., Navarro, R.E. and Aguirre, J. Mol. Gen. Genet. 258 (1998) 89.
Sanford, J.C., Smith, F.D. and Russell, J.A. (1993) Methods Enzymol 217, 483.
Schiestl, R.H. and Petes, T.D. Proc. Natl. Acad. Sci. USA. 88 (1991) 7585.
Schmidt-Dannert, C., Pleiss, J. and Schmid, R.D. Ann. NY Acad. Sci. 864 (1998) 14.
Schwer, B., and Shuman, S. Gene Expr. 5 (1996) 331.
Scott-Craig, J.S., Cheng, Y.Q., Cervone, F., De Lorenzo, G., Pitkin, J.W. and Walton, J.D. Appl. Environ. Microbiol. 64 (1998) 1497.
Scott-Craig, J.S., Panaccione, D.G., Cervone, F. and Walton, J.D. Plant Cell 2 (1990) 1191.
Seip, E.R., Woloshuk, C.P., Payne, G.A. and Curtis, S. E. Appl. Environ. Microbiol. 56 (1990) 3686.
Selker, E.U. Annu. Rev. Genet. 24 (1990) 579.
Shalon, D., Smith, S.J. and Brown, P.O. Genome Res. 6 (1996) 639.
Sharp, P.M. and Matassi, G. Curr. Opin. Genet. Dev. 4 (1994) 851.
Shi, Z., Christian, D. and Leung, H. Phytopathology 85 (1995) 329.
Singer, M.J. and Selker, E.U. Curr. Top. Microbiol. Immunol. 197 (1995) 165.
Singsit, C., Adang, M.J., Lynch, R.E., Anderson, W.F., Wang, A., Cardineau, G. and Ozias-Akins, P. Transgenic Res. 6 (1997) 169.
Sivadon, P., Peypouquet, M.F., Doignon, F., Aigle, M. and Crouzet, M. Yeast 13 (1997) 747.
Smith, A.W., Collis, K., Ramsden, M., Fox, H.M. and Peberdy, J.F. Curr. Genet. 19 (1991) 235.
Soni, R., Carmichael, J.P. and Murray, J.A. Curr. Genet. 24 (1993) 455.
Spellig, T., Bottin, A. and Kahmann, R. Mol. Gen. Genet. 252 (1996) 503.
St. Leger, R.J., S., Shimizu, S., Joshi, L., Bidochka, M.J., and Roberts, D.W. FEMS Microbiol. Lett. 131 (1995) 289.
Steiner, S., and Philippsen, P. Mol. Gen. Genet. 242 (1994) 263.
Stinchcomb, D.T., Struhl, K. and Davis, R.W. Nature 282 (1979) 39.
Straffon, M.J., Hynes, M.J. and Davis, M.A. Curr. Genet. 29 (1996) 360.
Sugita, T. and Nakase, T. Syst. Appl. Microbiol. 22 (1999) 79.
Sugiyama, H., Ohkuma, M., Masuda, Y., Park, S. M., Ohta, A. and Takagi, M. Yeast 11 (1995) 43.
Tanaka, K. Biochem. Biophys. Res. Commun. 247 (1998) 537.
Templeton, M.D., Sharrock, K.R., Bowen, J.K., Crowhurst, R.N. and Rikkerink, E.H. Gene 142 (1994) 141.
Timberlake, W.E. and Marshall, M.A. Science 244 (1989) 1313.
Todd, R.B. and Andrianopoulos, A. Fungal Genet. Biol. 21 (1997) 388.
Tsuge, T., Nishimura, S. and Kobayashi, H. Gene 90 (1990) 207.

Tsukagoshi, N. Nippon Ishinkin Gakkai Zasshi 39 (1998) 85.
Tsukuda, T., Carleton, S., Fotheringham, S. and Holloman, W. K. Molec. Cell. Biol. 8 (1988) 3703.
Tudzynski, B., Mende, K., Weltring, K. M., Kinghorn, J. R. and Unkles, S. E. Microbiol. 142 (1996) 533.
Turgeon, B.G., Garber, R.C. and Yoder, O.C. Molec. Cell Biol. 7 (1987) 3297.
Turner, G. Prog. Ind. Microbiol. 29 (1994) 641.
Unkles, S.E., Campbell, E.I., de Ruiter-Jacobs, Y.M.J.T., Broekhuijsen, M., Macro, J.A., Carrez, D., Contreras, R., van den Hondel, C.A.M.J.J. and Kinghorn, J. Molec. Gen. Genet. 218 (1989) 99.
Upchurch, R.G., Ehrenshaft, M., Walker, D.C. and Sanders, L.A. Appl. Environ. Microbiol. 57 (1991) 2935.
Usher, J.J., Hughes, D.W., Lewis, M.A. and Chiang, S.J. J. Ind. Microbiol. 10 (1992) 157.
van de Rhee, M.D., Graca, P.M., Huizing, H.J. and Mooibroek, H. Mol. Gen. Genet. 250 (1996) 252.
van den Hondel, C.A.M.J.J., Punt, P.J. and van Gorcam, R.F.M. (1991) Heterologous gene expression in filamentous fungi. In More Gene Manipulations in Fungi, J.L. Lasure, ed. (San Diego: Academic Press), pp. 396-428.
Van Dyck, E., Jank, B., Ragnini, A., Schweyen, R.J., Duyckaerts, C., Sluse, F. and Foury, F. Mol. Gen. Genet. 246 (1995) 426.
van Hartingsveldt, W., Mattern, I. E., van Zeijl, C.M., Pouwels, P.H. and van den Hondel, C.A. Mol. Gen. Genet. 206 (1987) 71.
van Heeswijck, R. and Hynes, M.J. Nucl. Acids Res. 19 (1991) 2655.
van Nistelrooy, J.G., van den Brink, J.M., van Kan, J.A., van Gorcom, R.F. and de Waard, M.A. Mol. Gen. Genet. 250 (1996) 725.
Van Veen, R.J.M., Hooykaas, P.J.J. and Schilperoort, R.A. (1988) Mechanisms of tumorigenesis by *Agrobacterium tumefaciens*. In Physiology and biochemistry of palnt-microbe interactions, N.T. Keen, T. Kosuge and L.L. Walling, eds. (Rockville, Maryland: American Society of Plant Physiologists), pp. 19-30.
VanEtten, H., Jorgensen, S., Enkerli, J. and Covert, S. F. Curr. Genet. 33 (1998) 299.
Vanwert, S.L. and Yoder, O.C. Cur. Genet. 25 (1994) 217.
Verdoes, J.C., Punt, P.J. and van den Hondel, C.A.M.J.J. Appl. Mibrobiol. Biotechnol. 43 (1995) 195.
Verdoes, J. C., van Diepeningen, A. D., Punt, P. J., Debets, A. J., Stouthamer, A. H. and van den Hondel, C. A. J. Biotechnol. 36 (1994) 165.
Vollmer, S. and Yanofsky, C. Proc. Nat. Acade. Sci. USA. 83 (1986) 4869.
Wach, A., Brachat, A., Pohlmann, R. and Philippsen, P. Yeast 10 (1994) 1793.
Wada, M., Beppu, T. and Horinouchi, S. Appl. Microbiol. Biotechnol. 45 (1996) 652.
Wang, J., Holden, D. W. and Leong, S. A. Proc. Natl. Acad. Sci. USA. 85 (1988) 865.
Wang, P. and Nuss, D.L. Proc. Natl. Acad. Sci. USA. 92 (1995) 11529.
Wang, Z., Fang, P. and Sachs, M.S. Mol. Cell. Biol. 18 (1998) 7528.
Wang, Z. and Sachs, M.S. Mol. Cell. Biol. 17 (1997) 4904.
Ward, M. (1989) Heterologous gene expression in *Aspergillus*. In Molecular Biology of Filamentous Fungi., HN. and M. Pentilla, ed.: Foundation for Biotechnical and Industrial Fermentation Research.), pp. 119-128.
Ward, M., Wilson, L.J., Kodama, K.H., Rey, M.W. and Berka, R.M. Biotechnology 8 (1990) 435.
Ward, P.P., May, G.S., Headon, D.R. and Conneely, O.M. Gene 122 (1992) 219.

Waring, R.B., May, G.S. and Morris, N.R. Gene 79 (1989) 119.
Watson, A.J., Fuller, L.J., Jeenes, DJ. and Archer, D.B. Appl. Environ. Microbiol. 65 (1999) 307.
Weidner, G., d'Enfert, C., Koch, A., Mol, P.C. and Brakhage, A.A. Curr. Genet. 33 (1998) 378.
Whiteway, M., Dignard, D. and Thomas, D. Y. Proc. Natl. Acad. Sci. USA. 89 (1992) 9410.
Wickens, M., Anderson, P. and Jackson, R.J. (1997) Curr. Opin. Genet. Dev. 7, 220.
Wickes, B.L. and Edman, J.C. Mol. Microbiol. 16 (1995) 1099.
Wnendt, S., Jacobs, M. and Stahl, U. Curr. Genet. 17 (1990) 21.
Wu, T.S., Skory, C.D., Horng, J.S. and Linz, J.E. Gene 182 (1996) 7.
Yang, G., Turgeon, B.G. and O.C. Yoder, Genetics 137 (1994) 751.
Yang, Z. and Bisson, L.F. Yeast 12 (1996) 1407.

Applied Mycology and Biotechnology
Volume 1. Agriculture and Food Production
G.G. Khachatourians and D.K. Arora (editors)

Aspergillus nidulans as a model organism for the study of the expression of genes encoding enzymes of relevance in the food industry

A.P. MacCabe, M. Orejas and D. Ramón

Departamento de Biotecnología; Instituto de Agroquímica y Tecnología de Alimentos (CSIC), PO Box 73, 46100-Burjassot, Valencia, Spain

1. INTRODUCTION

Filamentous fungi are of considerable importance in the food and beverage industries as producers of metabolites possessing either beneficial (food enzymes) or detrimental (mycotoxins) properties. Whilst these fungi are able to produce a very wide range of enzymatic activities only a restricted number of these enzymes are currently used in commercial food production. Nevertheless, this restricted number represents a large percentage of the total number of enzymes employed by the food industry. In general, enzymatic activities of industrial interest are obtained from fungal cultures as components in mixtures containing other non-required activities. Given the advances in the molecular characterisation of genes encoding enzymes of industrial relevance, a potentially useful alternative to screening programs for strain improvement may reside in the application of molecular genetic technology.

For the purposes of optimising the industrial production of enzymes by molecular means a logical first step is the study of the regulation of expression of the corresponding genes. The principal industrial scale fungal enzyme producers are predominantly members of the genera *Aspergillus* (*Aspergillus awamori, Aspergillus niger, Aspergillus oryzae*) and *Trichoderma* (*Trichoderma longibrachiatum, Trichoderma reesei*). Although not employed at the industrial level, *Aspergillus nidulans* is without doubt the organism of choice for carrying out such studies on gene regulation. Its ability to use a wide range of nutrient sources and the relative facility with which phenotypes can be studied by growth testing, combined with its amenability to genetic manipulation under laboratory conditions have resulted in its use as a genetic model organism for almost fifty years (1). To-date, a considerable number of *A. nidulans* mutants impaired in both structural and regulatory genes have been described (2), and during the last fifteen years the rapid application of recombinant DNA technology has resulted in the cloning and molecular characterisation of a large number of genes and hence a more detailed understanding of the biology of this micro-organism. By comparison to the industrially used species, *A. nidulans* is by far the filamentous fungus for which we have the greatest amount of genetic information.

Classical genetic studies have revealed the existence of two principal modes of genetic regulation in *A. nidulans*: pathway-specific regulation and wide domain regulation (3, 4). Pathway-specific regulation refers to the control of the synthesis of components (permeases and enzymes) of a single metabolic pathway whereas wide domain regulation refers to control exerted over a number of different pathways. For the majority of metabolic routes studied to-date both pathway-specific and wide domain regulatory mechanisms have been described. The most extensively studied examples of wide domain regulation are nitrogen metabolite repression, carbon catabolite repression and ambient pH regulation. In this chapter we will present details of these two models of regulation and subsequently focus on the control of the synthesis of enzymes involved in the degradation of xylan as an example of the interplay between specific induction and wide domain regulation.

2. WIDE DOMAIN REGULATORY SYSTEMS IN *A. nidulans*.

2.1. Nitrogen metabolite repression.

Nitrogen metabolite repression (also termed ammonium repression) refers to the repression of the synthesis of enzymes and permeases involved in the catabolism of nitrogen-containing compounds when readily assimilated nitrogen sources such as ammonium and L-glutamine are available as nutrients. Classical genetic studies in *A. nidulans* (5) identified two classes of mutation resulting in different nitrogen utilisation phenotypes that mapped to the same genetic locus on linkage group III. *areA*^{r} mutations were found to lead to loss of growth on a wide variety of nitrogen sources (nitrate, nitrite, purines etc) with the exception of ammonium and L-glutamine. The much rarer *areA*^{d} mutations resulted in derepression of one or more activities which in wild type (*areA*$^+$) are ammonium-repressible and hence represent a specificity class of mutation. The non-hierarchical heterogeneity of phenotypes (i.e. that the relative effects of one *areA* allele on a number of target systems are different to the relative effects exerted on the same targets by another allele) observed for both *areA*^{r} and *areA*^{d} classes of alleles and studies of dominance relationships between various alleles in diploids led to the conclusion that the functional product of the *areA* gene is a positively acting regulatory protein (5). Thus in *areA*$^+$ in the absence of ammonium or L-glutamine the gene product AreA would be functionally competent and available to positively regulate the expression of activities required for the obtention of nitrogen from less-readily assimilated sources. The particular catabolic activities stimulated would be determined by the presence of a specific inducer. By way of an example, the expression of the structural genes directly involved in nitrate utilisation (*crnA*, *niaD* and *niiA*) is induced in the presence of nitrate and repressed by ammonium (6). Induction by nitrate is mediated by the product (NirA) of the positive-acting pathway-specific regulatory gene *nirA* (7,8); repression is mediated by AreA when ammonium is present in the growth medium. Loss-of-function mutations in *areA* lead to either low or undetectable levels of nitrate utilisation activities in the presence of nitrate. Hence, high level expression of the activities required for growth on nitrate not only requires specific induction by NirA but also the function of the positively acting regulator AreA. Nitrogen metabolite repression is thus achieved by eliminating the positive regulatory effect of AreA.

Interestingly, nitrogen metabolite repression of certain other nitrogen assimilatory systems was found to be influenced by carbon source (5). Whereas no mycelial growth was observed for certain *areA*^{r} mutants (*areA*^{r}*1* and *areA*^{r}*2*) on proline/glucose medium, growth was observed when proline was present as the sole source of nitrogen and carbon. Mutants were generated which were able to suppress this effect allowing growth on proline as nitrogen source in the presence of glucose, and the suppressor mapped to a single locus, *creA* (see below).

Whereas many fungal genes have been cloned by transformation and phenotypic rescue of mutants, the molecular cloning of *areA* (9) was achieved by taking advantage of a genetic phenomenon in which potentially viable duplication-deficiency progeny may be generated in crosses in which a chromosome is pericentrically inverted with respect to its homologue. The mutation *xprD1*, originally selected as being derepressed for extracellular protease synthesis (10), was shown by Arst and Cove (5) to be an ammonium derepressed allele of *areA* (it is was also the most extreme *areA*^{d} allele, showing derepression of all ammonium repressible activities), and subsequent genetic analyses (11) showed that this allele resulted from a near terminal pericentric inversion of linkage group III. Duplication-deficiency progeny arising from crosses between an *xprD1* mutant and strains containing the non-inverted homologue (*xprD*$^+$) were found to be viable and lacked the *areA* gene and genomic sequences distal to it. Total DNA from both duplication-deficiency progeny and an *areA* wild type strain were isolated, radiolabelled and used to screen a wild type *A. nidulans* genomic library. Clones which hybridised to the wild type DNA but weakly or not at all to the DNA of the duplication-deficiency progeny were isolated and one of these was subsequently shown to detect banding patterns distinct from those of wild type in Southern blots of *areA* mutants which by classical genetics had been shown to have arisen from chromosomal translocations (*areA*^{r}*18* and *areA*^{d}*101*) and an inversion (*xprD1*). This latter clone was also shown to be able to rescue the *areA*$^+$ phenotype upon transformation of an *areA*^{r}*18* mutant.

DNA sequence analyses (12,13,14) have revealed a deduced *areA* translation product of 876 amino acid residues containing a highly acidic region of 70 amino acid residues (amino acids 487–557) possibly corresponding to a transcriptional activator function (15,16), a number of S(T)PXX motifs (17) and a single zinc finger DNA-binding motif of structure Cys-X_2-Cys-X_{17}-Cys-X_2-Cys (Figure 1). The 25 residues of the zinc finger lie within a 52 residue sequence (amino acids 671 – 722) of overall basic charge distribution (13) which exhibits very considerable similarity to the DNA-binding domains (DBD) of the eukaryotic GATA family of transcription factors (18) and nitrogen utilisation regulatory proteins from other fungi (19-24) which regulate the expression of target genes via binding to DNA motifs containing the core sequence 5'-GATA-3'. Sequencing of reverse transcription/PCR products revealed the presence of three introns, two occurring in the 5' untranslated region of the *areA* mRNA and the third located within the 5' coding sequence thus splitting the structural gene into two exons, the first of which encodes the 147 amino terminal amino acids of AreA (14).

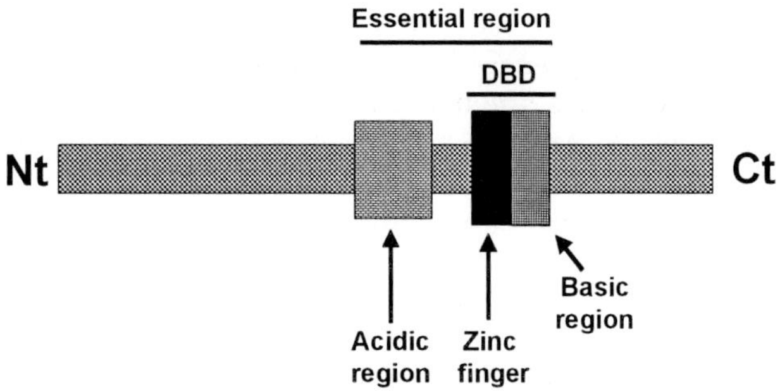

Figure 1. Structure of the *A. nidulans* Are A transcriptional factor.

Whilst the homologues of *areA* in other fungi (19-24) have also been cloned and characterised, including those of the industrially relevant *Aspergillus* species *A. niger* (25) and *A. oryzae* (26), the opportunities for sophisticated studies in *A. nidulans* are unsurpassed. The relative facility and sensitivity of nutritional growth testing in this fungus combined with the extensive distribution of classical genetic mutations (now largely molecularly characterised) throughout the *areA* locus, augmented by the site-specific introduction of precisely designed mutations obtained using PCR and gene transformation techniques, provides a powerful resource for the mutational analysis of *areA* expression and AreA function. Early studies (13,27) which molecularly characterised the classical mutations *xprD1* (the extreme *areA^d^* mutant) and an intracistronic revertant of the loss-of-function translocation mutant *areA^r^18* identified a minimal 224 amino acid segment (amino acids 500–723), lacking both an extensive amino terminal region and up to 153 carboxyl-terminal amino acids but containing most of the acidic region and all of the DBD, as being essential for the gene activation function of AreA. *xprD1*, which results in general nitrogen metabolite derepression, was shown to be due to the loss of 124 carboxyl-terminal amino acids, and the phenotypes of transformants carrying deletion constructs narrowed this to a region of 71 nucleotides essential for ammonium repression. Further analyses of the phenotypic effects of carboxyl-terminal-located mutations and their suppressors revealed that truncation of the AreA protein within the 52 amino acid conserved region resulted in loss-of-function phenotypes (*areA^r^146*, *areA^r^130*); some amino acid substitutions within the conserved region (e.g. *areA102*, *areA30*, *areA31*), or even its duplication (*areA300*), yielded specificity mutants showing different effects both positive and negative on different enzyme systems; truncations of AreA in the carboxyl region beyond the DBD (*areA1540*, *areA1600*, *areA1601*) whilst not affecting the utilisation of many nitrogen sources resulted in reduced expression of some genes and derepressed expression of others. This latter observation was taken to show that the region distal to the DBD may play an important role in modulating the level of AreA activity (27) in response to the physiological nitrogen status of the cell. An informative survey of mutations within the AreA DBD has recently been published (28).

Detailed molecular and phenotypic studies (29,30) have been conducted using, amongst others, specifically engineered deletion mutants to examine the role of regions within AreA responsible for the modulation of its activity. Apart from general effects on AreA function and altered specificity, amino acid substitutions in a short region of about 13 residues located within the conserved GATA-binding domain, but outside the zinc finger, result in derepression. Changes in the last five amino acid residues do affect AreA function but not its modulation. Beyond the DBD it has been shown that the region spanning amino acids 723–807 plays no role in modulating AreA activity as it can be deleted without resulting in derepression (independent deletion mutants lacking amino acids 723–738 and 738–807 showed near wild type phenotype for the expression of nitrate reductase under repressed and derepressed conditions). After demonstrating that the region responsible for modulation must lie within the carboxyl-terminal 69 amino acid residues, fine-point deletion mapping revealed that this function is wholly accounted for by the carboxyl-terminal 12 amino acids (the presence of only 9 terminal amino acids results in weak derepression), which in addition play an important role in the activation of a few of the genes regulated by AreA. New insights into the function of the carboxyl-terminal amino acids may be expected in the light of the recent cloning of the *nmrA* gene (31) which is the homologue of *nmr1* in *Neurospora crassa*. NMR1 has been shown to interact with the *N. crassa* equivalent of AreA (NIT2) in the basic region of the DBD and with the carboxyl-terminal residues (32).

A number of GATA motifs are located 5' to the *areA* coding region and *in vitro* binding of some to AreA has been shown (14). With regard to expression, in *areA*[+] under conditions of nitrogen starvation several *areA* mRNA species are detectable by northern blotting and the levels of these messengers are observed to be very considerably reduced in mycelia grown in the presence of ammonium. Several transcriptional start points have been identified and the specificity of use of some of these is determined by the growth conditions i.e. nitrogen repressing or derepressing. Studies of expression using various GATA-site mutants indicates that the smaller class of *areA* mRNA is autogenously regulated under nitrogen starvation conditions and the absence of growth phenotypes associated with these mutations indicates functional redundancy of the various *areA* mRNA species observed. With regard to translation, data concerning the phenotypes resulting from a number of 5'-located *areA* mutations (particularly frame-shifts) and their revertants has been compiled to provide an analysis of the translational properties of the *areA* mRNA (33). A minimal zone spanning some 329 nucleotides, and arguably larger, has been determined in which the initiation of translation can take place. Of the six AUG codons present in *areA*[+], only the first (nucleotide position +1) and third (nucleotide position +116 and hence out of frame) are strong initiation codons.

The 3' untranslated region of the *areA* transcript has been shown to play a role in the modulation of AreA since certain deletions within it result in derepression (29). Northern analysis has revealed effects on the stability of the *areA* mRNA. Whereas the half-life of *areA*[+] mRNA is about 40 minutes under derepressing conditions, it is just 7 minutes in the presence of ammonium. Deletion of a sequence in the 3' UTR resulted in an *areA* mRNA half-life of about 25 minutes in derepressing conditions and no reduction of it under repressing conditions. That double mutants containing both the 3' UTR deletion and a deletion of the 9 carboxyl-terminal amino acids (see above) showed additive levels of derepression indicates that the modulation activities of the 3' UTR and the amino terminal amino acids of AreA function independently. The total derepression phenotype of *xprD1* can now be understood since this mutation is an inversion, hence yielding a mutant product lacking both the 12 carboxyl-terminal amino acids and the 3' UTR of *areA*.

Currently, studies are being undertaken of the complexes formed between various physiologically relevant members of the target consensus site 5'-HGATAR-3' (where H = A, T, C; R = A, G) and wild type and mutant AreA DBDs (34,35,36). Correlating the differences observed between various AreA DBD-DNA target complexes with *in vivo* data regarding the phenotypes of corresponding mutants should provide considerable insight into the relationship between the structure and function of AreA. In this regard, *A. nidulans* may well be a valuable tool for the analysis of function of metazoan GATA factors from systems where the extent of genetic and phenotypic analysis at a level comparative to that of *A. nidulans* is not possible (37).

2.2. Carbon catabolite repression.

Carbon source utilisation by *A. nidulans* is strictly regulated. In the presence of preferred carbon sources, such as glucose, the expression of structural genes required for the utilisation of alternative carbon sources is repressed. This phenomenon is termed carbon catabolite repression (CCR) and there are a number of systems under its control the most studied being the ethanol regulon and the proline gene cluster (Table 1). Different strategies have been used to isolate *A. nidulans* CCR mutants. As noted above, one was based on the selection of suppressors of an *areA* loss-of-function mutant unable to use proline as a nitrogen source in the presence of a repressing carbon source such as D-glucose. Using this strategy, Arst and Cove (5) and Hynes and Kelly (38) isolated revertants that mapped to three genes named *creA, creB* and *creC*. A second strategy was based on the use of a *pdhA* mutant lacking a functional pyruvate dehydrogenase. This mutant is unable to produce acetyl CoA from pyruvate and requires an alternative carbon source such as ethanol or acetamide, the utilisation of which is under CCR. Several phenotypic suppressors were isolated which permitted growth on ethanol in the presence of glucose, mapping to the *creA* gene (39). The most extreme *creA* mutant isolated so far, named *creA^{d}30*, was isolated as a spontaneous mutation showing resistance to the toxicity of D-mannitol in a *frA1* background (40). This mutant produces a compact colony morphology and exhibits an altered flow through glycolysis leading to an increase in the formation of polyols which are subsequently secreted (41). Finally, a mutant named *creD* was selected as a suppressor of the effect of a *creC* mutant allele on acetamidase expression (42).

SYSTEM	REFERENCE
Araban degrading enzymes	43
Arabinose catabolism	44
Ethanol regulon	45
β-glucosidase production	46
Glycerol metabolism	47
Pectin degradation	48
Penicillin biosynthesis	49
Proline gene cluster	50
Rhamnosidase production	51
Xylan degradation	52

Table 1. Some *A. nidulans* systems under CCR.

The *creA* gene is located on linkage group I and early studies which noted the non-hierarchical heterogeneity of phenotype of different *creA* mutations indicated that this gene encodes a negatively acting regulatory protein (53). The genes *creB, creC* and *creD* are located on linkage group II. Unlike *creA, creB* and *creC* alleles are recessive to the wild type allele in diploids and exhibit hierarchical heterogeneity indicating that their functional role is likely to be indirect or via interaction with the *creA* gene product (54,55). As regards *creD*, there is only one mutant allele, named *creD34*, that suppresses the effects of different alleles of the *creA*, *creB* and *creC* genes suggesting some interaction between the products of all four *cre* genes (54).

The *creA* gene was cloned using a complementation strategy based on the sensitivity of *creA* mutants to media containing allyl alcohol and a repressing carbon source such as glucose (56). Under the same conditions the wild type strain is resistant due to the repression of alcohol dehydrogenase which otherwise converts allyl alcohol into the toxic compound acrolein. The *creA* mutant was transformed with a plasmid-based wild type *A. nidulans* genomic library and plated on minimal medium containing sucrose (a repressing carbon source) and allyl alcohol. Plasmids were rescued from transformants and the *creA* gene was located on a 2.3 kb *Bam*HI-*Xba*I fragment. Sequence data from this fragment revealed that the *creA* gene contains no introns and encodes a protein of 415 amino acids containing several features typical of DNA-binding proteins (Figure 2): two putative zinc finger structures of the Cys_2-His_2 type belonging to the same class as a number of mammalian early growth response genes, an alanine rich region and frequent S(T)PXX motifs (57). The zinc finger region of CreA is located in the amino terminal part of the protein and is very similar to the zinc fingers of MIG1, the main glucose repressor protein in *Saccharomyces cerevisiae.* The carboxyl-terminal region of CreA contains a stretch of 42 amino acids with 81% similarity to a region of the *S. cerevisiae RGR1* gene product. The functionally equivalent genes to the *A. nidulans creA* gene have been cloned from *A. niger, Trichoderma harzianum* and *T. reesei* (59,60,61). The overall structure of the encoded proteins is the same as that of *A. nidulans* suggesting that the essential mechanisms mediating CCR are very similar in related filamentous fungi.

Sequence analysis of different *A. nidulans creA* alleles indicates that the mutations are missense mutations in the zinc finger region or frame-shift or nonsense mutations occurring between the zinc finger domain and the carboxyl terminus of the protein (55). In order to generate a null mutant, gene replacement of *creA* was carried out leaving the flanking sequences intact (62). The resulting mutant is viable and has a severe phenotype very similar to that previously described for *creA*^{d}*30*, the latter having arisen from a pericentric inversion having one breakpoint within the *creA* gene on the left arm of linkage group I and the other between *bin*G and *y*A on the right arm (50).

Transcription of *creA* produces a mRNA of about 2 kb expressed at low level under conditions of carbon depletion. The addition of monosaccharides to depleted cultures produces a strong increase in the transcription of *creA* within a few minutes (63). Prolonged incubation with repressing monosaccharides such as glucose leads to autorepression, whereas incubation with derepressing monosaccharides such as arabinose maintains the levels of *creA* mRNA. When carbon sources other than monosaccharides (ethanol, acetate or glycerol) are used, the pattern of transcription is different showing a slow continuous increase in *creA* mRNA appearance (63). Using 6-deoxglucose it has been possible to demonstrate that *creA* transcription is triggered by monosaccharide transport or a reaction associated with sugar uptake (63) The recent cloning of several *A. nidulans* genes coding for hexose transporters may be of considerable importance in gaining an understanding of this phenomenon (our unpublished results).

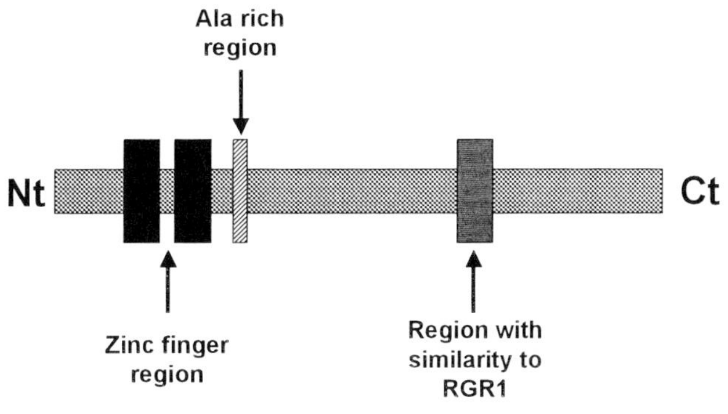

Figure 2. Structure of the *A. nidulans* CreA transcriptional factor.

The sequence of CreA clearly indicates that it is a DNA binding protein. Using DNaseI and methylation protection or methylation and depurination interference assays with fusion proteins comprising an amino terminal portion of CreA (including the two zinc fingers and the alanine rich region) and the *Shistosoma japonicum* glutathione transferase (GST::CreA), it has been possible to determine that the consensus sequence bound by the zinc finger domain of CreA is 5'-SYGGRG-3' (64,65), a sequence very similar to the 5'-GCGGAG-3' binding site previously described for the *S. cerevisiae MIG1* repressor. Interestingly, the binding in *A. nidulans* of at least some of the sequences derived fron the consensus depends on the presence of an upstream AT rich sequence (65,66). Genes under the control of CreA possess a number of consensus binding sites. The *in vivo* function of some of them has been formally demonstrated in the cases of the *alcA, alcR, prnB* and *xlnA A. nidulans* gene promoters (64,65,67), and also in the case of the *T. reesei xyn1* gene promoter using the homologous repressor Cre1 (68). Taking into account previous crystallographic work with other zinc fingers and statistical considerations, a proposed model of binding of the CreA zinc fingers to its target site has been suggested (65,69). In this model there are only three positions where amino acids make specific hydrogen bond contacts with bases, all of them being on the same strand. These amino acids are in the α-helical portion of the zinc finger or in the preceding shoulder, and the binding of the CreA zinc finger to the DNA is antiparallel. In all cases analysed so far in which *in vivo* CreA binding has been demonstrated there are two closely spaced CreA sites (64,65,67,68,70). This may suggest that CreA binds as a dimer, however there are no dimerization motifs in the sequence of CreA. In this regard, it is important to note that these experiments have used GST::CreA fusion proteins and GST forms a dimer. Consequently, as some authors have suggested, the existence of a dimerization artefact cannot be excluded (71).

What is the mode of action of *creA*? At present there are only two cases in which detailed knowledge of the action of CreA has been obtained. The first concerns the ethanol utilisation pathway which is subject to CCR. In this regulon, the *alcR* gene encodes an activator that is necessary for the induction of *alcA* and *aldA*, two genes coding for an alcohol dehydrogenase and an aldehyde dehydrogenase, respectively. When glucose is present, CreA completely prevents the transcription of the *alcR* gene and consequently blocks the induction of transcription of *alcA* and *aldA*. Moreover, CreA also exercises direct repression on the promoters of the *alcA* and *aldA* genes indicating the existence of a double-lock mechanism of repression (72,73). In the *alcA* promoter, the binding sites for AlcR and CreA overlap and competition for binding of both proteins has been demonstrated. The second well studied case is that of the proline gene cluster comprising the genes coding for the enzymes involved in the utilisation of proline as sole carbon and/or nitrogen source (*prnD* coding for proline oxidase and *prnC* coding for L-Δ^1-pyrroline carboxylate dehydrogenase), the specific permease of proline (*prnB*), the transcriptional factor mediating specific induction of the system (*prnA*), and a gene of unknown function (*prnX*). PrnA mediates the induction of transcription of *prnB, C* and *D* by proline which is also under nitrogen metabolite repression and CCR. The expression of *prnA* gene is constitutive and not repressed by glucose. Interestingly, the transcription of *prnB* and *D* is divergent and the intergenic region contains seven CreA binding sites two of which are *in vivo* targets for CreA, indicating only one level of repression (65).

At present there is a good deal of information about the mechanism of repression by CreA. However, little is known about the mechanism involved in glucose signalling in *A. nidulans*. Data indicate that glucose uptake is required for this signalling and some authors suggest that one of the glucose transporters could act as a sensor (68). In contrast to the *S. cerevisiae* model, hexokinase is dispensable for glucose repression in *A. nidulans* (71). With respect to cAMP, apparently it could play a role in a signalling pathway distinct from CreA (54,71). In this regard it is interesting to note that in three systems, the *ipnA* gene, the phenylacetic acid uptake system, and α-L-rhamnosidase production (51), several observations indicate the existence of a CCR mechanism independent of CreA (66,74). In conclusion, continued effort is required to improve our understanding of CCR in *A. nidulans*.

2.3. Ambient pH regulation.

In the natural environment micro-organisms can encounter considerable variations in ambient pH and thus require not only an efficient pH homeostatic system but also a regulatory mechanism which ensures that metabolic products exposed to the environment (e.g. permeases, exported metabolites and secreted enzymes) are only produced under the appropriate pH conditions. Hence, alkaline phosphatase is produced in alkaline growth conditions and acid phosphatase in acidic environments (75). A number of activities have been described in *A. nidulans*, the genes for which are subject to pH regulation. The mechanism of pH regulation in *A. nidulans* has been analysed both genetically and molecularly and is considered to be the paradigm for ambient pH regulation in fungi, from moulds to yeasts.

In *A. nidulans* eight genes (named *palA*, *B*, *C*, *F*, *H* and *I*, and *pacC* and *M*) have been identified, mutations in which can result in mimicry of growth at a pH other than the ambient pH (75-86). Mutations in any of the six *pal* genes result in acid growth phenotypes, yielding among others elevated levels of acid phosphatase or a lack of penicillin production. In the *pacC* gene three types of mutations have been described. Those designated *pacC*$^{+/-}$ mimic acid growth conditions albeit less extremely than *pal* mutations, whereas those mutations designated *pacC*^{c} mimic the effects of growth at alkaline pH, resulting for example in elevated levels of alkaline phosphatase and penicillin overproduction but low levels of acid phosphatase. A third type of mutation mimics neutrality and exhibits aspects of both acidity- and alkalinity-mimicking phenotypes. Mutations in the recently characterised gene *pacM* result in higher levels of acid phosphatase at alkaline pH, a feature they share with the other acidity-mimicking mutations. It is now well established that PacC is a transcriptional regulator (see below). The six *pal* gene products are components of a pH signal transduction pathway which are necessary for the conversion of inactive PacC to its functional form in response to alkaline ambient pH (75,78,79,87,88). The function of PacM it is as yet unknown.

The *A. nidulans pacC* gene has been cloned and characterised (79). The derived protein sequence of 678 residues contains three amino terminally located zinc fingers of the Cys_2-His_2 class and additional features in common with other transcription factors such as alanine, tyrosine, proline-glycine and serine-threonine rich regions, and an acidic region. A region rich in acidic and glutamine residues is also present in the carboxyl-terminal part of the protein (Figure 3). Sequence analysis of the two main categories of mutations found in *pacC* reveal that *pacC*^{c} mutations remove between 100 to 214 amino acids from the carboxyl terminus whilst mutations designated *pacC*$^{+/-}$ remove between 299 to 505 carboxyl-terminal residues. The phenotype of a null allele of *pacC* resembles that of mutations mimicking acidic growth pH (i.e. *pal*$^{-}$ and *pacC*$^{+/-}$). This result indicates that *pacC*^{c} mutations are gain-of-function mutations whereas *pacC*$^{+/-}$ are loss-of-function mutations. Using GST::PacC fusions containing the entire zinc finger region of PacC sequence-specific DNA binding was demonstrated to the core hexanucleotide 5'-GCCARG-3'. Replacement of the A at the fourth position of the consensus binding site by T prevents PacC binding (79). Further studies have shown that in the target sequence 5'-GCCAAG-3' any base substitution results in substantial or complete loss of binding, excepting the A at the fifth position. A model for PacC binding to its target site has been proposed recently (89).

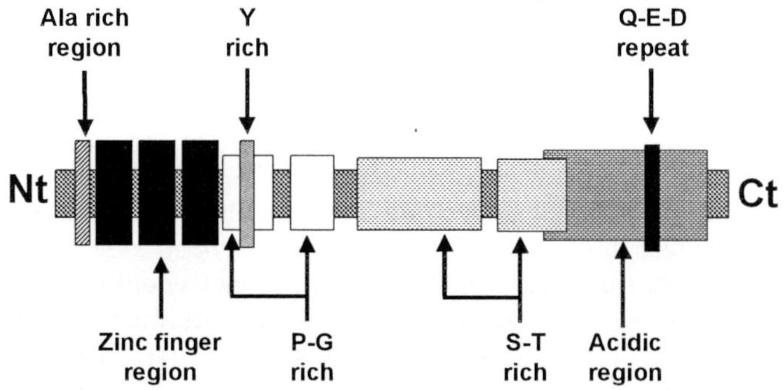

Figure 3. Structure of the *A. nidulans* PacC transcriptional factor.

What is the mechanism by which PacC operates? Current experimental data is consistent with a model of pH regulation in *A. nidulans* in which PacC proteolysis is an essential and pH-sensitive step in the regulation of gene expression by ambient pH. Under alkaline conditions the products of the six *pal* genes transduce the signal necessary for the conversion of the PacC to its functional form which is subsequently able to activate the transcription of alkaline-expressed genes and prevent the transcription of acid-expressed genes.

At acidic ambient pH, or in acidity-mimicking *pal* mutants, the PacC primary translation product is maintained in an inactive form via intramolecular interactions involving the carboxyl terminus. At alkaline ambient pH, the *pal* gene products transmit a signal the final consequence of which results in the disruption of these interactions and proteolytic removal of about 60% of the carboxyl terminus. PacC is thus converted into its functional truncated form. *pacC*^{c} mutations reduce the intramolecular interactions sufficiently to allow proteolysis even in the absence of the *pal* mediated signal and consequently functional PacC is synthesised irrespective of ambient pH, resulting in the repression of acid expressed genes and activation of alkaline expressed genes. Loss-of-function mutations in any of the signal pathway (*pal*) genes prevent modification of the intramolecular interaction yielding perpetually inactive PacC and the consequent derepression of acid expressed genes and the lack of activation of alkaline expressed genes (79,86,87).

In agreement with the proposed model, the promoters of alkaline-expressed genes contain several consensus PacC binding sites: for example three *in vitro* binding sites are present in the intergenic region *acvA-ipnA* (79) and two sites are present in the *xlnA* promoter (90). Only in the case of *ipnA* has the physiological relevance of these sites been determined by mutational analysis which demonstrated that PacC activates gene transcription at alkaline pH by their direct binding (88). Little is known regarding the molecular mechanism by which the cleaved version of PacC prevents the transcription of acid-expressed genes (82). Interestingly, whereas some acid-expressed genes such as *abfB* contain several PacC recognition sites, other genes such as *xlnB* and *gabA* present just single sites (90,91). A third class of acid-expressed genes does not contain consensus PacC target sites in the promoter (82). This heterogeneity raises the question of whether the PacC-mediated mechanism for repression acid-expressed genes varies between different set of such genes and is different from the mechanism of activation of alkaline-expressed genes. A further indication of differences between the mechanisms of regulation of acid-expressed and alkaline-expressed genes is provided by fact that *pacM* mutations suppress a number of characteristics associated with reduced expression of acid-expressed genes in *pacC*^{c} mutants whilst not affecting characteristics associated with alkaline-expressed genes (82).

PacC homologues have been described in yeasts and in other filamentous fungi such as the industrially relevant species *A. niger* (92) and *Penicillium chrysogenum* (93), both of which have been shown to complement *A. nidulans pacC*^{c} and null mutations, respectively. The *A. niger* and *P. chrysogenum* PacC proteins contain most of the sequence features described for the *A. nidulans* protein. The greatest divergence seems to occur in finger 1 which according to the model previously mentioned (89) is not involved in base-specific contacts. Greater differences are seen at the carboxyl terminus, the alternating stretch of glutamine and acidic residues being present only in PacC of *A. nidulans*. (Figure 4). The *S. cerevisiae* regulator of meiosis and invasive growth Rim101p, which shares a region of homology to the PacC three-zinc-finger DNA-binding domain, is also proteolytically activated in response to alkaline ambient pH (94). In *Yarrowia lipolytica* the transcription regulator YlRim101p also mediates pH regulation (95), and a carboxyl-terminally truncated form is able to activate transcription of a gene encoding an alkaline seryl protease (*XPR2*) regardless of pH. In pathogenic micro-organisms such as *Candida albicans* it has been noted that virulence is also controlled by pH via the PacC-related transcription factor encoded by *PRR2* (96). The existence of this range of homologues suggests that a regulatory system involving PacC-like transcription factors is conserved among a wide variety of micro-organisms.

With regard to the pH signal transduction pathway, although the integrity of the six *pal* genes is necessary for conversion of the inactive form of PacC to the functional form, no *pal* gene product directly effects such conversion. Sequence analysis of the genes has revealed that *palB* encodes a protein exhibiting similarities to the calpain family of calcium-activated cysteine proteases, though it is not responsible for the final proteolytic step that activates PacC (80). In *S. cerevisiae* there is a single gene that encodes a calpain-like protease which has significant amino acid sequence similarity to PalB and it has been suggested that it is implicated in processing Rim101p (97). The PalI polypeptide, which has homology with the meiotic signal transduction protein Rim9p of *S. cerevisiae* (a putative membrane protein acting upstream of Rim101p) contains four putative membrane-spanning domains and hence might be a membrane component involved in sensing ambient pH (81). PalH contains seven putative transmembranal domains and a long carboxyl-terminal hydrophilic region (98). PalA, the role of which in pH transduction is unknown, possesses putative SH3-domain binding motifs and putative homologues are present in *Caenorhabditis elegans* and possibly in *Schizosaccharomyces pombe* and *S. cerevisiae*, (84). Finally, functions have yet to be assigned to the *palF* (83) and *palC* translation products (98). *palF* has a functional homologue, *PPR1*, in *C. albicans*, which is involved in the alkaline induction of the regulator PRR2 (96). None of the six *pal* mRNA levels seem to be regulated by ambient pH via the PacC regulatory circuit. The existence of isofunctional homologues of the *A. nidulans pal* genes in yeasts (94,95,96,97,99,100) suggest that both the signalling pathway for proteolytic processing and perhaps the processing protease itself might be conserved. To address the latter, Mingot *et. al.* (86) constructed transgenic *S. cerevisiae* strains expressing various lengths of the *A. nidulans* PacC coding region under the control of the *GAL1* promoter. They showed that appropriate PacC processing occurred provided that a carboxyl-terminal gain-of-function truncation bypassed the need for the modification effected by the pH signal transduction pathway. Hence, the presumed signal transduction processes of *S. cerevisiae*, if present, either were not properly activated or cannot recognise the *A. nidulans* PacC protein. Moreover, the fact that PalA has a probable homologue in *C. elegans*, opens the possibility of conservation of pH signal transduction between fungi and animals (84).

The PacC-mediated pH regulatory system is, without doubt, of considerable biotechnological importance. pH not only controls penicillin biosynthesis in *A.*

nidulans (76,77,101) but also in *P. chrysogenum* (93). Moreover, the synthesis of mycotoxins in *A. nidulans* is also under PacC regulation (102). Finally, filamentous fungi are used as host organisms for heterologous expression, and extracellular protease production plays an important role in the quality of the final product. Whereas in *A. nidulans*, alkaline protease encoded by *prtA* is activated by PacC under alkaline conditions (79), in *A. niger* acidic proteases appear to be the main extracellular proteolytic activities involved in the degradation of polypeptides. Transcription of the genes encoding the acidic proteases *pepA* and *pepB in A. niger* is also controlled by ambient pH which overrides the effects of carbon and nitrogen derepression (103). The same is true for *pepF* which exhibits high levels of expression at low pH and no expression at high pH (104). As we will show later, the synthesis in *A. nidulans* of extracellular enzymes of relevance to the food and feed industries is also subject to control by the PacC regulatory system.

3. PATHWAY SPECIFIC REGULATORY SYSTEMS IN *A. nidulans*.

In fungi, most of the known pathway-specific transcription factors that control metabolic pathways where a specific inducer is involved, are activator proteins in which the DNA-binding domain shares the $Zn(II)_2Cys_6$ binuclear cluster motif typified by the yeast transcription factor GAL4 (105,106). In this motif, the six conserved cysteine residues form two α-helical structures, separated by a proline-associated loop, and co-ordinate two Zinc(II) ions to form a cloverleaf structure. These proteins are typically transcriptional activators and with some exceptions, such as the *S. cerevisiae* LEU3, they have dual roles acting as both activators and repressors (107). There are more than 80 $Zn(II)_2Cys_6$ proteins described so far, all of them having being identified in yeast or filamentous fungi, mainly in the ascomycete class (108). Searching the *S. cerevisiae* genomic database reveals the presence of 56 ORFs containing the $Zn(II)_2Cys_6$ motif. To-date ten of these regulators have been described in *A. nidulans* (Table 2). However, due to both the metabolic versatility of *A. nidulans* and the size of its genome, we would expect to find a greater number of these transcriptional factors in this filamentous fungus.

All of the $Zn(II)_2Cys_6$ transcriptional factors contain at least two functionally separable domains, one of which is responsible for sequence-specific DNA binding and the other(s), namely the transcriptional activation domain(s), directly or indirectly interact(s) with components of the basic transcription apparatus (118,119). Studies of the *S. cerevisiae* GAL4 factor demonstrate that at the DBD, which is generally located at the amino terminus, the binuclear zinc cluster is Cys-X_2-Cys-X_6-Cys-X_6-Cys-X_2-Cys-X_6-Cys, where the six cysteine residues tetrahedrally co-ordinate two zinc ions (105,106). A similar structure has been demonstrated for the *A. nidulans* AlcR gene (120,121). Comparisons between all the available motifs yields the following consensus: Cys-X_2-Cys-X_6-

$Cys-X_{5-16}-Cys-X_2-Cys-X_{6-8}-Cys$. Mutations introduced at several positions within the $Zn(II)_2Cys_6$ domains of the *A. nidulans* AlcR, AmdR and FacB proteins have evidenced the physiological relevance of Cys for binding and function (112,121,122). In $Zn(II)_2Cys_6$ transcription factors, the transcriptional activation domain is generally correlated with acidic regions which are usually located at the carboxyl terminus or within the middle part the protein. The so-called acidic activating regions, first identified in GAL4 (123), are also present in the *Aspergillus* proteins (reviewed in 124). The carboxyl-terminal regions of AmdR and FacB contain several acidic regions required for full activation of gene expression (112,122). Downstream of the $Zn(II)_2Cys_6$ region, a leucine zipper like heptad repeat motif is usually present (117,112). The region between the $Zn(II)_2Cys_6$ and the heptad repeat, named the linker, is involved in DNA binding specificity. The heptad motif has been shown to form coiled-coil structures involved in protein-protein interactions. As a consequence, homodimerization has been demonstrated for several of these proteins such as NirA (125). Interestingly, some of $Zn(II)_2Cys_6$ proteins such as AlcR lack an obvious heptad repeat motif and can act as monomers (126,127). Finally, the motif RRRLWW, first noted in UaY, is frequently found in these proteins (117).

PROTEIN	FUNCTION	REFERENCE
AflR	Regulator of sterigmatocystin pathway genes	109
AlcR	Specific activator of genes in ethanol regulon	110
AmdR	Control of genes involved in lactam and acetamide catabolism	111
AmyR	Putative regulator of amylase and α-glucosidase genes	NA
FacB	Regulation of genes for acetamide and acetate utilisation	112
NirA	Specific activator of genes involved in nitrate utilisation	113
PrnA	Transcription activator of genes involved in proline utilisation	114
QutA	Positive activator of the *qa* gene cluster	115
TamA	Positive role together with AreA controlling expression of genes subject to nitrogen repression	116
UaY	Mediates uric acid induction of genes encoding permeases and enzymes for purine utilisation	117

Table 2. *A. nidulans* $Zn(II)_2Cys_6$ transcriptional factors (NA: not available).

In the case of *A. nidulans*, the binding sites for some of these factors have been determined (Table 3). Most of the fungal $Zn(II)_2Cys_6$ proteins bind as dimers to inverted repeat sequences of the type 5'-CGG-N_x-CCG-3' where N_x represents a variable spacing between the two triplets, located one or more times in the 5' upstream region of the genes whose transcription they control. In some cases however, such as AflR and AlcR, direct repeats of the trinucleotide have been reported. Moreover, asymmetrical binding sites has also been detected for AlcR and NirA.

TRANSCRIPTION FACTOR	BINDING SITE	REFERENCE
AflR	TCG-N_5-CGA	109
	TCG-N_{10}-TCG[1]	
AlcR	CCGCA/T-$N_{2\text{-}8}$-CCGCA/T	126,127
	CCGCA/T-$N_{2\text{-}8}$-A/TGCGG	
	CCGCA/T[2]	
FacB	TCC/G-$N_{8\text{-}10}$-C/GGA	128
	GCC/A-$N_{8\text{-}10}$-G/TGC	
NirA	CTCCGHGG	125
UaY	TCGG-N_6-CCGA	117

Table 3. Binding sites of *A. nidulans* $Zn(II)_2Cys_6$ binuclear cluster transcription factors ([1]=possible site, [2]only detected *in vitro*).

4. GENE REGULATION OF XYLANASE PRODUCTION IN *A. nidulans.*

After cellulose, xylan is the most abundant renewable polysaccharide in nature. This polysaccharide is comprised of β-(1,4)-linked D-xylosyl residues to which several substituents are attached. Complete breakdown of branched xylan requires the action of several enzymes, mainly endo-(1,4)-β-xylanases which attack the polysaccharide backbone, β-xylosidase which hydrolyses xylooligosaccharides to D-xylose, and debranching enzymes such as α-L-arabinofuranosidase, acetyl esterase, α-glucuronidase and ferulic or cumaric acid esterase (129). Despite their ecological importance, the use of xylanases is of relevance for different industrial applications, mainly in the pulp and paper sector and also in the food and feed industries (Table 4).

Many filamentous fungi produce xylanolytic enzymes and whilst most of the corresponding genes have been cloned and sequenced, little is known about the molecular mechanisms controlling their expression. Recently, a positive-acting regulatory gene controlling xylanase gene expression has been identified in *A. niger* (130). This gene, named *xlnR*, encodes a typical $Zn(II)_2Cys_6$ protein and regulates not only the expression of the genes coding for xylan degrading enzymes but also the transcription of two cellulose degrading enzymes by binding to the asymmetrical sequence 5'-GGCTAAA-3' (131). In addition, regulatory elements involved in xylan-specific induction have been identified in the promoters of different fungal xylanase encoding genes (130,132,133).

INDUSTRIAL APPLICATION	REFERENCE
Clarification of juices and wine	128,140
Extraction of coffee, plant oils and starch	141, 142
Improvement of nutritional values	143
Increase in bread volume	144
Increase in fruity aroma in wines	145
Production of food thickeners	146

Table 4. Some industrial applications of fungal xylanases.

The availability of a number of regulatory mutants in *A. nidulans* makes this organism a convenient model for the basic study of the regulation of xylan degrading enzymes. When grown in liquid culture using xylan as sole carbon source, wild type *A. nidulans* secretes three endoxylanases named X_{22}, X_{24} and X_{34} (subscripts refer to their molecular masses in kDa), a β-xylosidase and at least one α-L-arabinofuranosidase (134,135). The X_{22} and X_{34} xylanases are major proteins in fungal culture filtrates whereas X_{24} is a minor protein. The purification and biochemical characterisation of the *A. nidulans* xylanases, the β-xylosidase and the α-L-arabinofuranosidase have been carried out using standard biochemical methods (135,136,137,138,139) and the genes encoding these enzymes have been cloned and sequenced (Table 5).

In *A. nidulans*, xylanase synthesis is induced by xylan, xylose and xylooligosaccharides such as xylobiose, xylotriose or xylotetraose (52,113, our unpublished data). Analysis of the promoters of the *A. nidulans* xylanase genes reveals the presence of sequences resembling the previously described *A. niger* XlnR consensus site in some of them. Very recently we cloned the equivalent *A. nidulans* XlnR gene. This gene also encodes a $Zn(II)_2Cys_6$ transcriptional factor (manuscript in preparation). Using reporter constructs comprising deletions of the *xlnA* promoter (*xlnA$_p$*) fused to the *goxC* gene, which encodes glucose oxidase, two regions that act positively in induction have been identified: one that mediates specific induction by xylose and another that includes two PacC binding sites and mediates the influence of ambient pH (see below).

GENE	ENZYME	REFERENCE
abfA	α-L-arabinofuranosidase	147
xlnA	Xylanase X_{22}	148
xnB	Xylanase X_{24}	148
xlnC	Xylanase X_{34}	149
xlnD	β-xylosidase	150
xlnR	Transcriptional factor	Unpublished

Table 5. Genes coding for xylan degrading enzymes in *A. nidulans*.

Previous work in our laboratory indicated that xylanase synthesis in *A. nidulans* is under CCR, and using different *cre* mutants we suggested a role for CreA (52). The promoters of the *xlnA, B* and *C* genes contain several CreA binding sites (Figure 4) and it has been possible to determine the CreA *in vitro* binding of some of them using GST::CreA fusion proteins. Northern analyses of gene expression using wild type and the *creA*^{d}*30* mutant have confirmed the role of CreA in the CCR of the expression of the *xlnA, B* and *C* genes (67, unpublished results). Using the previously described *xlnA*$_{p}$::*goxC* reporter and point mutational analysis we have identified a single site named *xlnA*.C1 that is responsible for direct CreA mediated repression *in vivo* (67). Comparison of the levels of expression in a *creA*^{d}*30* background of an allele mutated at the *xlnA*.C1 site indicates the existence of an indirect mechanism of repression by CreA, probably mediated at the level of *xlnR* expression and/or the synthesis of specific permeases or the inducer. Interestingly, this site is flanked by putative XlnR binding motifs but its inactivation exerts little or no influence on inducibility suggesting that induction and CCR of *xlnA* expression are two quite distinct molecular events (67). For the *xlnB* gene the situation is different. Mutational analysis of all four CreA binding sites present in its promoter (see Figure 4) reveals their lack of function *in vivo* (unpublished results). These results could indicate that *xlnB* CCR is exercised by CreA indirectly.

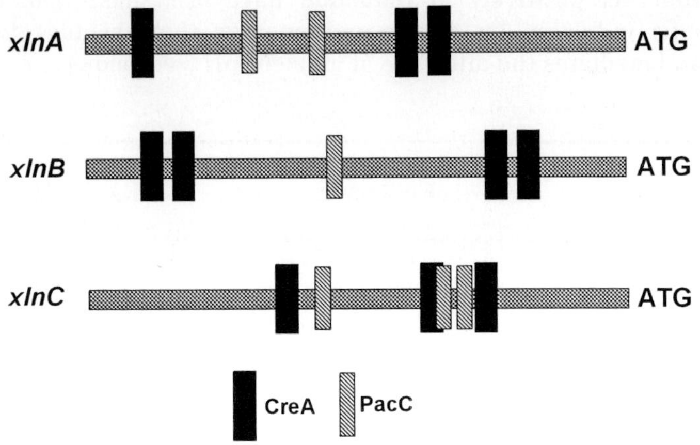

Figure 4. Presence of putative CreA and PacC binding sites in the promoters of the *A. nidulans xlnA, B* and *C* genes.

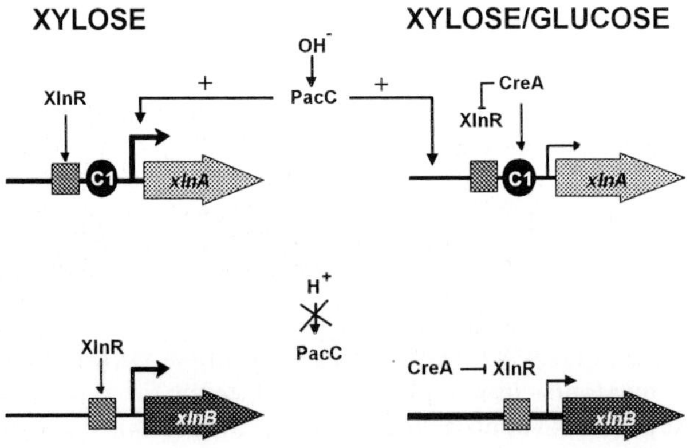

Figure 5. Model for the expression of the *A. nidulans xlnA* and *B* genes. Interactions between specific induction by XlnR, CCR and control by ambient pH are presented.

It has been formally demonstrated that *xlnA* and *xlnB* are subject to regulation by ambient pH via the transcriptional factor PacC (90). By northern analysis we have demonstrated that in the presence of D-xylose, *xlnA* is expressed mainly at alkaline ambient pH while *xlnB* is mainly expressed at acidic ambient pH. These results have been confirmed by using acidity- and alkalinity-mimicking *A. nidulans* mutants. By contrast, the transcription of *xlnC* is apparently not under ambient pH control (unpublished results). The promoters of the *xlnA* and *B* genes contain two and one PacC consensus sites, respectively (Figure 4). In the case of the *xlnA* gene promoter, and as mentioned above, both PacC sites map within a region involved in induction by xylose at alkaline ambient pH. Experiments are now in progress in order to confirm this direct role of PacC in *xlnA* induction. Due to the lack of information concerning the molecular mechanism for the negative action of PacC in the regulation of acid expressed genes, the presence of one PacC site in the promoter of the *xlnB* gene is of interest. Point mutational analysis of this site is also being undertaken.

Based on these results, a putative model for xylanase expression in *A. nidulans* is presented in Figure 5. From a biotechnological point of view, both cloning of monosaccharide transporter genes and increasing *xlnR* gene dosage are targets for the future. At present, by using genetic engineering we have constructed some *A. nidulans* strains able to overproduce and/or specifically produce xylanases (138, unpublished results). Due to the similarities between the *A. nidulans, A. niger* and *T. reesei* xylanolytic systems, it is evident that information gained in *A. nidulans* will be applied to these other fungal species which are used in the industrial production of enzymes.

REFERENCES

1. S.D. Martinelli in *Aspergillus*: 50 years on, S.D. Martinelli and J.R. Kinghorn (eds.), Elsevier, Amsterdam, (1994) 33.
2. http://www.genome.ou.edu/asper.html
3. H.N. Arst Jr. and C. Scazzocchio in Gene Manipulations in Fungi, J.W. Bennet and L.L. Lasure (eds.), Academic Press, Orlando, (1985) 309.
4. M.A. Davis and M.J. Hynes in More Gene Manipulations in Fungi, J.W. Bennet and L.L. Lasure (eds.), Academic Press, San Diego, (1991) 151.
5. H.N. Arst Jr. and D.J. Cove, Mol. Gen. Genet., 126 (1973) 111.
6. A.B. Tomsett and D.J. Cove, Genet. Res., 34 (1979) 19.
7. G. Burger, J. Tilburn and C. Scazzocchio, Mol. Cell Biol., 11(1991) 795.
8. G. Burger, J. Strauss, C. Scazzocchio and B.F. Lang, Mol. Cell. Biol., 11 (1991) 5746.
9. M.X. Caddick, H.N. Arst Jr., L.H. Taylor, R.I. Johnson and A.G. Brownlee, EMBO J., 5 (1985) 1087.

10. B.L. Cohen, J. Gen. Microbiol. 71 (1972) 293.
11. H.N. Arst Jr., Mol. Gen. Genet., 188 (1982) 490.
12. H.N. Arst Jr., B. Kudla, N. Martinez-Rossi, M.X. Caddick, S. Sibley and R.W. Davies, Trends Genet., 5 (1989) 291.
13. B. Kudla, M.X. Caddick, T. Langdon, N.M. Martinez-Rossi, C.F. Bennett, S. Sibley, R.W. Davies and H.N. Arst Jr., EMBO J., 9 (1990) 1355.
14. T. Langdon, A. Sheerins, A. Ravagnani, M. Gielkens, M.X. Caddick and H.N. Arst Jr. Mol. Microbiol., 17 (1995) 877.
15. E. Giniger and M. Ptashne, Nature, 330 (1987) 670.
16. I.A. Hope, S. Mahadevan and K. Struhl, Nature, 333 (1988) 635.
17. M. Suzuki, J. Mol. Biol., 207 (1989) 61.
18. S.H. Orkin, Blood, 80 (1992) 575.
19. Y.H. Fu and G.A. Marzluf, Mol. Cell. Biol., 10 (1990) 1056.
20. H. Haas, B. Bauer, B. Redl, G. Stoffler and G.A. Marzluf, Curr. Genet., 27 (1995) 150.
21. E.H. Froeliger and B.E. Carpenter, Mol. Gen. Genet., 251 (1996) 647.
22. S.E. Screen, A.M. Bailey, K. Charnley, R. Cooper and J.M. Clarkson, Gene, 221 (1998) 17.
23. S. Gente, N. Poussereau and M. Fevre, FEMS Microbiol. Lett., 175 (1999) 291.
24. B. Tudzynski, V. Homann, B. Feng and G.A. Marzluf, Mol. Gen. Genet., 261 (1999) 106.
25. A.P. MacCabe, S.A.S. Vanhanen, M. Sollewijn Gelpke, P. Van de Vondervoort, H.N. Arst Jr. and J. Visser, Biochim. Biophys. Acta., 1396 (1998) 163.
26. T. Christensen, M.J. Hynes and M.A. Davis, Appl. Environ. Microbiol., 64 (1998) 3232.
27. M. Stankovich, A. Platt, M.X. Caddick, T. Langdon, P.M. Shaffer and H.N. Arst Jr., Mol. Microbiol., 7 (1993) 81.
28. R.A. Wilson and H.N. Arst Jr., Microbiol. Mol. Biol. Rev., 62 (1998) 586.
29. A. Platt, T. Langdon, H.N. Arst Jr., D. Kirk, D. Tollervey, J.M. Mates-Sánchez and M.X. Caddick, EMBO J., 15 (1996) 2791.
30. A. Platt, A. Ravagnani, H.N. Arst Jr., D. Kirk, T. Langdon and M.X. Caddick, Mol. Gen. Genet., 250 (1996) 106.
31. A. Andrianopoulos, S. Kourambas, J.A. Sharp, M.A. Davis and M.J. Hynes, J. Bacteriol., 180 (1998) 1973.
32. X. Xiao, Y.H. Fu and G.A. Marzluf, Biochemistry, 34 (1995) 8861.
33. H.N. Arst Jr. and A. Sheerins, Mol. Microbiol., 19 (1996) 1019.
34. A. Ravagnani, L. Gorfinkiel, T. Langdon, G. Diallinas, E. Adjadj, S. Demais, D. Gorton, H.N. Arst Jr. and C. Scazzocchio, EMBO J., 16 (1997) 3974.
35. M.R. Starich, M. Wikstrom, H.N. Arst Jr., G.M. Clore and A.M. Gronenborn, J. Mol. Biol., 277 (1998) 605.
36. M.R. Starich, M. Wikstrom, S. Schumacher H.N. Arst Jr., A.M. Gronenborn and G.M. Clore, J. Mol. Biol., 277 (1998) 621.

37. M.X. Caddick, D.G. Peters, P. Hooley and A. Nayler, Genes Funct., 1 (1997) 37.
38. M.J. Hynes and J.M. Kelly, Mol. Gen. Genet., 150 (1977) 193.
39. C. Bailey and H.N. Arst Jr., Eur. J. Biochem., 51 (1975) 573.
40. H.N. Arst Jr., D. Tollervey, C.E.A. Dowzer and J.M. Kelly, Mol. Microbiol., 4 (1990) 851.
41. P. Van der Veen, G.J.G. Ruijter and J. Visser, Microbiology, 141 (1995) 2301.
42. J.M. Kelly and M.J. Hynes, Mol. Gen. Genet., 156 (1977) 87.
43. P. Van der Veen, H.N. Arst Jr., M.J.A. Flipphi and J. Visser, Arch. Microbiol., 162 (1994) 433.
44. P. Van der Veen, PhD Thesis, Wageningen Agricultural University, Wageningen, 1995.
45. R.A. Lockington, C. Scazzocchio, D. Sequeval, M. Mathieu and B. Felenbok, Mol. Microbiol., 1 (1987) 275.
46. L. Lee, K.S. Kwon and Y.C. Hah, FEMS Microbiol. Lett., 135 (1996) 79.
47. D.H.A. Hondmann, R. Bussink, C.F.B. Witteveen and J. Visser, J. Gen. Microbiol., 137 (1991) 629.
48. R.A. Dean and W.E. Timberlake, Plant Cell, 1 (1989) 275.
49. B. Pérez-Esteban, M. Orejas, E. Gómez-Pardo and M.A. Peñalva, Mol. Microbiol., 9 (1993) 881.
50. V. Sophianopoulou, T. Suárez, G. Diallinas and C. Scazzocchio, Mol. Gen. Genet., 236 (1993) 209.
51. M. Orejas, E. Ibáñez and D. Ramón, Lett. Appl. Microbiol., 28 (1999) 383.
52. F. Piñaga, M.T. Fernández-Espinar, S. Vallés and D. Ramón, FEMS Microbiol. Lett., 115 (1994) 319.
53. H.N. Arst Jr. and C.R. Bailey in Genetics and Physiology of *Aspergillus nidulans*, J.E. Smith and J.A. Pateman (eds.), Academic Press, London, (1977) 131.
54. J.M. Kelly in *Aspergillus*: 50 years on, S.D. Martinelli and J.R. Kinghorn (eds.), Elsevier, Amsterdam, (1994) 355.
55. B. Felenbok and J.M. Kelly in The Mycota III. Biochemistry and Molecular Biology, R. Brambl and G. Marzluf (eds.), Springer-Verlag, Heidelberg, (1996) 369.
56. C.E.A. Dowzer and J.M. Kelly, Curr. Genet., 15 (1989) 457.
57. C.E.A. Dowzer and J.M. Kelly, Mol. Cell. Biol., 11 (1991) 5701.
58. J. Östling, M. Carlberg and H. Ronne, Mol. Cell. Biol., 16 (1996) 753.
59. M.R. Drysdale, S.E. Kolze and J.M. Kelly, Gene, 130 (1993) 241.
60. J. Strauss, R.L. Mach, S. Zeillinger, G. Stöffler, M. Wolschek, G. Hartler and C.P. Kubicek, FEBS Lett., 376 (1995) 103.
61. M. Ilmen, C. Thrane and M. Penttilä, Mol. Gen. Genet., 251 (1996) 451.
62. R.A. Shroff, S.M. O'Connor, M.J. Hynes, R.A. Lockington and J.M. Kelly, Fungal Gen. Biol., 22 (1997) 28.
63. J. Strauss, H.K. Horvath, B.M. Abdallah, J. Kindermann, R.L. Mach and C.P. Kubicek, Mol. Microbiol., 32 (1999) 169.

64. P. Kulmburg, M. Mathieu, C. Dowzer, J. Kelly and B. Felenbok, Mol. Microbiol., 7 (1993) 847.
65. B. Cubero and C. Scazzocchio, EMBO J., 13 (1994) 407.
66. E.A. Espeso and M.A. Peñalva, FEBS Lett., 342 (1994) 43.
67. M. Orejas, A.P. MacCabe, J.A. Pérez-González, S. Kumar and D. Ramón, Mol. Microbiol., 31 (1999) 177.
68. R.L. Mach, J. Strauss, S. Zeillinger, M. Schindler and C.P. Kubicek, Mol. Microbiol., 21 (1996) 1273.
69. C. Scazzocchio, V. Gavrias, B. Cubero, C. Panozzo, M. Mathieu and B. Felenbok, Can. J. Bot., 73 (1995) S160.
70. W.E. Hintz and P.A. Lagosky, Bio/technology, 11 (1993) 815.
71. G.J.G. Ruitjer and J. Visser, FEMS Microbiol. Lett., 151 (1997) 103.
72. M. Mathieu and B. Felenbok, EMBO J., 13 (1994) 4022.
73. C. Panozzo, E. Cornillot and B. Felenbok, J. Biol. Chem., 273 (1998) 6367.
74. J.M. Fernández-Cañón and J.M. Luengo, J. Antibiot., 50 (1996) 45.
75. M.X. Caddick, A.G. Brownlee and H.N. Arst Jr., Mol. Gen. Genet., 203 (1986) 346.
76. A.J. Shah, J. Tilburn, M.W. Adlard and H.N. Arst Jr., FEMS Microbiol. Lett., 61 (1991) 209.
77. E.A. Espeso, J. Tilburn, H.N. Arst Jr. and M.A. Peñalva, EMBO J., 12 (1993) 3947.
78. H.N. Arst Jr., E. Bignell and J. Tilburn, Mol. Gen. Genet., 245 (1994) 787.
79. J. Tilburn, S. Sarkar, D.A. Widdick, E.A. Espeso, M. Orejas, J. Mungroo, M.A. Peñalva and H.N. Arst Jr., EMBO J., 14 (1995) 779.
80. S.H. Denison, M. Orejas and H.N. Arst Jr., J. Biol. Chem., 270 (1995) 28519.
81. S.H. Denison, S. Negrete-Urtastun, J.M. Mingot, J. Tilburn, W.A. Mayer, A. Goel, E.A. Espeso, M.A. Peñalva and H.N. Arst Jr., Mol. Microbiol., 30 (1998) 259.
82. S. Sarkar, M.X. Caddick, E. Bignell, J. Tilburn and H.N. Arst Jr., Biochem. Soc. Trans., 24 (1996) 360.
83. J. Maccheroni, G.S. May, N.M. Martínez-Rosi and A. Rossi, Gene, 194 (1997) 163.
84. S. Negrete-Urtastun, S.H. Denison and H.N. Arst Jr., J. Bacteriol., 179 (1997) 1832.
85. K. Then Bergh and A.A. Brakhage, Appl. Environ. Microbiol., 64 (1998) 843.
86. J.M. Mingot, J. Tilburn, E. Díez, E. Bignell, M. Orejas, D.A. Widdick, S. Sarkar, C.V. Brown, M.X. Caddick, E.A. Espeso, H.N. Arst Jr. and M.A. Peñalva, Mol. Cell. Biol., 19 (1999) 1390.
87. M. Orejas, E.A. Espeso, J. Tilburn, S. Sarkar, H.N. Arst Jr. and M.A. Peñalva, Genes Dev., 9 (1995) 1622.
88. E.A. Espeso and M.A. Peñalva, J. Biol. Chem., 271 (1996) 28825.
89. E.A. Espeso, J. Tilburn, L. Sánchez-Pulido, C.V. Brown, A. Valencia, H.N. Arst Jr. and M.A. Peñalva, J. Mol. Biol., 274 (1997) 466.

90. A.P. MacCabe, M. Orejas, J.A. Pérez-González and D. Ramón, J. Bacteriol., 180 (1998) 1331.
91. H. Hutchings, K.P. Stahmann, S. Roels, E.A. Espeso, W.E. Timberlake, H.N. Arst Jr. and J. Tilburn, Mol. Microbiol., 32 (1999) 557.
92. A.P. MacCabe, J.P.T.W. van den Homberg, J. Tilburn, H.N. Arst Jr. and J. Visser, Mol. Gen. Genet., 250 (1996) 367.
93. T. Suárez and M.A. Peñalva, Mol. Microbiol., 20 (1996) 529.
94. W. Li and A. Mitchell, Genetics, 145 (1997) 63.
95. M. Lambert, S. Blanchin-Roland, F. Le Louedec, A. Lépingle and C. Gaillardin, Mol. Cell. Biol., 17 (1997) 3966.
96. A.M. Ramón, A. Posta and W.A. Fonzi, J. Bacteriol., 181 (1999) 7524.
97. E. Futai, T. Maeda, H. Sorimachi, K. Kitamoto, S. Ishiura and K. Suzuki K, Mol. Gen. Genet., 260 (1999) 559.
98. S. Negrete-Urtastun, W. Reiter, E. Díez, S.H. Denison, J. Tilburn, E.A. Espeso, M.A. Peñalva and H.N. Arst Jr., Mol. Microbiol., 33 (1999) 994.
99. S.S.Y. Su and A.P. Mitchell A.P., Nucleic Acids Res., 21 (1993) 3789.
100. H. Sorimachi, S. Ishiura and K. Suzuki, Biochem. J., 328 (1977) 721.
101. O. Litzka, K. Then Bergh, J. Van den Brulle, S. Steidl and A.A. Brakhage, Anton. Leeuw. Inst. J. G., 75 (199) 95.
102. N.P. Keller, C. Nesbitt, B. Sarr, T.D. Phillips and G.B. Burow, Phytopathology, 87 (1997) 643.
103. G. Jarai and F. Buxton, Curr. Genet., 26 (1994) 238.
104. J.P.T.W. van den Homberg, A.P. MacCabe, P.J.I. van de Vondervoort and J. Visser, Mol. Gen. Genet., 251 (1997) 542.
105. T. Pan and J.E. Coleman, Proc. Natl. Acad. Sci. USA, 87 (1990) 2077.
106. R. Marmorstein, M. Carey, M. Ptashne and S.C. Harrison, Nature, 356 (1992) 408.
107. J.Y. Sze, E. Remboutsika and G.B. Kolhhaw, Mol. Cell. Biol., **13** (1993) 5702.
108. R.B. Todd and A. Andrianopoulos, Fungal Genetics and Biology, 21 (1997) 388.
109. M. Fernandes, N.P. Keller and T. Adams, Mol. Microbiol., 28 (1998) 1355.
110. P. Kulmburg, T. Prangé, M. Mathieu, D. Sequeval, C. Scazzocchio and B. Felenbok, FEBS Lett., 280 (1991) 11.
111. A. Andrianopoulos and M.J Hynes, Mol. Cell. Biol., 10 (1990) 3194.
112. R.B. Todd, R.L. Murphy, H.M. Martin, J.A. Sharp, M.A. Davis, M.E. Katz and M.J. Hynes, Mol. Gen. Genet., 254 (1997). 495.
113. G. Burger, J. Strauss, C. Scazzocchio, and B.F. Lang, Mol. Cell. Biol., 11 (1991) 5746.
114. B. Cazelle, A. Pokorska, E. Hull, P.M. Green, G. Stanway and C. Scazzocchio, Mol. Microbiol., 28 (1998) 355.
115. R.K. Beri, H. Whittington, C.F. Roberts and A. Hawkins, Nucleic Acids Research, 15 (1987) 7991.
116. M.A. Davis, A.J. Small, S. Kourambas and M.J. Hynes, J. Bacteriology, 178 (1996) 3406.

117. T. Suárez, M.V. de Queiroz, N. Oestreicher and C. Scazzocchio, EMBO J., 14 (1995) 1453.
118. M. Ptashne, Nature, 335 (1988) 683.
119. R. Tjian and T. Maniatis, Cell, 77 (1994) 5.
120. D. Sequeval and B. Felenbok, Mol. Gen. Genet., 242 (1994) 33.
121. I. Ascone, F. Lenouvel, D. Sequeval, H. Dexpert and B. Felenbok, Biochim. Biophys. Acta.. 1343 (1997) 211.
122. L.M. Parsons, M.A. Davis and M.J. Hynes, Mol. Microbiol.,6 (1992) 2999.
123. J. Ma, and M. Ptashne, Cell, 48 (1997) 847.
124. P. Schjerling and S. Holmberg, Nucleic. Acids Research, 24 (1996) 4599.
125. J. Strauss, M.I. Muro-Pastor and C. Scazzocchio, Mol. Cell. Biol., 18 (1998) 1339.
126. F. Lenouvel, I. Nikolaev and B. Felenbok, J. Biol. Chem. 272 (1997) 15521.
127. C. Panozzo, V. Capuano, S. Fillinger and B. Felenbok, J. Biol. Chem., 272 (1998) 22859.
128. R.B. Todd, A. Andrianopoulos, M.A. Davis and M. Hynes, EMBO J., 17 (1998) 2042.
129. P. Biely, Trends Biotechnol., 3 (1985) 286.
130. N.N.M.E. Van Peij, J. Visser and L.H. de Graaff, Mol. Microbiol., 27 (1998) 131.
131. N.N.M.E. Van Peij, M.M.C. Gielkens, R.P. de Vries, J. Visser and L.H. de Graaff, Appl. Environ. Microbiol., 64 (1998) 3615.
132. L.H. de Graaff, H.C. van der Broeck, A.J.J. van Ooijen and J. Visser, Mol. Microbiol., 12 (1994) 479
133. S. Zeillinger, R.L. Mach, M. Schindler, P. Herzog and C.P. Kubicek, J. Biol. Chem., 271 (1996) 25624.
134. M.T. Fernández-Espinar, D. Ramón, F. Piñaga and S. Vallés, FEMS Microbiol. Lett., 91 (1992) 91.
135. D. Ramón, P. v.d. Veen and J. Visser, FEMS Microbiol. Lett., 113 (1993) 15.
136. M.T. Fernández-Espinar, F. Piñaga, P. Sanz, D. Ramón and S. Vallés, FEMS Microbiol. Lett., 113 (1993) 223.
137. M.T. Fernández-Espinar, F. Piñaga, L. de Graaff, J. Visser, D. Ramón and S. Vallés, Appl. Microbiol. Biotechnol., 42 (1994) 555.
138. M.T. Fernández-Espinar, S. Vallés, F. Piñaga, J.A. Pérez-González and D. Ramón, Appl. Microbiol. Biotechnol., 45 (1996) 338.
139. S. Kumar and D. Ramón, FEMS Microbiol. Lett., 135 (1996) 287.
140. C.I. Beck and D. Scott, Adv. Chem. Ser., 136 (1974) 1.
141. K.Y. Ken and J.N. Saddler. *Trichoderma* xylanases, their properties and applications. In: Xylans and xylanases (J. Visser, G. Beldman, M.A. Kusters-van Someren and A.G.J. Voragen eds.), Elsevier, Amsterdam, (1992) 171.
142. P. Biely, ACS Symp. Ser., 460 (1991) 408.
143. D. Petterson and P. Aman, British J. Nutr., 62 (1989) 139.

144. J. Maat, M. Roza, J. Verbakel, H. Stam, M.J. Santos da Silva, M. Bosse, M.R. Egmond, M.L.D. Hagemans, R.F.M. van Gorcom, J.G.M. Hessing, C.A.M.J.J. van den Hondel and C. Van Rotterdam. Xylanases and their application in bakery. In: Xylans and xylanases (J. Visser, G. Beldman, M.A. Kusters-van Someren and A.G.J. Voragen eds.), Elsevier, Amsterdam, (1992) 349.
145. M.A. Ganga, F. Piñaga, S. Vallés, D. Ramón and A. Querol, Int. J. Food Microbiol., 47 (1999) 171.
146. J.G. Zeikus, C. Lee, Y.E. Lee and B.C. Saha, ACS Symp. Ser., 460 (1991) 36.
147. M. Gielkens, L. González-Candelas, P. Sánchez-Torres, P. van de Vondervoort, L. de Graaff, J. Visser and D. Ramón, Microbiology 145 (1999) 735.
148. J.A. Pérez-González, L.H. de Graaff, J. Visser and D. Ramón, Appl. Environ. Microbiol., 62 (1996) 2179.
149. A.P. MacCabe, M.T: Fernández-Espinar, L.H. de Graaff, J. Visser and D. Ramón, Gene, 175 (1996) 29.
150. J.A. Pérez-González, N. van Peij, A. Bezoen, A.P. MacCabe, D. Ramón and L.H. de Graaff, Appl. Environ. Microbiol., 64 (1998) 1412.

Applied Mycology and Biotechnology
Volume 1. Agriculture and Food Production
G.G. Khachatourians and D.K. Arora (editors)

Detection of food-borne toxigenic molds using molecular probes

Marianne E. Boysen, Anders R. B. Eriksson and Johan Schnürer

Department of Microbiology, Swedish University of Agricultural Sciences, Box 7025, SE-750 07 Uppsala, Sweden

1. INTRODUCTION

Unwanted fungal growth cause general food spoilage (reduced technical quality, formation of off-odors, loss of texture, nutritional losses) and might lead to the formation of mycotoxins and potentially allergenic spores (1,2). Food-borne fungi either grow as unicellular yeast, which do not produce mycotoxin, or as filamentous fungi, which may or may not form mycotoxins. The filamentous fungi, commonly known as molds, produce conidia (vegetative spores) from conidiophores, specialized aerial structures. Large differences in metabolic activity exists with different parts of their mycelium, and older parts might even be completely devoid of cytoplasm (3). Certain food-borne fungi produce sexual spores, predominantly ascospores, while others might produce thick-walled chlamydospores and other kinds of resting structures. Ascospores of several food-borne fungi have increased heat resistance compared to vegetative cells. This cause particular problem in the production of heat-processed canned fruit, fruit-juices, jams and other berry based food products.

A large number of fungal species, at least several hundred and possibly more than one thousand, can potentially damage food, making the correct identification of important species a daunting task. However, Filtenborg *et al.* (1) has suggested that only a specific, very limited group of fungal species, the associated funga (=mycobiota), is responsible for spoilage of a particular food type. This association of certain fungal species with each individual food is due to the selective effect of intrinsic factors (nutrient composition, pH, water-activity (a_w), inhibitory compounds etc.) in combination with extrinsic factors (temperature, RH, gaseous atmosphere) and processing techniques (slicing, heating, preservative additions). Apple rot is thus mainly caused by *Penicillium expansum*, *P. crustosum*, *P. solitum* and *Alternaria alternata*. Wheat and rye grain is commonly contaminated in the field by *Fusarium culmorum*, *F. graminearum*, *Alternaria alternata* and *A. infectoria*. Cereal grain stored in temperate climates is mainly spoiled by *Penicillium* species such as *P. verrucosum*, *P. hordei* and members of the *P. aurantiogriseum* complex. At lower water activities *Aspergillus* spp. (*A. candidus*, *A. versicolor*, *A. flavus*) and *Eurotium* spp. dominate on stored cereals. Spoilage of hard, semi-hard and semi-soft cheese without preservatives is mainly caused by *Penicillium commune* and

P. nalgiovense. Besides causing spoilage, several of the species mentioned above are also important mycotoxin producers in specific food commodities (1). Patulin is produced by *P. expansum*, ochratoxin A by *P. verrucosum*, deoxynivalenol by *F. culmorum* and *F. graminearum* and aflatoxin by *Aspergillus flavus*. This emerging knowledge of the associated funga makes a more targeted approach possible for analyzing fungal food contamination.

Methods able to detect and identify fungi at different levels of specificity are needed to evaluate the mycological quality of food. The main focus of this review is on the use of fungal volatiles and molecular techniques, i.e. methods based on nucleic acid sequences or antigens, to detect the presence of toxigenic food-borne fungi. In addition, the determination of colony forming units (CFU) and ergosterol will be covered briefly, as all new assays have to be related to classical methods.

Before selecting a method for detection of food-borne fungi it is important to specify the level of information that is needed. Methods can thus be suited for detection of: i) all fungi (i.e. general fungal presence), ii) certain genera or species, iii) mycotoxin producers, or iv) specific spoilage groups. Further considerations are whether the assay should be able to provide quantitative or only qualitative data and whether information on activity is required. Specifying time or sensitivity constraints is also important. Within the food industry there is often a wish for answers within minutes, as compared to the 5-7 days required for CFU determinations. In the commercial environment the degree of sensitivity required of an assay is often related to legislation, e.g. a need to know that a specified limit of 10^3 fungal CFU g^{-1} is not exceeded.

2. TRADITIONAL METHODS FOR THE DETECTION AND QUANTIFICATION OF FOOD-BORNE FUNGI

2.1. Colony forming units

The determination of numbers of colony forming units on general or selective agar substrates remain the standard method in food microbiology. Although reasonably well suited to quantify numbers of viable unicellular organism, i.e. bacteria and yeast, it is not well adapted for quantification of filamentous fungi and results of fungal CFU determinations are highly influenced by the degree of fungal sporulation (3,4). The use of selective substrates for various ecological groups can increase the precision and accuracy of CFU determinations. Although mycology has lagged behind food bacteriology with regard to development of specific substrates, substantial progress has been made in the development of selective and indicative growth media for food-borne fungi (5). The International Commission of Food Mycology (ICFM) (http://www.iums.vir.gla.ac.uk/mycology/comcofs.html) has been the driving force behind the international collaborative studies needed for validation of these new substrates. Xerotolerant fungi, e.g. *Eurotium* spp., *Aspergillus penicilloides*, and *Wallemia sebi* can be detected on e.g. dichloran glycerol 18 % agar (DG18), water activity (a_w) of 0.95. Organic acid tolerant fungi, e.g. *Penicillium roqueforti* and preservative resistant yeast, can be isolated on malt extract agar with 0.5 - 1.0 % acetic acid (5). Color changes of the colony reverse (backside of agar plate), based on reactions between substrate components and secondary metabolites, indicate the presence of certain mycotoxigenic fungi, e.g. *Aspergillus flavus* on *Aspergillus flavus* and *parasiticus* agar (AFPA) (6). It is likely that a further development of selective and indicative media will parallel the development of molecular techniques.

2.2. Ergosterol

Ergosterol is a fungal membrane lipid that can be used as a specific marker for quantification of fungi. Seitz *et al.* (7) originally proposed determination of ergosterol for estimating the degree of fungal infestation of grains. Ergosterol is a dominant membrane sterol in all eumycota fungi except chytrids, rusts, and some yeast (8). It is not found in bacteria, plants and animals, but has been found in certain algae and protozoa (8). Double bonds at C-5 and C-7 of the ergosterol molecule leads to a highly specific ultraviolet absorption spectrum with a maximum at 282 nm. Quantification is possible within one hour using high pressure liquid chromatography (HPLC) (7,9,10). The ergosterol assay has become a commonly used method for the quantification of fungal mass in cereal grain (11). Seitz and coworkers have also recently shown that near-infrared spectroscopy (NIR) can be used to predict ergosterol levels in single wheat kernels within a minute (12). This technology may provide a means of rapid screening of samples for fungal related food safety and quality problems. However, it has to be remembered that the ergosterol assay, whether run as a HPLC or a NIR method, is non-specific and does not provide any information on fungal species composition.

3. FUNGAL VOLATILES

Fungi produce volatile compounds, both during primary and secondary metabolism. Volatile metabolites from mainly *Aspergillus*, *Fusarium* and *Penicillium* spp. have been characterized with gas chromatography (GC), mass spectrometry and sensory analysis and can be used for detection and identification. Common volatiles are 2-methyl-1-propanol, 3-methyl-1-butanol, 1-octen-3-ol, 3-octanone, 3-methylfuran, ethyl acetate and the malodorous 2-methyl-isoborneol and geosmin (2). Given the occurrence of off-odorous fungal volatiles it is maybe not surprising that the most commonly used method for detecting fungal growth in food and feed is sensory analysis, i.e. the human nose. Sensory analysis is used to detect mold and other objectionable odors of grain in all international and in most national trade (13,14). This procedure for evaluating mycological quality, although fast and sensitive, has a number of drawbacks. The method is subjective (15), the inhalation of fungal spores from moldy grain is hazardous (16), and furthermore fungal volatiles may cause damage to human respiratory organs (17,18). Efficient alternative detection methods for fungal volatiles that are not too expensive and preferably as fast as human sensory analysis are thus needed.

Developments in sensor technology have led to the construction of "electronic noses", also known as volatile compound mappers (15). Exposure of different non-specific sensors to volatile compounds produces characteristic electrical signals. These are collected by a computer and processed by multivariate statistical methods or in artificial neural networks (ANN). Such systems can grade cereal grain with regard to presence of molds as efficiently as sensory panels evaluating grain odor. Volatile compound mapping can also be used to predict levels of ergosterol and fungal colony forming units in grain (2). Further developments should make it possible to detect individual fungal species as well as the degree of mycotoxin contamination of food and animal feeds.

4. MOLECULAR DETECTION AND QUANTIFICATION OF MYCOTOXIGENIC FOOD-BORNE FUNGI

As mentioned earlier the main focus of the review of molecular techniques are the immunochemical and nucleic acid based techniques, or more specifically the antigen-antibody Enzyme-Linked-Immunosorbent Assays (ELISA) and the polymerase chain reaction (PCR). PCR relies on the enzymatic amplification of a specific target sequence *in vitro* using specific oligonucleotide primers (19).

4.1. Immunochemical assays

A crucial step in developing an immuno assay for detection of molds is the selection of the fungal antigens. Mycelial fragments, e.g. glycoproteins in the cell wall, spores, soluble proteins, extra-cellular polysaccharides, or specific cloned genes used for monitoring gene-modified organisms could all be used.

Simple immunochemical tests have been introduced such as the mold latex agglutination (20) and the dip-stick (21) assays. However, most immunoassays developed for detecting molds in food and feed are sandwich ELISA. Recent examples of the development of ELISA assays for the detection of aflatoxigenic molds are the work of Tsai and Yu (22) detecting mycelial proteins from *Aspergillus flavus* and that of Shapira *et al.* (23) who directed the antibodies against proteins from different parts of the aflatoxin biosynthesis pathway. However, none of these assays were tested on natural samples. An ELISA kit for detection of more than 30 common field and storage fungi (e.g. *Fusarium*, *Penicillium*, and *Aspergillus* species) has been introduced commercially by Adgen Agrifood Diagnostics, Auchincruive, Scotland (http://www.adgen.co.uk) and is currently validated for grain. Non-specific cross reactivity is a general problem with the immuno-based assays when used on natural samples, but on request Adgen Agrifood Diagnostics provide detailed data on this. For further discussion on the general use and specificity of immunochemical assays for the detection of molds from foodstuff refer to a recent review by Girardin (24).

As a major concern in terms of fungal contamination of food and feed is the potential contamination with mycotoxins, analyzing for the toxin as such provides a different approach. The chemical methods typically used for detection and quantification of mycotoxins from food and feed extracts are based on HPLC and GC analysis as demonstrated by Scudamore *et al.* (25), when detecting and estimating 22 mycotoxins from maize products. However, faster and more user-friendly assays based on the immuno-detection and subsequent spectroscopic quantification of some of the major mycotoxins (e.g. ochratoxin, aflatoxins, and several trichothecenes) are available commercially from e.g. Rhône-Poulenc, France (http://www.rhone-poulenc.com) and Adgen Agrifood Diagnostics, Auchincruive, Scotland.

4.2. Polymerase chain reaction (PCR)

PCR for detection

For diagnostic purposes, amplification reactions have been developed to shorten the time required for diagnosis. Table 1 provides an overview of genes that have been used for the detection of some of the common food and feed spoilage fungi and to what extent the method has been applied for detection from natural samples. Numerous studies have been conducted on the PCR detection and differentiation of pathogenic fungi in clinical specimens (e.g. 26,27,28) and, more recently, on a wide variety of industrially important field and storage fungi (29-31). Amplification targets have been genes coding for specific proteins (30,32), the small ribosomal gene (18S rDNA) (26-29), the internal transcribed spacer (ITS) region of the ribosomal genes (31,33), or specific regions of the mitochondrial DNA (34).

For diagnostic purposes, the ribosomal RNA genes might contain too little variation to fulfil the requirements of the specific primers needed. An alternative approach is to identify species-specific random amplified polymorphic DNA (RAPD) fragments, determine their sequence, and hope to be able to identify appropriate primers for the given species. Schilling *et al.* (33) used this procedure after finding too few polymorphisms in the ITS regions to distinguish *Fusarium culmorum* and *Fusarium graminearum*. However, a disadvantage of this method is not knowing what gene is detected. As the amount of DNA sequences being published is growing almost exponentially, an alternative approach could be to look for biosynthetically important genes which would provide more and/or better information in terms of detecting groups of fungi. Examples of this were shown by Niessen and Vogel (35) in detecting a group of fungi producing the trichothecene toxins and Geisen (36) and Shapira *et al.* (37) detecting aflatoxigenic fungi by identifying specific sequences of the enzymes involved in the biosynthetic pathway of aflatoxin. For further discussion on the use of PCR methods for the detection of mycotoxin producing fungi the reader is referred to Geisen (38).

Adgen Agrifood Diagnostics (Auchincruive, Scotland) now also offers a service for PCR detection and quantification of a number of cereal pathogens, e.g. some of the *Fusarium* species that cause problems to the brewing industry.

Detection of PCR amplicons

Traditionally DNA fragments have been detected through electrophoretic gel-separation and visualization by UV transillumination after staining with the

Table 1
Common food and feed spoilage fungi detected by PCR

Mold	Target	Specificity	Application	Reference
Penicillium roqueforti, *P. carneum*	ITS	Species specific	Detection from soft cheeses	(31)
Penicillium species	ITS	Genus specific	Detection from soft cheeses	(31)
Aspergillus flavus/ *parasiticus*	Aflatoxin biosynthesis genes (*ver-1*, *omt-A*, *apa-2*, *nor-1*)	Species specific	Detection in corn, wheat grain, and figs	(36,37, 39,40)
Fusarium graminearum	Galactose oxidase gene (*gaoA*)	Species specific	Identification	(30)
Trichothecene produing fusaria	Trichothecene biosynthesis gene (*tri5*)	Group specific	Detection in wheat grain	(35,41)
Fumonisin producing fusaria	ITS	Group specific	n.a.[a]	(42)
Fusarium poae, *F. graminearum*, *F. culmorum*, *F. avenaceum*	Specific RAPD fragments	Species specific	Detection and quantification in cereals	(43-46)
Fusarium moniliforme	Specific polymorphic DNA fragments	Species specific	Detection in infected maize plants	(47)
Fusarium solani, *F. moniliforme*, *F. oxysporum*	Specific RAPD fragments	Species specific	Detection in blood and tissues	(48)
Alternaria linicola, *A. solani*	ITS	Species specific	Detection in linseed	(49)
Stachybotrys chartarum	ITS	Species specific	n.a.[a]	(50)
Mucor racemosus, *Neurospora crassa*	Elongation factor genes (EF-Tu, EF-1α)	Mold specific?	Detection in milk	(51)
Colletotrichum acutatum	ITS	Species specific	Detection from infected strawberry	(52)
Zygosaccharomyces and *Candida* species	Microsatellite DNA	Species specific	Quality control of mayonnaise and salad dressing production	(53)
Zygosaccharomyces bailii	Microsatellite DNA	Isolate specific	Quality control of mayonnaise and salad dressing production	(53)
Saccharomyces diastaticus	Glycoamylase gene	Species specific	Detection in beer products	(54)

[a] n.a.: not applied

double stranded DNA intercalating dye ethidium bromide (EtBr). Depending on what degree of resolution is needed the choice of gel would be either agarose or poly-acrylamide. The latter allows separation of fragments with as little as one single base differences in size. The dsDNA-specific dye SYBR® Green I (FMC Bioproducts, Rockland, ME, USA; Roche Molecular Biochemicals, Mannheim, Germany) offers an alternative to EtBr (55). SYBR® Green I is 25-100 times more sensitive for detecting DNA in agarose gels than EtBr (56).

A number of other methods can also be used to detect the PCR amplicons. Although it requires more expensive equipment, HPLC of the DNA provides another way of separating and quantifying PCR products by electrophoresis (57,58). HPLC does also allow a higher degree of automation than gel-electrophoresis. Yet another method which has been used for separation of PCR products is the capillary electrophoresis (59).

Reverse dot blot (60) can be used when screening for a large number of fungi. Various poly(dT)-tailed probes, specific to internal sequences of the PCR products, are immobilized as dots on a nylon membrane and the PCR products hybridized to the probes. During amplification the PCR product can be labeled with radioisotopes or small molecules i.e fluorescein, digoxigenin or biotin allowing for subsequent visualization of the specific hybridization product. Both fluorescein and digoxigenin can be bound to either primer or nucleotides (Roche Molecular Biochemicals, Mannheim, Germany), whereas biotin usually is primer-bound. Fluorescein is detected directly (61), digoxigenin is detected by antibodies conjugated to fluorescein, or alkaline phosphatase or horseradish peroxidase exposed to suitable substrate (62), and biotin is detected with a streptavidin-enzyme conjugate (63).

Restriction endonuclease digestion of the PCR product with subsequent separation by electrophoresis or HPLC, like reverse dot blot, adds an extra level of specificity to the analysis as it detects sequence variation within the PCR product.

Direct sequencing of the PCR product offers yet another way of detecting sequence differences within the amplicon. However, although automation is possible, it is more time consuming and expensive than the restriction endonuclease digestion and is therefor recommended only if the actual sequence is needed.

In the Scintillation Proximity Assay (Amersham Pharmacia Biotech, Sweden) a scintillant is integrated into microspheric beads, designed to bind specific molecules e.g. biotin. If a radioactive molecule (e.g. tritium labeled PCR product) binds to the bead, tritium is brought into close enough proximity of the scintillant to release sufficient energy to excite it and cause light emission (64,65).

During the last decade microwell based assays for either post- or real-time detection of PCR products have been introduced. This often involves hybridization of the PCR product to an immobilized probe followed by detection, or direct amplification of the PCR product using the immobilized probe as one of the PCR primers (66). The use of microwell assays is further discussed later.

Quantification by PCR

The simplest approach to quantify numbers of gene copies in a sample by PCR is to measure the amount of product in the exponential phase compared to an external standard (67). However, accuracy is often a problem, as minor variations during sample preparation will multiply during the amplification process. Wang *et al.* (68) introduced competitive PCR (cPCR) to overcome variations between samples. This method is based on the competitive co-amplification of the specific target DNA with a known concentration of an internal standard. The internal standard (competitor DNA) shares primer-binding sites with the target and is distinguished from the target by a different size. The easiest way to identify the products of target and competitor is by gel-separation. Figure 1 outlines the

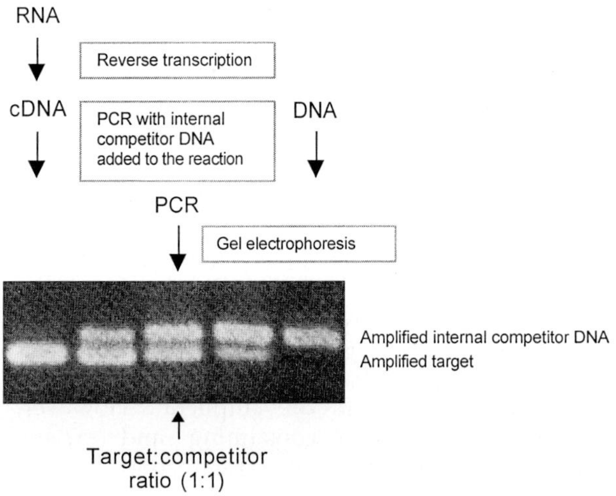

Figure 1. Principle of quantification by competitive PCR. A known number of internal competitor DNA copies is added to the PCR reaction and PCR products of target and competitor DNA are subsequently separated by gel electrophoresis. The products can be radioactively marked during reaction and quantified using a phospho-imager and appropriate software (available from e.g. Molecular Dymanics, USA). Alternatively, the intensity of fluorescence from bands on an EtBr stained gel can be quantified be scanning photos of the gel and using appropriate software (available from e.g. Amersham Pharmacia Biotech, Uppsala, Sweden). Extracts of RNA are converted into complementary DNA (cDNA) by reverse transcription prior to quantification.

principle of competitive PCR assays, starting from either DNA or RNA. The amount of target DNA in the initial sample is calculated from the ratio of the products expressed as:

$$\frac{[\text{Target DNA before PCR}]}{[\text{Target DNA after PCR}]} = \frac{[\text{Competitor DNA before PCR}]}{[\text{Competitor DNA after PCR}]}$$

It has been demonstrated that quantitative measurements are most accurate at equal ratios of competitor and target templates as indicated by the 1:1 ratio in Figure 1, and when the PCR process is in the exponential phase (69). However, if the competitor DNA is properly chosen and constructed, the amplification efficiency for both target and competitor should be identical throughout the entire PCR process (70). Construction and validation of the competitor DNA is on the other hand one of the most crucial and time-consuming parts of developing a competitive PCR assay. According to Jansson and Leser (71) there are various ways of constructing the competitor DNA. These involve construction of homologous competitors by either deleting or inserting internal fragments of the target sequence, construction of non-homologous competitors by linking primers to non-homologous DNA or generating fragments through low-stringency annealing of the primers.

Though not likely to be reported, unequal amplification efficiency of target and competitor DNA is a common experience for researchers working with the development of competitor DNA. Heteroduplex formation is presumably the major contributor to these problems and is often caused by the use of homologous competitors (56,71). On the other hand, it is possible to obtain equal amplification efficiencies by using non-homologous competitors with a G+C content similar to that of the target (71).

New products for the detection and quantification of PCR products are rapidly finding their way into the market. The use of fluorophores is a common theme for most of the approaches. Perkin Elmer (PE Applied Biosystems, Branchburg, USA) introduced the linear TaqMan® oligonucleotide with both a reporter and a quencher dye attached (Figure 2). The probe is designed to anneal between the forward and reverse PCR primers. When the probe is cleaved during the extension cycle by the 5′ nuclease activity of the *Taq* DNA polymerase, the reporter dye is separated from the quencher dye and a characteristic fluorescence is emitted.

Another kind of probe, first described by Tyagi and Kramer (72), is the molecular beacon (Figure 3). It resembles the TaqMan probe, though designed to form a hairpin with a fluorophore and a non-fluorescent quencher at opposite ends of the oligonucleotide. The ends have complementary sequences whereas the intervening region is designed to be specific to the target sequence. When free in solution the beacon forms a hairpin bringing the reporter and quencher close together. However, at loop-target recognition the binding is spontaneous and the stem-loop conformation opens. This brings the reporter and quencher sufficiently far apart for the beacon to fluoresce.

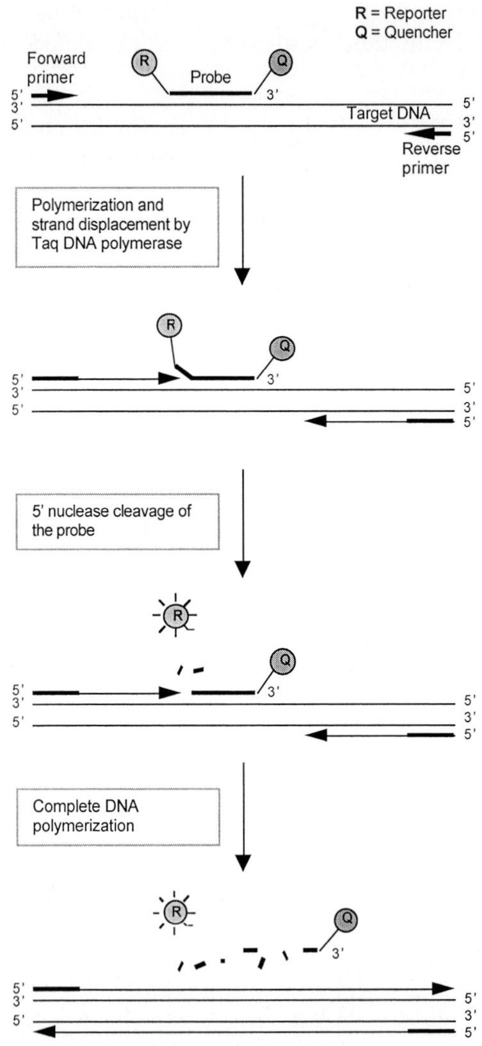

Figure 2. Principle of real-time PCR quantification using the TaqMan® probe. Two fluorescent dyes, a reporter (R) and a quencher (Q), are attached to an oligonucleotide probe and kept sufficiently close to each other to quench emission from the reporter dye. During the extension step of the PCR reaction the probe is displaced and cleaved by the *Taq* DNA polymerase leaving the reporter free in solution. Once separated from the quencher a characteristic fluorescence is emitted and can be monitored and quantified by fluorescence spectroscopy. The threshold cycle (C_T) (the cycle at which the emission is strong enough for detection) is used for quantification of the initial amount of target DNA. Redrawn from PE Applied Biosystems brochure (PE Applied Biosystems, Branchburg, USA).

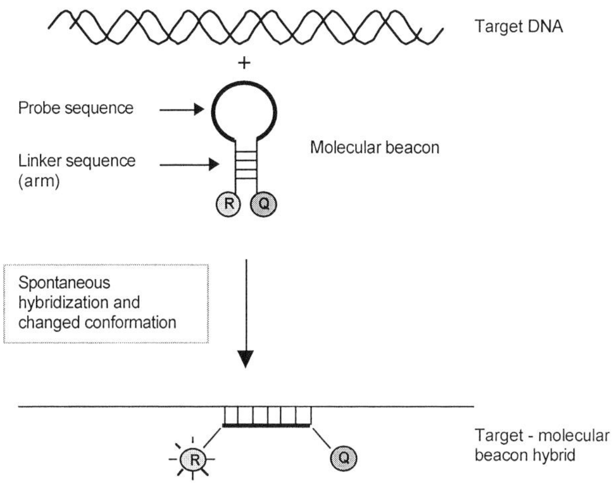

Figure 3. Principle of the molecular beacon. A fluorescent reporter (R) and a quencher (Q) at opposite ends of the molecular beacon are brought close to each other by a hairpin stem-loop conformation formed by hybridization of complementary 5 - 8-nucleotide-long arms. The hairpin stem-loop conformation cannot coexist with the rigid double helix that is formed at probe-target hybridization and consequently the reporter is moved away from the quencher. A 15-, 25-, or 35-nucleotide long probe maximizes the distance between the reporter and the quencher as the arms of the molecular beacon are arrayed in a *trans* configuration in relation to the probe-target double helix (72) and a characteristic fluorescence is emitted by the reporter. The light emission correlates to the amount of probe-target hybridization and can be monitored and quantified by fluorescence spectroscopy.

Finally, it is possible to add DNA intercalating dyes, e.g. EtBr or SYBR® Green I (FMC Bioproducts, Rockland, ME, USA; Roche Molecular Biochemicals, Mannheim, Germany) directly to the PCR reaction mixture and measure the increase in fluorescence.

An advantage of using a probe for detection rather than staining with intercalating fluorescent dyes is that the probe adds a level of specificity to the analysis by detecting internal sequence variation. Wittwer *et al.* (55) compared three different methods for real-time fluorescence monitoring of DNA amplification using the dsDNA specific dye SYBR® Green I and two probe-based methods. They found that the probe-based methods offered an increased level of specificity, however, using SYBR® Green I allowed real-time monitoring of denaturation, annealing, and extension within each cycle. In terms of probes the hairpin-structured beacon has advantages compared to the linear probes. The hairpin shape of the beacon has been reported to increase discrimination of single base-pair mismatches over linear probes (73,74). By using different fluorophores as reporters on different molecular beacons Tyagi *et al.* (74) showed that multiple targets can be detected and distinguished in the same solution, even if they differ from each other by as little as one nucleotide.

In all cases the detection and quantification of amplified PCR products can be done post amplification. Provided the PCR machine is equipped for fluorescence detection the amplification can also be monitored and quantified in real time. Some alternatives presently on the marked include the ABI PRISM 7700 (PE Applied Biosystems), the iCycler (Bio-Rad Laboratories, Herts, UK), and the LightCycler capillary system (Roche Diagnostics, Mannheim, Germany). One large advantage with the real-time quantification is that the plateau effect of the PCR reaction is overcome as the reaction is monitored cycle by cycle. This was exemplified by Higuchi *et al.* (75) by showing that the initial target copy number can be quantified based on the threshold cycle (the cycle at which the emission is strong enough for detection) during real-time PCR.

The real-time methods were developed for clinical use and so far there have been no publications using these techniques for the detection of toxigenic fungi in food. However, as the time and cost required for these kinds of analysis go down, they will potentially become useful for routine analysis of food and feed.

5. OTHER MOLECULAR METHODS

The reverse transcription-PCR (RT-PCR) technique makes it possible to discriminate between living and dead fungal cells due to the high turnover rate of the mRNA (sometimes less than a minute). The first step is to run a reverse transcription to make complementary DNA (cDNA) from the mRNA which is followed by a normal PCR reaction. The RT-PCR has for example been used for the detection of bacteria, molds and yeasts in milk (51). The method can be used as a semi quantitative method by adding an internal competitive DNA standard to the PCR reaction after completion of the RT step (RT-cPCR) (Figure 1). This is based on the assumption of a constant efficiency of the reverse transcription reaction (76). However, it is more likely that the data produced by RT-cPCR is highly subjective due to the different efficiency of the reverse transcription reaction. Yet, though difficult to handle due to the fast degradation of RNA, this problem can be overcome by including a competitive mRNA to the RT reaction tube (77,78). Although somewhat complicated the RT-PCR has a potential for use in detection and quantification of active/living food-borne fungi as recently described by Doohan *et al.* (79) when studying the expression of the *tri5* (the first gene in the trichothecene biosynthetic pathway) by *Fusarium* species in planta.

The starting point for Immuno-PCR was the development of a linker molecule with specific affinity for both DNA and antibodies. The idea is to performed a standard PCR reaction after the formation of a conjugate between antigen-specific antibody-linker-DNA (80). However, the fact that the linker molecule was not commercially available and quite expensive promoted the development of the universal immuno-PCR for target protein detection. The special linker molecule was substituted with commercially available biotinylated secondary antibodies and free streptavidin. The principle, as shown in Figure 4, is that the components: target-antigen (proteins or DNA), specific antibody, biotinylated

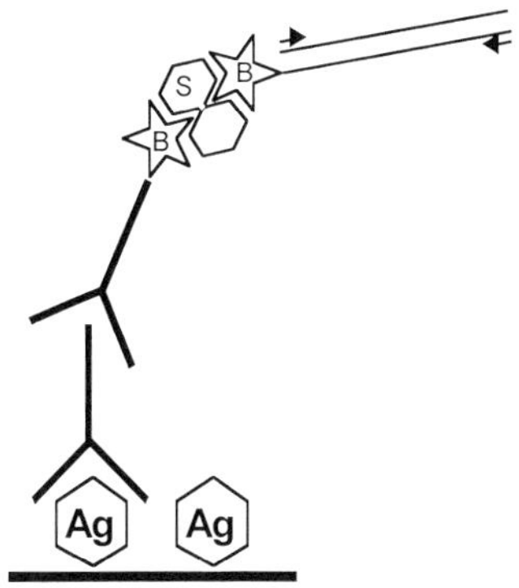

Figure 4. Description of Universal Immuno-PCR. The method relies on an initial specific antigen-antibody detection of the target DNA or protein (antigen) and a subsequent PCR amplification of a signal DNA to amplify the signal and increase the sensitivity of the reaction. Antigen, specific primary antibody, biotinylated secondary antibody, free streptavidin and biotinylated signaling DNA are stepwise immobilized to the PCR tube. After that, PCR amplification is performed on the immobilized signal DNA and the products are visualized on agarose gel. Antigen (Ag), Biotin (B) and Streptavidin (S).

secondary antibody, streptavidin and biotinylated signaling DNA are stepwise immobilized to the PCR tube. A normal PCR reaction is then performed and the PCR product is subsequently visualized on an agarose gel (81). The easy to use universal immuno-PCR is a promising method for evaluation of the mycological quality of food. It could to a certain extent also be expected to be able to distinguish between living and dead fungal cells depending on the selection of the target protein.

Solid phase PCR (66,82) is based on the separation of the primer pair into one solid and one liquid phase. One of the primers is immobilized to the plastic of a microwell by a covalent phosphoramidate binding (83). The solid phase PCR method has been further developed to DIAPOPS (detection of immobilized, amplified product in a one phase system) (66,82). Here, a normal PCR is performed, and the DNA strand polymerized to the 3' terminus of the solid phase primer becomes trapped to the microwells. Detection by hybridization with a specific internal oligonucleotide probe coupled to a marker (e.g. enzyme based) is carried out in the same well. DIAPOPS has been used for detection and estimation of *Fusarium graminearum* DNA from a mixture of fungal and plant DNA. The total processing time required for the DIAPOPS assay was estimated to seven hours using ELISA equipment (84).

Northern blot analysis is a sensitive method for identification and quantification of the transcript of a specific target gene. The RNA (total or mRNA) has to be isolated and separated on an agarose gel, transferred to a nylon membrane and fixated. The transcript is detected by hybridization with a specific internal probe and visualized either by incorporation of a radioactive labeled compound or by an enzymatic reaction. The northern blot analysis includes many different steps and is expensive, laborious and time consuming. Therefor it has no future as a routine method for detection and quantification of mold in food and feed samples. On the other hand it can certainly be a useful complement when there is a special need for a sensitive method for identification and quantification of low abundant mRNA.

In situ-PCR is used to provide information on target localization of RNA and DNA within cells or tissues, e.g. in cancer research or molecular plant pathology as it combines the high sensitivity of PCR with the anatomic localization of *in situ* hybridization (85). However, in terms of fungal contamination of food and feed it is not likely to be a method of choice unless a very specialized diagnosis is requested.

6. SAMPLE PREPARATION

One concern with both diagnostic- PCR and ELISA (24) is the reliability of the assay. DNA amplification efficiency is highly dependent on sample preparation since various components from both the sample preparation and processing have been shown to inhibit the rate or extent of the reaction. Lantz *et al.* (86) observed inhibition of PCR reactions with cheese samples and Rossen *et al.* (87) tested great numbers of components present in food and from DNA extractions. They reported inhibition in samples containing high levels of fat and protein, as well as some inhibition caused by substances used for DNA extraction (e.g. detergents, NaOH, alcohols). In addition, factors from plant tissue, soil and sediments have been reported to be inhibitory (reviewed by 88). High concentrations of DNA may also be inhibitory to the reaction, and Farber *et al.* (40) observed a ten-fold reduction in sensitivity of an aflatoxigenic-specific PCR when DNA from figs was introduced to the reactions. Nevertheless, PCR assays are finding their way into routine analysis laboratories, and in general sensitivity and ease of detection are improving. Fletcher *et al.* (27) tested a microtitre plate detection assay of PCR products of *A. fumigatus* and found it as sensitive as detection by Southern blot hybridization. The sensitivity of the PCR may be additionally increased by using nested PCR (89,90).

Quantitative DNA extraction from infected food and feed (e.g. cereal grain) is likely to be influenced by the degree and age of infection. Fungal biomass from the surface of the product is easier to extract than fungal biomass that has grown into the matrix. The cell lysis efficiency pose another problem. If the method used does not result in lysis of both mycelia and spores a quantitative method will never be achieved. Olsson *et al.* (91) recently developed a fast and efficient method for DNA extraction from fungal mycelia as well as conidia of a number of food- and feed-borne fungi using heated NaOH and Sodium Dodecyl Sulfate (SDS) for cell lysis. Also, Haugland and coworkers (92) published a work testing different methods (e.g. grinding, sonication, glass bead milling) for cell lysis of fungal conidia. Results were evaluated by cPCR and showed that glass bead milling was the most efficient method in terms of absolute copy numbers recovered as well as extract to extract variation. However, both studies were done with extracts of pure cultures and might therefor only to a certain extent reflect extraction from natural samples.

As mentioned earlier, a number of components from natural samples have been shown to inhibit the PCR reaction. Möller and Jansson (93) suggested that, when using PCR for quantification, variations in DNA extractions can be overcome by using co-extraction of a competitive internal standard for subsequent quantification by cPCR. In this way any inhibition or loss of DNA during the extraction steps will affect DNA amplification of the standard and target proportionally. Nazar *et al.* (94) demonstrated that applying competitor DNA to an environmental sample (e.g. soil) will eliminate artifacts resulting from inhibitors in the extract as losses in target and competitor DNA will be proportional. However, this will not eliminate failures in DNA extraction efficiencies. In a present study procedures for fungal DNA extraction from infected grain were evaluated, testing e.g. lyophilization and ultra sound sonication or grinding for cell lysis as well as various extraction buffer components i.e. proteinase K, NaCl, EDTA, Hexadecyl Trimethyl-Ammonium Bromide (CTAB), and SDS (95). Results were evaluated by statistical experimental design to find the optimal extraction procedure, which includes lyophilization, sonication and high concentrations of SDS. Further use of the optimized extraction procedure has shown good correlation between traditional quantification by CFU counts and cPCR at mold levels higher than 10^2 CFU per gram grain, i.e. at the same level of detection as with surface plating for CFU (Boysen *et al.*, unpublished).

7. CONCLUDING REMARKS

The distribution of mycotoxigenic fungi in a food matrix can show a high degree of temporal and spatial variation. Neither cultivation methods nor molecular methods can provide a "true" answer of the mycological status of a given food commodity unless a correct sampling strategy has been used. In reality this often involves processing a large number of sub-samples, implying the handling of large quantities of food material. An example can be found in a European Commission directive (98/53/EG) for official control laboratories on how to sample and analyze for aflatoxin contamination of foodstuff (96). The directive states that total sample sizes of dried fruit, peanuts, pistachios, nuts in general, and cereal grain should be 30 kg (collected from 100 sub-samples), unless lots are smaller than 15 ton (50 ton for cereal grain). This sample should then be divided into three 10-kg laboratory samples for milling. This situation of course is not very familiar for the molecular biology laboratories developing new assays and normally dealing with mg samples. However, for successful implementation of new molecular techniques the sampling problems and handling of large sample quantities have to be thought through at an early stage of assay development.

In a critical review on the possibilities and limitations of nucleic acid amplification technology in diagnostic microbiology Vaneechoutte and van Eldere (97) state that routine bacteriology laboratories will continue to rely on plate counts of CFU as the preferred quantification method for most diagnostic applications. This is a prediction based on the drawbacks of the PCR techniques in terms of false-positives and false-negatives as well as the difficulties of getting quantitative results. A similar future, relying on cultivation of fungi on agar plates will most definitely be the case also for food mycology. The problems for the PCR techniques, e.g. inhibitory compounds, are likely to be overcome in a not-too-distant future, but other barriers will more seriously delay the use of new molecular techniques in food mycology. One has to do with education of staff in the food industry and related laboratories. Most corporations have down-sized their organizations to such an extent that very limited time is available for learning and development of new skills. Another and perhaps more serious barrier is that the food industry will not use new techniques unless prompted to by legislation. In judging whether tomato products have been properly prepared to eliminate rot and decay by molds, the US Food and Drug Administration uses the Howard mold-count test. If mold filaments are present in excess of amounts stated in the Food Defect Action Levels, FDA refuses admission to import shipments and considers enforcement action against domestic shipments. The Howard mold-count test was first published 1911 (Howard, BJ. 1911. USDA 68 Bur Chem Circ) but still remain the food mycology standard for the tomato-based food industry. In order for a new molecular technique to replace the Howard mold-count it first has to be compared with the existing test in a lengthy number of inter-laboratory calibrations. Then translations of the semi-quantitative scale used in the Howard mold-count test into molecular data has to be incorporated into Food Defect Action Levels and other regulations/legislation before education and implementation can take place within the food industry. Similar legislation related problems are likely to be found when comparing plate count CFU and result from molecular techniques. Although the traditional techniques will continue to form the basis for permissible levels and guideline levels of legislation and directives for a long time yet, the molecular techniques will surely supplement these. The combination of extremely high sensitivity, specificity and short analyze time will make them an increasingly important complement in the food mycology laboratory.

REFERENCES

1. Filtenborg, O., Frisvad, J.C., and Thrane, U. (1996) Moulds in food spoilage. Int. J. Food Microbiol. 33, 85-102.
2. Schnürer, J., Olsson, J., and Börjesson, T. (1999) Fungal volatiles as indicators of food and feeds spoilage. Fung. Genet. Biol. 27, 209-217.
3. Schnürer, J. (1993) Comparison of methods for estimating the biomass of three food-borne fungi with different growth-patterns. Appl. Environ. Microbiol. 59, 552-555.
4. Pitt, J.I. (1984) The significance of potentially toxigenic fungi in foods. Food Technol. Aust. 36, 218-219.

5. Samson, R.A., Hoekstra, E.S., Frisvad, J.C., and Filtenborg, O. (1995) Introduction to Food-Borne Fungi. 4th ed. Centraalbureau voor Schimmelcultures, Baarn, The Netherlands.
6. Pitt, J.I., Hocking, A.D., and Glenn, D.R. (1983) An improved medium for the detection of *Aspergillus flavus* and *A. parasiticus*. J. Appl. Bacteriol. 54, 109-114.
7. Seitz, L.M., Mohr, H.E., Buroughs, R., and Sauerr, D.B. (1977) Ergosterol as an indicator of fungal invasion in grains. Cereal Chem. 54, 1207-1217.
8. Newell, S.Y. (1992) Estimating fungal biomass and productivity in decomposing litter. In The Fungal Community. Its Organization and Role in the Ecosystem (G.C. Carroll and D.T. Wicklow Eds.), pp. 521-561. 2nd ed. Marcel Dekker Inc., New York.
9. Seitz, L.M., Sauer, D.B., Burroughs, R., Mohr, H.E., and Hubbard, J.D. (1979) Ergosterol as a measure of fungal growth. Phytopathology 69, 1202-1203.
10. Newell, S.Y., Arsuffi, L.T., and Fallon, R.D. (1988) Fundamental procedures for determining ergosterol content in decaying plant material by liquid chromatography. Appl. Environ. Microbiol. 54, 1876-1879.
11. Schwadorf, K. and Müller, H.M. (1989) Determination of ergosterol in cereals, mixed feed components, and mixed feeds by liquid chromatography. J. Assoc. Off. Anal. Chem. 72, 457-462.
12. Dowell, F.E., Ram, M.S., and Seitz, L.M. (1999) Predicting scab, vomitoxin, and ergosterol in single wheat kernels using near-infrared spectroscopy. Cereal Chem. 76, 573-576.
13. Börjesson, T., Eklöv, T., Jonsson, A., Sundgren, H., and Schnürer, J. (1996) An electronic nose for odor classification of grains. Cereal Chem. 73, 457-461.
14. Smith, E.A., Chambers, E., and Colley, S. (1994) Development of vocabulary and references for describing off-odors in raw grains. Cereal Food. World 39, 495.
15. Dickinson, T.A., White, J., Kauer, J.S., and Walt, D.R. (1998) Current trends in "artificial-nose" technology. TIBTECH. 16, 250-258.
16. Rylander, R. (1986) Lung diseases caused by organic dusts in the farm environment. Am. J. Ind. Med. 10, 221-227.
17. Larsen, F.O., Clementsen, P., Hansen, M., Maltbaek, N., Larsen, T.O., Nielsen, K.F., Gravesen, S., Skov, P.S., and Norn, S. (1998) Volatile organic compounds from the indoor mould *Trichoderma viride* cause histamine release from human bronchoalveolar cells. Inflamm. Res. 47, S5-S6.
18. Wålinder, R., Norbäck, D., and Johanson, G. (1998) Pulmonary reactions after exposure to 3-methylfuran vapour, a fungal metabolite. Int. J. Tuberc. Lung. Dis. 2, 1037-1039.
19. Mullis, K.B. and Faloona, F. (1987) Specific synthesis of DNA *in vitro* via a polymerase-catalyzed chain reaction. Methods Enzym. 155, 335-350.
20. Stynen, D., Meulemans, L., Goris, A., Braendlin, N., and Symons, N. (1992) Characteristics of a latex agglutination test based on monoclonal antibodies for the detection of fungal antigens in foods. In Modern Methods in Food Mycology (R.A. Samson, A. Hocking, J.I. Pitt, and A.D. King Eds.), pp. 213-219, vol. 31. Elsevier, Amsterdam.
21. Dewey, F.M., MacDonald, M.M., Phillips, S.I., and Priestley, R.A. (1990) Development of monoclonal-antibody-ELISA and DIP-STICK immunoassays for *Penicillium islandicum* in rice grains. J. Gen. Microbiol. 136, 753-760.
22. Tsai, G.J. and Yu, S.C. (1997) An enzyme-linked immunosorbent assay for the detection of *Aspergillus parasiticus* and *Aspergillus flavus*. J. Food Protect. 60, 978-984.
23. Shapira, R., Paster, N., Menasherov, M., Eyal, O., Mett, A., Meiron, T., Kuttin, E., and Salomon, R. (1997) Development of polyclonal antibodies for detection of aflatoxigenic

molds involving culture filtrate and chimeric proteins expressed in *Escherichia coli*. Appl. Environ. Microbiol. 63, 990-995.
24. Girardin, H. (1997) Detection of filamentous fungi in foods. Sci. Aliment. 17, 3-19.
25. Scudamore, K.A., Nawaz, S., and Hetmanski, M.T. (1998) Mycotoxins in ingredients of animal feeding stuffs: II. Determination of mycotoxins in maize and maize products. Food Addit. Contam. 15, 30-55.
26. Bretagne, S., Costa, J.M., Marmoratkhuong, A., Poron, F., Cordonnier, C., Vidaud, M., and Fleuryfeith, J. (1995) Detection of *Aspergillus* species DNA in bronchoalveolar lavage samples by competitive PCR. J. Clin. Microbiol. 33, 1164-1168.
27. Fletcher, H.A., Barton, R.C., Verweij, P.E., and Evans, E.G.V. (1998) Detection of *Aspergillus fumigatus* PCR products by a microtitre plate based DNA hybridisation assay. J. Clin. Pathol. 51, 617-620.
28. Hopfer, R.L., Walden, P., Setterquist, S., and Highsmith, W.E. (1993) Detection and differentiation of fungi in clinical specimens using polymerase chain reaction (PCR) amplification and restriction enzyme analysis. J. Med. Vet. Mycol. 31, 65-75.
29. Kappe, R., Okeke, C.N., Fauser, C., Maiwald, M., and Sonntag, H.G. (1998) Molecular probes for the detection of pathogenic fungi in the presence of human tissue. J. Med. Microbiol. 47, 811-820.
30. Niessen, M.L. and Vogel, R.F. (1997) Specific identification of *Fusarium graminearum* by PCR with gaoA targeted primers. System. Appl. Microbiol. 20, 111-123.
31. Pedersen, L.H., Skouboe, P., Boysen, M., Soule, J., and Rossen, L. (1997) Detection of *Penicillium* species in complex food samples using the polymerase chain reaction. Int. J. Food Microbiol. 35, 169-177.
32. Crampin, A.C. and Matthews, R.C. (1993) Application of the polymerase chain reaction to the diagnosis of candidosis by amplification of an HSP 90 gene fragment. J. Med. Microbiol. 39, 233-238.
33. Schilling, A.G., Möller, E.M., and Geiger, H.H. (1996) Polymerase chain reaction-based assays for species-specific detection of *Fusarium culmorum, F. graminearum*, and *F. avenaceum*. Phytopathology 86, 515-522.
34. Smith, O.P., Peterson, G.L., Beck, R.J., Schaad, N.W., and Bonde, M.R. (1996) Development of a PCR-based method for identification of *Tilletia indica*, causal agent of Karnal bunt of wheat. Phytopathology 86, 115-122.
35. Niessen, M.L. and Vogel, R.F. (1997) A molecular approach to the detection of potential trichothecene producing fungi. Cereal Res. Commun. 25, 245-249.
36. Geisen, R. (1996) Multiplex polymerase chain reaction for the detection of potential aflatoxin and sterigmatocystin producing fungi. System. Appl. Microbiol. 19, 388-392.
37. Shapira, R., Paster, N., Eyal, O., Menasherov, M., Mett, A., and Salomon, R. (1996) Detection of aflatoxigenic molds in grains by PCR. Appl. Environ. Microbiol. 62, 3270-3273.
38. Geisen, R. (1998) PCR methods for the detection of mycotoxin-producing fumgi. In Applications of PCR in Mycology (P.D. Bridge, D.K. Arora, C.A. Reddy, and R.P. Elander Eds.), pp. 243-266. CAB International, Wallingford, UK.
39. Geisen, R., Mulfinger, S., and Niessen, L. (1998) Detection of *Aspergillus flavus* in wheat by PCR. J. Food Mycol. 1, 211-218.
40. Farber, P., Geisen, R., and Holzapfel, W.H. (1997) Detection of aflatoxinogenic fungi in figs by a PCR reaction. Int. J. Food Microbiol. 36, 215-220.
41. Niessen, M.L. and Vogel, R.F. (1998) Group specific PCR-detection of potential trichothecene-producing *Fusarium* species in pure cultures and cereal samples. System. Appl. Microbiol. 21, 618-631.

42. Grimm, C. and Geisen, R. (1998) A PCR-ELISA for the detection of potential fumonisin producing *Fusarium* species. Lett. Appl. Microbiol. 26, 456-462.
43. Doohan, F.M., Parry, D.W., and Nicholson, P. (1999) Fusarium ear blight of wheat: the use of quantitative PCR and visual disease assessment in studies of disease control. Plant Pathol. 48, 209-217.
44. Nicholson, P., Doohan, F., Rezanoor, H.N., Simpson, D., Smith, P.H., Turner, A., and Weston, G. (1997) Detection and quantification of individual fungal species in *Fusarium* disease complexes of cereals by polymerase chain reaction (PCR). Cereal Res. Commun. 25, 477-482.
45. Nicholson, P., Simpson, D.R., Weston, G., Rezanoor, H.N., Lees, A.K., Parry, D.W., and Joyce, D. (1998) Detection and quantification of *Fusarium culmorum* and *Fusarium graminearum* in cereals using PCR assays. Physiol. Molec. Plant Pathol. 53, 17-37.
46. Parry, D.W. and Nicholson, P. (1996) Development of a PCR assay to detect *Fusarium poae* in wheat. Plant Pathol. 45, 383-391.
47. Murillo, I., Cavallarin, L., and Segundo, B.S. (1998) The development of a rapid PCR assay for detection of *Fusarium moniliforme*. Eur. J. Plant Pathol. 104, 301-311.
48. Hue, F.X., Huerre, M., Rouffault, M.A., and deBievre, C. (1999) Specific detection of *Fusarium* species in blood and tissues by a PCR technique. J. Clin. Microbiol. 37, 2434-2438.
49. McKay, G.J., Brown, A.E., Bjourson, A.J., and Mercer, P.C. (1999) Molecular characterisation of *Alternaria linicola* and its detection in linseed. Eur. J. Plant Pathol. 105, 157-166.
50. Haugland, R.A. and Heckman, J.L. (1998) Identification of putative sequence specific PCR primers for detection of the toxigenic fungal species *Stachybotrys chartarum*. Mol. Cell. Probes 12, 387-396.
51. Vaitilingom, M., Gendre, F., and Brignon, P. (1998) Direct detection of viable bacteria, molds, and yeasts by reverse transcriptase PCR in contaminated milk samples after heat treatment. Appl. Environ. Microbiol. 64, 1157-1160.
52. Sreenivasaprasad, S., Sharada, K., Brown, A.E., and Mills, P.R. (1996) PCR-based detection of *Colletotrichum acutatum* on strawberry. Plant Pathol. 45, 650-655.
53. Couto, M.M.B., Hartog, B.J., Veld, J.H.J.H.I., Hofstra, H., and van der Vossen, J.M.B.M. (1996) Identification of spoilage yeasts in a food-production chain by microsatellite polymerase chain reaction fingerprinting. Food Microbiol. 13, 59-67.
54. Yamauchi, H., Yamamoto, H., Shibano, Y., Amaya, N., and Saeki, T. (1998) Rapid methods for detecting *Saccharomyces diastaticus*, a beer spoilage yeast, using the polymerase chain reaction. J. Am. Soc. Brew. Chem. 56, 58-63.
55. Wittwer, C.T., Herrmann, M.G., Moss, A.A., and Rasmussen, R.P. (1997) Continuous fluorescence monitoring of rapid cycle DNA amplification. Biotechniques 22, 130-138.
56. Schneeberger, C., Speiser, P., Kury, F., and Zeillinger, R. (1995) Quantitative detection of reverse transcriptase-PCR products by means of a novel and sensitive DNA stain. PCR Meth. Applic. 4, 234-238.
57. Palma, F., Potenza, L., Amicucci, A., Fiorani, M., Labella, D., Di Biase, S., and Stocchi, V. (1998) HPLC and CE analysis of PCR products: A comparative study. J. Liq. Chrom. & Rel. Technol. 21, 1527-1540.
58. Maslow, J.N., Slutsky, A.M., and Arbeit, R.D. (1993) Application of pulsed-field gel electrophoresis to molecular epidemiology. In Diagnostic Molecular Microbiology. Principles and Applications (D.H. Persing, T.F. Smith, F.C. Tenover, and T.J. White Eds.), pp. 563-572. American Society for Microbiology, Washington D.C.
59. Martin, F., Vairelles, D., and Henrion, B. (1993) Automated ribosomal DNA fingerprinting by capillary electrophoresis of PCR products. Anal. Biochem. 214, 182-189.

60. Saiki, R.K., Walsh, P.S., Levenson, C.H., and Erlich, H.A. (1989) Genetic analysis of amplified DNA with immobilized sequence-specific oligonucleotide probes. Proc. Natl. Acad. Sci. USA 86, 6230-6234.
61. Gibellini, D., Zauli, G., Re, M.C., Furlini, G., Lolli, S., Bassini, A., Celeghini, C., and Laplaca, M. (1995) In-situ polymerase chain reaction technique revealed by flow cytometry as a tool for gene detection. Anal. Biochem. 228, 252-258.
62. Holmstrøm, K., Rossen, L., and Rasmussen, O.F. (1993) A highly sensitive and fast nonradioactive method for detetion of polymerase chain reaction products. Anal. Biochem. 209, 278-283.
63. Mansfield, E.S., Worley, J.M., McKenzie, S.E., Surrey, S., Rappaport, E., and Fortina, P. (1995) Nucleic-acid detection using nonradioactive labelling methods. Mol. Cell. Probes 9, 145-156.
64. Bosworth, N. and Towers, P. (1989) Scintillation proximity assay. Nature 341, 167-168.
65. Kenrick, M.K., Jiang, L.X., Potts, C.L., Owen, P.J., Shuey, D.J., Econome, J.G., Anson, J.G., and Quinet, E.M. (1997) A homogeneous method to quantify mRNA levels: a hybridization of RNase protection and scintillation proximity assay technologies. Nucleic Acids Res. 25, 2947-2948.
66. Chevrier, D., Rasmussen, S.R., and Guesdon, J.L. (1993) PCR product quantification by nonradioactive hybridization procedures using an oligonucleotide covalently bound to microwells. Mol. Cell. Probes 7, 187-197.
67. Cross, N.C.P. (1995) Quantitative PCR techniques and applications. Br. J. Haemat. 89, 693-697.
68. Wang, A.M., Doyle, M.V., and Mark, D.F. (1989) Quantitation of mRNA by the polymerase chain reaction. Proc. Natl. Acad. Sci. USA 86, 9717-9721.
69. Arnold, B.L., Itakura, K., and Rossi, B.T. (1992) PCR-based quantitation of low levels of HIV-1 DNA by using an external standard. Genet. Anal. Tech. Appl. 9, 113-116.
70. Zimmermann, K. and Mannhalter, J.W. (1996) Technical aspects of quantitative competitive PCR. Biotechniques 21, 268.
71. Jansson, J.K. and Leser, T. (1996) Quantitative PCR of environmental samples. In Molecular Microbial Ecology Manual (A.D.L. Akkermans, J.D. van Elsas, and F.J. de Bruijn Eds.), pp. 2.7.4: 1-19. Kluwer Academic Publisher, Dordrecht.
72. Tyagi, S. and Kramer, F.R. (1996) Molecular beacons: Probes that fluoresce upon hybridization. Nat. Biotech. 14, 303-308.
73. Bonnet, G., Tyagi, S., Libchaber, A., and Kramer, F.R. (1999) Thermodynamic basis of the enhanced specificity of structured DNA probes. Proc. Natl. Acad. Sci. USA 96, 6171-6176.
74. Tyagi, S., Bratu, D.P., and Kramer, F.R. (1998) Multicolor molecular beacons for allele discrimination. Nat. Biotech. 16, 49-53.
75. Higuchi, R., Fockler, C., Dollinger, G., and Watson, R. (1993) Kinetic PCR analysis. Real-time monitoring of DNA amplification reactions. BioTechn. 11, 1026-1030.
76. Hall, L.L., Bicknell, G.R., Primrose, L., Pringle, J.H., Shaw, J.A., and Furness, P.N. (1998) Reproducibility in the quantification of mRNA levels by RT-PCR-ELISA and RT competitive-PCR-ELISA. Biotechniques 24, 652-657.
77. Riedy, M.C., Timm, E.A., and Stewart, C.C. (1995) Quantitative RT-PCR for measuring gene expression. Biotechniques 18, 70-76.
78. Tsai, S.J. and Wiltbank, M.C. (1996) Quantification of mRNA using competitive RT-PCR with standard-curve methodology. Biotechniques 21, 862-866.
79. Doohan, F.M., Weston, G., Rezanoor, H.N., Parry, D.W., and Nicholson, P. (1999) Development and use of a reverse transcription-PCR assay to study expression of *Tri5* by *Fusarium* species in vitro and in planta. Appl. Environ. Microbiol. 65, 3850-3854.

80. Sano, T. and Cantor, C.R. (1991) A streptavidin-protein A chimera that allows one-step production of a variety of specific antibody conjugates. BioTechn. 9, 1378-1381.
81. Zhou, H., Fisher, R.J., and Papas, T.S. (1993) Universal immuno-PCR for ultra sensitive target protein detection. Nucleic Acids Res. 21, 6038-6039.
82. Rasmussen, S.R., Rasmussen, H.B., Larsen, M.R., Hoff-Jørgensen, R., and Cano, R.J. (1994) Combined polymerase chain reaction-hybridization microplate assay used to detect bovine leukemia virus and *Salmonella*. Clin. Res. 40, 200-205.
83. Rasmussen, S.R., Larsen, M.R., and Rasmussen, S.E. (1991) Covalent immobilization of DNA onto polystyrene microwells - The molecules are bound at the 5' end. Anal. Biochem. 198, 138-142.
84. Niessen, L., Klunsmann, J., and Vogel, R.F. (1998) Quantitative estimation of *Fusarium graminearum* DNA using a solid phase PCR assay (DIAPOPS). J. Food Mycol. 1, 73-84.
85. Bagasra, O. and Hansen, J. (1997) In Situ PCR Techniques. Wiley-Liss, New York.
86. Lantz, P.G., Tjerneld, F., Borch, E., Hahn-Hägerdal, B., and Rådström, P. (1994) Enhanced sensitivity in PCR detection of *Listeria monocytogenes* in soft cheese through use of an aqueous two-phase system as a sample preparation method. Appl. Environ. Microbiol. 60, 3416-3418.
87. Rossen, L., Nørskov, P., Holmstrøm, K., and Rasmussen, O.F. (1992) Inhibition of PCR by components of food samples, microbial diagnostic assays and DNA-extraction solutions. Int. J. Food Microbiol. 17, 37-45.
88. Wilson, I.G. (1997) Inhibition and facilitation of nucleic acid amplification. Appl. Environ. Microbiol. 63, 3741-3751.
89. Niepold, F. and Schöber-Butin, B. (1997) Application of the one-tube PCR technique in combination with a fast DNA extraction procedure for detecting *Phytophthora infestans* in infected potato tubers. Microbiol. Res. 152, 345-351.
90. Ibeas, J.I., Lozano, I., Perdigones, F., and Jimenez, J. (1996) Detection of *Dekkera brettanomyces* strains in sherry by a nested PCR method. Appl. Environ. Microbiol. 62, 998-1003.
91. Olsson, J., Schnürer, J., Pedersen, L.H., and Rossen, L. (1999) A rapid and efficient method for DNA extraction from fungal spores and mycelium for detection using PCR. J. Food Mycol., in press.
92. Haugland, R.A., Heckman, J.L., and Wymer, L.J. (1999) Evaluation of different methods for the extraction of DNA from fungal conidia by quantitative competitive PCR analysis. J. Microbiol. Meth. 37, 165-176.
93. Möller, A. and Jansson, J.K. (1997) Quantification of genetically tagged cyanobacteria in Baltic Sea sediment by competitive PCR. Biotechniques 22, 512-518.
94. Nazar, R.N., Robb, E.J., and Volossiouk, T. (1996) Direct extraction of fungal DNA from soil. In Molecular Microbial Ecology Manual (A.D.L. Akkermans, J.D. van Elsas, and F.J. de Bruijn Eds.), pp. 1.3.6: 1-8. Kluwer Academic Publisher, Dordrecht.
95. Boysen, M.E. (1999) Molecular identification and quantification of the *Penicillium roqueforti* group. Ph.D. Thesis. Swedish University of Agricultural Sciences, Uppsala, Sweden.
96. European Commission (1998) Kommissionens Direktiv 98/53/EG av 16 juli 1998 om fastställande av provtagnings- och analysmetoder för officiell kontroll av högsta tillåtna halt av vissa främmande ämnen i livsmedel [The Commission's Directive no. 98/53/EG]. Europeiska Gemenskapens Officiella Tidning (Sweden) [Official Journal of The European Commission] L 201, 17.7.98, 93-101.
97. Vaneechoutte, M. and van Eldere, J. (1997) The possibilities and limitations of nucleic acid amplification technology in diagnostic microbiology. J. Med. Microbiol. 46, 188-194.

Applied Mycology and Biotechnology
Volume 1. Agriculture and Food Production
G.G. Khachatourians and D.K. Arora (editors)

Strain improvement in filamentous fungi-an overview

K.M.H. Nevalainen

Department of Biological Sciences, Macquarie University, Sydney, NSW 2109, Australia

Filamentous fungi are used for the commercial production of various metabolites. Early strain development programs involved the improvement of product yield by random mutagenesis and screening and protoplast fusion. Currently, molecular techniques such as genetic transformation and protein engineering are being applied for the further refinement of fungal products and production systems. As eukaryotic microorganisms capable of post-translational modification of gene products and of secreting them into the culture medium, filamentous fungi are attractive hosts for products of high value or volume. Functional genomics and the emerging proteome studies will generate an invaluable information base paving the way for a holistic approach in fungal strain improvement.

1. INTRODUCTION

Filamentous fungi are widely employed in industrial processes for the production of enzymes, antibiotics, organic acids and other bioactive compounds as well as in the more traditional industries of food fermentation. Strain improvement is an elementary part of process development, generally aiming at reduction of production costs. Properties such as increased enzyme and biomass yields, augmented physiological characteristics and effective utilisation of a variety of industrially relevant substrates are targets for genetic strain improvement.

The main genera of filamentous fungi used in industry are *Penicillium* (antibiotics, cheese), *Cephalosporium* (antibiotics), *Aspergillus* (enzymes, organic acids, fermented foods), *Trichoderma* (enzymes), *Mucor* (enzymes) and *Rhizopus* (enzymes). Examples of the growing list of fungi cultivated as food include *Fusarium venenatum* (fungal protein Quorn®), *Agaricus bisporus* (common mushroom), *Pleurotus ostreatus* (oyster mushroom), *Lentinus edodes* (shiitake) and *Tuber* spp. (truffles). Fungi are increasingly being employed to produce therapeutic metabolites and to carry out various biotransformations requiring particular specificity such as product isomerization and steroid hydroxylation. Filamentous fungi (*Trichoderma*, *Verticillium*, *Metarhizium*) are also used and developed as biocontrol agents to replace environmentally recalcitrant toxic compounds in pest control.

The practical options for fungal strain improvement can vary considerably depending on the nature of the product application (secreted enzyme/human consumption/biological control), the level of knowledge and information on the genetics and physiology and the availability of molecular biology tools concerning a particular organism. In case of secreted industrial enzymes, the prevailing technique is genetic engineering. For the improvement of edible products and biocontrol agents for field release the non-recombinant approach involving traditional mutagenesis and screening or sexual and parasexual recombination (when applicable) may be chosen.

2. TECHNIQUES FOR FUNGAL STRAIN IMPROVEMENT

Genetic approaches to fungal strain development generally feature one of more of the following: (i) the use of chemical and physical mutagens to induce random mutations in the genome, (ii) application of parasexual and sexual reproduction to obtain novel recombinants and (iii) genetic engineering to introduce novel material in the fungal genome or to inactivate unvanted genes (Table 1). These methods can be applied either separately or in different combinations. The work of Dunn-Coleman *et al.* (1) where the production of calf chymosin in *Aspergillus niger* was brought up to a commercial level is an example of successful combination of serial mutagenesis and screening and gene cloning. A fifteen-fold increase in the production of the enzyme glucoamylase in *A. niger* has been achieved by multiplication of the copy number of the glucoamylase gene followed by somatic crossing of two multicopy transformants (2).

Table 1
Genetic approaches to fungal strain improvement

Technique	Effect on the genome
Sexual crossing	Fusion of specialized cells, diploid formation and meiotic recombination involving multiple cross-overs throughout the whole genome
Somatic crossing hyphal anastomosis protoplast fusion	Fusion of vegetative cells, karyogamy of somatic nuclei and mitotic recombination (parasexual cycle); involves the whole genome but there is normally only one cross-over per pair of chromosomes
Random mutagenesis and screening	Mutations are induced randomly in a (preferably haploid) genome by the use of chemical and physical mutagenic agents
Genetic engineering	Targets a particular (known) genetic determinant, can be applied to introduce novel properties into an organism

Although some genera (e.g. *Aspergillus*, *Claviceps* and *Emercicellopsis*) of industrially utilized filamentous fungi accommodate strains that possess a complete sexual cycle, for the majority of economically important fungi, sexual breeding is not available. However, sexual crossing remains as an important research tool with a view to applications within fungal genera, such as *Aspergillus*, that accommodate both sexually and asexually reproducing species.

In fungi which do not have a convenient natural mechanism for recombination of genetic material, protoplast fusion provides a method to facilitate heterokaryon formation, potentially leading to fusion of vegetative nuclei and mitotic recombination. Protoplast fusion can be conducted between strains belonging to same species (intraspecific recombination) or with some restrictions, between strains representing different species (interspecific recombination). Protoplast fusion has successfully been applied for the improvement of penicillin and cephalosporin yields (3,4,5).

2.1. Random mutagenesis

Before the advent of recombinant DNA technology, random mutagenesis and screening was the main technique applied for fungal strain improvement. A number of industrially relevant filamentous fungi (e.g *Trichoderma reesei*) produce haploid conidia that provide ideal material for mutagenic treatment: in the absence of a complementing set of genes mutations will be readily discovered by the use of a suitable screen and the mutant stability is generally good.

Mutation is the ultimate source of all genetic variation. The efficiency of induced mutation depends on the type of damage (base pair substitution, insertion, deletion etc.) a given mutagen causes on DNA and the cellular mechanism(s) involved in the repair of that damage. Among the most popular mutagens used for fungal strain improvement are DNA alkylating NTG (*N'*-methyl-*N'*-nitro-*N'*-nitrosoguanidine) which typically produces a variety of point mutations and UV irradiation which causes the formation of pyrimide dimers leading to point mutations and deletions (reviewed in 6 and 7). In addition to effectiveness, an important aspect in the application of chemical and physical mutagens is user safety. In this respect, rather than using long-lived, very potent chemical mutagens such as NTG that require the use of protective gloves, respiratory mask and neutralization upon disposal, physical mutagens such as UV that represent a lower risk to the user, may be chosen. In any case, a successful mutagenesis strategy should be based on the repeated (and rotated) application of one or more mutagenic agents of which the effectiveness to a particular organism has been tested to cover a broad range of mutant types.

Mutagenesis programs have been extremely successful in the improvement of secreted enzyme yields in *Aspergillus* (8, 9) and *Trichoderma* (reviewed in 10), and penicillin production in *Penicillium chrysogenum* (11). The published family trees for *T. reesei* high cellulase-producing mutants (12) and *P. chrysogenum* penicillin-producing mutants (11) span several decades, feature the rotational use of both chemical and physical mutagens and involve academic and industrial partners. As a result of the penicillin enhancement program, the yield was increased from 60 mg/L to 7000 mg/L and the best *T. reesei* cellulase producers today are capable of secreting up to 40 g/L of protein in their culture medium, most of which consists of cellulases (13). To obtain the best results, genetic strain improvement is routinely accompanied by process optimization which often contributes significantly to the productivity of novel strains. Currently, many high-secreting fungal mutant strains are used as hosts for recombinant protein production.

In the strain improvement programs discussed above that utilize fungal conidia as the material for mutagenic treatments, the genetic base of the high-yielding mutants lies in the genes of a single nucleus of the original parent strain. Filamentous fungi have proved to be surprisingly tolerant to strong mutagenic treatments retaining their vigour and productivity even after radical rearrangements of their chromosomes (14).

2.2. Genetic engineering

Recombinant DNA technology provides the tools for increasing the gene dosage and gene expression from strong promoters, deletion of unwanted genes from the fungal genome, manipulation of metabolic pathways and developing fungal strains for the production of heterologous proteins. Genetic transformation systems have been established for both ascomycetous and basidiomycetous fungi including gilled basidiomycetes (15). A suitable starting point for the development of a transformation protocol for any fungus can be found amongst the published methods. These include PEG-mediated transformation of protoplasts, electroporation, transformation of germinated conidia by incubation in the presence of lithium salt and microprojectile bombardment of intact conidia (reviewed in 16). The most recently described high frequency transformation of both fungal conidia and protoplasts uses the *Agrobacterium* T-DNA (17). "Global" transformation selection markers such as the *A. nidulans amdS* gene (18) and bacterial antibiotic resistance markers, e.g. *hph* (hygromycin B resistance; 19) have proven to be functional over a wide range of fungi. Methods for gene targeting and replacement have been developed (20,21) and a series of expression cassettes featuring strong promoters such as the *A. niger glaA* (glucoamylase; 22) and *T. reesei cbh1* (cellobiohydrolase I; 23) as well as transcription termination sequences have been constructed for the effective expression of a variety of proteins in filamentous fungi.

The majority of present day microbial industrial bulk enzymes are made using genetically engineered strains. Among these enzymes are glucoamylases for starch processing, cellulases for animal feed applications and the textile industry, xylanases for pulp bleaching and proteases for detergent use (24). In addition to high yield, the enzyme profiles of the strains may have been recombinantly modified to better suit a particular application (10, 25). For example, a fungal strain producing a xylanase preparation aimed for paper bleaching will overexpress xylanase enzymes but lack cellulase activity and a strain used for the manufacture of an enzyme preparation for animal feed improvement typically secretes elevated amounts of ß-glucanases but little cellobiohydolase.

Recombinant DNA technology can also be used to introduce site-directed or random mutations in fungal genes by the application of REMI (restriction enzyme mediated integration) and transposon mutagenesis or by PCR (polymerase chain reaction). PCR approach has proved successful in protein engineering studies that address the structure-function relationships in proteins and aim at evolving novel enzymes. Random evolution (26, 27) of industrial enzymes where mutant enzyme gene libraries are shuffled to increase genetic diversity in a population, and subsequent robotized screening for particular properties of the gene products, will lead to the isolation of enzymes that meet the specific needs of industrial processes. For example, an ideal alpha-amylase enzyme for detergent application should feature an oxidation-stable backbone, exhibit calcium independent performance and show stability and activity at highly alkaline pH 10.5 in the temperature range of 25-60°C (28). An enzyme fulfilling these requirements may not exist in nature or may have been eliminated in the course of natural evolution. However, by increasing genetic diversity and recombination *in vitro*, such an enzyme may be in the reach of an industrial geneticist. After initial screening using *E. coli*, evolved novel proteins would need to be produced on a large scale for industrial application.

At the moment, the most popular filamentous fungal hosts for the synthesis of homologous, heterologous and engineered gene products are found among genus *Aspergillus* (*A. niger*, *A. awamori*, *A. oryzae*), followed by *T. reesei*. Examples of fungal systems under development for heterologous gene expression include *Fusarium oxysporum*, *F. graminearum*, *Topylocladium geodes* (29), and *Penicillium chrysogenum* (30).

2.3. Screening methods

An effective screening regime is a prerequisite for the successful identification of mutants, novel recombinants and transformant cells exhibiting desired characteristics. Traditional screening on cultivation plates is based, for example, on the growth of fungal colonies on a substrate that the sought after enzyme(s) will degrade. On plate screening, excessive enzyme production is typically indicated by large clearing zones (halos) around the growing colony or by the formation of a coloured or fluoresceing product. Examples of substrates used for the screening of various hydrolytic enzyme activities are listed in Table 2.

Table 2
Substrates added into growth plates to indicate production of a particular hydrolase activity. All percentages are w/v with the exception of olive oil which is 2.5% (v/v)

Enzyme activity	Carbon source	Reference
Protease	2% glucose, 1% skim milk	31
Lipase	2.5% olive oil, 0.001% rhodamine B	32
Cellulase	0.5% Walseth cellulose	33
β-mannanase	0.5% locust bean gum	34
Xylanase	0.5% birch xylan	35
β-glucanase	0.1% AZCL-pachyman	35
Chitinase	0.3% chitin	36
Pectinase	0.1% pectin	37

Today, automated systems are often preferred over the more traditional and labour-intensive manual screening. Automated screening systems are typically based on the use of microtitre plates that are capable of accommodating 96 or more microbial cultures of which the activity can be assayed *in situ*. Inoculation of the microtiter plates is carried out by a robot. The enhancement of the production of calf chymosin in *A. niger* by Genencor International was one of the first published examples featuring robotized screening of mutant and transformant fungal colonies (1). Recently, a similar automated system has been applied to the screening of over 50,000 fungal strains produced by REMI mutagenesis. This automated screening successfully resulted in the isolation of transformant strains producing increased amounts of a heterologous lipase enzyme (38). In a subsequent investigation, a mutation in the *palB* gene encoding a cysteine proteinase enzyme was detected in the best lipase producing transformants.

Another approach into large scale screening, so far mainly applied for the analysis of non-filamentous organisms is the use of flow cytometry. Flow cytometers are employed to quantitatively measure the optical characteristics of particles, such as cells, passing in front of the light (usually laser) in single file. These instruments are capable of analyzing cells at rates of 40,000 per second. Modern flow cytometers can sort and physically collect individual particles based on predetermined light scattering and fluorescent characteristics. To enhance the ability of the instrument to discriminate between populations of cells, various fluorescent labelling subtrates are available (Molecular Probes Inc). Substrates such as βgalactoside resorufin that produces a fluorescent product when cleaved can be used for the detection of enzymatically active cells. Fluorescently labelled antibodies can also be utilized. Flow cytometry has been successfully applied for example, for the screening of pathogenic eukaryotic protozoa in water samples (39) and novel hybrids of baker's yeast (40).

The filamentous nature of many industrial fungi may not present an insurmountable problem for flow cytometry. We have shown that *T. reesei* can be succesfully separated into defined populations seven hours after germination (Figure 1, Bradner *et al.*, unpublished). Provided that the induction and synthesis of a given enzyme (or other protein of interest) can be achieved in that time frame, flow cytometry presents an intriguing possibility for the screening and sorting of filamentous fungal strains. Identification of novel fungal promoters functional under particular physiological conditions such as low or high pH or oxygen pressure could also be performed using flow cytometry.

3. FILAMENTOUS FUNGI AS HOSTS FOR HETEROLOGOUS GENE PRODUCTS

Filamentous fungi can satisfy even a most demanding biotechnologist as production hosts for large-scale industrial processes. Fungi thrive on relatively simple and cheap media, even on waste from other industrial processes such as surplus whey from cheese manufacture. High secreting mutants as well as recombinant strains have proven to be genetically stable in production processes. Furthermore, fungal cells are able to carry out post-translational modifications of proteins often relevant for their performance. However, while many fungi can secrete gram amounts of endogenous and heterologous fungal gene products in the culture medium, attempts to produce high yields of proteins of bacterial, plant and mammalian origin have, so far, been less successful. Some factors limiting foreign protein yields are listed in Table 3.

Table 3
Factors limiting the yield of heterologous products synthesized in filamentous fungi

Factor	Effect
Codon usage	No transcript/premature ending of transcription
Proteolytic processing	Incomplete proteolytic cleavage for activity
Protein folding	Incomplete or inaccurate folding leading to product degradation
Glycosylation	Incorrect glycosylation affecting structure, stability, activity and secretion
Proteolytic degradation	Intra-and extracellular host proteases attack the foreign gene product
Missorting	Incorrect or incomplete address tags result in mislocalization and subsequent degradation of a foreign gene product

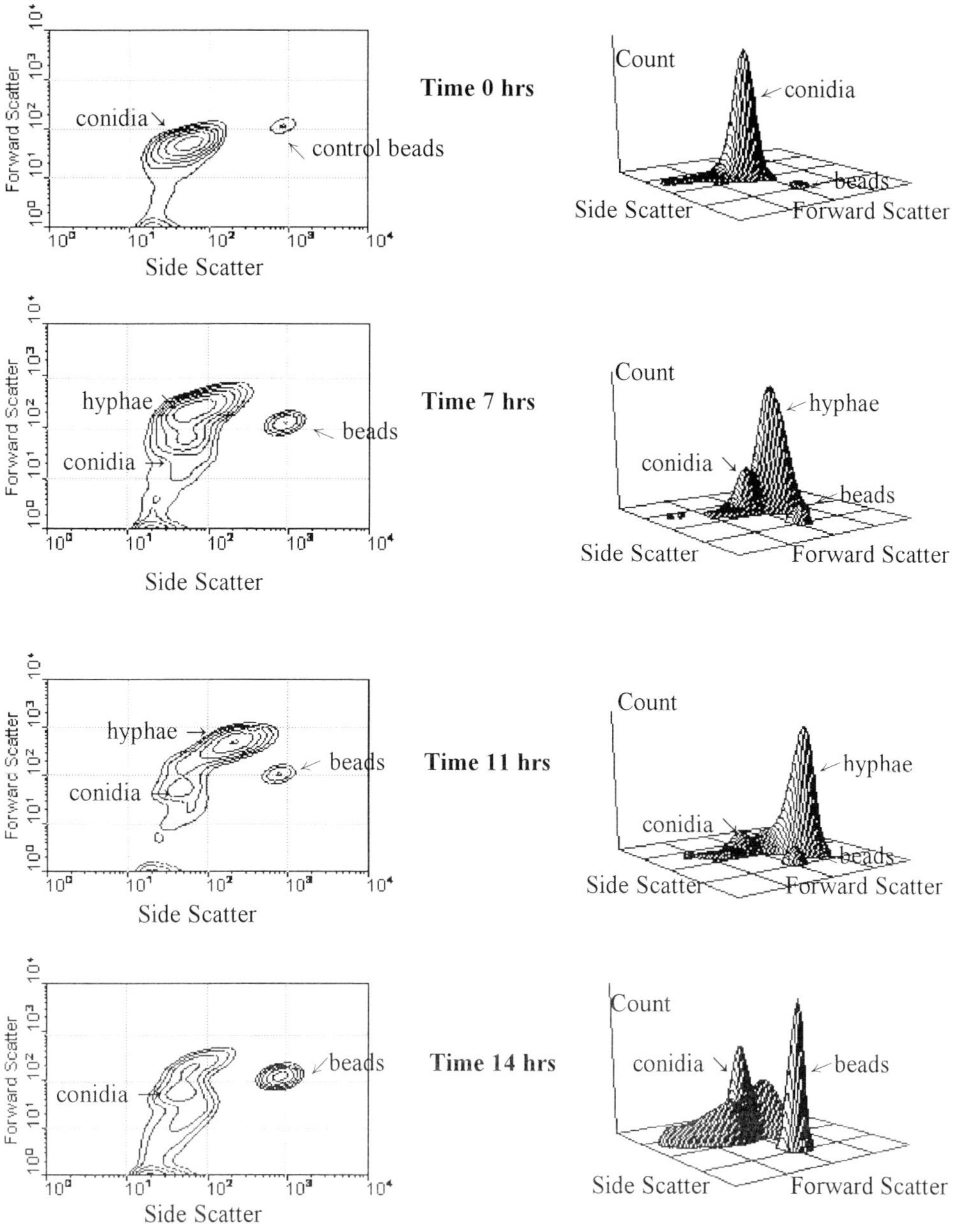

Figure 1. Progressive changes observed in *Trichoderma reesei* grown in liquid culture and viewed by flow cytometry. Contour plots (LHS) and 3-D images (RHS) clearly identify the conidial and hyphal populations. The developing population exhibiting hyphal growth can be clearly distinguished after 7 hours incubation. After 14 hours, hyphal growth has exceeded the limit of detection of the instrument.

3.1 Codon usage

Effectively expressed genes of filamentous fungi, such as cellulase-encoding genes of *Trichoderma reesei* exhibit a strong bias against NTA in their codons, where N can be any nucleotide (41). This bias is especially notable in the *cbh1* gene that codes for the main cellobiohydrolase enzyme CBHI which represents approximately 50% of all protein secreted by *T. reesei*. Different organisms and genes exhibit different preferences and therefore mismatched codon usage between the incoming gene and the expression host can lead to problems like incorrect processing of pre-mRNA and low mRNA stability. These have not yet been studied in great detail in filamentous fungi (reviewed in 42).

It has been shown that in addition to proteins encoded by particular bacterial genes such as *hph, lacZ* and *gusA* (43), gene products following yeast, plant, murine and mammalian codon usage can be expressed in filamentous fungi (e.g. 44, 45, 46, reviewed in 47). Even though codon optimisation has not been a major focus in the expression of heterologous genes in filamentous fungi, it has proved necessary where the codon usage between two organisms differs significantly. For example, in the xylanase B (*xynB*) gene of thermophilic bacterium *Dictyoglomus thermophilum*, the overall AT content is 61% whereas in a typical *T. reesei* cellulase gene it is under 40%. In addition, *D. thermophilum* favors A or T at the wobble position but the effectively expressed *T. reesei* genes discriminate against A and T at the third codon position. In order to achieve expression of the *xynB* product in *T. reesei*, the entire 1014 bp long *xynB* gene was reconstituted by primer extension PCR according to the codon preference of *Trichoderma* (48). To our knowledge, *xynB* is the first synthetic bacterial gene for which the codon usage has been optimized to better suit expression in a fungal host.

3.2 Protein processing, folding and localization

A widely applied strategy for the synthesis of heterologous non-fungal proteins in filamentous fungi is to fuse the foreign gene to the 3' end of a highly expressed endogenous gene that encodes an efficiently secreted product. Fungal genes successfully used as fusion partners include *A. niger* and *A. awamori* glucoamylase (*glaA*) and *T. reesei* cellobiohydrolase 1 (*cbh1*). Both glucoamylase and cellobiohydrolase I are composed of an N-terminal catalytic domain and a C-terminal substrate binding domain joined together by a linker region. In most fusion experiments the C-terminal binding domain has been replaced by the foreign gene to be expressed in the fungal host. The N-terminal fungal fusion partner has been suggested to stabilize the recombinant mRNA, serve as a carrier to facilitate the translocation of foreign proteins in the secretory pathway and protect the heterologous part from degradation. Heterologous products synthesized using the gene fusion approach include bovine chymosin (49, 50), porcine pancreatic phospholipase (51), human interleukin-6 (52, 53, 54), hen egg-white lysozyme (55), human lactoferrin (56) and murine antibody Fab fragments (46). The fusion proteins may be cleaved by autocatalytic processing (49, 50), yet unidentified fungal protease(s) (51, 46) or a KEX2 like protease, for which a recognition site can be introduced in the expression vector (52, 53, 56). A schematic presentation of an expression vector designed for protein fusions is shown in Figure 2. Even when synthesized as fusion products, the yields of non-fungal heterologous proteins have remained 10-100 times lower when compared to effectively secreted fungal proteins indicating the presence of other bottlenecks at post-translational level.

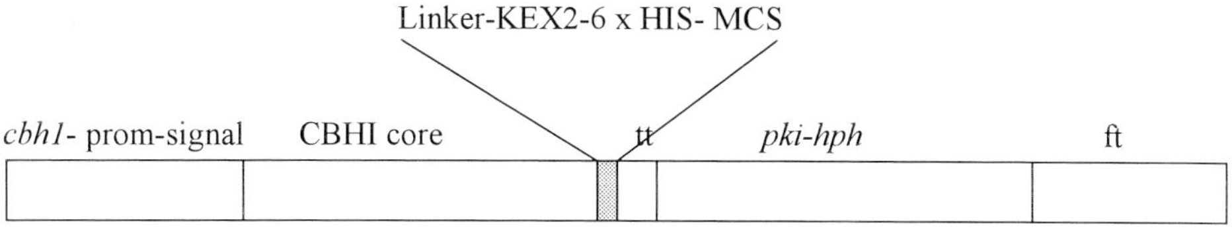

Figure 2. Schematic presentation of an expression vector designed for the synthesis of heterologous gene products, based on the use of parts of the *T. reesei cbh1* gene and protein. Prom- signal denotes the *cbh1* promoter and leader sequence; tt is a truncated *cbh1* terminator; 6 x HIS tag is inserted for product purification; MCS is a multicloning site for genes to be expressed; *pki-hph* denotes the bacterial hygromycin resistance gene under the *T. reesei pki* (pyruvate kinase) promoter and ft is the full length *cbh1* terminator.

The intracellular quality control, initiated by inefficient translation, ineffective or incorrect protein translocation, incomplete or inaccurate folding, proteolytic processing or incomplete glycosylation evidently works effectively in clearing the system of defective gene products. In addition to the gene-fusion approach discussed above, other strategies have been proposed to improve the yield of heterologous proteins from filamentous fungi. These include isolation and overexpression of cellular foldases such as disulphide isomerase (PDI; *pdiA* gene family) and chaperones like BiP (*hsp70* gene family) that has proved successful in increasing the secretion of intracellularly accumulated foreign proteins in *S. cerevisiae* (57, 58). However, attempts to apply a comparative strategy for filamentous fungi, especially in black *Aspergilli*, have so far failed to produce expected results (59, 60, 61). It seems likely, that cooverexpression of several foldases/chaperones would be needed to facilitate folding and secretion of foreign proteins from mycelial fungi. Some prokayotic enzyme proteins, especially lipases, possess enzyme-specific chaperones that are required for the correct folding of a particular protein and are located adjacent to the protein encoding sequences (62). Such specifc chaperones have not been found in fungi so far, however, it has been suggested that N-terminal sequences of a *Rhizopus delemar* lipase enzyme are essential for the correct folding and processing of the protein into its active form (62) thus carrying out a chaperone-type function.

3.3 Glycosylation

Studies into the detailed mode of glycosylation of proteins produced by filamentous fungi represent a relatively new research area; consequently, the complete oligosaccharide composition is known for only a limited number of fungal proteins. In general, filamentous fungi produce high mannose type N-glycans (Man_{5-9} $GlcNac_2$, reviewed in 63) and are also capable of effective O-glycosylation. Recently, NMR spectroscopy has been applied for detailed structural characterisation of the N-linked oligosaccharides on the *Trichoderma reesei* main cellobiohydrolase I enzyme (64, 65). Advances in glycoprotein microsequencing facilitating the sequencing through highly O-glycosylated regions (66) have resulted in the characterization of the complete glycosylation of the CBHI protein of *T. reesei* (67) that features an N-glycosylated catalytic core and a heavily O-glycosylated linker peptide. There is evidence that O-glycosylation of the linker may protect the peptide against proteolysis (68) and that the O-linked glycosylation (but not the N-linked) is necessary for the secretion of two endoglucanases, EGI and EGII, from *T. reesei* (69).

Authentic glycosylation of gene products is of particular importance when contemplating the expression of heterologous glycosylated human proteins in filamentous fungi for therapeutic uses. Potential problems may arise from nonauthentic glycosylation leading to recognition and clearing of foreign proteins from the blood, antigenicity provoked by noncustomary glycosidic substituents and the yet unknown effects of mannose linked phosphate groups in fungally produced proteins. An example of ongoing approaches towards modification of fungal glycosylation is the cloning of the UDP-GlcNAc:α-3-D-mannoside ß-1,2 acetylglucosaminyltransferase I into the glycosylation pathway to allow for the synthesis of more complex glycans (70,71) without compromising further processing and correct translocation of proteins.

Many secreted fungal proteins such as *Trichoderma* cellulases and chitinases, *Aspergillus* invertase and and *Fusarium oxysporum* pectinases are synthesized as isoforms varying in the glycan composition and length (reviewed in 63). This variable glycosylation does not, in general, affect enzyme activity. It has also been shown that strains of *T. reesei* glycosylate their endogenous cellobiohydrolase I (64, 65, 67) and a heterologous *A. niger* phytase differently (72) pointing to the existence of genetic differences in the glycosylation pathway. Learning more about the molecular biology of fungal glycosylation will assist in modification of sugar moieties and promote successful matching of suitable strains with particular heterologous proteins that may require certain type of glycosylation. Finally, recombinant introduction of an N-glycosylation site to a protein which is not naturally glycosylated may increase its secretion as shown with calf chymosin produced in *A. niger* (20) and an unsuitable location of the glycosylation may interfere with the specific activity (20) or processing (73) of a given heterologous protein.

3.4 Proteolytic degradation

Protein folding, processing, glycosylation and translocation are functions of the secretory pathway. Proteolytic degradation occurs in the intracellular environment, the cell wall and outside the cell, in the growth medium. Thus, in order to overcome generally recognised bottlenecks in the expression and secretion of foreign proteins from filamentous fungi, a better understanding of the secretion kinetics and the protease makeup will be necessary. It has been shown that fungal proteases are responsible for truncated forms of cellulase enzymes in the cultivation medium (74) and proteases also degrade heterologous proteins produced in filamentous fungi. Studies into the proteolytic system of *Aspergillus niger* have revealed a series of proteases and their genes including metallo-, serine- and aspartyl proteases (75).

A majority of problems related to degradation of heterologous proteins in *Aspergillus* are caused by acid (aspartyl) proteases, the production of which is often induced by the growth medium used for the synthesis of foreign proteins. Some strategies into strain improvement aimed at reduction or elimination of the harmful effects of fungal proteases feature the isolation of protease negative mutants by random mutagenesis and screening (76, 77, 31), disruption of the major extracellular protease gene (*pepA* in *Aspergillus*, *pep1* in *Trichoderma*; 78,31), isolation of multiple protease deficient strains by disruption of several protease genes, isolation of wide domain regulatory mutations and modification of growth medium to reduce protease synthesis (reviewed in 75). It has been demonstrated that disruption of the major protease gene *pepA* in *A. awamori* (78) and *A. niger* (76) reduced the amount of extracellular PEPA protein to about one fifth of the wild type level. Recently, a *T. reesei* mutant entirely deficient of extracellular acid protease activity has been isolated by

UV-mutagenesis (Nevalainen *et al.*, unpublished). The use of low protease strains as expression hosts has led to significant improvement in the yields of heterologous proteins (79, 80, 51, 53).

Recently, it has been proposed that a cell wall-associated protease is involved in the authentication of secreted proteins and clearing of the misfolded proteins from the vicinity of the translocation complex in *Bacillus subtilis* (81). Such a mechanism would have direct implications in the yield of heterologous proteins synthesized in any given microbial host. The observation that a mammalian calf chymosin, when produced without fusion to a native *Trichoderma* protein, is retained (and possibly degraded) in the cell wall may suggest the presence of a functionally similar protease in the fungal cell wall (Nykänen *et al.*, in preparation). The 2-dimensional separation and characterization of proteins associated with the fungal cell wall would provide the means of targeting the proteins of specific interest and enable the subsequent cloning of the genes which encode them (82). Novel genes coding for cell wall proteins will provide excellent tools for basic research and for the development of fungal hosts more effectively secreting both homologous and heterologous proteins.

4. FUNGI IN FOOD AND ENVIRONMERNTAL APPLICATIONS

The use of recombinantly prouced enzymes in food and animal feed applications such as baking (xylanase, α-amylase, lipase), brewing (decarboxylase), and poultry feed (enzyme preparations enriched with β-glucanase activity) is a general practice today. The tests and methods used in a safety evaluation of industrial enzymes are in general based on the guidelines formulated by internationally recognized authorities such as Organization for Economic Cooperation and Development (OECD), Joint FAO/WHO Expert Committee on Food Additives (JECFA), European Economic Community (EEC) and the Food and Drug Administration of the USA (FDA) (reviewed in 83). Regulations concerning genetically modified microorganisms are documented in international agreements, directives and recommendations and, at the national level, by legislation or guidelines concerning such aspects as the labelling of recombinant products and deliberate release of recombinant microorganisms. However, some regulations concerning particular food applications and the deliberate release of genetically modified microorganisms in the environment may vary in between countries. Also, the public perception concerning these issues may be somewhat sensitive. Thus, strain development programs targeting fungal products intended for dissemination in the environment (biocontrol and bioremediation agents) and direct human or animal consumption (fodder yeast, mushrooms and fungal cell protein) will require the consideration of these regulations and consultation with authorities as an integral part of the process.

4.1 Mushroom breeding

Button mushrooms are among the oldest cultivated crops and their economical importance as health food high in protein, low in fat, is continuously increasing. Notwithstanding the commercial importance, the approaches to genetically improve the strains used for commercial production have been limited. Traditional breeding of *Agaricus bisporus* has suffered from the natural barriers in the mushroom's life cycle and a low level of genetic diversity (84). As with many other industrially exploited fungi, the system would benefit from the introduction of novel gene material originating from the wild. This could be achieved, for example, by producing intraspecies hybrids of *Agaricus* by protoplast fusion (85). Studies on transposable *Agaricus* mitochondrial plasmids and the application of genetic transformation techniques represent other avenues for mushroom strain improvement (reviewed in 86).

4.2 Biological control

Main emphasis in the development of fungal biocontrol products has so far been in product formulation. Fungal strains presently used for the biocontol of insects (e.g. *Metarhizium anisopliae*) and plant pathogenic fungi (e.g. *Trichoderma harzianum*) are those isolated from the nature and have not been subjected to intensive strain development programs. Recent research into understanding the molecular mechanisms of antagonistic interactions and isolation of genes encoding lytic enzymes, such as chitinases and proteases, suggested to have a role in biological control, have paved the way for genetic engineering of novel biocontrol strains (87). For example, overexpression of selected genes for synergistic combinations of lytic enzymes in a given fungal strain without compromising the vigour and ecological competence of the fungus is technically possible and will present a considerable leap forward in the war against pathogens causing significant economical losses. A value added effect of biological control, also attractive to the general public, is the reduction of environmental pollution by the replacement of toxic chemicals by benign biological agents.

5. FUTURE DEVELOPMENTS

The next advancements in fungal strain improvement will emerge from research into the bottlenecks in heterologous protein production and from fungal genome and proteome studies. The rapid progress of genome scale sequencing programs has spawned a new field of functional genomics. Projects on functional genomics on filamentous fungi are well underway at several universities and commercial organizations, involving, for example, *Aspergillus*, *Neurospora, Fusarium*, *Magnaporthe* and *Metarhizium* genomes. Rapid determination of the entire catalogue of genes that constitute the physiological capacity of a particular fungus will lead to the identification of novel genes for various regulatory functions, commercially interesting products and metabolic pathways. Proteome studies addressing the protein complement of a genome by the application of two-dimensional gel electrophoresis and mass spectrometry for the identification of proteins complement the genome studies by adding a new dimension in the network of biological information (88). The wealth of data resulting from these research programs will provide an enormous information base for holistic approaches into fungal strain improvement.

REFERENCES

1. N.S. Dunn-Coleman, P. Bloebaum, R.M. Berka, E. Bodie, N. Robinson, G. Armstrong, M. Ward, M. Przetak, G.L. Carter, R. LaCost, L.J. Wilson, K.H. Kodama, E.F. Baliu, B. Bower, M. Lamsa and H. Heinsohn, Bio/Technol., 9 (1991) 976.
2. D.B. Finkelstein, J. Rambosek and M.S. Crawford, In C.L. Hershberger, S.W. Queener and G. Hegeman (eds.), Genetics and Molecular Biology of Industrial Microorganisms, American Society for Microbiology, Washington (1989) 295.

3. C. Ball, In Z. Vanek, Z, Hostalek and J. Cudlin (eds.), Genetics of Industrial Microorganisms, Elsevier Publishing Company, Amsterdam (1973) 227.
4. G.B. Bradshaw and J.F. Peberdy, Enzyme Microb. Technol., 6 (1984) 121.
5. P.F. Hamlyn and C. Ball, In O.K. Sebek and A.I. Laskin (eds.),Genetics of Industrial Microorganisms, American Society for Microbiology, Washington DC (1979) 185.
6. R.T. Rowlands, Enzyme Microb. Technol., 6 (1984) 3.
7. R.T. Rowlands, In D.B. Finkelstein and C. Ball (eds.), Biotechnology of Filamentous Fungi, Butterworth-Heineman, USA (1992) 41.
8. Y.K. Park and M.S.S. De Santi, Induction of high amyloglucosidase-producing mutant from *Aspergillus awamori*, J. Ferment. Technol., 55 (1977) 193.
9. D.B. Finkelstein, Antonie van Leeuwenhoek, 53 (1987) 349.
10. A. Mäntylä, M. Paloheimo and P. Suominen , In G.E. Harman and C.P Cubicek (eds.), *Trichoderma* and *Gliocladium*, Basic Biology, Taxonomy and Genetics, Vol.2, Taylor and Francis Ltd., UK (1998) 291.
11. Y. Aharonowitz and G.Cohen, Sci. Amer. Sep., (1981) 141.
12. H. Nevalainen, P. Suominen and K. Taimisto, J. Biotechnol., 37 (1994) 193.
13. H. Durand, M. Clanet and G. Tiraby, Enzyme Microb. Technol., 10 (1988) 341.
14. A.L. Mäntylä, K.H. Rossi, S.A. Vanhanen, M.E. Penttilä, P.L. Suominen and K.M.H. Nevalainen, Curr.Genet., 21(1991) 471.
15. D.B. Finkelstein, In D.B. Finkelstein and C. Ball (eds.), Biotechnology of Filamentous Fungi, Technology and Products, Butterworth-Heinemann, USA (1992) 113.
16. R. Mach and S. Zeiliger, In C.P Cubicek and G.E. Harman (eds.), *Trichoderma* and *Gliocladium*, Basic Biology, Taxonomy and Genetics, Vol.1, Taylor and Francis Ltd., UK (1998) 225.
17. M.J.A. de Groot, P. Bundock, P.J.J. Hoykaas and A.G.M. Beijersbergen, Nature Biotechnol., 16 (1998) 839.
18. J. M. Kelly and M.J. Hynes, EMBO J., 4 (1985) 475.
19. P. Punt, R.P. Oliver, M.A. Dingemanse, P.H. Pouwels and C.A.M.J.J. van den Hondel, Gene, 56 (1987) 117.
20. M. Ward, In H. Nevalainen and M. Penttilä (eds.), Molecular Biology of Filamentous Fungi, Foundation for Biotechnical and Industrial Fermentation Research, Vol. 6 (1989) 119.
21. T. Karhunen, A. Mäntylä, H. Nevalainen and P. Suominen, Mol.Gen.Genet., 241 (1993) 515.
22. M. Ward, In C.L. Hershberger, S.W.Queener and G. Hegeman (eds.), Genetics and Microbiology of Industrial Microorganisms, American Society for Microbiology, Washington DC (1989) 288.
23. H. Nevalainen, M. Penttilä, A. Harkki, T. Teeri and J. Knowles, In . S. A. Leong and R. M. Berka (eds.), Molecular Industrial Mycology, Marcel Dekker Inc., New York (1991) 129.
24. http://www.novo.dk/backgrou/position/list.htm
25. A. Harkki, A. Mäntylä, S. Muttilainen, R. Bühler, P. Suominen, J. Knowles and H. Nevalainen, Enzyme Microb.Technol., 13 (1991) 227.
26. K. Chen and F.H. Arnold, Proc. Nat. Acad. Sci. USA 90 (1993), 5618.
27. F.H. Arnold, In IBC Directed Enzyme Evolution, San Diego, CA (1998).
28. S. Carlsen, In IBC Enzyme Technologies, San Francisco, CA (1999).
29. M. Baron and G. Tiraby, J. Biotechnol., 24 (1992) 253.

30. S. Graessle, H. Haas, E. Friedlin, H. Kürnsteiner, G. Stöffler and B. Redl, Appl. Environ. Microbiol., 63 (1997) 753.
31. A. Mäntylä, R. Saarelainen, R., Fagerström, P. Suominen, and H. Nevalainen, In 2nd European Conference on Fungal Genetics, Lunteren, The Netherlands (1994), Abstract B52.
32. Kouker, G. and K. Jaeger, Appl. Environ. Microbiol., 53 (1987) 211.
33. C.S. Walseth, Tappi J., 35 (1952) 228.
34. M. Rättö and K. Poutanen, Biotechnol. Lett., 10 (1988) 661.
35. J. R. Bradner, M. Gillings and K. M. H. Nevalainen, World. J. Micobiol., (1999) in press.
36. M. O'Brien and R.R. Colwell, Appl. Environ. Microbiol., 53 (1987) 1718.
37. A.E. Hagerman, D.M. Blau and A.L. McClure, Anal. Biochem., 151 (1985), 334.
38. D. Yaver, In The 20th Fungal Genetics Conference, Asilomar, Pacific Grove, CA (1999), Abstract 119.
39. G. Vesey, P. Hutton, A. Champion, N. Ashbolt, K.L. Williams, A. Warton and D. Veal, Cytometry 16 (1994)
40. P.J. L. Bell, D. Deere, J. Shen, B. L. Chapman, P. V. Attfield, P. H. Bissinger, and D. Veal, Appl. Environ. Microbiol., 64 (1998).
41. T. Teeri, P. Lehtovaara, S. Kauppinen, I. Salovuori and J. Knowles, Gene, 51 (1987), 43.
42. R.J. Gouka, P.J. Punt and C.A.M.J.J. van den Hondel, Appl. Microbiol. Biotechnol., 47 (1997) 1.
43. S. Menne, M. Waltz and U. Kück, Appl. Microbiol. Biotechnol., 42 (1994), 57.
44. R.J. Gouka, P.J. Punt, J.G.M. Hessing and C.A.M.J.J. van den Hondel, Appl. Environ. Microbiol., 62 (1996) 1951.
45. R. Saarelainen, A. Mäntylä, H. Nevalainen and P. Suominen, Appl. Environ. Microbiol., 63 (1997) 4938.
46. E. Nyyssönen, M. Penttilä, A. Harkki, A. Saloheimo, J. Knowles and S. Keränen, Bio/Technol., 11 (1993) 591.
47. R. Radzio and U. Kück, Process Biocem., 32 (1997) 529.
48. V.S. Te'o, A. Ciferszky, P.L. Bergquist and H. Nevalainen, In The 20th Fungal Genetics Conference, Asilomar, Pacific Grove, CA (1999), Abstract 369.
49. A. Harkki, J. Uusitalo, M. Bailey, M. Penttilä and J.K.C. Knowles, Bio/Technology, 7 (1989) 596.
50. M. Ward, L.J. Wilson, K.H. Kodama, M.W. Rey and R.M. Berka, Bio/Technology, 8 (1990) 43.
51. I.N. Roberts, D.J. Jeenes, D.A. MacKenzie, A.P. Wilkinson, I.G. Summer and D.B. Archer, Gene, 122 (1992) 155.
52. R. Contreras, D. Carrez, J.R. Kinghorn and C.A.M.J.J. van den Hondel, Bio/Technology 9, (1991) 378.
53. M.P. Broekhuijsen, I.E. Mattern, R. Contreras, J.R. Kinghorn and C.A.M.J.J. van den Hondel, J. Biotechnol., 31 (1993) 135.
54. J. Demolder, X. Saelens, M. Penttilä, W. Fiers and R. Contreras, In 2nd European Conference on Fungal Genetics, Lunteren , The Netherlands, Abstract B38.
55. D.J. Jeenes, D.A. Marckzinke, D.A. MacKenzie, and D.J. Archer, FEMS Microbiol. Lett., 107 (1993) 267.
56. P.P. Ward, C.S. Piddington, G.A. Cunningham, X. Zhou, R.D. Wyatt and O.M. Conneely, Bio/Technology, 13 (1995) 784.
57. K.D. Wittrup and A.S. Robinson, WO 94/08012 (1994).

58. A.S. Robinson, V. Hines and K.D. Wittrup, Bio/Technology, 12 (1994) 444.
59. P.J. Punt, I.A. van Gemeren, J. Drint-Kuijvebhoven, J.G.M. Hessing, G.M. van Muijlwijk-Harteveld, A. Beijersbergen, C.T. Verrips and C.A.M.J.J. van den Hondel, Appl. Environ. Microbiol., 50 (1998) 447.
60. I.A. van Gemeren, A. Beijersbergen, C.A.M.J.J. van den Hondel and C.T. Verrips, Appl. Environ. Microbiol., 64 (1998) 2794.
61. H. Wang, In The 20th Fungal Genetics Conference, Asilomar, Pacific Grove, CA (1999).
62. K.E. Jaeger, S. Ransac, B.W. Dijkstra, C. Colson, M. van Heuvel and O. Misset, FEMS Microbiol. Rev., 15 (1994) 29.
63. D.B. Archer and J.F. Peberdy, Crit. Rew. Biotechnol., 17 (1997), 273.
64. M. Maras, A. deBruyn, J. Schraml, P. Herdewijn, M. Claeyssens, W. Fiers and R. Contreras, Eur. J. Biochem., 245 (1997) 617.
65. K. Klarskov, , J. Piens, J. Ståhlberg, P. B. Høj, J. van Beeumen and M. Claeyssens, Carbohyd. Res., 304 (1997) 143.
66. A.A. Gooley and K.L. Williams, Nature, 385 (1997) 557.
67. M. Harrison, A.S. Nouwens, D.R. Jardine, N.E. Zachara, A.A. Gooley, H. Nevalainen and N.H. Packer, Eur. J. Biochem. 256 (1998), 119.
68. H. Nevalainen, M. Harrison, D. Jardine, N. Zachara, M. Paloheimo, P. Suominen, A. Gooley and N. Packer, In M. Claeyssens, Nerinckx and K. Piens (eds.), Carbohydrases from *Trichoderma reesei* and Other Microorganisms. Structures, Biochemistry, Genetics and Applications, The Royal Society of Chemistry, Thomas Graham House, Cambridge, UK (1998), 335.
69. C. P. Kubicek, T. Panda, G. Schreferi-Kunar, F. Gruber and R. Messner, Can. J. Bot., 33 (1987) 698.
70. W.E. Hintz, Kalsner, I., Plawinski, E., Guo, Z.M. and P.A. Lagosky, Can. J. Bot., 73 (1995) Suppl. 1E-H: S876.
71. M. Maras, A De Bruyn, J. Schraml, P. Hederwijn, K. Piens, M. Claeyssens, J. Uusitalo, M. Penttilä, W. Fiers and R. Contreras, In M. Claeyssens, Nerinckx and K. Piens (eds.), Carbohydrases from *Trichoderma reesei* and Other Microorganisms. Structures, Biochemistry, Genetics and Applications, The Royal Society of Chemistry, Thomas Graham House, Cambridge, UK (1998), 323.
72. H. Nevalainen, M. Paloheimo, A. Miettinen-Oinonen, T. Torkkeli, M. Cantrell, C. Piddington and J. Rambosek, PCT WO94/03612 (1994).
73. M. Nykänen, R. Saarelainen, M. Raudaskoski, H. Nevalainen and A. Mikkonen, Appl. Environ. Microbiol., 63 (1997) 4929.
74. K. Hagspiel, D. Haab and C. Kubicek, Appl. Microbiol. Biotechnol., 32 (1989)
75. J.P.T.W. van den Homberg, P.J.I. van de Vondervoort, L. Fraissinet-Tachet and J. Visser, TIBTECH, 15 (1997) 256.
76. I.E. Mattern, J.M. van Noort, P. Berg, D.B. Arst, I.A. Roberts and C.A.M.J.J. van den Hondel, Mol. Gen. Genet., 234 (1992) 332.
77. J.P.T.W. van den Homberg, P.J.I. van de Vondervoort and N.C.B.A.van der Heijden, Curr Genet., 28 (1995) 299.
78. R.M. Berka, M. Ward, L.J. Wilson, K.J. Hayenga, K.H. Kodama, L.R. Carlomagno and S.A. Thompson, Gene, 86 (1990) 153.
79. R.M. Berka, F.T. Bayliss, P. Bloebaum, D. Cullen, N.S. Dunn-Coleman, K.H. Kodama, K.J. Hayenga, R.A. Hitzeman, M.A. Lamsa, M.M. Przetak, M.W. Rey, L.J. Wilson and

M. Ward, In L.W. Kelly and T.O. Baldwin (eds.), Applications of Enzyme Biotechnology, Plenum Press, New York (1991) 273.
80. R. Berka, K. Hayenga, V.B. Lawlis and M. Ward, EP No.0 429 490 B1 (1991).
81. K. Stephenson and C.R. Harwood, Appl. Environ. Microbiol., 64 (1998) 2875.
82. H. Nevalainen, In Luiz Pereira Ramos (ed.), Proceedings of the Fifth Brazilian Symposium on the Chemistry of Lignins and Other Wood Components, Curitiba, Brazil, (1997) 275.
83. H. Nevalainen and D. Neethling, In C.P Kubicek and G.E. Harman (eds.), *Trichoderma* and *Gliocladium*, Basic Biology, Taxonomy and Genetics, Vol.1, Taylor and Francis Ltd., UK (1998) 193.
84. C.A. Raper, R.E. Miller and J.A. Raper, Mycologia, 64 (1972) 1088.
85. A.J. Castle, P.A. Horgen and J.B. Anderson, Appl. Environ. Microbiol., 54 (1988) 1643.
86. P.A. Horgen and J.B. Anderson, In D.B. Finkelstein and C. Ball (eds.), Biotechnology of Filamentous Fungi, Technology and Products, Butterworth-Heinemann, USA (1992) 447.
87. I. Chet, N. Benhamou and S. Haran, In G.E. Harman and C.P Kubicek (eds.), *Trichoderma* and *Gliocladium*, Basic Biology, Taxonomy and Genetics, Vol.2, Taylor and Francis Ltd., UK (1998) 153.
88. W.P. Blackstock and M.P. Weir, TIBTECH, 17 (1999) 121.

Applied Mycology and Biotechnology
Volume 1. Agriculture and Food Production
G.G. Khachatourians and D.K. Arora (editors)

Fungal solid state fermentation - an overview

M.K. Gowthaman[a], Chundakkadu Krishna[b] and M.Moo-Young[a]

[a] Department of Chemical Engineering, University of Waterloo, Waterloo, ON, N2L 3G1, Canada

[b] Department of Biosystems & Agricultural Engineering, University of Kentucky, Lexington, KY, 40546-0276, USA

1. INTRODUCTION

Fungal fermentation metabolites have an enormous impact on the food, pharmaceutical and agricultural industries [1, 2]. The basis of production of fungal products has largely been by liquid-based Submerged Fermentation (SmF). At present, interest in Solid State Fermentation (SSF) in the USA is small compared to the oriental countries where fungi are exploited for their natural affinity for moist environments. Japan has always been a trendsetter in the area of SSF, and has some of the largest SSF production plants in the world. Recently, the West is trying to copy Japan in this respect as it has been done in the fields of microelectronics and automobile manufacturing.

SSF can be defined as the growth of microorganisms on moist, water-insoluble solid substrates in the absence or near-absence of free liquid [3]. In addition to the above mentioned benefits for fungi, SSF has several potential advantages over SmF such as low capital cost, low energy expenditure, less expensive downstream processing, low wastewater output and potential higher volumetric productivity. For convenience, FSSF will be used as an abbreviation for fungal solid state fermentation in this chapter. An overview of the various factors which affect the development and application of FSSF systems with reference to the growing body of literature will be given.

2. PROCESS PROTOCOL FOR FSSF

General guidelines for FSSF methodologies have been given by Lonsane et al [4] as follows:

1) Inoculum preparation - generally spores raised on the actual substrate
2) Substrate preparation including size reduction, nutrient addition and pH adjustment

• MKG acknowledges the grant of a Biotechnology Overseas Associateship by the Department of Biotechnology (DBT), India.

3) Autoclaving to sterilize/pasteurize and cook the medium for increased amenability to fungal growth
4) Inoculation of the moist solid medium
5) Incubation under near optimal conditions in suitable reactor systems
6) Drying of the solids and extraction of product(s).
7) Final steps involving filtration, concentration/purification.

3. FACTORS AFFECTING FSSF:

3.1 Inoculum type

Several advantages of using spores have been cited for the common use of spore rather than vegetative cells for inoculum. These are convenience, ease of inoculum preparation, prolonged storability for subsequent use and higher resistance to mishandling during transfers [5]. They do have however, a few disadvantages such as longer lag time [6], different optimal conditions for spore germination and vegetative growth and larger inoculum size requirement. Mycelial inoculum has been found to yield higher protein content in a fermentation system of wheat straw and *Chaetomium cellulolyticum*, due to instant availability of the necessary enzymes [7]. Inoculum density is another important factor in FSSF (Table 1).

Table 1
Effect of inoculum density in FSSF

Organism	Substrate	Spores/cells per g substrate	Effect	Reference
R. oryzae	Cassava flour	10^7	Poor mycelial devt	8
		10^5	Heavy sporulation	
		10^6	Maximum yield	
A. niger	Cassava meal	10^6 - 10^7	High growth rates	9
		10^8	Low growth rates	
P. chrysosporium	Lignin	2 X 10^7	Longer lag	10
A. koppan		10^7	Shorter lag & rapid growth	11
S. cerevisiae	Sweet sorghum	4 X 10^6	Max. EtOH production	12
R. oryzae	Rice bran	10^2 - 10^7	No effect on protease	13

3.2. Water activity

Fungi are well known to favor moist environments for their growth. An optimum moisture level however has to be maintained since, lower moisture tends to reduce nutrient diffusion, microbial growth, enzyme stability and substrate swelling [3,4]. Higher moisture levels lead to particle agglomeration, gas transfer limitation and competition from bacteria. Different fungi have different optimal moisture requirement and this characteristic also varies with the substrate nature. As an example, the optimum moisture level for growth of *A. oryzae* on rice was 40% [14] while it was 80% on coffee pulp [15]. A more appropriate term, water activity is used to describe the moist environment. This is defined by the ratio of the vapour pressure of the water in the substrate (p) to the vapor pressure of pure water (p_0) at the same temperature [16].

i.e., $aw = p / p_0$; Also, $a_w = \% RH / 100$

Most fungi capable of SSF have minimum a_w values between 0.8 and 0.9 [17, 18].

Table 2 gives typical values of water activity in FSSF systems. The optimum water activity may be different for growth and product formation [19] therefore, this condition offers the possibility of manipulating the water activity during fermentation [14]. The moisture content of the substrate largely depends on the water activity and small changes in the latter could have a great effect on the former [20]. Fermentation also leads to increase in water activity of the substrate due to co-production of water but still, may not help to maintain the moisture levels in the substrate due to evaporative losses [21]. Hence, control of moisture levels assumes great significance. Laukevics et al [22] suggest the use of relative humidity of air for moisture regulation.

Table 2
Typical levels of water activity in FSSF systems

Microorganism	Substrate	Product	Water activity	Reference
Trichoderma viride	Sugar beet pulp	beta-xylanase	0.995	19
Candida rugosa	Coconut oil cake	lipase	0.92	23
P. roquefortii	---	growth & sporulation	0.96	24
A. oryzae	Various Substrates	protease	0.982 - 0.986	25
R. oligosporus	Chickpea	--	0.92	26

3.3. pH requirements of fungal growth

Filamentous fungi have reasonably good growth over a broad range of pH (2 - 9) with an optimal range of 3.8 to 6.0. On the other hand, yeasts have a pH optimum between 4 and 5 and can grow over a large pH range of 2.5 to 8.5 [27, 28]. This typical pH versatility of fungi can be beneficially exploited to prevent or minimize bacterial contamination, especially choosing a lower pH. Yang [29], reported a bacterial content of only 10^3 cells per g dry mass during the fermentation of sweet potato by yeast with the final pH being around 4.0. A pH of 3 to 3.5 in the fermentation of fodder beet pulp by *S. cerevisiae* [30], 3.5 for fermentation of Jerusalem artichoke by *K. fragilis* [31] and 3.5 to 4.5 in amylase production [32] have reportedly been effective for prevention of bacterial contamination. Also, the optimum pH for mycelial growth (pH 3-7) may be different from that required from protease production (pH 3.55- 4.3) as in the case of *A. niger* [33]. The important thing to remember in FSSF is that the values mentioned above are the initial pH values and they are not necessarily the same as the fermentation proceeds. In FSSF, unlike the homogenous three phase system of submerged fermentation, pH control is practically impossible due to the heterogeneous three phase system. The pH change will occur according to the nitrogen source selected as well as the growth characteristics. Table 3 lists pH values for some FSSF systems.

Table 3
Typical pH ranges in FSSF systems for growth and product formation

Microorganism	Substrate	Product	pH	Reference
Aspergillus carbonarium	canola meal	phytase	4.7	34
Aspergillus flavus	sugarcane press mud	degradation products	7.2	35
Streptomyces rimosus	sweet potato residue	oxytetracycline	6.0	36
Monascus fulginosus	-	lipase/esterase	3.0 to 5.5	37
Coprinus sp.	Wheat straw	-	9.0	37

3.4. Temperature variation and control

Probably the most important of all the physical variables affecting FSSF performance is temperature. Similar to that for pH, fungi can grow over a wide range of temperatures from 20-55°C and the optimum temperature for growth could be different from that for product formation [38]. Table 4 shows the temperature ranges for some FSSF systems. The critical criteria in FSSF is temperature control which is quite difficult to effect. This is because the very high heat output [39] due to the aerobic nature of the process is not easily dissipated owing to the poor thermal conductivity of the solid matrix as well as the predominantly static nature of FSSF. In combination with local moisture content and availability of void spaces this could lead to thermal gradients. Therefore, the key issue in FSSF is heat removal and hence, most studies on reactor designs are focused on maximizing heat removal [40, 41]. The problem is aggravated in large scale systems where heat generation leads to serious moisture losses and maintaining constant temperature and moisture content is not easy [42]. In systems like composting, very large temperature gradients can be encountered [43], as well as in tempeh fermentation [44]. The most appropriate way in dealing with heat build-up is by means of forced aeration, which plays multiple roles in FSSF. The critical step would be the flow rate [45] and to humidify or dehumidify the air since low or high moisture content brings about adverse effects [21].

Table 4
Typical temperature ranges in some important FSSF production systems.

Microorganism	Substrate	Product	Temp °C	Reference
Rhizopus oligosporus	Wheat bran	acid protease	25	46
Taralomyces sp.	Wheat bran	avicelase	45	47
A. niger	Wheat bran	glucoamylase	30	48
Penicillium capsulatum	Beet pulp	poygalacturonase	30	49
Humicola sp.	Agri-residues	alpha galactosidase	45	50
Penicillium candidum	Wheat bran	lipolytic enzyme	29	51
A. carbonarium	Canola meal	phytase	53	34
T. viride		cellulase	30-50	52

4. SUBSTRATES FOR FSSF

One of the main advantages of FSSF is the ease with which fungi can grow on complex natural solid substrates like agro-industrial wastes, without much pretreatment. Table 5 shows various substrates that are used in FSSF.

Table 5
Substrates commonly used for FSSF

Lignocellulosics	Starchy Materials
Wheat straw	Wheat bran
Corn stover	Rice
Rice stover	Cassava
Sugar beet pulp	Corn
Feedlot waste	Banana meal
Wood	Buckwheat seeds
	Rye meal

There are several factors involved in the selection of a suitable substrate for the desired fungi to grow. These are the macromolecular structure, particle size and shape, porosity and particle consistency [53]. The complexity of the solid substrate arises from the presence of macromolecular compounds such as cellulose, starch, lignin and even smaller sugar molecules. Fungi cannot directly absorb these macromolecules, they are induced by low molecular weight compounds to synthesize and secrete the enzymes to hydrolyze the macromolecules into smaller metabolizable compounds [54].

As far as particle size is concerned, a larger surface area to volume ratio is preferred for high yields especially when the fungi lack sufficient penetrative ability, so that the substrate molecules are more easily accessible for the hydrolytic enzymes. If the surface area is less, then the enzyme diffusion tends to become rate-limiting. The particle size is also important in obtaining favourable physical conditions in the bed. Smaller particles result in bed compaction affecting gas exchange while, larger particles limit substrate accessibility (because of decreased surface to volume ratio). Different substrate particles have different particle shapes, which have a direct bearing on the void fraction when packed into the reactor system. It has been indicated that spherical and long thin particles support sufficient void spaces than cubical and thin slab-like ones [55].

The substrate particle would also need to have sufficiently large pore openings for easy mycelial entry and passage. Porosity can be further improved by pretreatment. Other considerations in substrate selection is the compressibility of the particles especially in large scale systems where the bottom layers have to resist deformation by the weight of the upper layers. Also, stickiness and other surface properties will decide the outcome of microbial growth.

5. GROWTH PATTERNS IN FSSF

Literature on the growth patterns of fungi are very few and these have been collated [55]. A clear understanding of fungal morphology and growth is impaired by the complex orientation of the mycelium with the substrate types, substrate heterogeneity and the lack of techniques for the direct estimation of viable biomass. A few workers have characterized the fungal growth patterns on flat surfaces such as agar plates. These studies focus on the hyphal mode of growth [56], growth information based on the development of nutrient gradients [57], and the advantage of overculture technique to achieve uniform growth (single point) in contrast to colony growth (58). However, the picture becomes more complex when solid substrates are involved. Two stages of *Aspergillus niger* development on cassava and potato flour were reported by Aufeuvre and Raimbault [59] viz., swelling of spores, extension of germ tubes and development into a loose hyphal network; then, intense growth and branching forming a dense mycelial network knitting together the substrate particles. In another study [60] with *R. oligosporus*, similar results were obtained with additional production of aerial and penetrative hyphae. Hyphal extension and penetration depend on the extent of cracks and fissures available for hyphal contact: the greater their number, the better the penetration, in which case efficient gas exchange is facilitated [20]. It is therefore an additional benefit when sterilization by steam is effected that the substrate surface undergoes modification by increased cracks and fissures, rendering it more susceptible for the desired hyphal penetration. This is especially important in the fermentation of recalcitrant lignocellulosics [61]. Table 6 gives a list of important fungi in SSF and where they find applications.

6. BIOREACTOR SYSTEMS USED IN FSSF

The bioreactor is the critical stage of the FSSF process since, all the earlier and later stages are dependent on the its capabilities. Unlike submerged fermentation, in FSSF the choice of a suitable reactor system is difficult, given the heterogeneity of the substrate matrix. The selection will depend on factors such as substrate type, process variables, extent of control required, etc. In the classical FSSF koji process, tray reactors have been traditionally employed. Other major types of reactors are the packed column, rotating drum, and fluidized bed.

6.1. Tray systems

The general features of tray systems are:

1) The substrate is evenly spread in layers about 2 to 4 cm in height in trays with perforated bottom to provide maximum surface availability for gas exchange.

2) The trays after autoclaving, are cooled, inoculated and stacked one above the other in trolleys and moved to the fermentation chamber where conditions such as relative humidity, aeration, temperature are regulated.

3) After the fermentation is done, the solids are harvested from the trays and recovery operations executed.

Table 6
Some examples of fungi and fungal species used in FSSF (Doelle et al, 1992)

Fermented foods	**Secondary Metabolites**
Rhizopus oligosporus	*Aspergillus* sp.
Aspergillus oryzae	*penicillium* sp.
Pediococcus halophilus	*Fusarium* sp.
Monascus sp.	*Monascus* sp.
Cephalosporium sp	*Streptomyces* sp.
Fusarium sp.	*Gibberella* sp.
Neurospora sitophila	**Composting**
Exoenzymes	*Aspergillus* sp.
Aspergillus sp.	*Chetomium* sp.
Rhizopus sp.	*Geotrichum* sp.
Mucor sp.	*Humicola* sp.
Penicillium sp.	*Trichoderma* sp.
Trichoderma sp.	**Lignocellulosics**
Fusarium sp.	*Chaetomium cellulolyticum*
Organic acids	*Trichoderma* sp.
Aspergillus sp.	*Cellulomonas* sp.
Saccharomycopsis sp.	*Coriolus versicolor*
Penicillum sp.	*Sporotrichum pulverulentum*
Candida sp.	**Bioremediation**
Edible Mushrooms	*Phanerochaete chrysosporium*
Agaricus bisporus	*Trametes versicolor*
Lentinus edodes	*Polyporus*
Volvariella volvacea	*Peurotus* sp.
Pleurotus sp.	

Tray systems have some very special features that distinguish them from other systems:
1) Simultaneous study of different process variables in different trays is possible, e.g., the effects of media components, moisture, packing density, inoculum type and size, etc can be selectively investigated.
2) Tray systems are simple as far as reactor operation is concerned, and the fermentation can be observed in all the trays any time the operator desires, since all one has to do is to simply walk (after taking safety measures) inside the fermentation chambers and take a good look at each tray and check out any trays that are of concern. This eliminates the need for sophisticated instrumentation, monitoring and control systems to track down the fermentation. This will be especially important for those processes where the economics is a critical factor, e.g., where there is no valuable product synthesis such as for bioremediation, composting, etc.

Tray reactors have been successful in laboratory, pilot and large scale operations [62 - 65] . However, one major disadvantage is that it can be very labour-intensive unless a reliable automation system is present. Also, heat transfer limitations restrict the height of the substrate bed.

6.2. Packed column bioreactors

Reactors in which the substrate is generally loose-packed to larger heights (than trays) and air is forced through the bed from the bottom fall into this category. Closed reactors where air is passed over the surface of the substrate bed are not packed bed reactors since, the mass transfer occurs in this case by diffusion instead of forced convection as in the actual packed bed bioreactors. The advantages of packed beds are:

1) Substrate bed height can be larger since, the forced aeration helps to minimize temperature gradients [66] and gas concentration gradients [67], compared to tray systems.
2) Being a closed system they maintain more aseptic conditions.
3) Humidified air also helps in regulating moisture.
4) After the fermentation, the column reactor can be used as a trickle bed extractor for product recovery.

The packed bed reactor lends itself as a versatile tool to study various effects. Gowthaman et al [68] estimated the oxygen transfer coefficient in FSSF of amyloglucosidase using a packed bed bioreactor under different conditions. Heat transfer studies were conducted in a packed bed bioreactor for citric acid production with *A.niger*, where the authors found that the convective and evaporative mechanisms were more prominent than conductive heat transfer [40]. Recently, a multi-layer packed bed reactor for FSSF production of citric acid from *A. niger* was found to be more efficient for mass transfer and allowed precise measurements of various gradients [69] compared to a single-layer packed bed reactor.

Packed beds at increasing scales of operation lead to non-uniform growth and transport gradients. In spite of this, it is a convenient tool to study at the laboratory and pilot scales and many references have been cited by [70].

6.3. Rotating drum bioreactors

In this configuration, a horizontal or inclined cylinder rotates around its own longitudinal axis such that the substrate within the reactor is in constant motion to promote gentle inter-mixing. This is in contrast to the packed bed where the substrate is in static conditions. The tumbling action promotes very efficient heat and mass transfer. The problem with this bioreactor is that particles tend to agglomerate over a period of time and also the rotational speed becomes an important parameter if the mycelia are shear-sensitive. Baffles along the inner wall of the rotating drum facilitate mixing by the lifting and dropping of the substrate [71] and increased the maximum oxygen uptake rate by 60% over unbaffled runs [72]. Amyloglucosidase production was affected by the time at which the rotation was first initiated during the fermentation in a *A.niger*/wheat bran rotating drum fermenter [73].

6.4. Other configurations

There are also a few other bioreactor types such as stirred bioreactors [74] where vertical screws mounted on a conveyor at the top of the reactor to agitate the bed at 22 rpm while moving across the bed. Chamielec et al [75] reported the performance of a 50L bioreactor using a planetary type agitator. In another case, a Z-blade mixer was modified as a bioreactor where the mixing quality was found to be dependent on the number of revolutions rather than the speed [76]. Air-solid fluidized beds [77] are another type of reactors where the bed is kept in a fluidized state and eliminates several drawbacks found with the other bioreactors [78]. In one specific study with *Aspergillus oryzae* the operating costs were compared and found to be just one-sixth of the other types of SSF bioreactors [79]. A novel design of a large scale solid state fermenter, called Zymotis --the condensed term based on Greek word "Zymothiras ," meaning 'fermenter', offers efficient control of various fermentation parameters such as temperature, moisture, and aeration of the fermenting moist solids. A large quantity of metabolic heat is claimed to be easily removed by the novel cooling system employed. The unit can be operated at different capacities simply by adding or removing the compartments [80]. For more comprehensive information on FSSF bioreactors the reader is referred to the articles by Lonsane et al [4] and Mitchell et al [70].

7. DOWNSTREAM PROCESSING

Some FSSF products do not require any downstream processing. Some examples of these are koji starter, composts, ensiled grasses, upgraded agro-industrial residues, enzymes in crude form, poultry and animal feed supplements. In the downstream processing of FSSF bioproducts, the major difference with that of submerged fermentation systems is that, an additional step of leaching is involved, but liquid removal is less important. This stage ensures the transfer of product(s) from the solid to liquid which then is akin to the submerged fermentation broth.

In some other cases, simple operations are all that is required:

1) Seed koji can be dried under controlled conditions, bagged and stored for later use [81].

2) The Japanese fermented food Miso, is blended, pasteurized and packed [82].

Very few products such as citric acid and aflatoxin may require additional steps to meet the purity specifications depending on the application of the product. Since FSSF finds more applications for the production of enzymes, a brief outline of the possibilities for downstream processing is given below:

1) Direct use of enzyme as in 'sake' preparation with 'koji'.

2) Drying: The moist fermented solids can be dried under controlled conditions to about 8% moisture content at 50-60°C or even lower, depending on product stability; they are then pulverized, screened and stored in suitable containers for future use [83].

3) Leaching: The leaching step is probably the most important step in FSSF downstream processing since, the economics of the process will be dictated by the leached product concentration. Hence, it is imperative to select the right leaching technique and conditions such as solvent type, concentration, etc. An ideal solvent should be able to leach out the product effectively at room temperature and other desirable conditions. Solvents that can be used include tap water, distilled water, dilute salt/glycerine solution, buffer solutions and dilute ethanol solution [84]. Other factors in leaching are diffusivity of solute and solvent, solvent retention by solids, extraction efficiency of solvent, solid/solvent ratios, contact time, temperature and pH. One of the following leaching techniques may be selected:

1) Percolation: In this technique, moldy bran is first charged into cylindrical vessels. The solvent is then regulated to flow through the bed and the extract is collected from the bottom or if required it is recycled or passed into another stage depending upon the level of leaching desired/achieved.

2) Pulsed plug flow extraction: This is a variation of the percolation technique in which the solvent is pulsed at appropriate flow rates, instead of a continuous feed. Better contact time with this technique has been reported [85].

3) Counter-current extraction. Very common in the chemical industries, this technique involves the movement of solute and solvent in a direction opposite to each other. This is more efficient than the other techniques and has been widely tried.

4) Hydraulic pressing: Recovery of cellulases from solid state cultures of *T. harzianum* can be efficiently achieved by hydraulic pressing. Pressing of fermented solids yielded carboxymethyl-cellulase (CMCase) extraction efficiency of 71% and a ratio of leachate to fermented solids of 0.58 (v/w) [86].

5) Clarification/Concentration/Purification: Once leaching is effected, the extract may be clarified either by filtration/centrifugation to remove suspended cells/spores/solid residues. Generally, the extract after leaching contains very low biomass (cells, mycelia, spores) because most of the biomass strongly adheres to the solid substrate and not easily removed which is a hassle-saving advantage. This biomass can now be easily filtered off by simple filtration techniques because of their relatively larger size than bacteria. After this, the product may be either recovered in solid form by concentration and/or precipitation or aqueous two phase separation (ATPS) or the broth concentrated by membrane processes. ATPS has been attempted for the recovery of amyloglucosidase from moldy bran [87] as well as its comparison with recovery using ultrafiltration [88]. In another study, Tirumurthy and Gowthaman [89] were able to achieve a 30-fold concentration of

amyloglucosidase from the dilute aqueous extract using a Millipore Minitan ultrafiltration system.

8. EFFLUENT DISPOSAL

The problems of waste treatment such as handling of large volumes of waste water in submerged fermentation, is obviously absent in FSSF. However, in the case of those products which require to be leached and recovered from the extract the problem may have to handled the same way as in submerged fermentation. However, the BOD/COD are expected to be less than in SmF since, the major chemical constituents remain in the spent solid after the leaching process and can also be treated by simple waste treatment methods. The spent solid residue may then be sterilized and dried and serves as a valuable animal feed, fertilizer or raw material for biogas production [90]. Other uses of the spent residue are in the manufacture of bricks, boards, paper, as fuel and for land-filling.

9. APPLICATIONS OF FSSF

There are several applications of FSSF viz., fermented foods, exoenzymes, organic acids, lignocellulosics, grass ensiling, composting, biocontrol, etc. Recent interest in solid state fermentation has placed a special emphasis on FSSF as alternative fermentation route simply because of its simple and non-intensive capital nature. The following sections will give an overview about these aspects.

9. 1. Fermented foods

FSSF has found traditional application in the classical oriental foods believed to have existed over 20 years ago. Most prominent among them being Koji, soysauce and Tempeh. Koji is a mixture of various hydrolytic enzymes and is the first step in soysauce production. It is prepared by first growing *Aspergillus oryzae* on a mixture of wheat and soybeans at 30°C for 5 days; drying, mixing with roasted wheat and soybeans before being added to cooked soybeans to effect the fermentation [91]. Tempeh is an Indonesian fermented whole soybean product. A mucoraceous fungus is used for inoculum, and inoculated soybean is fermented by *Rizopus oligosporus, R. oryzae, R. arrhizus*, etc. until the beans are covered with white mycelium as a cake [92]. In Indonesia, peanut is also used for making tempeh. Ontjom is a tempeh-like fermented peanut product. However, fermentation of ontjom is done with *Neurospora sitophilia*, and unlike tempeh, the cake is pink in color. Tempeh is consumed daily as a side dish or snack in Indonesia.

Furu is a Chinese traditional product having Foiegras-like texture and salty taste which is made from soybean curd (Tofu) by mold fermentation and brining process. Typical fungi involved are *Actinomucor elegans, Mucor hiemalis, Mucor silvaticus, Rhizopus chinensis* while *Monascus purpureus, Monascus anka* and the like are used as the occasion demands for coloring [93].

9.2. Extracellular enzymes

Because of their biocatalytic efficiency at mild reaction conditions [94], enzymes find a host of applications viz., food, feed, brewing, distillery, beverages, fats, oils, textile, pharmaceutical, leather, laundry, etc New enzymes/applications are being developed constantly. The break-up of global industrial consumption of enzymes are 30% for detergent, 30% for starch, 15% for dairy and the balance is distributed between all other applications. With the exception of thermostable alkaline proteases, almost all commercial enzymes (high-volume, low value) are derived from fungi. Table 7 shows the range of enzymes and the sources derived from. Table 8 shows the market for fungal enzymes.

Table 7
Exoenzymes produced by FSSF [95]

Enzyme	Microorganism	Substrate	Application
Soy sauce koji	*A. oryzae, A.sojae*	Soybean + wheat	Fermented food
Miso-dama	*A. oryzae, A.sojae*	Soybean	-do-
Soybean koji	Natural microflora	Soybean alone or with wheat	-do-
Sake rice koji	*A. oryzae*	Rice	-do-
Ang-kak	*Monascus* sp.	Rice	-do-
Miso rice koji	*A. oryzae*	Rice	-do-
Tane koji	*A. flavus - oryzae*	Rice	-do-
Pectinases	*A.carbonarius* *A. sojae, A. saito* *A.niger*	Wheat bran	Fruit processing
Fungal alpha-	*A. oryze* *Rhizopus, Mucor* *A. oryzae* *A. niger*	Wheat bran; Rice grains Barley, soybeans Cassava meal Wheat bran	Baking, brewing, Distilliery, Pharma
Glucoamylase	*A. oryzae*	Rice, soybeans	liquid glucose, dextrose,
	A. niger	Wheat bran alone or + Corn flour	HFCS, brewing, distillery
	Rhizopus sp.	Bran	
Rennet	*Mucor pusilus* *Mucor meihei* *R. oligosporus*	Wheat bran Wheat bran Wheat bran + rice bran	Cheese making
Proteases	*A. oryzae* *A. oryzae*	Wheat bran Wheat bran + rice bran + Soybean cake	baking, brewing, pharma, protein-hydrolysis, soybean processing
	Aspergillus sp.	Cereal grnaules, bran, oat hulls, straw	
Cellulase	*T. reesei* *Pleurotus sojor-caju*	Wheat bran Agro-industrial wastes	digestive aids, animal feed, textile

Table 7 (contd.)
Exoenzymes produced by FSSF

	T. koningii	Wheat bran + saw dust	
Lipase	*A. Niger*	Wheat bran	digestive aids, hydrolysis
	A. luchuensis	Wheat bran	of fats & oils, flavor modification, prodn. of fatty acids & glycerol
Alpha-galactosidase	*A. awamori*	Wheat bran	sugar refining; soybean milk procesing
Beta-galactosidase	*A.niger; A. oryzae* *Fusarium* sp.	Wheat bran	digestive aids; lactose hydrolysis
Catalase	*R. niveus*	Wheat bran	food industries to
	A. oryzae	Wheat bran	remove hydrogen peroxide; for controlled release of oxygen
Xylanase	*A. niger* *S. pulverulentum* *A. terreus*	Wheat bran + rice straw	Conversion of hemicellulose into pentose sugars
Alpha-glucosidase	*A. oryzae*	Wheat bran	hydrolysis of cellulose into glucose
Beta-D-glucosidase	*A, ustus,.* *S. pulverulentum* *A. terreus,* *Trichoderma* sp	Wheat bran + rice straw	-do-
Phytase	*R.oligosporus*	Rape seed meal	Hydrolysis of phytic acid in feeds
Linamarse	*Aspergillus* sp. *Penicillium* sp. *Fusarium* sp.	Cassava	Detoxification in cassava tubers
Chitinase	*A. niger*	Wheat bran	food processing and microbial cell lysis
Invertase	*Aspergillus* sp.	Wheat bran	sucrose inversion

Table 8
Estimated world market for various fungal enzymes [96]

Enzyme	% of total microbial enzymes
Amyloglucosidase	14
Pectinases	10
Rennet	5
Alpha-amylase	4
Proteases	4

9.2.1 Amyloglucosidase

Glucoamylase or amyloglucosidase hydrolyses aplha-1-4-linkages from the non-redcing ends of starch and dextrin molecules, thereby splitting one molecule of glucose of the chain at a time. The enzyme also can split the alpha-1-6-linkage in amylopectin and convert the whole starch molecule directly into glucose [96]. It is produced by FSSF in many Asian countries. The enzyme from SSF is generally free of transglucosidases which if present, convert starch into alpha-1-6-oligosaccharides instead of glucose. In the FSSF of amyloglucosidase from wheat bran, transglucosidase co-production is suppressed while, it is co-produced in SmF [97]. Many new research findings are reported and one of these is the enzyme being produced by *A. niger* in SSF, SmF and an aqueous, two-phase system of polyethyleneglycol (PEG) and salt. In SSF, a fed-batch mode of operation gave a yield of 64 U/ml compared with 44 U/ml in batch mode. Similar trends were observed for SmF, where fed-batch cultivation gave a yield of 102 U/ml compared with 66 U/ml in batch [98].

9.2.2. Proteases

The commercial protease complexes produced by *A. oryzae* or *A. sojae* by FSSF contain acid, neutral, and alkaline-proteases in addition to exopeptidases [99], whereas, those produced by SmF contains only neutral proteases [100]. More then half the bread made in the USA involves the use of proteases derived from *A. oryzae*. The pH of the medium and the substrate markedly affected protease production by *Aspergillus oryzae* [101]. Very interesting research is being carried out on proteases. For instance, acid protease production reached its highest level, a 43 % increase over air, with a mixture of 4 % CO_2 and 21% O_2. The protease production was strongly related to the mold metabolic activity as represented by the total CO_2 evolved [102]. A decrease of O_2 concentration from 21 % to 0.5 % did not alter protease production by *R. oligosporus* but retarded amyloglucosidase production [103]. An acid protease was produced by *Mucor bacilliformis* under SSF and purified by diafiltration [104]. The production of an alkaline protease was carried out using the microorganism *Rhizopus oryzae* and the product tested for unhairing of skins. The enzyme was stable from pH 3-11 and at temperature up to 80°C [105].

Proteases are among the easiest of enzymes to be produced by fungi. Although, the production is desirable when it is the principal product, it is actually something of a liability, because the co-produced protease hydrolyses the desired enzymes thus, rendering it reducing its final yield. Hence, current research on protease reduction is of intense interest around the world. For example, xylanase and beta-xylosidase production in corn cob and sugar cane bagasse did not result in high protease production [106]; similarly in the heterologous production of lysozyme (HEWL) in SSF using *A,niger*/wheat bran, Gowthaman et al [107] reported low production of protease activity compared to SmF.

9.2.3. Cellulases

These enzymes cleave the endo-beta-1-4 glucanases at random positions in the chain and exo-beta-1-4 glucanases, which split cellobiose or less commonly glucose residues from the chain ends of cellulose molecules [108]. Cellulase from FSSF showed three times more activity than that from SmF [109]. A number of references giving excellent information on this subject can be cited [110-114]. Some of the recent and interesting studies are as follows: Solid substrate fermentation of leached beet pulp with *T. aureoviride* yielded cellulase of 29 FP Units/g solids [115]. *Penicillium citrinum* using rice husks produced maximum cellulase yields (37 Units/g) after 12 days with a cellulose utilization of more than 70%. Enzyme yields were three times higher than in shake-flask cultures [116]. Optimal cellulase and xylanase levels of 4 IU/g dry weight (DW) and 180 IU/g DW, respectively, were achieved in 120 h-fermentation when *T. reesei* was inoculated into sweet sorghum silage at 0 h, followed by the inoculation of *A. niger* at 48h [117]. Coconut pith, available in plenty in tropical countries, has been found to be an excellent new substrate for the production of cellulase enzyme by solid substrate cultivation of *T. viride* [118]. The fungus *Hymenoscyphus ericae* has been reported [119] to produce a complete cellulase enzyme complex (cellulase, cellobiohydrolase and beta -D-glucosidase) along with components of the hemicellulolytic (endoxylanase, beta -D-xylosidase, alpha -L- and beta -L-arabinosidase and glucoronidase) and mannanolytic (mannanase, beta -D-mannosidase, alpha -D- and beta -D-galactosidase) complexes. A cellulase system, showing enhanced hydrolytic potential and beta -glucosidase under SSF, was obtained by co-cultivation of *A. ellipticus* and *A. fumigatus* [120]. Cellulase production by *T. viride* conducted in a column, was not affected with periodic changes in temperature in the range of 30-50°C [52]. Very recently, high levels of cellulase and hemicellulase production at 49°C by the thermophilic fungus *Thermoascus aurantiacus* has been reported in SSF by [121].

9.2.4. Pectinases

Pectinases hydrolyse pectin, by means of a complex of pectin methylesterases, transelliminases, endo- and exo- polyglacturonases. They cause de-esterification, chain-splitting and glycoside-bond cleavage. The major part of pectinase production is by FSSF [122]. There are quite a few differences between pectinases produced by SSF and SmF [123], the SSF often preferred because of development of unique properties. In SSF, all pectinase activities were more stable at extreme pH and temperature values but the K_m values of endo-pectinase and pectin lyase were higher with respect to those activities obtained by the submerged-culture technique [124]. Production of pectinases by *A. niger* in solid state fermentation at high initial glucose concentrations were studied by [125] and they found that there was no drastic effect of the sugar concentrations on the pectinase activity. The pectin lyase activity obtained by the submerged-culture technique showed substrate inhibition but the enzyme obtained by solid-state fermentation did not. Comparison of pectinesterase and polygalacturonase by *A. niger* in submerged and solid state systems showed four and six times higher activity respectively in a solid state system than in a submerged fermentation system [126].

9.2.5. Lipases

Lipases catalyse the hydrolysis of ester bonds a the interface between water and insoluble fatty acid ester or glyceride phase and the aqueous phase containing the enzyme [108]. Lipase are curently produced both by SSF and SmF, although the former is preferred for some reasons [127]. Rivera-Munoz et al [128] compared the activity of lipase produced by *P. candidum* in SSF and SmF and found that the solid system gave the highest titers and a stable production. An ideal lipase especially for lanudry applications should have good pH (generally alkaline)and temperature stability. Bhushan et al [129] produced lipase from an alkalophilic yeast species (*Candida*) at pH 8.5 by solid state fermentation using rice and wheat brans as alternative cheap solid substrates. Liu et al [37] reported a thermostable lipase and esterase produced by *Monascus fulginosus* in solid-state fermentation. The yield of lipase as 207 u/g. The enzyme almost retained 100% activity after treatment at 55°C for 1 hour and 24 hours at 45°C. Kamini et al [130] found that lipase produced by *A. niger* using gingelly oil cake was stable between pH 4.0-10.0 and 4-50°C and also showed remarkable stability in the presence of detergents. A lipase activity of 363.6 U/g of dry substrate was obtained at 72 h under optimum conditions. In another study, a lipase yield of 118.2 Units per gram of dry fermented substrate (U/gds] at 72 h from *Candida rugosa* in a mixed-solid substrate containing coconut oil cake and fine and coarse wheat bran was reported [131].

9.2.6 Phytases

Phytases hydrolyse phytic acid into phosphoric acid and myo-inositol. Most of the phosphorus contained in typical feedstuffs such as cereals and oilseeds exists as the plant storage form called phytate or phytic acid. Pigs and poultry cannot digest the phosphorus contained within these complex phytate structures as they lack the right enzymes to break phytate down and release the phosphorus. Phytases are used to supplement feed for the hydrolysis of phytates and phytic acid in soyben, rapeseed meal, protein isolates and other related products. They represent one of the fastest-growing markets for feed enzymes today. Hence, the hydrolysis of phytates and phytic acid in soybean, rapeseed meal, protein isolates and other related products is of major concern. The earliest attempts to develop phytases began with a microorganism of the *Aspergillus* family. Commercial phytases from *A. niger* are currently produced by SmF. High Phytase production was possible from *A. ficuum* [132] which reduced phytic acid content in canola in 48 h. Al-Asheh and Duvnjak [133] reported that the production of phytase by *A. carbonarius* from canola meal (the residue from canola oil processing in Canada) was growth-associated and that the homogenization of the inoculum effected the maximum biomass and minimum phytic acid. The same workers also improved phytase production by the addition of surfactants such as Na-oleate and Tween-80 to the canola medium [134].

In the United States, there is considerable industrial interest for phytase production by FSSF where laboratory results have yielded activity of about 1,000 Units/g dry substrate along with co-activities of many desirable enzymes [135]. Mycelial morphology and inoculum quality have been cited important in phytase production by *A. niger*, for which a response surface methodology was applied to predict the effect of inoculum age, media composition, and fermentation duration [136]. All the treatments had significant main effects on phytase production in SSF. Recently, Novo Nordisk has derived a new phytase called Bio-Feed® Phytase from the fungus *Peniophora lycii* growing on the trunk of a tree in Denmark [137]. This phytase differs from the others in that it is a 6-phytase and releases the phosphorus in the sixth position, whereas *Aspergillus* phytases are 3-phytases. However, this is believed to be produced by SmF.

9.2.7. Xylanase

This enzyme hydrolyses beta-1,4-linkages in xylan which has great significance in the hemicellulase conversion to pentose sugars. One of the main advantages of xylan produced by FSSF is that it is more thermostable [95]. Xylanases from *Thermomyces lanuginosus* and *Thermoascus aurantiacus* were most active at 70°C, but at pH 6.0 and 5.0, respectively. Both xylanases displayed remarkable pH (5.0 to 11.0) and thermal stabilities by retaining most of their activities even after having been subjected to temperatures much higher than their optimal [138]. Srivastava [139] while studying the thermostability of three hemicellulolytic enzymes from a *Trichoderma* strain, found that all three enzymes were stable up to 80°C; but it was only the xylanase which had the longest half-life of 540 h at 40°C. On the contrary, Brustovetskaya et al [140] reported that the thermostability of xylanases did not depend on the type of fermentation (whether SmF or SSF) while studying the enzyme production and stability of cellulases and xylanases from two fungal strains, Wiacek-Zychlinska [141] studied the production of xylanases by *Chaetomium globosum* and *Aspergillus niger* on several media combinations of different carbon sources. The best xylanolytic activities from both the strains were obtained by growth on a medium containing wheat bran (75%), beet pulp (20%) and apple pomace (5%). In another study, *Melanocarpus albomyces* a thermophilic fungus was used for the production of xylanase on various agroresidues and the author found that pretreatment of rice straw and rice husk enhanced xylanase production, while pretreatment of wheat straw and bagasse decreased it. Also, the xylanse produced by SSF was more than that by SmF [142]. Cellulase-free xylanase production from *Thermomyces lanuginosus* in SSF has been carried out by Purkarthofer et al [143]. A multivariant statistical approach to predict the effect of initial moisture content in the medium, cultivation time, inoculum level, and bagasse mass on xylanase production using *Thermoascus aurantiacus* was developed by Souza et al. [144].

9.2.8. Other enzymes

Several other enzymes have also been tried by FSSF and some of the recent reports are tabulated below (Table 9):

Table 9
Other important enzymes recently studied by FSSF

ENZYME	Organism	Substrate	Reference
Chitinase	*Beauveria bassiana*	Prawn waste	145
alpha-	*Humicola* sp.	Soy flour	146
galactosidase	*A. niger*	wheat bran + lactose + guar flour	147
beta-galactosidase	*Rhizomucor* sp	--	148
alpha-amylase	*Pycnoporus sanguineus*	Wheat bran	149

9.3. Mushrooms

Nothing can be a more prominent SSF application than the cultivation of mushrooms. The multiple advantages of mushroom farming have been extolled by Zadrazil et al [150]:
* Industrial plant residues and by-products can be efficiently recycled into the ecosystem through natural processes.
* Bioconversion of solid and liquid wastes is directly accomplished.
* Formation of fruit bodies take place by breaking down of recalcitrant lignocellulosics.
* The apparent biomass that can be harvested free of substrate, in the form of fruiting bodies merits the usefulness of such culturing of mushrooms in solid state fermentation [151].
.* Inexpensive separation of fruiting bodies from the fermented substrate.

There is a growing trend worldwide in the production of mushrooms and there are quite a few species cultivated (Table 10).

Table 10
Production of edible fungi [150]

Species	Common Name	Quantity (in 1000 tons)
Agaricus bisporus *A.bitorquis*	White mushroom	1227
Lentinus edodes	Shii-take	314
Volvariella volvacea	Straw mushroom	178
Pleurotus sp.	Oyster mushroom	169
Auricularia sp.	Jewish ear	119
Flammulina velutipes	Winter mushroom	100
Tremella fuciformis		40
Pholioto nameko	Nameco	25
Other sp.		10
TOTAL		2182

Fruit and vegetable wastes are proving to be excellent substrates for mushroom cultivation. The technical feasibility of using mixtures agricultural wastes (mango and date industry wastes) as a substrate for the cultivation of *Pleurotus ostreatus* was evaluated by Jwanny et al [152]. They selected these wastes since, they were good sources of nonstarchy carbohydrate (67%) and protein (27.44%), containing amounts of essential amino acids, especially lysine and low RNA (3.81%). Comparison of two mushrooms, *Flammulina velutipes* and the white-rot fungus *Trametes versicolor* cultivated separately on sugarcane bagasse were reported by Pal et al [153]. An increase in the relative abundance of aromatic carboxylic acids suggested that the oxidative transformation of lignin unit side-chains was more prominent with *Flammulina velutipes* than with *T. versicolor*. Straatsma et al [154] emphasize the significance of *S. thermophilum* in compost preparation where, naturally occurring strains of *S. thermophilum*, present in ingredients, readily colonize compost during preparation. This finding is of relevance for the environmentally controlled production of high-yielding compost.

9.4. Secondary metabolites

A diverse range of compunds synthesised as secondary metabolites mainly by fungi and *Streptomyces,* have great commercial value. Antibiotics, growth promoters and mycotoxins are typical examples. Traditionally secondary metabolites have been produced by SmF but more workers are resorting to SSF because of some special advantages [155].

1) the mycelial morphology is suited to penetrative growth on solid substrate
2) scale up of mycelial cultures poses major difficulties in SmF - high viscosity, non-Newtonian broths and heavy foam production, thus, demanding more power, and
3) product quality is superior in SSF.

An acute need for mycotoxin existed in the US during the seventies for toxicity studies, and sufficient quantity could be obtained only by FSSF [72, 156]. Monascus pigments synthesised by *Monascus pupureus* is traditionally produced in Asia to impart flavour and colour for foods including red rice (ang-kak), fish and soybean cheese. Cepholosporins have been produced from barley using *S. cluvuligerus* [157]. High concentrations of the secondary aryl metabolites, veratryl alcohol and 3-chloro-p-anisaldehyde white rot fungus were produced by a *Bjerkandera* sp. during solid state fermentation of lignocellulosic substrates [158]. The enhanced production of these secondary metabolites compared to defined liquid cultures is suggested to be due to the release of lignin degradation products serving as alternative precursors for their biosynthesis. Bandelier et al [159] used a fed-batch technique in an aseptic pilot-scale reactor for production of gibberellic acid by *Gibberella fujikuroi*. However, FSSF technology for secondary metabolites has not yet fully realized commercial reality since, the major hurdle is the lack of appropriate instrumentation and control systems, more so essential, because of the extended duration of such fermentations.

9.4. Organic acids

There is at least one report on SSF production of itaconic acid by *A.terreus* [160] and the production of gluconic acid using spores of *A.niger* by SSF on buckwheat seeds [161]. However, citric acid has been in the forefront among the organic acids and fairly large amount of literature is available on this subject. Gutierrez-Rojas et al [165] studied the citric acid production in SSF by *A. niger* using high glucose concentration on an amberlite inert support and found that citric acid production was free of heavy metals inhibition. Wheat bran [163], sugar-cane press mud [164], coffee husk [165], pineapple waste [166], and carrot processing waste [167] have been found to be good substrates for citric acid production. By addition of 6 % w/w methanol in the medium, citric acid production could be enhanced from carob pod by SSF [168]. Application of a packed bed reactor [169], multi-layer packed-bed reactor to citric acid production with *A. niger* [170] and heat transfer studies [41] indicate the importance of reactor and engineering aspects for citric acid production by SSF.

9.5. Lignocellulosic substrates

Lignocellulosics are by far the most abundant natural raw material present on earth accounting for more than 60% of total plant biomass produced, and thus form the main structural component of plant cell walls. The net photosynthetic production of the dry biomass by plants on earth has been estimated to be 155 billion tons per year [171]. Regardless of the source, lignocellulosic materials contain cellulose, hemicellulose and lignin as major components. This inexhaustible resource is not yet fully exploited, because of the resistance of lignocellulose to chemical and biological transformations [172]. Microbial utilization of the lignocellulosic biomass for the production of industrial chemicals, liquid fuels, protein-rich food and feed, and preparation of cellulose polymers to help meet energy and food demands has received great attention in the past few decades.

Higher fungi, with white-rot fungi and mushrooms mainly reported as the major degraders of lignocellulosic material. Fungi infect wood by the aid of spores or hyphae, which secrete specific enzymes for attacking the cell wall of wood fibers. The complex structural features of lignocellulosic materials dictate unusual constraints on biodegradation systems. Fungi grow in nature on insoluble lignocellulose, with atmospheric moisture content, which then was the inspiration for FSSF development.

9.5.1. Structure of lignocellulosic biomass

The plant cell wall is a multilayered structure - S3 closest to the lumen, S2 the middle layer, and S1 the outermost layer [173] - and consisting of cellulose microfibrils embedded in an amorphous mixture of different hemicelluloses and lignin. The concentration of cellulose is highest in sub-layer S2 and decreases toward the middle lamella, where it is present only in small amounts. The S3 layer is rich in hemicelluloses, while lignin is the dominant compound in middle lamellae [174]. There are several detailed reports on this subject [175 - 179].

9.5.2. Degradation of lignocellulose by fungi

The fungi belong to *Ascomycetes, Deuteromycetes or Basidiomycetes* groups are able to produce necessary enzymes for the degradation of lignocellulosic material. Fungi living on dead wood that preferentially degrade one or more of the wood components cause three types of wood rot i.e., soft rot, brown rot and white rot [180, 181].

Important fungi causing soft-rot include *Chaetomium cellulolyticum, Aspergillus niger, Trichoderma viride (reesei), Fusarium oxysporum, Thielavia terrestris, Penicillium jenthillenum, Dactylomyces crustaceous* and different species of *Paecilomyces, Papulaspora, Monodictys, Allescheria, Hypoxylon, Xylaria and Graphium* [181, 182]. All these fungi efficiently attack wood carbohydrates but modify lignin only to a limited extent. Fungi causing brown-rot type of decay include, *Poria placenta, Tyromyces balsemeus, Gloeophyllum trabeum, Lentinus lepidius, Lenzites trabeum, Coniophora puteana, Laetiporus sulphureus and Fomitopsis pinicola* [181-184]. These also exhibit preference for cellulose and hemicelluloses, lignins being degraded only to a limited extent. White-rot fungi are the only wood degraders, to any extent, can attack all the components of plant cell walls. The most studied fungi of this group are *Phanerochaete chrysosporium, Trametes versicolor, Dichomitus squalens, Phlebia radiata, Heterobasidium annosum, Phellinus pini, Cyathus stercoreus, Pleurotus ostreatus, Ceriporiopsis subvermispora, Polyporus anceps and Ustulina vulgaris* [181, 185, 186]. The normal pattern of wood decay by these fungi involves simultaneous attack on both polysaccharides and lignin [181]. Fungal hyphae enter the cell lumen, first colonize the ray parenchyma cells, and then penetrate from cell to cell via pits or by the development of boreholes directly through the cell walls [181, 187, 188].

9.5.2.1. Cellulose hydrolysis

Cellulose hydrolysis is a complex process, which requires the synergistic action of several cellulases. Research on cellulase has progressed very rapidly in the past few decades, emphasis being on enzymatic hydrolysis of cellulose to glucose [189]. The synergistic effect of a combination of three enzymes shown below, the dependency of enzyme activity upon the degree of polymerization of cellulose and substrate inhibition effects in cellulose degradation have been studied using a mechanistic model (190). The enzymatic hydrolysis of cellulose requires the use of cellulase [1,4-(1,3:1,4)-β-D-glucan glucanohydrolase, EC 3.2.1.4], a multiple enzyme system consisting of three type of activities:

(1) Endo 1,4-β-D-glucanase
(2) Exo 1,4-β-D-glucanase (Cellobiohydrolase)
(3) Cellobiase

The most active cellulases are produced by fungi, although cellulases are also produced by a range of bacteria. However, relatively few are capable of producing all the necessary enzymes for degradation of crystalline cellulose [191]. Because of the resistance of native cellulose to enzymatic hydrolysis due to its intrinsic properties, pretreatment is necessary for the effective utilization of lignocellulosic material. Different pretreatment methods employed are: physical (mechanical), thermal (autohydrolysis, steam explosion, hydrothermolysis), chemical (alkali, acid, oxidizing agents, gases, cellulose solvents, solvent delignification) or by biological treatment (whole cells or by isolated enzymes), or by a combination of theses methods, which in principle cause breakdown of the structure, thereby creating large surface area on which enzymes can act. The main intention of pretreatment is to open the structure of the lignocellulosic materials, making it accessible to the cellulolytic enzymes [192]. This can be accomplished by (1) increasing the specific surface area, (2) removing lignin, or (3) by solubilizing hemicellulose. The pretreatment methods and its effect on lignocellulosics was reviewed by Ghosh and Singh [193].

9.5.2.2 Lignin degradation

Phanerochaete chrysosporium is the most efficient and best characterized lignolytic organism available that posses characteristic features of rapid growth, extensive degradation of lignin, relatively high temperature optima, formation of asexual spores (conidiospores) in abundance, and rapid completion of the sexual cycle [194]. For lignin metabolism, the lignolytic system of the fungus is produced constitutively, expressed only during secondary metabolism, triggered by carbon, sulfate, and nitrogen limitation and markedly influenced by oxygen concentration [195, 196]. Lignin peroxidase is the major enzyme involved in lignin degradation [197]. White-rot fungi also produce two other types of extracellular enzymes, Mn (II) dependent peroxidases (MnPs) and glyoxal oxidase, an enzyme involved in extracellular H_2O_2 production. Both types of peroxidases are usually involved in lignin degradation, although they differ in their catalytic mechanisms [198].

Different pretreatment methods are used to disrupt the structure and make it easily accessible to enzymatic hydrolysis. Particle size reduction by milling the substrate is generally employed. Acid and alkali pretreatment are generally used to promote cellulose utilization and are not used when lignin degradation is desired [61]. Autoclaving increases the availability of growth-limiting nutrients and favors growth and non-lignin conversions while depressing lignin utilization [10]. However, a hot water treatment promoted lignin utilization in SSF [199]. Addition of sugars in small quantities induces lignin degradation, since lignin is co-metabolized rather than being used as a sole carbon source of carbon and energy [200].

9.5.3. Use of lignocellulose as substrate for SSF process

Numerous workers have reported on different lignocellulosic substrates. However, due to space limitations, only few references are shown. Wheat straw [201], corn and rice stover [62], wheat bran [202], sugar beet pulps [203], feedlot waste fibres [204] and wood [205]. The aim of these studies was usually either to increase the protein content for feeding to ruminants, or for the production of cellulases or other enzymes.

9.5.3.1. Fungal growth on lignocellulose in SSF

Fungal degradation of lignocellulosic material is a complex process. Typically, the fungal attack starts at a weak point of the plant cell wall, into which an active hyphal tip rich in the enzyme complex penetrates. Filamentous fungi grow on the substrate surface in the absence of free water, utilizing the bound water of the substrate [3, 43]. Fungal growth on lignocellulosic substrate is linear rather than logarithmic rate, limited by steric hindrance and substrate accessibility [172]. The solid substrate surface is inoculated with the fungus (spore or hypha) must first colonize the substrate surface by adhesion, then spread from one particle to another by branching. The fungal hyphae penetrate into the solid particle and the mycelium releases enzymes which diffuses into the substrate and hydrolyze the polysaccharide. The soluble sugars released then diffuse back to the mycelium. And oxygen diffuses from the interparticle spaces through a stagnant gas film into the mycelial layer on the particle surface. The mycelium simultaneously consumes the soluble sugars and oxygen [206]. In the core, lignin surrounds and protects the cellulose from enzyme hydrolysis, thus partial delignification is neccessary for cellulose hydrolysis in plants having higher than 10 % lignin content [172]. A coordinate attack of cellulases and ligninases is hard to achieve, due to their substrate specificity and complexity, making lignocellulose degradation a low-efficient process [43, 196, 207, 208]. This low efficient and comparatively slow process, which take 3-8 weeks for maximum degradation of hardwood and 3-4 weeks for softwood, is uneconomical for high technology industrial fermentation, where a maximum of 3-4 days would be required for the profitable operation [172]. Pretreatment of the lignocellulosic substrate causes partial delignification, increasing the separation of lignin from cellulose causing the removal of the outer cementing components of the cell wall, and therefore increases the efficiency and rate of the hydrolysis [209].

9.5.4. Applications of lignocellulose conversion by SSF

The increasing interests for the bioconversion of lignocellulosic materials as a renewable feedstock for the microbial production of food, feed and chemicals, and the applications of fungi in biotechnology have led to further exploitation of solid state fermentation (SSF) systems. The inherent steric limitations of fungal growth on solid substrate leads to lower substrate utilization and poor biomass conversion compared to liquid cultures. From the lab-scale lignocellulose conversion studies by liquid and solid state fermentation, it is evident that liquid culture gives greater efficiency of conversion and higher biomass (protein) yield; however, solid culture has a greater volumetric productivity [172]. The economic picture for lignocellulosic degrading enzyme production is quite favorable [3, 43] for scale-up. The high volumetric productivity and reduced recovery costs in SSF may assure viability for such high value products. Most of the cell wall degrading enzymes are inducible by lignocellulosic substrates, thus producing profitable enzymes by SSF [172]. The production of enzymes by solid state fermentation is detailed in a former section of this chapter.

The presence of lignin along with cellulose and hemicellulose decreases the cellulase activity and the feed value of lignocellulose wastes [210]. The white-rot fungi, belonging to the *Basidiomycetes*, are the most efficient lignin degraders [207], and their biodegradation ability has been extensively reviewed by Leonowicz et al. [211]. They are capable of depolymerizing lignin and metabolizing it to CO_2 and H_2O. The lignin degrading system of this fungi have great potential applications in improving the digestibility of wood or straw for animal feed [212 – 215], biobleaching and biopulping [216 – 221] and the biodegradation of xenobiotic pollutants [222 – 224]. Some species can remove lignin faster than cellulose and hemicellulose, and thus could most useful for the pretreatment of wood and straw for enzymatic saccharification to increase digestibility for feed purposes or for biological pulping [225]. Lignin-degrading system of the white-rot fungi have potential applications in the area of lignocellulose bioconversion by SSF [226]. Progress in this area has been made with white-rot fungi [227 – 229], and the recent reports shows the production of oxidative enzymes in solid state fermentation systems that have not been reported in liquid cultures previously [230]. The most widely studied of these is *Phanerochaete chrysosporium* [231 – 234]. Studies have been focused on the physiology, enzymology, and molecular genetics of the biodegradation process. The key enzyme for polymer fragmentation by *P. chrysosporium* is lignin peroxidase [231, 232, 235].

Lignocellulose degradation and activities related to lignin utilization were studied in the solid state fermentation system of cotton stalks by comparing two white-rot fungi *P. chrysosporium* and *Pleurotus ostreatus* [236]. *P. chrysosporium* grew vigorously, with rapid nonselective degradation of 55% of the organic components of the cotton stalks within 15 days. Whereas, *P. ostreatus* grew more slowly with obvious selective lignin degradation of only 20% of the organic matter after 30 days incubation. The relationship between CO_2 evolution and the extent of ligninolytic and proteolytic activities of *Phanerochaete chrysosporium* grown on sugarcane bagasse pith using SSF was investigated [237]. CO_2 evolution profile showed that MnP activity was expressed during the idiophase when glucose was limited then decreased due to the protease activity. An inert carrier (nylon sponge), a non-inert carrier (barley straw) and the addition of veratryl alcohol or manganese (IV) oxide to the cultures were found to induce the production of ligninolytic enzymes by *Phanerochaete chrysosporium* during semi-solid state fermentation conditions [238].

Effect of supplementary carbon sources and exogenous phenolic compounds on the lignocellulolytic system of *Cerrena unicolor* during the solid state fermentation of grapevine cutting wastes was reported by Elisashvili et al. [239]. During the process lignocellulolytic system of the higher Basidiomycete *Cerrena unicolor* could be regulated by supplementary carbon sources and exogenous phenolic compounds. The addition of glucose or cellobiose to the fermentation medium with grapevine cutting wastes enhanced laccase and ligninase activities and decreased cellulase and xylanase activities of *C. unicolor* over the first week of SSF. Conversely, Avicel increased the polysaccharase activity of the culture, decreased laccase activity, and fully inhibited ligninase activity. They concluded that the relatively lower activities of cellulase and laccase in the presence of phenolic compounds might be due to the retarded growth of the fungus rather than to the suppressed synthesis of these enzymes. Soyhull a lignocellulosic material produced in large amounts during soybean processing, was utilized for cellulase production through FSSF [240]. Of five known fungi, *Phanerochaete chrysosporium* gave maximum yields of the enzyme. The white rot fungus *Pleurotus ostreatus* was grown in a chemically defined solid stale fermentation system amended with cotton stalk extract [241]. Treated cultures exhibited increased laccase activity as well as enhanced lignin mineralization. The *basidiomycetous* fungus *Nematoloma frowardii* produced manganese peroxidase (MnP) as the predominant ligninolytic enzyme during SSF of wheat straw [242]. They stated a partial extracellular mineralization of lignin on the basis high levels of organic acids in the fermented straw. The maximum concentrations in the water phases of the straw cultures were 45 mM malate, 3.5 mM fumarate, and 10 mM oxalate, which rendered MnP effective and therefore a partial direct mineralization of lignin can be achieved.

9.5.4.1. Lignocellulose conversion into fungal biomass

Processing of agricultural wastes in solid state fermentation for microbial protein production was reviewed [225, 243]. The production of fungal biomass from lignocellulosic substrates has been investigated widely at laboratory or pilot scale by several groups [225]. Lignocellulosic material can be converted to microbial biomass by some yeasts, fungi and bacteria by direct cultivation and utilization [244 – 248].

Important factors affecting lignin degradation are temperature, pH, water content [249], carbon dioxide and oxygen levels [10, 249 – 253]. In general carbon dioxide is inhibitory [249] and oxygen stimulates lignin degradation, which may be best with a pure oxygen atmosphere [10, 251], although specific effects of oxygen on lignin degradation depend on the particular fungus [250]. The presence of added nitrogen source will, in general favor growth and non-lignin conversions while depressing lignin utilization [249, 252]. Two groups of fungi can be distinguished on the basis of the effect of added nitrogen on lignin degradation, with one group being stimulated and the other depressed [253]. Manganese-mediated lignin degradation by *Pleurotus pulmonarius* solid state fermentation of [C^{14}] lignin wheat straw was reported by Camarero et al. [254].

Chaetomium cellulolyticum was used to upgrade lignocellulosic wastes to microbial biomass [204, 255 – 258] and *Trichoderma lignorum* for animal feed production [257]. Conversion of cellulosic and other organic matter into microbial proteins by fungi has been reported for usage as microbial proteins a highly desirable feed for intensive aquaculture [259]. The fibrous residue of the Napier grass (*Pennisetum purpureum Schumaker*) remaining after protein extraction was converted into a protein-enriched animal feed supplement by solid-state fermentation with *Phanerochaete chrysosporium*

[260]. Best improvement in feed value of wheat straw was obtained using young mycelium in the form of small pellets of *Cyathus stercoreus* [261]. Biological upgrading of wheat straw by the white-rot fungi *Phanerochaete chrysosporium*, *Pleurotus eryngii*, *Phlebia radiata*, and *Ceriporiopsis subvermispora* was monitored during 60 day SSF [262]. The degradation of lignocellulosic biomass of banana pseudostem during SSF by *P. ostreatus* and *P. sajor-caju* and their potential use in the conversion into mycelial protein-rich fermented animal feed was investigated [263].

Technological improvements in bioreactor design and fermentation process may lead to further improvements in biomass productivity. Use of fed-batch or continuous operation of SSF can improve efficiency, because sporulation and feedback inhibition of enzyme production by catabolic repression may be avoided [7]. By combining solid (*Trichoderma viride* or *Coriolus versicolor*) and liquid cultivation (*Endomycopsis fibuliger* or *Candida utilis*) in a percolator bioreactor, where the mold grew on wheat straw and the yeast in the continously percolated liquid, the biomass yields and conversion efficiency of wheat straw has been improved, mainly because of substrate utilization and avoidance of catabolite repression by reducing sugars [264]. Controlling temperature and moisture by evaporative cooling and water replenishment in a rotating drum bioreactor increased biomass productivity by 50% in SSF of *Rhizopus oligosporus* on corn grit [265].

Integrated processes, sequential culture by different microbes or mixed culture processes in a micro-ecosystem show promise for future SSF application. It has been reported that 48 h anaerobic fermentation suppressed indigenous microflora and permitted colonization of *Pleurotus ostreatus* on non-sterile straw equal to that on autoclaved straw [266]. Straw delignified with white-rot fungi can be anaerobically digested to methane more efficiently than untreated straw [267]. Anaerobic lignocellulose degrading fungi perform delignification efficiently in the rumen in synergism with other rumen microorganism [268, 269]. An alternative process to aerobic fungal SSF is a new biological process combines traditional ensiling (anaerobic lactic acid fermentation) with enzymatic hydrolysis of the plant fiber (ENLAC), which has shown improved digestibility of the fibers and increased recovery of protein [172].

Reid [21] studied the optimization of solid-state fermentation for selective delignification of aspen wood with *Phlebia tremellosa*. The white-rot fungus *Phlebia tremellosa* can delignify aspen (Populus) wood and increase the accessibility of its polysaccharides through enzymatic hydrolysis (prior to use in pulping, for animal feed, or as sugars). Forced aeration was needed for delignification of wood layers more than a few millimeters thick but air circulation between the particles was not blocked by mycelium. Bruder [270] reported on the solid-state fermentation of sugarbeet pulp by filamentous fungi and concluded that the increase of feed value in this way was technically successful but uneconomical.

Possible biotechnological methods of treating beet pulp are ensiling, submerged fermentation to produce biogas, or protein enrichment by submerged or solid-state fermentation using *Fusarium oxysporum* and fungi imperfecti [271]. Mathot and Brakel [272] reported on the microbial up-gradation of feed by solid state fermentation. *Aspergillus niger* used in simple, non-aseptic fermentation of barley, wheat and dehydrated beet pulp increased true protein content by 20% after 48 h fermentation. Grujic et al. [273] studied the possibilities of the solid-state fermentation of new and of fermented alkali pretreated grape pomace, and of the alkali pretreated ground maize stover into protein-enriched animal feed by *Chaetomium cellulolyticum* alone or in the mixed culture with *Candida utilis*. Fermentations were performed in tray bioreactors and in an aerated bioreactor with air humidification. Solid-state fermentation of alkali-pretreated maize stover, resulted in protein increases of 52-71%, and average growth rates of 0.092-0.246/d were obtained, depending on the pretreatment conditions.

A thermophilic strain *Thermoascus aurantiacus* produced alpha-L-arabinofuranosidase on a medium based on leached sugarbeet pulp [274]. The thermostability of this enzyme would be useful in the manufacture of pelleted animal feeds. SSF of sugar beet pulp yielded dietary fibers, including soluble fibers, and an enzyme (alpha-L-arabinofuranosidase) produced by *Trichoderma reesei* and *Aspergillus niger* which increased pentosan digestibility and is useful in feedstuffs for monogastric animals [275].

Zadrazil and Puniya [276] reported on the effect of particle size of sugarcane bagasse into animal feed using white-rot fungi *Pleurotus sp.*, *Agrocybe aegarita*, *Pleurotus eryngii* and *Kuehneromyces mutabilis*. The digestibility of all fractions was increased by *P. eryngii* but decreased by *A aegarita;* the other fungi had inconsistent effects but all 3 increased the digestibility of the coarsest fraction. Hence, the extent of breakdown and separation can affect the culture of white-rot fungi to enhance the nutritive value as animal feed (and presumably also in the culture of edible fungi). A two-fold enrichment of protein content of sugar beet pulp, wheat bran and citrus waste by SSF using *Neurospora sitophila* was reported [277]. A strain of white-rot fungus *Lentinus edodes* resulted in about 50 % improvement in dry matter digestibility [278].

Scerra et al. [279] reported on the influence of *Penicillium* sp. on nutrient content of citrus fruit peel by SSF with 3 strains of *Penicillium camembertii* and 7 strains of *Penicillium roquefortii*. The activity of these moulds could prove a cost-effective way of recycling citrus fruit peel in animal feed, thus increasing its nutritional value by single cell protein. An efficient way of utilizing lignocellulosic wastes is in the cultivation of edible fungi which is discussed in an earlier section.

9.6. Composting

Composting has been used as a treatment process for solid wastes and slurries of organic origin for many centuries. The aerobic decomposition of high solid organic wastes through composting is a solid state fermentation process [46, 280] that is receiving considerable attention in the recent years due to its environmental significance. Recently, composting is also been used as a treatment procedure for hazardous materials [281]. Compost (the final product of the composting process) has been shown to suppress plant pathogens [282] and has been used successfully as a medium for biofiltration and bioremediation [283 – 287].

Composting accelerates the biodegradation process of organic wastes and produces a stable material compost, which can be used to improve the soil properties. The composting process is conditioned by temperature, moisture and aeration, the latter

influence the other two factors [288]. Composting is a controlled biooxidative process, which involves the following stages:
(1) a heterogeneous organic substrate in the solid state, (2) degradation process passing through a mesophilic stage initially, followed by a thermophilic phase and a final mesophilic stage (3) and leads to the production of CO_2, water, minerals and stabilized organic matter (compost). The process and parameters involved in composting was detailed by Stentiford and Dodds [289]. The composting process generally results in a 25-35% reduction (dry basis) of the waste material, due to the release of CO_2 and H_2O by the growing microorganisms [46].

9.6.1 Role of fungi in composting

Compost is a rich source of nutrients and thus harbors a diversified microbial flora includes bacteria, fungi, actinomycetes, viruses, helminths and protozoa [290 – 293]. Microbial flora differs in number and type to some extent depending on the compost starting material, and the geographical location [294]. Temperature is considered as the key parameter in composting both for process control and for sanitation provided the moisture content and the aeration do not become limiting parameters [295]. Generally a microbial progression from mesophilic to thermophilic was observed during composting. Subsequently, as substrate becomes limiting or when the process conditions fall below optimum level, the temperature of the compost mass falls leading to re-establishment of a mesophilic population. Other conditions such as moisture content, hydrogen ion concentration (pH) and microbial interaction can also influence the diversity of the microbial population [289].

Fungi flourish where conditions are suboptimal for other microorganisms, for example at lower moisture contents [293, 296] (RH 90%) and at lower pH values. Both mesophilic and thermophilic species are isolated from the compost [289]. A wide range of fungi exist in composting masses [296 – 298]; as with bacteria the diversity decreases as the temperature increases, no fungi have been isolated at temperatures greater than 60°C [291, 296]. Fungi play an important role in the decomposition of cellulose and other complex carbonaceous materials [293, 297, 299]. Lignin decomposition is limited to higher fungi (*Basidiomycetes*) [293, 297]. The decomposition of these complex materials takes place later in the composting process, after the thermophilic stage.

In the initial stage, mesophilic organisms, mainly bacteria begin to breakdown soluble sugars, starches, some proteins and some hemicelluloses. Metabolic heat generation leads to a rapid rise in temperature, which is further enhanced by the insulating properties of the composting mass [290, 300]. This results in a change in the microflora from mesophilic to thermophilic in nature and less diverse [290, 291, 296, 299, 301], temperature continues to rise and optimum degradation rates are obtained at temperatures in the range 45-55°C [297]. In uncontrolled conditions the composting mass will continue to release metabolic heat and temperatures rise as high as 80°C, partially due to chemical reactions [296, 302]. This excessive temperature is not conducive to composting, as only a few bacterial species are viable at temperatures above 70°C [302]. With some hazardous materials, however the temperature will not reach 50°C [281]. If other conditions are optimum, the temperature declines when the readily available substrates become exhausted [46], allowing the recolonization of the material by mesophilic organisms (from heat resistant spores or by re-invasion from outside) [290, 291].

The composting process is conditioned by temperature, moisture and aeration, the latter of which influences the other two factors [288]. Moisture content varies during composting, linked to metabolic water production during aerobic respiration [296], and to interstitial evaporative cooling followed by subsequent moisture loss from the composting mass [301]. However, the metabolic water produced by the fungi eventually leads to the recolonization of bacteria in the composting material [296]. The composition of gas phase can be used to monitor how effective the process is in achieving optimum development of the microbial population [303]. In general, a minimum of 5% O_2 (v:v) is required in the composting material [304]; excessive aeration does not increase the decomposition rate but leads rather to excessive drying [290]. On the other hand, the optimum temperature for O_2 uptake and CO_2 release is between 55°C and 65°C. High relative concentration of CO_2 in the composting material is considered to reflect the intense microbial development required for compost formation [288].

C/N ratio and pH also alters throughout composting and influence the microbial population. The initial C:N ratios (52:1, 35:1, 48:1, 47:1) for four different compost heaps were significantly reduced to 24:1, 14:1, 18:1 and 12:1, respectively, after 60 days of composting, resulting in the production of a stable humus that was suitable for crop production [305]. A trial run on composting process residuals of seaweed (*Ascophyllum nodosum*) was described in which composting may be used to produce materials suitable for land application, such as tree planting or mulching, from seaweed-based decanter waste otherwise unsuitable for any horticultural purpose [306]. This organic material self-heated readily, suggesting potential as a biological compost activator. Phytotoxicity of the waste was reduced, but not eliminated, by composting. It was concluded that organic nature of material is an inadequate sole indicator of desirability as a compost ingredient. Most recently, the feasibility of adjusting individual composting processes to produce the required composition of desired compost mass for mushroom cultivation was investigated [307].

9.6.2. Composting of toxic compounds

Composting has been used as a means of treating soil contaminated with chlorophenols [281, 308] and TNT [309 – 311]. A field study has shown that the concentrations of TNT, RDX and HMX in contaminated sediments placed into composts degraded to a marked extent [312]. Laboratory scale test of the composting of TNT suggests that an initial anaerobic treatment before the aerobic phase of composting is desirable for the complete destruction of the explosive [313]. Potter et al. [314] reported extensive degradation of PAHs in bench scale studies.

9.7. Bioremediation

Bioremediation refers to the productive use of biodegradative processes to remove or detoxify pollutants in the environment. Solid phase treatments are widely used for petroleum contaminated soils, which is constantly finding new uses [315]. FSSF may play a crucial role in the development of mass culturing and delivery of biopesticides [172] for biocontrol application and for bioremediation of xenobiotic compounds, which has been receiving considerable attention in the recent years. Some of the recent advances in this field are summarized below.

9.7.1. Biocontrol agents

Biological pest control is increasingly receiving attention due to the search for eco-friendly pesticides and consistent with the goal of sustainable agriculture. The knowledge about the potential use of fungi as biocontrol agents dates back to Louis Pasteur. The first practical attempt was made to control wheat chafer, *Anisoplia austriaca*, and the sugar beet curculio, *Cleonis punctiventris*, using the fungus *Metarhizium* during 1880s [316]. Fungi constitute the largest group of more than 500 species that can parasitize insects. Most of the taxonomic groups contain entomopathogenic fungi. Genera such as *Metarhizium, Beauveria, Verticillium, Nomuraea, Entomophthora, Neozygites and Pythium* are commonly found in nature [317]. Although sufficient knowledge has been accumulated on the use of biocontrol agents in agriculture, an effective delivery system and the way to monitor the fate of the delivered microbial agents in the environment is lacking [172].

Bacterial pesticides are already commercially in use. Among fungi biocontrol agents are *Trichoderma sp* that parasitize soil borne pathogenic fungi and stimulate plants for more vigorous growth [318]. The main impediment in their widespread use is lack of economical means for the mass production of spores or mycelial biomass, and their proper formulation in effective delivery systems to the plant root zone. Special plant protective fungi, such as *Trichoderma sp* or mycorrhizae (fungal associations with plant roots), may be mass produced by SSF systems, resembling or combined with composting [172]. Both liquid and solid state fermentation systems can be used for biopesticide production. SSF is having potential advantages over liquid cultures for biopesticide production, the spores produced as living propagules in SSF tend to be more tolerant to desiccation and more stable as a dry preparation compared with spores produced in submerged fermentation. Large scale production of infective propagules in a two-stage cultivation system in which fungus is first grown in submerged fermentation and then allowed to sporulate as a surface culture in either semisolid medium or still culture is also widely used. For stable biopesticide formulation, the fungal propagules in biopesticides must be alive at the time of their use and expected to have a minimum shelf life of 18 months at 20°C [317].

Tengerdy [172] reported on the laboratory investigation of *T. harzianum* sp in a multi-tray semi-continuous SSC reactor. After 6-7 days cultivation, 75% of the fungal biomass converted to chlamydospores and they can survive ten times longer than the conidia in the soil. The bioreactor content was directly mixed with potting soil and used as a biocontrol agent. *Entomophthora virulenta* is pathogenic to aphids that can produce zygospores in surface and submerged fermentation [319], which can be used in biopesticide formulations. The entomopathogenic fungus, *Beauveria bassiana*, was cultivated by SSF for the production of a new bioinsecticide against European corn borer, which shows a field efficiency of 80%, which persists for 3 weeks [320]. Mycoparasitic *T. harzianum* strains develop chlamydospores in submerged fermentation and conidia on solid culture [317].

Biosynthesis of flaviolin and 5,8-dihydroxy-2,7-dimethoxy-1,4-naphthoquinone by *Aspergillus niger* and 2,7-dimethoxynaphthazarin by *Streptomyces* using SSF technology was reported [321]. Radioactivity from shikimic acid was effectively incorporated into flaviolin; this conversion, however, proceeded by way of acetic acid. The latter stages of biosynthesis of 2,7-dimethoxynaphthazarin by the Streptomycete follows the pathway: flaviolin to mompain initially and then to 2,7-dimethoxynaphthazarin. Mutagenicity and antibacterial activity of mycotoxins produced by *Penicillium islandicum* and *Penicillium rugulosum* in solid-state fermentation on grains was reported by Stark *et al.* [322].

SSF of *Lathyrus sativus* seeds using *Aspergillus oryzae* and *Rhizopus oligosporus* to eliminate the neurotoxin beta-ODAP without loss of nutritional value was evaluated [323]. The nutritional qualities were improved in the fermented seed meal with increased content of protein, higher amino acid content for sulphur-containing and aromatic amino acids. Better resistance to high temperature and oxidation, and drastic reduction of the flatulence factors was also noticed. Daigle et al. [324] reported on the production of fungal biocontrol agents by SSF plus extrusion as an effective process for encapsulating hardy fungal propagules in granular products for the biocontrol of agricultural pests. *Colletotrichum truncatum*, an *Alternaria sp.*, *Paecilomyces fumosoroseus*, or a toxigenic *Aspergillus flavus* and *Aspergillus parasiticus*, was mixed with wheat flour, kaolin, and water and extruded into granules grown on rice flour in plastic bags. The inoculum survived extrusion and fluid bed drying at 50°C and was better than the inoculum produced in liquid fermentation. Depending on the agent, the high level of flour infestation permitted a 1:9 to 1:1600 dilution to yield the $1X10^6$ cfu/g in the final product, which is usually needed for biocontrol efficacy.

Development of a suitable medium and technological improvements in the SSF technology is necessary for the economic feasibility for large-scale mycopesticide production. *Metarhizium anisopliae, Beauveria tenella,* and *Aschersonia aleyrodes* produce aerial spores when cultivated on mixtures of organic bases and porous materials within 4 weeks [325]. Spore production of *Coniothyrium minitans* on laboratory scale SSF using several natural media, extracts and agar media were reported [326 – 328]. A chemically defined solid medium was developed with starch and trace elements, found to be inducing spore production of *Coniothyrium minitans* [329]. Use of lignocellulosic substrate to increase water-retention capacity and forced aeration for dissipation heat can lessen the technological problems related to SSF. It was reported that sporulation of *M. anisopliae* was significantly improved with the forced aeration in a column bioreactor [330]. An experimental ecosystem was simulated using a packed-bed, continuous-flow column reactor for the interactions between *Fusarium culmorum*, a causative agent of foot rot in cereal plants in the soil, and its potential biocontrol agent, *Trichoderma harzianum* [331].

9.7.2. Bioremediation of toxic compounds

Though biodegradation of waste is a centuries old technology, only recently that serious attempt have been made to use biodegradative process with the goal of large scale technological applications for effective and affordable environmental restoration. Soils are often treated by solid-phase technologies, which has been receiving considerable interest in the recent years [281, 315, 332]. Solid-phase bioremediation describes the treatment of the contaminated soil or other solid materials (eg. dredged sediment or sludge) in a contained, aboveground system using conventional soil management practices (i.e., tilling aeration, irrigation, fertilization) to enhance the microbial degradation of organic contaminants. Three treatment strategies are considered 1) landfarming, (2) composting, and (3) engineered soil cells [315].

Landfarming describes the process of treating contaminated surface soil on-site via conventional agricultural practices. The goal of landfarming is to cultivate catabolically relevant, indigenous microorganisms and facilitate their aerobic catabolism of organic contaminants. Composting is a variation of solid-phase treatment that involves adding large amounts of readily degradable organic matter to a contaminated material, followed by incubation, usually aerobic lasting several weeks or months. The technology is very similar to landfarming, exception that the material is not necessarily aerated or tilled. The engineered soil cells is a hybrid of the landfarming and the composting processes. The greatest advantage of the solid-phase bioremediation is the on-site operation of the process. Compared with *in situ* technologies, the kinetics of biodegradation of toxic chemicals in *ex situ* bioremediation are more favorable and economical [315].

White-rot fungi, the cultures of *Phanerochaete chrysosporium*, the most widely studied fungi for its xenobiotic bioremediation, which has the ability to degrade various compounds in soil [333, 334] and a number of PAHs, PCBs, 2,3,7,8-tetracholorodibenzo-p-dioxin, DDT, lindane, chlordane, and explosives such as TNT, RDX and HMX [281]. Pleurotus is another lignocellulolytic fungus that has been studied for bioremediation properties [335]. Xenobiotic breakdown of white-rot fungi is thought to be related to oxidative enzymes secreted by fungi to effect lignin depolymerization [336, 337], as most of the recalcitrant environmental pollutants have the carbon skeleton similar to lignins [333]. Exploitation of the ligninolytic properties of *basidiomycetes* is becoming increasingly important in developing bioremediation processes [338 – 340].

Degradation of PCP by *Phanerochaete* sp was tested on soil [341, 342]. Enhanced reductive dehalogenation of polychlorinated biphenyls (PCBs) was accomplished by SSF in reactors of 15 cubic meters of volume by natural attenuation process [343]. The suitability of solid-state fermentation techniques to dispose of atrazine (purity 99.0%) and carbofuran (purity 99.0%) formulations was evaluated [344]. They developed a SSF simple less expensive technique to dispose pesticide waste. The biotransformation of atrazine added to a mixture of cotton and wheat straw (CWS) and inoculated with the white-rot fungus, *Pleurotus pulmonarius*, was studied, as a proposed system for bioremediation [345]. The concentration of methanol-extractable atrazine was reduced, due to both biological transformation and physical-chemical adsorption to the straw. Only 32% of the total radioactivity added as ^{14}C-ring-labeled atrazine to pasteurized CWS inoculated with *Pleurotus* was extracted two weeks after fungal colonization, and less than 70% from non-inoculated CWS. The reduction in extractable radioactivity increased with time of incubation. No mineralization of the triazine ring was found during six weeks of incubation, but transformation to two groups of atrazine metabolites, chlorinated and dechlorinated, occurred, as a result of the activity of the fungus inoculated and natural bacterial population.

Zafar et al. [346] reported the influence of nutrient amendment on the biodegradation of wheat straw during SSF with *Trametes versicolor*. Kuilman and Fink [347] reported on a co-operative study between the European countries to find the safety of three decontamination processes of contaminated peanut meal based on solid-state fermentation using biological degradation of the mycotoxins by non-toxicogenic moulds. The suitability of SSF as a means to dispose of carbofuran (2,3-dihydro-2,2-dimethyl-7-benzofuranyl methylcarbamate) and the fate of [U-ring-^{14}C] labeled carbofuran was determined in incubation chambers containing either sphagnum peat moss (limestone-amended) or a peanut hulls/peat moss/ steam-exploded wood (1:1:1) mixture (PPW) [348]. The metabolite carbofuran-7-phenol, detected in the ethyl acetate extracts of the peat and PPW mix following 4 and 8 week of incubation, was susceptible to oxidative coupling in the presence of horseradish peroxidase (HRP) and H_2O_2, suggesting that this metabolite had become covalently bound to organic matter during humic substance formation. These results may provide the basis for the development of an effective and economical method for detoxification or containment of waste pesticide residue.

Two strains of white-rot fungi, *Dichomitus squalens* and a *Pleurotus* sp., were tested for their effect on mineralization of ^{14}C-labelled pyrene [349]. *Pleurotus* sp. was highly resistant to microbial attack and had the ability to penetrate the soil. *D. squalens* was less competitive and did not colonize the soil. The resistance of the fungus was dependent on the duration of fungal pre-incubation. Mineralization of [^{14}C] pyrene by mixed cultures of *D. squalens* and soil microorganisms was higher than by the fungus or the soil microflora alone when soil was added after 3 weeks of incubation or later. With *Pleurotus* sp., the mineralization of [^{14}C] pyrene was enhanced by the soil microflora irrespective of the time of soil application. With *D. squalens*, which in pure culture mineralized less [^{14}C] pyrene than did *Pleurotus* sp., the increase of [^{14}C] pyrene mineralization caused by soil application was higher than with *Pleurotus* sp. Both dechlorination and mineralisation of the xenobiotic compound were effected by solid substrate cultures of *Lentinula edodes* when inoculated sterilized and non-sterilized soils contaminated with pentachlorophenol (PCP) [340].

Solid-state fermentation of eucalyptus wood with several fungal strains was investigated as a possible biological pretreatment for decreasing the content of compounds responsible for pitch deposition during C^{l2}-free manufacture of paper pulp [350]. The results showed that some *Basidiomycetes* that decreased the lipophilic fraction also released significant amounts of polar extractives, which were identified by thermochemolysis as originating from lignin depolymerization. *Phlebid radiata, Funalia trogii, Bjerkandera adusta*, and *Poria subvermispora* strains were identified as the most promising organisms for pitch biocontrol, since they degraded 75 to 100% of both free and esterified sterols, as well as other lipophilic components of the eucalyptus wood extractives. *Ophiostoma piliferum*, a fungus used commercially for pitch control, hydrolyzed the sterol esters and triglycerides, but it did not appear to be suitable for eucalyptus wood treatment because it increased the content of free sitosterol, a major compound in pitch deposits. *Phanerochaete chrysosporium* grown on sugarcane bagasse pith, a lignocellulosic residue, is proposed as a bioremediation agent for aromatic contaminated soils [351]. To investigate the use of pith for the development of a fungal inoculum, the effect of culture conditions on fungus survival and microbial respiration under solid state fermentation were studied.

10. CONCLUDING REMARKS

Fungal solid state fermentation (FSSF) can be a viable alternative to submerged fermentation (SmF). The need for recognition of FSSF when SmF is usually viewed as a better way to produce biological compounds stems from the simple principle that nature has evolved efficient systems. Worldwide, nations have started looking for eco-friendly technologies in all aspects of life. FSSF would be an ideal technique to utilize the abundant quantities of agro-industrial and forest residues beckoning to be converted into useful products, by exploiting the natural capabilities of fungi. There is hope as industries have once again started considering FSSF as a potential way to harness fungal capabilities. FSSF is eco-friendly, simple, economical and flexible to both developing and developed countries. The future is going to witness great strides in this dynamic area.

REFERENCES

1. J.E. Smith and D.R. Berry, Eds. The filamentous fungi, Edward Arnold, London, vol. 1, (1975).
2. C.P. Kurtzman, Mycologia. 75 (1983) 374.
3. M. Moo-Young, M., A.R. Moreira and R.P. Tengerdy, In: The Filamentous Fungi, vol..4., ed. J.E. Smith, D.R. Berry and B. Kristiansen, Edward, Arnold, London (1983) 117.
4 B.K. Lonsane, N.P. Ghildyal, S. Budiatman and S.V. Ramakrishna, Enzyme Microb. Technnol. 7 (1985) 258.
5 D.A. Mitchell, In: Solid Substrate Cultivation. Eds: H.W. Doelle, D.A. Mitchell and C.E. Rolz, Elsevier Applied Science, London, 17, 1992.
6 H.J. Kronenberg, Econ. Bot. 38 (1984) 433.
7 A.L. Abdullah, R.P. Tengerdy and V.G. Murthy, Biotechnol. Bioeng. 27 (1985) 20.
8 W.E. Trevelyan, Tropical Science. 16 (1974) 179.
9 M. Raimbault and D. Alazard, Eur. J. Appl. Microbiol. Biotechnol. 37 (1980) 203.
10 R.E. Mudgett and A.J. Paradis, Enzyme Micorb. Technol. 7 (1985) 150.
11 S.Y. Huang, H.H. Wang, C-J. Wei, G.W. Malaney and R.D. Tanner, In: Topics in Enzyme and Fermentation Biotechnology, ed. A. Wiseman, Ellis Horwood, Chichester, England, 10 (1985) 88.
12 W.R. Gibbons and C.A. Westby, Appl. Environ. Microbiol. 52 (1986) 960.
13 L. Ikasari and D.A. Mitchell, World J. Microbiol. Biotechnol. 10 (1994) 320
14 H. Narahara, Y. Koyama, T. Yoshida, S. Pichangkura, R. Ueda and H. Taguchi, J. Ferment. Technol. 60 (1982) 311.
15 W. Penaloza, M.R. Molina, R.G. Brenes and R. Bressani, Appl. Environ. Microbiol. 49 (1985) 388.
16 W.J. Scott, Adv. Food Res. 7 (1957) 83.
17 J.E.L. Corry, Progr. Ind. Microbiol. 12 (1973) 73.
18 L. Leistner and W. Rodel, In: Inhibition of and Inactivation of Vegetative Microbes. ed. F.A. Skinner and W.B. Hugo, Academic Press, London (1976) 219.
19 W. Grajek and P. Gervais, Enzyme Microb. Technol. 9 (1987) 658.
20 R.E. Mudgett, In: Manual of Industrial Microbiology and Biotechnology, ed. A.L. Demain and H.A. Solomon. American Society for Microbiology, Washington, D.C, (1986), 66
21 I.D. Reid, Enzyme Microb. Technol. 11 (1989) 786.
22 J.J. Laukevics, A.F. Apsite, U.E. Viesturs and R.P. Tengerdy, Biotechnol. Bioeng. 26 (1984) 1465
23 S. Benjamin and A. Pandey, Acta Biotechnol. 18 (1998) 315.
24 C. Larroche, M. Theodore and J.B. Gros, Appl. Microbiol. Biotechnol. 38 (1992) 183.
25 R.A. Battaglino, M. Huergo, A.M.R. Pilosof and G.B. Bartholomai, Appl. Microbiol. Biotechnol. 35 (1991) 292.
26 O. Parades-Lopez, J. Gonzalez-Casteneda, and A. Carabez-Trejo, 71 (1991) 58.

27 P.J. Vandemark and B.L. Batzing, The Microbes. An Introduction to their Nature and Importance. (1987) 162.
28 C.L. Cooney. In: Biotechnology, Vol. 1 ed. H.J. Rehm and G. Reed, velag-Chemie Weisheim (1981) 73.
29 S.S. Yang, Biotechnol. Bioeng. 32 (1988) 886.
30 W.R. Gibbons and C.A. Westby, J. Ferment. Technol. 64 (1986) 179.
31 W.R. Gibbons, J. Ferment. Technol. 67 (1989) 258.
32 L.A. Underkofler, In: Industrial Fermentations ed. L.A. Underkofler and R. Hickey, Chemical Publications, NewYork (1954) 97.
33 S.S. Yang and W.F. Chiu, Chinese J. Microbiol. Immunol. 19 (1986) 276.
34 S. Al-Asheh and Z. Duvnjak, Acta Biotechnologica. 14 (1994) 223.
35 S.Xavier and B.K. Lonsane, Appl. Microbiol. Biotechnol.41 (1994) 291.
36 S.S. Yang, L. Chiu and S.S. Yaun, World J. Microbiol. Biotechnol. 10 (1994) 215.
37 G, Liu, S. Lu, Y. Jiang and Y. Wu, Acta Microbiol.Sin. 35 (1995) 109.
38 J.S. Yadav, Biotechnol. Bioeng. 31 (1988) 414.
39 B.A. Prior and J.C. Du Preez, In: Solid Substrate Cultivation, ed. H.W. Doelle, D.A. Mitchell and C.E. Rolz, Elsevier Applied Sciences, London (1992) 65.
40 J.E. Smith and K.E. Aidoo, In: Physiology of Industrial Fungi. ed. D.R. Berry, Blackwell Scientific Publications, Oxford (1988) 249.
41 M. Guiterrez-Rojas, S.A.A. Hosn, R. Auria, S. Revah and E. Favela-Torres, Proc. Biochem. 31 (1996) 363.
42 N.P. Ghildyal, M.K. Gowthaman, KSMS. Raghava Rao and N.G. Karanth, Enzyme microb. Technol. 16 (1994) 253.
43 R.P. Tengerdy, Trends in Biotechnol. 3 (1985) 96.
44 S.M. Finger, R.T. Hatch and T.M. Regan, Biotechnol. Bioeng. 18 (1976) 1193.
45 B.L. Rathbun and M.L. Shuler, Biotechnol. Bioeng. 25 (1983) 929.
46 E. Cannel and M. Moo-Young, Proc. Biochem. 15 (1980) 24.
47 H.L. Wang, J.B. Vespa and C.W. Hesseltine, Appl. Microbiol. 27 (1974) 906.
48 N. Nishio, H. Kuriso and S. Nagai, J. Ferment. Technol. 59 (1981) 407.
49 N.P. Ghildyal, B.K. Lonsane, K.R. Sreekantiah and V.S. Murthy, J. Fd. Sci. Technol. 22 (1985) 171.
50 P.J. Considine, T.J. Hackett and m.P. Coughlam, Biotechnol. Lett. 9 (1987) 131.
51 S.M. Kotwal, M.M. Gote, S.R. Sainkar, M.I. Khan and J.M. Khire, Proc. Biochem. 33 (1998) 337.
52 E. Ortiz-Vazquez, M. Granados-Baegra and G. Riviera-Munaz, Biotechnol. Adv. 11 (1993) 409.
53 S. Tao, L. Beihmi, L. Deming, L. Zuohu, Biotechnol. Lett. 19 (1997) 171.
54 D.A. Mitchell, Z. Targonski, J. Rogalski and A. Leonowicz, In: Solid Substrate Cultivation, ed. H.W. Doelle, D.A. Mitchell and C.E. Rolz, Elsevier Applied Sciences, London (1992) 29.
55 J.S. Knapp and J.A. Howell, In: Topics in Enzyme Ferment. Biotechnol. ed. A. Wiseman, Vol. 4, Ellis Horwood Ltd. Chichester, England (1980) 85.
56 D.A. Mitchell, In: Solid Substrate Cultivation, ed. H.W. Doelle, D.A. Mitchell and C.E. Rolz, Elsevier Applied Sciences, London (1992) 87.

57 A.T. Bull and A.P.J. Trinci, Adv. Microbiol. Physiol. 15 (1977) 1.
58 S.G. Oliver and A.P.J. Trinci, In: Comp. Biotechnol. Vol.1. ed-in-chief M. Moo-Young, Pergamon Press, New York (1985) 159.
59 G. Georgiou and M.L. Shuler, Biotechnol. Bioeng. 28 (1986) 405.
60 M.A. Aufeuvre and M. Raimbault, Compt. Rend. Sci. Series III., 294 (1982) 949.
61 D.A. Mitchell, P.F. Greenfield and H.W. Doelle, World J. Microbiol. Biotechnol. 6 (1990) 201.
62 D.S. Chahal, M. Moo-Young and D. Vlach, Mycologia, 75 (1983) 597.
63 L.A. Underkofler, A.m. Severson, K.J. Goering and L.M. Christensen, Cereal. Chem., 24 (1947) 1.
64 N.P. Ghildyal, S.V. Ramakrishna, P.N. Devi, B.K. Lonsane and H.N. Asthana, J. Fd. Sci. Technol. 18 (1981) 248.
65 V. Hecht, W. Rosen and K. Schugerl, Appl. Microbiol. Biotechnol. 21 (1985) 189.
66 P. Danbresse, S. Ntibashirwa, A. Gheysen and J.A. Meyer, Biotechnol. Bioeng. 29 (1987) 962.
67 N.P. Ghildyal, M.K. Gowthaman, K.S.M.S. Raghava Rao and N.G. Karanth, Enzyme Microb. Technol. 16 (1994) 253.
68 M.K. Gowthaman, K.S.M.S. Raghava Rao, N.P. Ghildyal and N.G. Karanth, J. Chem. Technol. Biotechnol. 56 (1993) 233.
69 M.K. Gowthaman, K.S.M.S. Raghava Rao, N.P. Ghildyal and N.G. Karanth, Proc. Biochem. 30 (1995) 9.
70 M.Y. Lu, I.S. Maddox and J.D. Brooks, Proc. Biochem. 33 (1998) 117.
71 D.A. Mitchell, B.K. Lonsane, A. Durand, R. Renaud, S. Almanza, J. Maratray, C. Desgranges, P.S. Crooke, K. Hong, R.D. Tanner and G.W. Malaney, In: Solid Substrate Cultivation, ed. H.W. Doelle, D.A. Mitchell and C.E. Rolz, Elsevier Applied Sciences, London (1992) 115.
72 L.A. Lindenfelser and A. Ciegler, Appl. Microbiol. 29 (1975) 323.
73 C.J. Fung and D.A. Mitchell, Biotechnol. Tech. 9 (1995) 295.
74 M.K.Gowthaman, P. Nandakumar, N.P. Ghildyal and N.G. Karanath, Unpublished Data (1997)..
75 A. Durand and D. Chereau, Bioechnol. Bioeng. 31 (1988) 476.
76 Y. Chamielec, R. Renaud, J. Maratray, S. Almanza, M. Diez and A. Durand, Biotechnol. Tech. 8 (1994) 245.
77 S.P. Ellis, K.R. Gray and A.J. Biddlestone, Trans. Inst. Chem. Eng. part C, Vol. 72, No.C3 (1994) 158.
78 T. Akao and T. Sakasi, US Patent No.4 046 921 (1977).
79 K. Hong, R.D. Tanner, P.S. Crooke and G.W. Malaney, Appl. Biochem. Biotechnol. 18 (1988) 3.
80 T. Akao and Y. Okomoto, Proc. 4th int. Conf. Fluid. Eng. Foundation, New York (1984) 631.
81 S. Roussos, M. Raimbault, J-P Prebois and B.K. Lonsane, Appl. Biochem. Biotechnol. 42 (1993) 37.
82 E. Cannel and M. Moo-Young, Proc. Biochem. 15 (1980) 2.

83 T. Yokotsuka, In: Microbiology of Fermented Food, ed. B.J.B. Wood, Vol.1. Elsevier Applied Science Publishers, London (1985) 197.
84 N. Lotong, In: Microbiology of Fermented Food, ed. B.J.B. Wood, Vol.2. Elsevier Applied Science Publishers, London (1985) 237.

85 B.K. Lonsane and M.M. Krishniah, In: Solid Substrate Cultivation, ed. H.W. Doelle, D.A. Mitchell and C.E. Rolz, Elsevier Applied Sciences, London (1992) 147.
86 N.P. Ghildyal, M. Ramakrishna, B.K. Lonsane and N.G. Karanth, Proc. Biochem. Int.25 (1990)
86 S. Roussos, M. Raimbault, G. Saucedo-Castaneda and B.K. Lonsane, Biotechnol. Tech. 6 (1992) 429.
87 M. Ramadas, O. Holst and B. Mattiasson, Biotechnol. Adv. 9 (1995) 901.
88 S. Tanuja, N.D. Srinivas, K.S.M.S. Raghava Rao, M.K. Gowthaman, Proc. Bioche. 32 (1997) 635.
89 K. Tirumurthy and M.K. Gowthaman, Unpublished Data (1994).
90 B.K. Lonsane, N.P. Ghildyal and V.S. Murthy, In: Technical Brochure, Symp, Fermented Foods, Food Contaminants, Biofertilisers and Bioenergy, Assoc. Microbiologists, India (1982) 12.
91 D.A. Mitchell and B.K. Lonsane, In: Solid Substrate Cultivation, ed. H.W. Doelle, D.A. Mitchell and C.E. Rolz, Elsevier Applied Sciences, London (1992) 1.
92 C. W. Hesseltine and H.L. Wang, Biotech and Bioeng., 9 (1967) 275.
93 H. L. Wang and C.W. Hesseltine, J. Agric Food Chem., 18 (1970) 572.
94 R. Borris, In: Biotechnology, ed J.F. Kennedy, Verlag Chemie, Weinheim, Vol. 7A (1987) 35.
95 B.K. Lonsane and N.P. Ghildyal, (1992). In: Solid Substrate Cultivation, ed. H.W. Doelle, D.A. Mitchell and C.E. Rolz, Elsevier Applied Sciences, London (1992) 191.
96 K. Aunstrup, Appl. Biochem. Bioeng. 2 (1979) 27.
97 S. Xavier, M. Ramakrishna, B.K. Lonsane, N.G. Karanth and M.M. Krishnaiah, Chem. Mikrobiol. Technol. Lebensm. 15 (1993) 25
98 M. Ramadas, G. Holst and B. Mattiasson, World. J. Microbiol. Biotechnol. 12 (1996) 267.
99 O.P. Ward, In: Microbial Enzymes and Biotechnology, ed. W.M. Fogarty, Applied Science Publishers, London (1983) 251.
100 B.F. Klapper, D.M. Jameson and R.M. Mayer, Biochim. Biophys. Acta. 304 (1973) 505.
101 R.A. Battaglino, M. Huergo, A.M.R. Pilosof and G.B. Bartholomai, Appl. Microbiol. Biotechnol. 35 (1991) 292.
102 E. Villigas, S. Aubauge, L. Alcantara, R. Auria and S. Revah, 11 (1993) 387.
103 L. Ikasari and D.A. Mitchell, Biotechnol. Lett. 20 (1998) 349.
104 H.M. Fernandez-Lahore, E.R. Fraile and O. Cascone, J. Biotechnol. 62 (1998) 83
105 S. Pal, R. Banerjee, B.C. Bhattacharya and R, Chakraborty, J. Am. Leather Chem. Asscoc. 91 (1996) 59.
106 G. Ferreira, C.G. Boer and R.M. Peralta, FEMS Microbiol. Lett. 173 (1999) 335.

107 M.K. Gowthaman, M. Gyamerah and M. Moo-Young, RAFT III Meeting, Sarasota, FL, .Nov 13-16, 1999.
108 G.M. Frost and D.A. Moss, In: Biotechnology, ed. J.F. Kennedy, Verlag-Chemie, Weingheim, Vol 7a (1987) 65.
109 D.S. Chahal, In: Foundations of Biochemical Engineering: Kinetics and Thermodynamics in Biological Systems. ed. H.W. Blanch and E.T. Poputsakis, ACS, Washington D.C (1983) 421.
110 N. Toyama, In: Cellulases and their applications, ed. R.F. Gould, American Chemical Society, Washington D.C, Adv. Chem. Series 95 (1969) 359.
111 F. Deschamps, C. Giuliano, M. Asther, M.C. Huet and S. Roussos, Biotechnol. Bioeng. 27 (1985) 1385. (1985).
112 N. Pamment, C.W. Robinson, J. Hilton and M. Moo-Young, Biotechnol. Bioeng. 20 (1978) 1735.
113 R.E. Mudgett, J. Nash and R. Rufner, Dev. Ind. Microbiol. 23 (1982) 397.
114 R.W. Silman, J.E. McGhee and R.J. Bothast, Biotechnol. Lett. 6 (1984) 115.
115 A. Illanes, G. Arora, L. Cabello and F. Acevedo, World J. Microbiol. Biotechnol. 8 (1992) 488.
116 R.C. Kuhad and A. Singh, World J. Microbiol. Biotechnol. 9 (1993) 100.
117 M.R. Castillo, M. Gutierrez-Correa, J.C. Linden and R.P. Tengerdy, Biotechnol. Lett. 16 (1994) 967.
118 P.K.A. Muniswaran and N.C.L.N. Charyulu, Enzyme Microb. Technol. 16 (1994) 436.
119 R.M. Burke and J.W.G. Cairney, Mycol. Res. 101 (1997) 1135.
120 A. Gupte and D. Madamwar, Biotechnol. Prog. 13 (1997) 166.
121 E. Kalogeris, G. Fountonkides, D. Kekos and B.J. Macris, Bioresource Technol. 67 (1999) 313.
122 F.M. Rombouts and W. Pilnik, In: Economic Microbiology, ed. A.H. Rose, Academic Press, London, Vol. 5 (1980) 227.
123 A. Ayres, J. Dingle, A. Phipps, W.W. Reid and C.L. Solomons, Nature 170 (1952) 834.
124 M.E. Acuna-Argueelles, M. Gutierrez-Rojas, G. Viniegra-Gonzalez and E. Favela-Torres, Appl. Microbiol. Biotechnol. 43 (1995) 808.
125 S . Solis-Pereyra, E. Favela-Torres, M. Gutierrez-Rojas, S. Roussos, G. Saucedo-Castaneda, P. Gunasekaran and G. Viniegra-Gonzalez, World J. Microbiol. Biotechnol. 12 (1996) 257.
126 M.C. Maldonado and AMStrasser de Saad, J. Ind. Microbiol. Biotechnol. 20 (1998) 34.
127 K. Yamada, Biotechnol. Bioeng. 19 (1977) 1563.
128 G. Rivera-Munoz, J.R. Tinoco-Valencia, S. Sanchez and A. Farres, Biotechnol. Lett. 13 (1991) 277.
129 B. Bhushan, N.S. Dosanigh, K. Kumar and G.S. Hoondal, Biotech. Lett. 16 (1994) 841
130 N.R. Kamini, J.G.S. Mala and R. Puvanakrishnan, Proc. Biochem. 33 (1998) 505.
131 S. Benjamin and A. Pandey, Acta Biotechnol. 18 (1998) 315.

132 V.C. Nair and Z. Duvnjak, Appl. Microbiol. Biotechnol. 34 (1990) 183
133 S. Al-Asheh and Z. Duvnjak, World J. Microbiol. Biotechnol. 11 (1995) 228.
134 S. Al-Asheh and Z. Duvnjak, Biotechnol. Lett. 16 (1994) 183.
135 K. Filer, Recent Advances in Fermentation Technology (RAFT III) Meeting, Sarasota, FL, Nov. 13-16, 1999.
136 C. Krishna and S.E. Nokes, Recent Advances in Fermentation Technology (RAFT III) Meeting, Sarasota, FL, Nov. 13-16, 1999.
137 Nova Nordisk, Biotimes, No.3, 9/99
138 M. Alam, I. Gomes, G. Mohiuddin and M. Hoq, Enzyme Microb. Technol. 16 (1994) 298.
139 K.C. Srivastava, Biotechnol. Adv. 11 (1993) 441.
140 Brustovetskaya, TP; Okunev, ON; Shul'Ga, AV, Prikl. Biokhim. Mikrobiol. 27 (1991) 577.
141 A. Wiacek-Zychlinska, J. Czakaj and R. Sawika-Zulouska, Bioresource Technol. 49 (1994) 13.
142 A. Jain, Proc. Biochem. 30 (1995) 705.
143 Purkarthofer, M. Sinner and W. Steiner, Enzyme Microb. Technol. 15 (1993) 677
144 M.C.de O. Souza, I.C. Roberto and A.M.F. Milagres, Appl. Microbiol. Biotechnol. 52 (1999) 768.
145 P.V. Suresh and M. Chandrasekaran. World J. Microbiol. Biotechnol.
146 S.M. Kotwal, M.M. Gote, S.R. Sainkar, M.I. Khan, and J.M. Khire, Process Biochem. 33 (1998) 337
147 M.R.S. Srinivas, S. Padmanabhan and B.K. Lonsane, Chem. Mikrobiol Technol. Lebensm. 15 (1993) 41.
148 S.A. Shaikh, J.M. Khire and M.I. Khan, J. Ind. Microbiol. Biotechnol. 19 (1997) 239.
149 E.M. De Almeida Siqueira, K. Mizuta and J.R. Giglio, Mycol. Res. 101 (1997) 188.
150 F. Zadrazil, D. Ostermann and G. Dal Compare, In: Solid Substrate Cultivation, ed. H.W. Doelle, D.A. Mitchell and C.E. Rolz, Elsevier Applied Sciences, London (1992) 283.
151 S. Rajarathnam, M.N.J. Shashirekha and Z. Bano, Crit. Rev. Biotechnol. 18 (1998) 91.
152 E.W. Jwanny, M.M. Rashad and H.M. Abdu, Appl. Biochem. Biotechnol. 50 (1995) 71.
153 M. Pal, A.M. Calvo, M.C. Terron and A.E. Gonzalez, World J. Microbiol. Biotechnol. 11 (1995) 541.
154 G. Straatsma, T.W. Olignsma, J.P.G. Gerrits, J.G.M. Amsing, H.J.M. Opden Camp and L.J.L.D. Van Griensven, Appl. environ. Microbiol. 60 (1994) 3049.
155 M.R. Johns, In: Solid Substrate Cultivation, ed. H.W. Doelle, D.A. Mitchell and C.E. Rolz, Elsevier Applied Sciences, London (1992) 341.
156 R.W. Silman, H.F. Conway, R.A. Anderson. and E,B. Bagley, Biotechnol. Bioeng. 21 (1979) 1799.
157 M.F.G. Jermini and A.L. Demain, Experientia 45 (1989) 1061.

158 T. Mester, R. Sierra-Alvarez and J.A. Field, Holzforschung 52 (1998) 351.
159 S. Bandelier, R. Renaud, A. Durand, Proc. Biochem. 32 (1997) 141.
160 M.C. Huang and Y.C. Tsai, 5th International Mycological Congress, Vancouver, BC (Canada), 14-21 Aug 1994.
161 C. Moksia, Larroche and J-B. Gros, Biotechnol. Lett. 18 (1996) 1025
162 M. Gutierrez-Rojas, J. Cordova, R. Auria, S. Revah and E. Favela-Torres, Biotechnol. Lett. 17 (1995) 219.
163 V.S. Shankaranand and B.K. Lonsane, Chem. Mikrobiol. Technol. Lebensm. 14 (1992) 33.
164 V.S. Shankaranand and B.K. Lonsane, World J. Microbiol. Biotechnol. 9 (1993) 377.
165 V.S. Shankaranand and B.K. Lonsane, World J. Microbiol. Biotechnol. 10 (1994) 165.
166 C.T. Tran and D.A. Mitchell, Biotechnol. Lett. 17 (1995) 1107.
167 N. Garg and Y.D. Hang, J. Food Sci. Technol. 32 (1995) 119.
168 T. Roukas, Enzyme Microb. Technol. 24 (1999) 54.
169 M.Y. Lu, J.D. Brooks and I.S. Maddox, Enzyme Microb. Technol. 21 (1997) 392.
170 M.Y. Lu, I.S. Maddox and J.D. Brooks, Proc. Biochem. 33 (1998) 117.
171 I.S. Goldstein, Organic chemicals from biomass, CRC Press, Boca Raton, 1981.
172 R.P. Tengerdy In: Solid Substrate Cultivation, ed. H.W. Doelle, D.A. Mitchell, C.E. Rolz, Elsevier Applied Sciences, London, (1992) 269.
173 E.B. Cowling and T.K. Kirk, Biotechnol. Bioeng. Symp., 6 (1976) 95.
174 L.T. Fan, Y.H. Lee and M.M. Gharpuray, Adv. Biochem. Eng., 22 (1982) 158.
175 R.L. Mitchell and M.A. Millet, TAPPI, 39 (1956) 571.
176 D.A. Applegarth and G.S. Dutton, TAPPI, 48 (1965) 208.
177 I.G. Gilbert and G.T. Tsao, Annu. Rep. Ferm. Proc., 6(1983) 323.
178 H. Nimz, Ange. Chem. Int. Ed., 86 (1974) 336.
179 W.G. Glasser and G. Glasser, Holzforschung, 28 (1974) 5.
180 T.K. Kirk and E.B. Cowling, Adv. Chem. Ser., 207 (1984) 455.
181 K. El. Eriksson, R.A. Blanchette and P. Ander, Microbial and enzymatic degradation of wood and wood components, Springer, Berlin Heidelberg, New York (1990).
182 R.A. Blanchette, T. Nilsson, G. Daniel and A. Abad, Adv. Chem. Ser., 225 (1990) 141.
183 O.V. Niemenmaa, A.K. Usui-Raua and A.I. Hatakka, In: Biotechnology in pulp and paper industry, ed. M. Kuwahara and M. Shimada, UNI, Tokyo (1992) 221.
184 M.A. Goni, B. Nelson, R.A. Blanchette, and J.I. Hedges, Geochin Cosmochim Acta., 57 (1993) 3985.
185 M. Akhtar, M.C. Attridge, G.C. Meyers and R.A. Blanchette, Holzforschung, 47 (1993) 36.
186 U. Tuor, K. Winterhalter and A. Fiechter, J. Biotechnol., 41 (1995) 1.
187 W.W. Wilcox, US Fox Serv Res Pap., FPL-70 (1968) p 46.
188 W.W. Wilcox, Bot. Rev., 36 (1970) 2.
189 C. Krishna, Bioresour. Technol., 69 (1999) 231.
190 M. Okazaki and M. Moo-Young, Biotechnol. Bioeng. 20 (1978) 637.
191 R.C. Kuhad, A. Singh, K-EL. Eriksson, Adv. Biochem. Eng., 57 (1997) 45.

192 T.K. Ghose and P. Ghosh, Process Biochem., 14 1979 20.
193 P. Ghosh and A. Singh, Adv. Appl. Microbiol., 39 (1993) 295.
194 V. Kaushik and V.S. Bisaria, J. Sci. Ind. Res., 48 (1989) 276.
195 T.K. Kirk, In: Biochemistry of microbial degradation, ed. D.T. Gibson, Dekker, NewYork (1983).
196 T.K. Kirk and M. Shimada, In: Biosynthesis and biodegradation of wood components, ed. T. Higuchi, Academic Press, San Diego (1985) 579.
197 J.A. Buswell and E. Odier, CRC Crit. Rev. Biotechnol., 6 (1987) 1.
198 P. Bonnarme and T.W. Jeffries, Appl. Environ. Microbiol., 56 (1990) 210.
199 O. Milstein, Y. Vered, A. Sharma, J. Gressel and H.M. Flowers, Biotechnol. Bioeng., 28 (1986) 381.
200 S.K. Saxena, V.S. Bisaria, J. Verma and K.S. Gopalakrishnan, J. Ferment. Technol., 63 (1985) 307.
201 J.S. Yadav, Biotechnol. Bioeng., 31 (1988) 414.
202 W. Grajek, Biotechnol. Lett., 8 (1986) 587.
203 P. Nigam and M. Vogel, Biotechnol. Lett., 10 (1988) 755.
204 D.C. Ulmer, R.P. Tengerdy and V.G. Murphy, Biotechnol. Bioeng. Sym., 11 (1981) 449.
205 C. Mishra and G.F. Leatham, J. Ferment. Bioeng., 69 (1990) 8.
206 P. Sangsurasak, M. Nopharatana and D.A. Mitchell, J. Sci. Ind. Res., 55 (1996) 333.
207 R.L. Crawford, Lignin biodegradation and transformation, Wiley-Interscience, New York (1981).
208 T.K. Kirk and R.L. Farrell, Ann. Rev. Microbiol., 41 (1987) 465.
209 B.E. Dale, Ann. Rep. Ferm. Proc., 8 (1985) 229.
210 S. Rajarathnam and Z. Bano, CRC Crit. Revs. Food Sci., Nutrition, 28 (1989) 31.
211 A. Leonowicz, A. Matuszewska, J. Luterek, D. Ziegenhagen, M. Wojtas-Wasilewska, N.S. Cho, M. Hofrichter and J. Rogalski, Fungal Genetics and Biology, 27 (1999) 175.
212 F. Zadrazil, Angew. Botanik, 59 (1985) 433.
213 M. Valmaseda, G. Almendros, and A.T. Martínez, Biomass Bioenergy, 1 (1991) 261.
214 T. Vares, M. Kalsi, and A. Hatakka, Appl. Environ. Microbiol., 61 (1995) 3515.
215 O.S. Isikhuemhen, F. Zadrazil and I.O. Fasidi, J. Sci. Ind. Res., 55 (1996) 388.
216 K.A. Onysko, Biotechnol. Adv., 11 (1993) 179.
217 I.D. Reid and M.G. Paice, Appl. Environ. Microbiol., 60 (1994) 1395.
218 K. Messner and E. Srebotnik, FEMS Microbiol. Rev., 13 (1994) 351.
219 S. Camarero, J.M. Barrasa, M. Pelayo and A.T. Martinez, J. PULP AND PAPER SCI., 24 (1998) 197.
220 A. Breen and F.L. Singleton, Current Opinion in Biotech., 10 (1999) 252.
221 M. Akhtar, Recent Advances in Fermentation Technology (RAFT III) Meeting, Sarasota, FL, Nov. 13-16, (1999).
222 E. Dejong, J.A. Field, H.E. Spinnler, J.B.P.A. Wijnberg and J.A.M. Debont, Appl. Environ. Microbiol., 60 (1994) 264.
223 D.P. Barr and S.D. Aust, Environ. Sci. Technol., 28 (1994) A78.
224 D.K. Cha, P.C. Chiu, S.D. Kim and J.S. Chang, Water Environ. Res., 71 (1999) 870.

225 A.S. El-Nawawy, In: Solid substrate cultivation, ed. H.W. Doelle, D.A. Mitchell, and C.E. Rolz, Elsevier Applied Sciences, London, (1992) 247.
226 M.M. Berrocal, J. Rodriguez, A.S. Ball, M.I. PerezLeblic, and M.E. Arias, Appl. Microbiol. Biotechnol., 48 (1997) 379.
227 S. Rios and J. Eyzaguirre, Appl. Microbiol. Biotechnol., 37 (1992) 667.
228 S. Lobos, J. Larraín, L. Salas, D. Cullen and R.I. Vicuña, Microbiology, 140 (1994) 2691.
229 M. Valmaseda, G. Almendros, and A.T. Martínez, Appl. Microbiol. Biotechnol., 35 (1994) 817.
230 T. Vares, O. Niemenmaa and A. Hatakka, Appl. Environ. Microbiol., 60 (1995) 569.
231 R.L. Farrell, K.E. Murtagh, M. Tien, M.D. Mozuch and T.K. Kirk, Enzyme Microb. Technol. 11 (1989) 322.
232 A. Datta, A. Bettermann and T.K. Kirk, Appl. Environ. Microbiol., 57 (1991) 1453.
233 E.L.F. Holzbaur, A. Andrawis and M. Tein, In: Molecular Industrial Mycology., ed. S.A. Leong and R.M. Berka, Marcel Dekker, Inc., New York (1991).
234 B. Lee, A.L. Pometto III, A. Fratzke and T.B. Bailey Jr., Appl. Environ. Microbiol., 57 (1991) 678.
235 T. Higuchi, Wood Sci. Technol. 24 (1990) 23.
236 Z. Kerem, D. Friesem and Y. Hadar, Appl. Environ. Microbiol, 58 (1992) 1121.
237 T. Cruz-Cordova, T.G. Roldan-Carrillo, D. Diaz-Cervantes, J. Ortega-Lopez, G. Saucedo-Castaneda, A. Tomasini-Campocosio and R. Rodriguez-Vazquez, Resources Conservation and Recycling., 27(1999) 3.
238 S.R. Couto and M. Ratto, Biodegradation, 9 (1998) 143.
239 V.I. Elisashvili, L.P. Daushvili, N.G. Zakariashvili, E.T. Kachlishvili, M.O. Kiknadze and K.A. Tusishvili, Microbiol., 67(1998) 33.
240 K. Jha, S.K. Khare and A.P. Gandhi, Bioresour. Technol., 54 (1995) 321.
241 O. Ardon, Z. Kerem and Y. Hadar, Can. J. Microbiol., 44 (1998) 676.
242 M. Hofrichter, T. Vares, M. Kalsi, S. Galkin, K. Scheibner, W. Fritsche and A. Hatakka, Appl. Environ. Microbiol., 65 (1999) 1864.
243 P. Nigam, and D. Singh, J. Sci. Ind. Res., 55 (1996) 373.
244 Y.W. Han and C.D. Callihan, Appl. Microbiol., 27 (1974) 159.
245 K.E. Erikkson and K. Larsson, Biotechnol. Bioeng., 17 (1975) 326.
246 A.E. Humphrey, A. Moreira, W. Amiger and D. Zabriski, Biotechnol. Bioeng. Suppl. Symp. 7 (1977) 45.
247 M. Moo-Young, A.J. Douglas, D.S. Chahal and D.C. McDonald, Process Biochem., 14 (1979) 38.
248 F. Zadrazil, Eur. J. Appl. Microbiol. Biotechnol., 9 (1980) 243.
249 I.D. Reid, Appl. Environ. Microbiol., 50 (1985) 133.
250 E. Agosin and E. Odier, Appl. Microbiol. Biotechnol., 21 (1985) 397.
251 D.N. Kamra and F. Zadrazil, Biotechnol. Lett., 7 (1985) 335.
252 E. Levonen-Munoz and D.H. Bone, Biotechnol. Bioeng., 27 (1985) 382.
253 F. Zadrazil, In: Production and feeding of single cell protein, ed. M.P. Ferranti and A. Fiechter, Applied Sci. Publ., London (1983) 76.

254 S. Camarero, B. Bockle, M.J. Martinez and A.T. Martinez, Appl. Environ. Microbiol., 62 (1996) 1070.
255 M. Moo-Young, D.S. Chahal and D. Vlach, Biotechnol. Bioeng., 20 (1978) 107.
256 D.S. Chahal, D. Vlach and M. Moo-Young, Adv. Biotechnol., 2 (1981) 327.
257 U.E. Viesturs, A.F. Apsite, J.J. Laukevics, V.P. Ose, M.J. Bekers and R.P. Tengerdy, In: Third Symposium on Biotechnology in Energy Production and Conservation: proceedings, Gatlinburg, Tenn., May 12-15, 1981. Wiley, New York (1981) 359-369.
258 O. Bravo, A. Ferrer, C. Aiello, A. Ledesma, and M. Davila, Biotechnol. Lett., 16 (1994) 865.

259 V.R. Srinivasan, In: Conference on Detrital Systems for Aquaculture, Bellagio, Como (Italy). (1985), 26-31.
260 V.M.M. Braga and J.R. Nicoli, Revista de Microbiologia, 18 (1987) 58.
261 S. Yuillet, C. Demarquilly, A. Durand, H. Blachere, H. Odier and F. Zadrazil, In: Treatment of lignocellulosics with white-rot fungi, ed. P. Reiniger, Elsevier Applied Science Publishers Ltd., Barking, Essex, UK (1988) 105.
262 J. Dorado, G. Almendros, S. Camarero, A.T. Martinez, T. Vares and A. Hatakka Enzyme Microb. Technol. 25 (1999) 605.
263 M. Ghosh, R. Mukherjee and B. Nandi, Acta Biotechnologica, 18 (1998) 243.
264 U.E. Viesturs, S.V. Strikauska, M.P. Leite, A.J. Berzincs and R.P. Tengerdy, Biotechnol. Bioeng., 30 (1987) 282.
265 L.M. Barstow, B.E. Dale and R.P. Tengerdy, Biotechnol. Techniques, 2 (1988), 233.
266 F. Zadrazil, and A. Peerally, Biotechnol. Lett., 8 (1986) 663.
267 H.W. Muller and W. Trosch, Appl. Microbiol. Biotechnol., 24 (1986) 180.
268 D.E. Akin and R.I. Brenner, Appl. Environ. Microbiol., 54 (1988) 1117.
269 E. Grenet, A. Breton, P. Barry and G. Fonty, Animal Feed Sci. Technol., 26 (1989) 55.
270 M. Bruder, In: Technical Conference - British Sugar plc., Eastbourne, UK, 25-28 June 1990.
271 M. Vogel, In: Technical Conference - British Sugar plc., Eastbourne, UK, 25-28 June 1990.
272 P. Mathot and J. Brakel, In: Mededelingen van de Faculteit Landbouwwetenschappen Rijksuniversiteit Gent (Belgium), (1992). v. 56(4b) p. 1611.
273 O. Grujic, S. Popov, and S. Gacesa, Microbiologija (Yugoslavia)., 29 (1992) 41.
274 N. Roche, C. Desgranges, A. Durand, J. Biotechnol., 38 (1994) 43.
275 N.W. Broughton, C.C. Dalton, G.C. Jones and E.L. Williams, International Sugar Journal., 97 (1995) 1154.
276 F. Zadrazil, and A.K. Puniya, Bioresource Technol. 54 (1995) 85.
277 S.A. Shojaosadati, R. Faraidouni, A. Madadi-Nouei, I. Mohamadpour, Resources Conservation and Recycling, 27 (1999) 73.
278 C.B. Pham, T.J. Ramirez, and S.A. Sedano, Philippine Journal of Biotechnology, 6 (1995) 61.

279 V. Scerra, A. Caridi, F. Foti and M.C. Sinatra, Animal Feed Sci. Tech., 78 (1999) 169.
280 J.S. VanderGheynst, L.P. Walker and G.B. VanderGheynst, J. Air & Waste Manage. Assoc., 47 (1997) 1041.
281 M. Alexander, In: Biodegradation and Bioremediation, 2nd edition, ed. M. Alexander, Academic Press, USA (1999) 325.
282 H.A. Hoitink and M.E. Grebus, Compost Science and Utilization, 2 (1994) 7.
283 R.L. Corsi and L. Seed, Environmental Progress, 14 (1995) 151.
284 L.F. Diaz, G.M. Savage and C.G. Golueke, In: The Science of Composting, ed. M.de Bertoldi, P. Sequi, B. Lemmes and T. Papi, Champman and Hall, London (1996).
285 K. Hupe, J.C. Luth, J. Heerenklage and R. Stegmann, Acta Biotechnology, 16 (1996) 19.
286 X. Liu and M.A. Cole, In: The Science of Composting, ed. M.de Bertoldi, P. Sequi, B. Lemmes and T. Papi, Champman and Hall, London (1996).
287 T.O. Williams and F.C. Miller, Biocycle, 33 (1992) 73.
288 L.J. Sikora and M.A. Sowers, J. Environ. Qual., 14 (1985) 434.
289 Stentiford and Dodds, In: Solid substrate cultivation, ed. H.W. Doelle, D.A. Mitchell, C.E. Rolz, Elsevier Applied Sciences, London, (1992) 269-282.
290 K.R. Gray, K. Sherman and A.J. Biddlestone, A review of composting –Part 1. Process Biochem., (1971) 32.
291 K.R. Gray and A.J. Biddlestone, The Chemical Engineer, 270 (1973) 71.
292 H.I. Shuval, C.G. Gunerson and S. Julius, Appropriate technology for water supply and sanitation nightsoil composting, World Bank, 10 (1981).
293 J.T. Pereira-Neto, On the treatment of municipal refuse and sewage sludge using aerated static pile composting – a low cost technology approach. Ph.D. thesis, Leed University, UK (1987).
294 O. Krogstad and R. Gudding, Acta Agriculturae Scandinavica, 25 (1975) 281.
295 V.L. McKinley and J.R. Vestal, Appl. Environ. Microbiol., 47 (1984) 933.
296 M.S. Finstein and M.L. Morris, Adv. Appl. Microbiol., 19 (1975) 113.
297 M. de Bertoldi, G. Vallini and A. Pera, Waste Management and Research, 1 (1983) 157.
298 P.S. Grewal, Annals of Applied Biology, 115 (1989) 299.
299 P.F. Storm, Appl. Environ. Microbiol., 50 (1985) 906.
300 E.I. Stentiford, S. Kelling and J.L. Adams, Refuse-derived fuel-improved profitability by composting fines residue. Institute of Mechanical Engineering, (1988) 31.
301 S.T. MacGregor, F.C. Miller, K.M. Psarianos and M.S. Finstein, Appl. Environ. Microbiol., 41 (1981) 1321.
302 P.J.L. Derikx, H.J.M. Op Den Camp, C. van der Drift, L.J.L.D. Van Griensven and G.D. Vogels, Appl. Environ. Microbiol., 56 (1990) 3029.
303 M.T. Baca, E. Esteban, G. Almendros and A.J. Sanchez-Raya, Bioresource Technol., 44 (1993) 5.
304 K.L. Schulze, Apparatus for control of aerobic decomposition, US Patent No. 3 010 801 (1961).
305 J.J. Thambirajah., M.D. Zulkali, M.A. Hashim, Bioresource Technol., 52 (1995) 133.

306 R.A.K. Szmidt, Compost Science & Utilization, 5 (1997) 78.
307 G. Straatsma, J.P.G. Gerrits, J.T.N.M. Thissen, J.G.M. Amsing, H. Loeffen and L.J.L.D. Van Griensven, Bioresource Technol., 72 (2000) 67.
308 R. Valo and M. Salkinoja-salonen, Appl. Microbiol. Biotechnol., 25 (1986) 68.
309 D.L. Kaplan and A.M. Kaplan, Appl. Environ. Microbiol., 44 (1982) 757.
310 J.D. Isbister, G.L. Anspach, J.F. Kitchens and R.C. Doyle, Microbiologica, 7 (1984) 47.
311 R.T. Williams, P.S. Ziegenfuss and W.E. Sisk, J. Ind. Microbiol., 9 (1992) 137.
312 P.S. Ziegenfuss, R.T. Williams and C.A. Myler, J. Hazard. Mater., 28 (1991) 91.
313 J. Breitung, D. BrunsNagel, K. Steinbach, L. Kaminski, D. Gemsa, and E. vonLow, Appl. Microbiol. Biotechnol., 44 (1996) 795.
314 C.L. Potter, J.A. Glaser, M.A. Dosani, S. Krishnan and E.R. Krishnan., In: In Situ and On site Bioremediation, Fourth Symposium, Vol. 2 p. 85-90, Battle Press, Columbus, OH (1997).
315 R.L. Crawford, In: Bioremediation Principles and Applications, ed. R.L. Crawford and D.L. Crawford, Biotechnology Research Series: 6 (1996), Cambridge University Press, UK.
316 R.J. Milner, Mycologist 21 (1987) 147.
317 M.V. Deshpande, CRC Crit. Revs. Microbiol., 25 (1999) 229.
318 I. Chet, In: Innovative Approaches to Plant Disease Control, ed. I. Chet, Wiley, New York (1987) 137.
319 J.P. Latge and R.S. Soper, Biotechnol. Bioeng., 19 (1977) 1269.
320 C. Desgranges, C. Vergoignan, A. Lereec, G. Riba and A. Durand, Biotech. Adv., 11 (1993) 577.
321 E.P. McGovern and R. Bentley, Biochemistry, 14 (1975) 3138.
322 A.A. Stark, J.M. Townsend, G.N. Wogan, A.L. Demain, A. Manmade and A.C., J. Environ. Pathol. Toxicol., 2 (1978) 313.
323 Y.H. Kuo, H.M. Bau, B. Quemener, J.K. Khan and F. Lambein, J. Sci. food Agric., 69 (1995) 81.
324 D.J. Daigle, W.J. Connick Jr, C.D. Boyette, M.A. Jackson and J.W. Dorner, Biotechnol. Techniques, 12 (1998) 715.
325 Katakura, Biotechnol. Abstr., 98-10909 E1 A2 (1998).
326 M.P. McQuilken and J.M. Whipps, Eur. J. Plant Pathol., 101 (1995) 101.
327 M.P. McQuilken, S.P. Budge and J.M. Whipps, Mycol. Res., 101 (1997) 11.
328 L.P. Ooijkaas, C.J. Ifoeng, J. Tramper and R.M. Buitelaar, Biotechnol. Letts., 20: (1998) 785.
329 L.P. Ooijkaas, E.C. Wilkinson, J. Tramper and R.M. Buitelaar, Biotechnol. Bioeng., 64 (1999) 92.
330 B. Dorta and J. Arcas, Enzyme Microb. Technol., 23 (1998) 501.
331 J.L. Cheetham, M.J. Bazin, and J.M. Lynch, Enzyme Microb. Technol., 21 (1997) 321.
332 R.T. Lamar, Appl. Environ. Microbiol., 56 (1990) 3093.
333 J.A. Bumpus, M. Tein, D.S. Wright and S.D. Aust, Science, 228 (1985) 1434.

334 P.K. Donnelly, J.A. Entry and D.L. Crawford, Appl. Environ. Microbiol., 59 (1993) 2642.

335 S. Masaphy, D. Levanon and Y. Henis, Biores. Technol., 56 (1996) 207.

336 J.A. Bumpus and S.D. Aust, Appl. Environ. Microbiol., 53 (1987) 2001.

337 G.D. Mileski, J.A. Bumpus, M.A. Jurek and S.D. Aust, Appl. Environ. Microbiol., 54 (1988) 2885.

338 G. Palmieri, P. Giardina, L. Marzullo, B. Desiderio, G. Nitti, R. Cannio and G. Sannia, Appl. Microbiol. Biotechnol., 39 (1993) 632.

339 B.C. Okeke, A. Paterson, J.E. Smith and I.A. Watsoncraik, Appl. Microbiol. Biotechnol., 41 (1994) 28.

340 B.C. Okeke, A. Paterson, J.E. Smith and I.A. Watsoncraik, Appl. Microbiol. Biotechnol., 48 (1997) 563.

341 R.T. Lamar, M.W. Davis, D.M. Dietrich and J.A. Glaser, Soil Biology & Biochem. , 26 (1994) 1603.

342 J.A. Glaser and R.T. Lamar, In: Bioremediation Science and Applications, ed. H.D. Skipper and R.F. Turco, Soil Science Society of America, Madison, WI (1995) 117.

343 F. Kastanek, K. Demnerova, J. Pazlarova, J. Burkhard and Y. Maleterova, Int. Biodeterior. Biodegrad., 44 (1999) 39.

344 D.F. Berry, R.A. Tomkinson, G.H. Hetzel, D.E. Mullins and R.W. Young, J. Environ. Quality, 22 (1993) 366.

345 S. Masaphy, D. LevAnon and Y. Henis, Bioresource Technol., 56 (1996) 207.

346 S.I. Zafar, N. Abdullah, M. Iqbal, and Q. Sheeraz, Int. Biodeterior. Biodegrad., 38 (1996) 83.

347 M. Kuilman and J. Fink-Gremmels, In: Proceedings Mycotoxin Workshop at Braunschweig (Germany). (1995) 80.

348 H.P.L. Willems, D.F. Berry, and D.E. Mullins, J. Environ. Quality, 25 (1996) 162.

349 C. in derWiesche, R. Martens and F. Zadrazil, Appl. Microbiol. Biotechnol., 46 (1996) 653.

350 A. Gutierrez, J.C. del Rio, M.J. Martinez, and A.T. Martinez, Appl. Environ. Microbiol., 65 (1999) 1367.

351 R. Rodriguez-Vazquez, T. Cruz-Cordova, J.M. Fernandez-Sanchez, T. Roldan-Carrillo, A. Mendoza-Cantu, G. Saucedo-Castana and A. Tomasini-Campocosio, Folia Microbiologica., 44 (1999) 213.

Applied Mycology and Biotechnology
Volume 1. Agriculture and Food Production
G.G. Khachatourians and D.K. Arora (editors)

ROLE OF FUNGAL ENZYMES IN FOOD PROCESSING

R.K. Saxena, Rani Gupta, Shashi Saxena and Ruchi Gulati

Department of Microbiology, University Of Delhi, South Campus, New Delhi- 110 021, India

ABSTRACT

At present many microbial, particularly fungal enzymes have been found to play a vital role in food processing industries and their impact is going to be felt much more in coming years. Many fungal enzymes are used either for enhancing the overall process or to generate additional flavors. Besides this, enzyme mediated processes are preferred due to their high specifictiy and least by-products. The food processed in this way is close to the natural products and are thus categorised as 'Green'. The applications of enzymes in food processing industry has been known for ages and the oldest known enzyme-mediated process is alcoholic fermentation involving yeasts. Today, enzymes are used in bakery, brewing, dairy, meat, sugar, fruit processing and other food industries. Various enzymes used are amylases, proteases, lipases, gluco-oxidases, pectinases and tannases and many others. The world wide application of these will increase by developing tailor-made enzymes through recombinant DNA technology which could enhance the yields and modify their specificity to meet consumer's requirement in terms of cost, calorie and taste. In this chapter, a comprehensive attempt has been made to elaborate the usage and present status of the applications of fungal enzymes in various food industries.

1. INTRODUCTION

It is generally believed that life began almost four billion years ago on this planet and since the inception of life on earth, food was the main necessity for all living beings. Therefore, efforts have continuously been made round the world to fulfill the food demand partially or completely using biological systems by exploiting the unutilised resources to produce human food or animal feed and to improve various food or feed products.

With the civilization which evolved with the age of mankind, man had started using different materials from both plants and animals for their needs especially with respect to food. He also developed ways and means by which the raw food could be processed, made more edible and had better taste. In this respect, roasting, addition of salts and different spices have been traditional practices of improving flavor and quality of food. Unknowingly in food processing, man had started using organic molecules which we know today as enzymes, and biologically speaking they are organic molecules which initiate and control biological reactions important in dairy, baking, brewing, fruit processing and meat industry (1,2,3). These enzymes are obtained from plant, animal and microbial sources. In the last three decades, microbial enzymes have comparatively achieved a paramount importance. They can be grown in a wide variety of conditions on simple media and growth period is much smaller as compared to animal and plant systems. These micro-organisms can also be induced to produce a variety of enzymes by manipulation of cultural conditions. Yet, out of 2500 enzymes known today, only a few are produced on a large scale and used industrially (4). These are mainly hydrolytic enzymes which degrade the naturally occuring biopolymers. These hydrolytic enzymes account for 80% of all the commerciallly produced enzymes. Approximately 60% of all enzymes used by industrial concerns are proteases. About 27% are carbohydrases (carbohydrate hydrolysers) which find applications in baking, brewing, starch and textile industries. The remainder of the enzymes are accounted for by lipolytic enzymes (3%) and oxidoreductases (10%) (4).

The importance of enzymes in food and drink manufacture can very well be realized as primitive food processing had made use of enzymes by using the whole organisms. Obvious examples are the use of yeasts in alcoholic fermentations, where a series of enzymes come into play to turn a variety of carbohydrates – mainly sucrose, glucose and fructose–into ethanol. The enzymes are conveniently packaged inside the yeast cells, but the processes are certainly enzyme mediated. This was subsequently followed by fermentations involving cell free extracts. This is well exemplified by the *Aspergillus*-mediated conversions of soyabeans to produce a whole range of fermented foods where several extracellular enzymes act in concert. Besides this, there are examples of using specific enzymes isolated from animals.The oldest example is probably rennet (also called 'chymosin'), a protease obtained from the calf stomach which is used in cheese manufacture (5).

There are, however, a number of attractive processes and many important bio-conversions which can be carried out efficiently and accurately, provided a desired enzyme is available in sufficient amounts. This is well exemplified from important conversions like starch to maltose, glucose to fructose, modifications of some gums to improve their properties, modification of lipids and protein *etc.* Before enzymes are exploited for food processing they must meet the basic demands of a viable commercial venture. The process should have the following advantages:

1. Yields better improved quality product than the traditional one
2. Has a wide range of potential applications
3. Is cheaper ,which of course may be achieved directly by saving the raw materials, cost of labour and /or machinery required and produce the enzyme in sufficient large quantities in short time periods.
4. Yields products that were not previously available or were available in limited quantities.
5. The process should not yield any toxic by-products.

Today, with the advent of biotechnology, all the demands, by and large could be met and the processes simplified and made efficient to a great extent. It is possible now, to produce virtually any enzyme potentially available in desirable quantites. In this respect, the scenario of metabolic engineering , recombinant DNA technology, protein engineering , gene modification and construction of transgenics has led to a revolution in the area of enzyme technology. Use of bio-catalysis methods is especially of advantage for the production of enzymes because, when carried out rigorously, such techniques allow them to be described as " natural" , and so command the premium prices conferred on them as a result of insistent consumer demand for "natural" or "green" food ingredients (6). It is possible to carry out any chemical or biochemical process earlier deemed impossible. Thus, today it is possible to develop a tailor-made enzyme to carry out the increased intrinsic enzyme performance and stability with required precise functions, boosting production yield, and optimising formulation and to modify the biochemical characteristics of the enzyme so that they can work in a wide range of physico-chemical conditions such as temperature, pH, salinity, solvent *etc.*

Considering the industrial scenario, currently many of the enzymes used in industry are used in food processing, for instance, three enzymes most commonly used in industry i.e α-amylase, glucoamylase, and glucose isomerase, are employed for the production of glucose and fructose syrups (7). This emphasis is hardly surprising, since most foods are derived from natural sources, which are synthesized enzymatically in the plant, animal, or micro-organism of origin. Enzymes are present and active in many foods, for instance, stored meat becomes more tender owing to the action of enzymes, particularly lysosomal proteases (cathepsins) and collagenases; and we use enzymes to break down our food in our digestive tracts. A good estimate of progress is the time-scale for the introduction of enzymes for commercial uses. Bacterial proteases were first widely marketed in the early 1960s, followed by a wide range of other proteases, amylases, cellulases and lipases (2).

Today, the estimated world enzyme market is approaching two billion dollars. Enzyme technology has made an impact in numerous industries and companies. However it appears that this is only the tip of the ice'berg . It is likely that in the not-too-distant future almost all facets of modern life will be touched by enzyme technology . A number of applications using novel enzymes have been introduced to the marketplace. These include the firming of fruit, biopreparation (biscouring) of cotton, production of potable and fuel alcohol and poultry feed:

1.1. The characteristics of industrial enzymes for food processing

A pre-requisite for any enzyme to be used in food is that its preparation must be "safe" for human consumption . In this light, for a long time, yeast and rennet from calf stomach have found extensive application . But now, after increased awareness and development of toxicity tests, new enzyme sources are also finding market place. Among the microbial sources, **'GRAS'** (generally regarded as safe) micro-organisms are recommended for food preparations (2).

Enzymes, unlike many other substances, are recognized and sold by their activity rather than their weight, so that the stability of enzyme preparations during storage is of prime importance. The enzymes used in industry are very rarely crystalline, chemically pure, or even single protein preparations. These impurities need not interfere with the activity of the enzyme. However, enzyme impurities can catalyze the formation of side-products, or present a toxicological hazard. For, instance, the potent carcinogen aflatoxin can contaminate extracts from *Aspergillus flavus*.

An enzyme that is commercially useful in food processing must be cheap and must be stable under the physical conditions prevalent during traditional food processing steps. That is, it is preferable to screen different enzymes for activity under these conditions rather than to manipulate an established process or product so as to accommodate a potentially useful new enzyme. The enzyme must also be stable – many industrially used enzymes operate at temperatures in excess of 50°C; it must be available in sufficient quantities, and it must be safe. Since the costs of petitioning regulatory authorities are so great, whenever possible, it is much easier to make use of an enzyme already approved for food use rather than to obtain legal status for a new enzyme. It is also noticeable that the same enzyme may be useful in very different applications; for instance, α-amylases are used in both brewing and baking, and proteases in brewing, baking, cheese-making, and in the tenderization of meat (1,2,3).

The majority of enzymes act during the preparation and processing of food rather than in the final product itself. Thus, the use of enzymes is advantageous because they operate under wide range of conditions such as pH, temperature, etc. that are consistent with the retention of the desired structure and other properties of the food, and minimize the energy requirements during processing, whereas the high temperatures and pressures associated with the use of chemical catalysts would often be detrimental to the product.

Frequently, the residues of enzymes remaining in the final product are very small indeed, and are regarded as a processing aid. Sometimes they are important to be removed as their presence may deteriorate the product on storage. In this respect, the use of immobilized enzymes offers additional advantage because it can be completely recovered from the final product without heating the product to denature the enzyme, a procedure which cannot always be used because of the temperature sensitivity of the product. Also, larger concentrations of immobilized rather than free enzyme can be utilized because the immobilized enzyme can be recovered and reused, resulting in shortening of reaction times and/or the size of vessel needed to carry out the reaction, virtual absence of enzyme from the final product such that it only has to be approved as a food-processing aid and not as a food additive.

In this chapter, the role of various enzymes in different sectors of industries such as baking, brewing, dairy, fruit processing, meat, oils and fats, starch and sugar are discussed.

2. BAKING INDUSTRY

Bread is one of the most common and low cost foods, associated with traditions in many countries and can be considered as one of the oldest biotechnology industry . Today, consumers have certain criteria for bread, including appearance, freshness, taste, flavour, variety and a consistent quality. It is a great challenge for the baking industry to fulfill these demands for several reasons. Firstly, the main ingredient of the bread *i.e* flour, varies due to varied wheat quantities and milling technology. Secondly, the baking industry needs various quantities of ingredients. The general requirement in bread making is that the dough should be acceptable for handling even in fast mechanical systems and should be able to initiate and sustain gas production that will raise the volume while retaining a good structure. Also, it should not collapse or otherwise degrade when baked or at some time after baking . There are always the requirements of taste and texture, crumb and crust color and overall keeping properties. It is well recognised that the starch component of the flour is the source of the sugars for the fermentation, and this may be supplemented with small amounts of added sugars. The creation of carbon dioxide is only important in the proof stage and its retention is critical to the bread quality. It is also established that the yeast activity contributes to flavor, although crust formation is probably more significant . Some new variety breads can be made by simple adjustment of the formulation or baking procedure. However, others require the bakers to develop new techniques. Therefore, both millers and bakers need agents or process aids such as chemical oxidants, emulsifiers or enzymes to standardize the quality of the products and diversify the product range (6). For decades, enzymes such as malt and microbial α -amylases have been used for bread making (8). Due to the changes in the baking industry and the demand for more varied and natural products, number of enzymes have gained more and more importance in bread formulations. Through new and rapid developments in biotechnology, a number of enzymes have recently been made available to the baking industry. Fungal amylases and sometimes bacterial amylases, proteinases and enzyme active soyaflour are well established contributors to the baking performance (9). Recent advances in our understanding of the dough forming and overall baking processes at the molecular level have focussed attention on further improvements that can be achieved by the application of more specially tailored enzymes to modify components of the flour that are neither starch nor protein.

There is a little fermentable sugar, mostly maltose in bread flour. It rarely exceeds 0.5% and this is insufficient for bread-making. In order that yeasts may generate carbon dioxide to raise the dough it requires a supply of fermentable sugar. Wheat and wheat flour contains endogenous enzymes, mainly α and β amylases which can activate the dough to produce more fermentable sugars. These may be unbalanced and also overactive in some cases. Ungerminated wheat has large amounts of β-amylase and low levels of α-amylase (10). The result is inadequate amounts of fermentable sugars. It is therefore quite usual to supplement the flour by adding fungal α-amylase at the mill.

Traditional methods of baking are based on the presence of endogenous enzymes which catalyse natural changes during growth, ripening and storage. They also carry out the saccharification of starch prior to fermentation and the degradation of the gluten which is a very important determinant of the rheological properties of the dough. The strength of the dough is related to the quantity and quality of the gluten present and the number of disulphide bonds which form when dough is mixed. Proteases and peptidases are normally present at very low levels in wheat but are increased by malting. However, wheat gluten is not susceptible to attack by flour proteinases and proteases, hence supplementation is usually essential (11). The following are the important enzymes used in baking industry as shown in Table1.

Table1

Enzymes currently available for use in baking

Enzyme	Type	Source
Amylase	Fungal	*Aspergillus* spp. *Rhizopus* spp.
	Bacterial	*Bacillus subtilis* *Bacillus licheniformis* Genetically modified organisms (GMOs)
Beta-glucanase	Fungal	*Aspergillus* spp. *Penicillium emersoni*
	Bacterial	*Bacillus subtilis*
Cellulase	Fungal	*Trichoderma* spp.
Glucose oxidase	Fungal	*Aspergillus niger*
Hemicellulase	Fungal	*Aspergillus* spp. *Trichoderma* spp.
Lipase	Fungal	*Aspergillus niger* *Candida cylindracea(rugosa)* *Mucor* spp. *Rhizopus* spp.
Pentosanase	Fungal	*Aspergillus* spp. *Trichoderma* spp.
Peptidase	Fungal	*Aspergillus* spp. *Rhizopus* spp.

Table1

Enzyme	Type	Source
Peroxidase	Fungal	*Saccharomyces* spp.
Phospholipase	Fungal	*Aspergillus* spp.
Proteinase	Fungal	*Aspergillus, Rhizopus* spp.
	Bacterial	*Bacillus subtilis*

2.1. Amylases

The amount of amylases in most sound, ungerminated wheat or rye flour is negligible (10). Cereal β-amylase yields maltose from damaged starch granules. This is slowly fermented and can only be fermented after the yeast has been induced by glucose molecules. Flours with low natural α-amylase activity to generate glucose will have a poor fermentation character.This can be corrected by the addition of another amylase, glucoamylase or direct addition of glucose. Commercial amylases used in the baking industry are generally α-amylases (mostly from *Aspergillus oryzae*) that specifically hydrolyse the α -1,4 glycosidic linkages of amylose and amylopectin molecules in starch to release glucose and soluble inter-mediate size dextrins of DP2- DP12 (12). Fungal amylases are effective in partially hydrolyzing damaged starch which varies depending on wheat sort and milling condition and are often added to flour as flour correction to develop desirable properties such as oven spring and brown color to the crust. Generally, flour made from hard winter wheat contains more damaged starch than soft wheat. These have the potential to adversely affect the dough and the finished product. The α-amylases provide fermentable sugar, which results in an increased volume, better crust color, and improved flavor. The dough character is altered by the hydrolysis of damaged starch granules which are rapidly and readily hydrated. It is generally thought that the dextrins produced by amylase action play a part in the overall hydration of the dough resulting in dough softening and to a small extent increase the shelf life of the product. They may also contribute to texture and "mouthfeel" in the final product (13). However, extensive degradation of the damaged starch due to over-dose of amylases leads to sticky dough. These amylases also have antistaling effect due to their limited thermostability and are for the most part inactivated before the onset of starch gelatinization during baking. It was reported by Si and Simonsen (12) that fungal amylase, although it reduces the crumb firmness, has no effect on reducing starch degradation.The overall pattern of amylase activity during the baking stage depends on the proportions of the different types of amylases involved and their respective deactivation temperatures. The fungal amylases are the predominant choice for baking supplementation but bacterial amylases are sometimes considered. The volume and crumb structure improve with increased dosage of fungal amylases. Although a high dosage can provide a larger volume increase , the dough would be too sticky to work. The optimum dosage is thus defined as the dosage with maximum reachable volume without a sticky dough (14). In cases where wheat has been subject to adverse conditions prior to harvest, there may be excess α-amylase. This is usually adjusted by blending the flour with other flours with much lower amylase content. Although a good number of fungal and cereal amylases have been used in baking industries, some of the bacterial amylases have also been tried . A brief description of these will also be helpful for readers to know bacterial amylases in baking industries.

Initially tried as replacements for expensive and rather variable malt extract sources of additional amylase, the bacterial amylases were found to have too great a heat tolerance for most conventional bread-making. The bacterial amylases are far lower in price than fungal amylases and they are very fast acting. These are potential benefits for the baker, but the control problem outweighs the benefit. The most current use in baking is to create a soft, sweet and sticky final product such as honey cake and malt loaf. The Catamyl range (Biocatalysts) are selected blends of conventionally produced bacterial amylases available for baking , while Novamyl (Novo Nordisk) is from a genetically modified strain of *Bacillus subtilis* (11). *A.oryzae* enzyme is most active at about 55°C while *B.subtilis* enzyme is active at about 75°C and is too stable and its activity remains after the baking process causing overhydrolysis of the starch resulting in a sticky product. So the ideal condition is a moderate hydrolyis of the starch, but with all the enzyme activity being lost during baking so that an α-amylase with an temperature optima of 70°C is required. Such an enzyme has been found to be produced by *A.niger* with no remaining activity left in the bread. Similar enzymes have been found in *B.megaterium* , *B.stearothermophilus* and *B.subtilis* for case of commercial production. Another trick is to optimize mixtures of different α-amylases so as to achieve best results for instance, enzyme bio-systems have developed a potential mixture of *Aspergillus niger* and *Bacillus stearothermophilus* α-amylases and claim that bread so produced keeps fresh for a week (15). Some work has also been carried out to treat wheat, removing phytic acid and antinutritional factor by fungal phytase form *A. ficcum* (16). In addition, presently, attempts have also been made to use a α-amylase obtained from *A.oryzae* which acts as an efficient antistaling agent (personal communication) (Fig. 1).

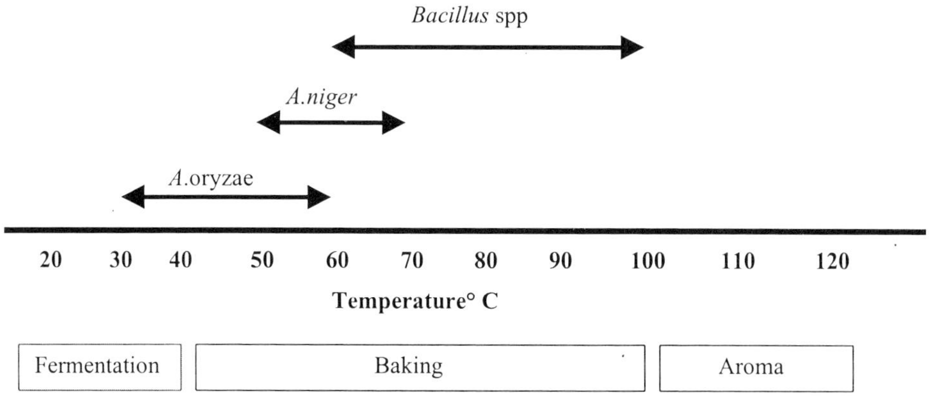

Fig: 1 Temperature profile with the use of Amylases in the baking process

2.2. Hemicellulases

Several enzyme companies offer hemicellulase / pentosanase enzyme preparations. These are usually fungal products and of extremely high quality . Individually these enzymes find use in the bakery mainly as components of prepared “bread improvers”. Works by Hamar (17,18) indicate that the use of pentosanases increased gluten coagulation in a diluted dough system. Si and Goddik (19) reported that a good baking xylanase increases the gluten strength. A study conducted by Jakobsen and Si (20) using four different pure xylanases concluded that the best enzyme in terms of baking performance is the xylanase that has a certain level of activities toward soluble and insoluble wheat arabinoxylans in a dough system. Using enzyme combinations for bread making is not new. It is well known that hemicellulase or xylanase used in combination with fungal α-amylase has synergistic effects. A high dosage of a pure xylanase may give some increase in volume but the dough with this dosage of xylanase would be too sticky to be handled in practice. When the xylanase is combined with even a very small amount of fungal α-amylase, a lower dosage of xylanase with the amylase provides a larger increase in volume and better overall score without the dough stickiness problems . The combination of xylanase and fungal α-amylase has its limitations because of the dough stickiness easily caused by these enzymes. In addition, lipase improved dough stability against over-fermentation because of its effect on gluten strengthening. With the addition of α-amylase alone or in combination with xylanase,the volume and overall scores of the bread improved. When α-amylase, xylanase and lipase were used together, the overall scores were further improved. The dough exhibited good stability on over-fermentation , which resulted in significantly better baguettes. Because lipase does not make the dough sticky and significantly improves dough stability and crumb structure, the synergistic effects between xylanase or amylase provide many possibilities for improved bread quality (6).

2.3. Glucose Oxidase

Glucose oxidase has good oxidising effect that results in a stronger dough. It can be used to replace oxidants such as bromate and ADA (azodicarbon–amide) in some baking fomulations such as with ascorbic acid , it is an excellent dough strengthener. Use of glucose oxidase with fungal α-amylase gives the dough extensibility. The commercial enzyme is normally produced from *A.niger* (21). The combination of these enzymes can therefore achieve a synergistic effect. When the two enzymes are used together with a smaller amount of ascorbic acid, the dough is not only very stable but also absorbs 1-2% more water, resulting in a greater volume increase and a crispier crust. Glucose oxidase combined with fungal α-amylase can replace the bromate in some bread formulations. The basic formulation contains both ascorbic acid and potassium bromate. By adding glucose oxidase and fungal α-amylase instead of bromate, the final bread has a much improved appearance and a volume increase of approximately 40 %.

2.4. Proteases and peptidases

It is estimated that two-thirds of the white bread baked in U.S. is treated with *A.oryzae* protease. Fungal proteases from *A.oryzae* are normally used in bread making due to their lower temperature stability and also their amylase content (22) . *A.oryzae* fungal protease contains both endo- and exo-peptidase. Endo-peptidase modifies the visco-elastic properties of the dough by hydrolyzing interior gluten peptide bonds. Exopeptidase releases amino acids from gluten which can react with the sugar through the Maillard reaction during baking, thereby contributing to flavor and acid crust colour. Improvements in the aroma of bread due to fungal protease have been observed (23,24). Both fungal and bacterial proteases have application in the manufacture of cracker and biscuit dough. Bacterial neutral protease improves the extensibility of cracker dough allowing it to be rolled very thickly without tearing and reducing bubbling during baking. This enzyme is also used in high protein flour doughs for cookies, pizza and biscuits.

3. BREWING INDUSTRY

Brewing refers to a process involving the production of beer, a beverage which is a dilute solution of ethanol containing a unique mixture of flavouring substances. The characteristic flavour arises from the use of malted barley hops and yeast's nutrients. The brewing process involves extracting and breaking down the carbohydrates from the malted barley to make a sugar solution and using this as a source of nutrients for anaerobic yeast to break down the simple sugars, releasing energy and producing ethanol as a metabolic byproduct (1). The major biological changes in the brewer process are catalysed by naturally produced enzymes from the barley and yeast respectively. The rest of the brewing process largely involves heat exchange, separation and clarification which only produces minor changes in chemical composition when compared with the enzyme catalysed reactions. Barley is able to produce all the enzymes needed to degrade starch, β-glucan (the principal component of the cell wall), pentosans, lipids and proteins which are the major compounds of concern to the brewer. Sometimes, brewers may find themselves using poorer quality malts then they would ideally like to supplement the malt enzymes of insufficient potency and in some cases to provide additional activities not inherently present in the malt. These various shortcomings can be tackled by the selection of appropriate mash enzymes.

3.1. Brewing with adjuncts

Adjuncts, which are any carbohydrate source other than malt, are used principally as an additional source of extract to replace or more usually supplement malt. Originally, adjunct may have been used to compensate for a shortage of suitable malt or as a cheaper source of extract. Adjuncts have been found, however, to contribute flavor and processing advantages which have become an integral part of many brewing formulations. Unmalted cereal adjuncts usually contain no active enzymes, and therefore usually rely on the malt or exogenous enzymes to provide the necessary enzymes for sugar conversion which are generally obtained from fungal or bacterial sources. Although some of the cereals may require milling before use, they all require to be gelatinized and liquified before they can undergo starch hydrolysis. Hydrolysis is often achieved by using the enzyme activity from the malt; however in some cases where high levels of starch are used it may dilute the malt enzymes to a limiting level and these require to be augmented or replaced by commercial exogenous enzymes as α-amylase, proteases (Fig 2). The requirements for processing in the brewery depends on the state and type of adjunct used and in particular the starch gelatinization or liquefaction temperatures. If the gelatinization temperature is above the normal malt saccharification temperature (usually 60°C), then the starch has to be gelatinized and liquefied before adding to the mash; there is a high risk of forming retrograded starch if the gelatinized starch is cooled down without being liquefied (11).

With poorly modified malts or when using high adjunct levels it is common practice to use fungal and bacterial α-amylases during cooking. Conventionally, α-amylase requires a stand of 20-30 min at 70-75°C to work, and requires high levels of calcium for stability; however an alternative is to use thermostable bacterial α-amylase from *Bacillus licheniformis* which requires a shorter contact time at a higher temperature, lower enzyme dose rate, and lower calcium ion concentrations.

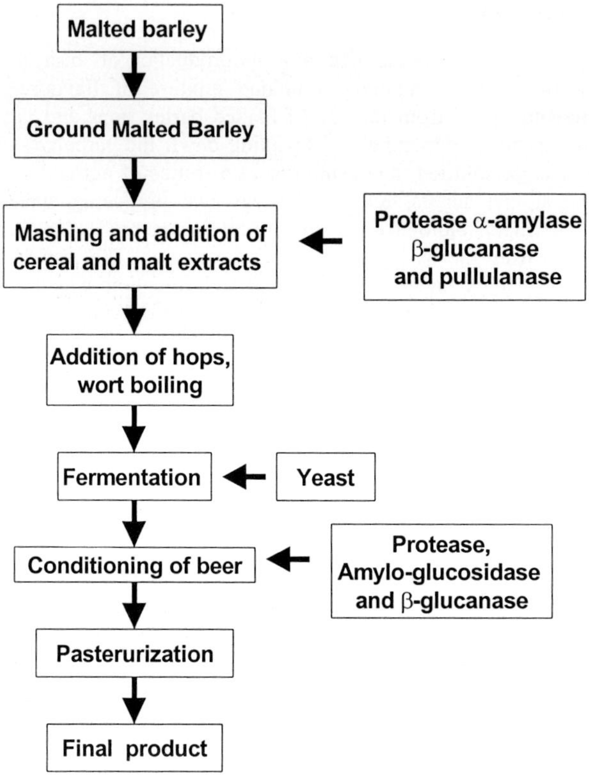

Fig: 2 A flow diagram for the beer manufacturing process

Another advantage of using heat stable α-amylase is that there is no risk of producing retrograded starch, which happens when insufficient starch breakdown has occurred at high temperatures and on cooling below 98°C, strong hydrogen bonds are formed within the amylase chains making it resistant to further enzyme degradation. The heat stable α-amylase ensures adequate starch breakdown during the stand at higher temperatures. Under these conditions, retrograded starch is stable, insoluble and can not be detected by the iodine test. The production of retrograded starch leads to a loss of extract (up to 3 per cent), and an undesirable increase in viscosity. It has also been implicated in some beer hazes.

Brewers however, often find it convenient to use exogenous microbial enzymes either to improve the saccharification performance when using poorer quality malts, or to allow the use of high levels of unmalted adjunct. Improvements in saccharification can be obtained by using microbial α-amylase added to the mash either in the mash mixing vessel or mash tun, usually at a rate of 1 Kg BAN (Bounded amino nitrogen) 120L (Novo) per tonne of malt. This treatment is particularly effective when dealing with set mashes caused by inadequate breakdown of starch and dextrins or for preventing carry forward of starch in the bright wort during run-off.

Malt generally contains sufficient free α-amino nitrogen (FAN>140 mg/L) to ensure a healthy fermentation. About 60 per cent of the α-amino nitrogen comes from hordein breakdown during malting, and by the end of a thick mash with prolonged protein rests, some 10-15 per cent of the barley protein gets solubilized. With thin mashes and high adjunct ratios, however very little additional α-amino nitrogen is released; additional exogenous proteolytic activity may be required when brewing with certain adjuncts or poor quality malts to modify the endosperm structure and to facilitate saccharification, particularly to release bound β-amylase, and to adjust the ratio of soluble nitrogen necessary for yeast growth. (25). The most suitable proteolytic enzyme is one containing only bacterial neutral protease produced from *Bacillus subtillis* used at 0.3-1.5 Kg per tonne of malt. This is particularly suited to the temperature and conditions of programmed mash stand between 52°C and 55°C where it raises the soluble nitrogen levels and improves the extract performance. Excessive proteolytic activity can have a detrimental effect on the beer foam stabilizing proteins, leading to poor haze formation in the beer. Other enzymes that have been used for chill-proofing include tannase from *A.niger* which degrades the polyphenol component of haze (26), this enzyme also having been proposed as a means of solubilising solids in instant tea (27). Increasing the soluble wort nitrogen too far can effect the beer ester formation and wort buffering capactiy leading to higher than usual wort pH values.

3.2. Unmalted cereal brewing

In some cases very high levels of unmalted cereals as in barley, sorghum are used, *i.e.* up to 100 per cent of total grist composition, and in these cases exogenous enzymes have to be used.

3.2.1. Barley Brewing

For barley brewing it is current practice to replace up to 50 per cent of the malt with barley and this is usually milled together with the malt, often using a wet mill. A new enzyme preparation for improving the steeping of barley has been developed by the Alko Co. of Finland, which contains cellulase and phytase (phosphatase) activities desired from an *Aspergillus* strain (28). Use of this enzyme is expected to facilitate the separation of starch from barley fibre and gluten so increasing starch yields, reducing steeping times, to produce phytic acid-free corn steep liquor. The grist is mashed at 50°C with a short proteolytic stand of 10 min before heating up for stands at 63°C, and then 66°C, with a final temperature rise to 78°C to halt enzyme action. Bacterial protease and bacterial α-amylase from *Bacillus subtilis* are added to the mash to assist with proteolysis and saccharification. Although the natural α-amylase from *Bacillus subtilis* may also contain bacterial β-glucanase (2), it is rather heat labile and therefore additional thermostable fungal β-glucanase, such as that produced from *Penicillium emersonii*, is usually also added to the mash. There are additional benefits from selecting more than one β-glucanase for the mash, as it has been shown that there are differences in the specificity between enzymes. Although not usually practised by brewers, it is perfectly possible to make a totally satisfactory wort from up to 100 per cent barley using exogenous enzymes as described above. The product is commonly supplied as barley syrup used as a liquid adjunct in brewing and for home brew kits.The typical range of enzymes that are used for barley brewing are described in Table 2

Table 2

Enzymes for barley brewing

Enzyme type	Example of product	Example of supplier	Dose rate (kg /tonne barley)
Bacterial neutral Proteinase	Neutrase 0.5L	Novo Nordisk	1.2 - 1.5
Bacterial α-Amylase	BAN 120L	Novo Nordisk	2.0 - 3.0
Bacterial β-Glucanase	BAN 120L	Novo Nordisk	2.0 - 3.0
Thermostable Fungal β - Glucanase	β-glucanase 150L	ABM (Rhone-Poulenc)	1.5- 1.0
Completed blended enzyme treatments for barley brewing			
Mixed bacterial Protease	Alphalase AP	ABM (Rhone-Poulenc)	0.5 – 1.5
α-amylase, β-glucans	Brewers flow	Gist-Brocades	1.0 – 2.0
and fungal	Promalt	Quest	1.0 – 2.0
β-glucanase	Ceremix	Novo Nordisk	1.0 - 2.0

Some enzyme manufacturers have produced a blend of enzymes to provide the brewer with a single addition to simplify the process. Unmalted barley contains a high level of β-amylase, some of which is bound to barley protein and will be released during mashing by proteolytic enzymes. β-amylase is unable to attack starch granules, and so has a limited action during the initial stages of mashing. During this period, amounts of soluble dextrins are produced by the action of exogenous α-amylase on starch granules. Temperatures in excess of 60°C are required to gelatinize the starch and it is only at this stage that the starch becomes susceptible to rapid degradation by β-amylase; as the temperature exceeds 64°C, the enzyme becomes rapidly deactivated.

3.2.2. Sorghum Brewing

Another cereal used in high proportion, particularly in Africa, is sorghum which can be used at up to 100 per cent as a replacement for malt. Although trials are underway to malt sorghum, most of the commercial beer is produced from raw sorghum with the use of exogenous enzymes. The best results have been achieved with short grained white sorghum. Red sorghums are not recommended due to their high tannin content which can produce excessive protein precipitation and astringent flavors. Sorghum brewing presents its own special problems as the starch gelatinizaiton temperature is 68-75°C which is above normal mashing temperatures and a special mashing regimen would involve a 50 per cent malt mash and a separate 50 per cent raw sorghum mash which had been finely milled and mashed in at

50°C with a liquor:sorghum ratio of 3:1 (11). A blend of neutral protease, thermostable α-amylase and a range of β-glucanases obtained from different microorganisms is required and have the following benefits:

- Increases FAN (soluble nitrogen) by 25-30 per cent. The addition of other enzymes such as pancreatin and carboxypeptidase result in the increase in (FAN) of up to 100 per cent.
- Improves extract yield
- Reduces viscosity and improves wort filtration
- Reduces the incidence of starch positive worts

After a 30 min stand, additional thermostable bacterial α-amylase is added and the mash is heated at 10°C per minute up to 80°C. There is then a 10 min rest after which it is heated to 90°C, held for 50 min before being combined with a standard malt mash, mashed in at 50°C and these have already been treated with α-amylase, β-glucanase, and bacterial neutral protease. The wort is finally separated. It is also possible to produce a wort from 100 per cent sorghum. The same initial mash profile as described earlier is followed, but after the mash stand at 90°C, the mash is cooled to 60°C and the pH is adjusted to 5.5 with the addition of HCl. The cooled mash is treated with a cocktail of enzymes comprising a blend of fungal α-amylase and amyloglucosidase to suit the attenuation limit required, and the mash is held at 60°C for 30 min, then 75°C for 20 min. The mash is then ready for filtration; however, because of the lack of husk material the worts are usually clarified in a mash filter. Sorghum worts are also a poor source of nitrogen and usually have to be supplemented with yeast nutrients.

3.3. Reduced fermentability beers

Recent developments involve the production of a range of low and no alcohol beers, and some techniques require the production of low fermentable worts. This can be achieved by mashing conventional malts at very high temperature and adding thermostable α-amylase and β-glucanase of microbial origin . The brew is mashed at 72-75°C with the addition of heat-stable fungal β-glucanase (at very high dose rates) and thermostable α-amylase (such as enzyme prepared from *Bacillus stearothermophilus*) to ensure adequate starch release and breakdown and to reduce wort viscosity. Under these conditions, starch saccharification and extraction can proceed, but the temperature is too high for the heat-labile malt β-amylase to survive, resulting in the production of much higher than normal unfermentable oligosaccharides. In this way the fermentability of the wort can be reduced to between 30 and 50 per cent of normal. Further reductions in wort fermentability can be achieved through the addition of very low fermentable adjuncts such as partially hydrolysed maize starch for maltodextrins (11).

3.4. Low Carbohydrate 'Lite' Beer

At the other extreme, superattenuated or 'lite', low carbohydrate beers are often produced by the addition of enzymes to the unpitched wort or during fermentation. The α-1,6 bonds in amylopectin are not hydrolysed by α and β-amylase, and although malt is able to produce all the enzymes needed to break down starch into fermentable sugars, the malt limit dextrinase is unable to attack intact starch granules, which require gelatinization and liquefacation by α and β-amylase before they can be degraded. Exogenous enzymes can be used to break the α-1,6 links and produce more fermentable sugars from the wort, with the following results:

- Added to the fermenting wort it will yield more fermentable sugars increasing the attenuation limit to produce low carbohydrate/'diabetic' beers (29).
- Added after primary fermentation ,it will produce simple sugars to give priming sweetness to a beer, or to fuel a secondary fermentation.

A small increase in fermentability of 1-2 degrees can often be achieved by adding fungal α-amylase to ensure the complete breakdown of all the α-1,4 links. A greater increase in fermentablity can be obtained by adding either amyloglucosidase which preferentially breaks the α-1,4 linkages, and also slowly hydrolyses the α-1,6 links to produce glucose, or pullulanase which breaks the α-1,6 bonds. The addition of pullulanase is usually accompanied by either α-or β-amylase to break the straight chain dextrins into smaller units. The different enzyme combinations produce a wort with a different spectrum of fermentable sugars. The pullulanase / β-amylase blend produces a wort which more closely represents a conventionally produced malt wort. Both enzymes, amyloglucosidase and pullulanase / β-amylase blend, can be added directly to the fermenter between 50 and 150 ppm. Amyloglucosidase is relatively stable in beer and will survive low levels of pasteuriztion. For this reason, some brewers add amyloglucosidase to the wort before wort boiling (30) (cooling to 60°C with a 30 min stand) to allow the enzyme to complete its task before ensuring it is destroyed during the boil, giving better control over the residual fermentability of the wort. An alternative is to use the pullulanase / β-amylase or α-amylase blends which are readily deactivated on storage and with low levels of pasteurization.

The degradation of mash β-glucans by the addition of exogenous β-glucanases has been demonstrated to decrease wort and beer viscosity and to improve filterability. The results show that the addition of β-glucanase decreased wort viscosity and improved the filtration performance. Not all commercial β-glucanase preparations have the same specificity.

- Bacterial β-glucanases have similar specificity to the naturally occurring malt β-glucanases, which would suggest that it should produce significant filtration improvements.
- The β-glucanase from *Aspergillus niger* has a similar activity spectrum to that of *Bacillus* species and with an equally poor performance in the mash.
- The β-glucanases from *Trichoderma* have a high β-1,4 (1,3) and β-1,4 glucanase activity and were able to show an 18 per cent improvement in filtration which compared well with the improvements observed for the β-glucanases from *Pencillium funiculosum* and *Penicillum emersonii.* If the enzymes are compared on a unit basis the *Trichoderma* performance would be poor, possibly because of its poorer heat stability which causes its activity to fall off towards the end of the mash stand ($t_{1/2}$ at 65°C is 14.9 min).
- The best performance for β-glucanase was from *Penicillum emersonii* which although it has a lower activity, has superior heat stability and shows a greater activity towards non-starch polysaccharides.

The β-glucanases from *Penicilium funiculosum* and *Penicillium emersonii* both have substantial pentosanase β-1,4 xylanase activity, which may also contribute to their improved performance. It also makes the enzymes particularly suitable for brewing with adjuncts where high pentosanase activity is required. The ideal solution now offered by some suppliers is a blend of β-glucanase enzymes, with the majority of β-glucanase activity coming from a more cost effective source such as *Trichoderma sp.* and being supplemented by heat-stable β-glucanase from *Penicillum funiculosum* and *Penicillum emersonii* .These provide the necessary enzyme activity at the end of the mash stand and any pentosanase activity that may be required. Although the addition of β-glucanase often gives a noticeable improvement in filtration performance, it is not the only polysaccharide which can hinder filter performance.

A two stage fermentation system has also been devised for the production of low carbohydrate beers (31), amylases (gluco and α-amylases) from *Schwanniomyces castelli* are produced in a highly inducing medium containing maltose. Subsequently, the cells are removed, the culture filterate is concentrated and added to wort previously inoculated with a genetically manipulated strain of *Sacchromyces diastaticus* or with a brewing production strain of *Saccharomyces uvarum (carlsbergensis)*. Recently Novo Nordisk have brought out a new enzyme which is an α-acetolactate decarboxylase (Maturex) from *Bacillus brevis* which is capable of converting the precursor α-acetolactate directly into acetoin (32), thus avoiding the slow oxidative decarboxylation and the subsequent yeast reduction stage. The enzyme must be added at the start of fermentation where it prevents the build-up of diacetyl thus avoiding the need for a long warm maturation or diacetyl stand after fermentation, enabling the beer to be ready more quickly and thus improving vessel utilization. At present the enzyme has only received approval in certain countries and its application is at an early stage.

The following Table 3 shows some of the problems which can occur during the brewing process and recommends some of the possible enzyme treatments that can be used to 'cure' the problem.

Table 3
A Brewing First-Aid Kit

Location	Symptom	Remedy
Cereal cooker	Glutenous starch	Add heat-stable bacterial α-amylase
Mash mixer	Retrograded starch Enzyme deficient malt starch in wort Set mashes: wort will not filter	Add bacterial α -amylase
	Filtration problems High malt glucans Poor extract recovery	Add heat-stable fungal β-Glucanase
	Adjunct brewing wheat/bariey	Add heat-stable fungal β-Glucanase and/or pentosanase
	Low wort nitrogen High adjunct brewing	Add neutral proteinase
Fermentation	Poor wort fermentability Starch in beer	Add fungal α-amylase
	Require highly attenuated beers	Add amyloglucosidase or Pullulanase and β-amylase
	Rapid diacetyl removal	Add α-acetolactate decarboxylase
Maturation and Filteration	Low sweetness Fuel for secondary fermentation	Add amyloglucosidase

Table 3
A Brewing First-Aid Kit

Location	Symptom	Remedy
	Chill haze	Add papain
	Poor filterability	Add fungal α-amylase – β-glucanase/pentosanase
	Haze from starch, glucans, Protein	Add fungal α-amylase – β-glucanase/pentosanase-Papain
Bottling	Poor resistance to staling in Bottle	Add immobilized glucose oxidase in the crown liner

4. DAIRY INDUSTRY

In the present day dairy industry, the most important process involving enzymes is the cheese processing . Here we have discussed in detail the process of cheese manufacture and the steps involving microbial enzymes during the flavor development and cheese ripening process.

4.1. Cheese manufacture

Cheese is one of the earliest known milk products. This milk product comprises the largest use of enzymes in the dairy industry. Traditionally, cheese was prepared by the use of chymosin obtained from the calves stomach. Certain cheese varieties are characterised by their distinctive flavor which is produced by the presence of the enzyme lipase which naturally occurs in milk. The demand for more consistent quality of cheese had led to the commercial development of standardized lipase enzyme preparations to add to cheese milk. Commercial pressures on cheese makers have resulted in the demand for the flavor of their cheese to develop with increasing speed. Research into the use of enzymes has led to the development of systems for the accelerated ripening of cheese to meet these demands. Besides cheese processing, whey, a by-product of cheese manufacture is also processed enzymatically.

Whey contains lactose, proteins, minerals and some lactic acid. It is particularly prone to spoilage, as a consequence of its low solids and contamination with cheese starter organisms. An alternative to spray drying to preserve its nutritional and functional value is to use a lactase enzyme to hydrolyse the lactose in whey to yield glucose and galactose, which are both more soluble and also sweeter than lactose. The advent of improved technology and the introduction of new proprietary products has boosted growth in the demand for enzymes. The use of enzymes in the dairy industry for cheese making and for the production of low lactose dairy products is the second largest market at 65 million dollars (21).

One of the major traditional applications of enzymes in the dairy industry is in the clotting of milk in cheese manufacture and in the maturation of the treated curd to yield the final product. Milk contains two broad classes of proteins, caseins, whey protein, and a sugar lactose. In raw milk, lipid is also present, predominantly as triglycerides sequestered from the aqueous environment by inclusion into milk-fat globule membranes.

In the initial stages in cheese manufacture, microbial starter cultures of *Lactococcus* and *Lactobacillus* ferment lactose to lactic acid, reducing the pH of the solution. At pH 6.0, proteolytic enzymes are added, traditionally used enzyme is chymosin rennet from the milkfed calf stomach. This enzyme specifically cleaves the bond between residues 150 and 106 of k-casein (33), releasing the glycosylated portion of the protein into solution as glycomacropeptide, and retaining the insoluble paracasein in the submicelle. Thus, there is no further barrier to polymerization of α-and β-casein which link, at a frequency dependent upon the calcium ion concentration, which after further mechanical, thermal and maturation process becomes the cheese. Hence, an alternative source preferably a microbial source of these enzymes have to be searched for.

During the maturation process, enzymes and some micro-organisms act on the proteins and lipids of the curd to develop the flavor and texture regarded as typical for that cheese type (34). Following are the enzymes which have been used in cheese processing.

4.2. Lactase

Whey, which is a by-product of cheese manufacturing industry is composed of two main components- lactose (70-75% of the whey solid) and whey protein (6%). Almost 70% of the world's population over the age of 3 years old are lactose-tolerant (35).The enzyme, lactase (beta-D-galactoside galacto-hydrolase; E.C 3.2.1.23) converts the whey into a potentially useful food ingredient (36), by hydrolysing it into its glucose and galactose subunits. Lactases are found in the intestines of young mammals, in plants, fungi, yeasts and bacteria. Lactase is being used increasingly in the dairy industry for several purposes. The industrially important lactase enzyme preparations are exclusively derived from yeasts, fungi and bacteria. The best commercial sources of this enzyme are yeast such as *Saccharomyces lactis*, moulds like *Aspergillus niger* or *A. oryzae*, or bacteria such as *Esherichia coli* . Pure lactase preparations have been derived from *E.coli* , *Saccharomyces* (or *Kluyveromyces*) species and fungi. Of the *Saccharomyces* lactases derived from the dairy yeast, *K.lactis* is at present the most widely used commercial lactase preparation. The second most well known commercial lactase preparation is a fungal lactase derived from *Aspergillus niger*. Its optimal process conditions are around 50°C and pH 3.5- 4.5. This lactase is therefore more suitable for the treatment of whey. Lactases find their application in several areas as :

Milk for people with milk tolerance problems especially infants.

Milk destined for concentrated milk-based products such as cheese and yoghurt making.

Whey or lactase for the production of sweetners and soluble hydrolysed whey syrups.

Concentrated milk products such as condensed milk, and for preventing the gritty texture of ice-cream made with lactose (2).

The *Aspergillus oryzae* enzyme is now the most commonly used enzyme for immobilized systems as it is stable over a wide pH range whereas the *Kluyveromyces marxianus var. lactis* enzyme is currently the most widely used commercial lactase preparation for batch applications. Lactase activity occurs in most starter culture strains used by cheesemakers and Smart *et al.* (37) suggested *Streptococcus salivarius* spp. *thermophilus* may in the future compete with yeasts as a more natural enzyme source for dairy products.

Although the potential for the use of lactases in dairy processing has been recognized for many years, it was not until the development of a commercial enzyme preparation from microbial sources in the 1960s that large scale applications of lactases began. The hydrolysis of lactose in cheese milk produces a slightly faster acidification rate which leads to higher final bacterial counts. This results in a high initial rate of proteolysis and this contributes to a more rapid development of flavor and texture.

4.3. Catalase

Hydrogen peroxide is an effective chemical sterilant, and it is used as a preservative for milk or whey in some countries where there is a lack of refrigeration or pasteurization equipments.

The enzyme catalase (E.C.1.11.1.6) is used to destroy any excess hydrogen peroxide remaining after treatment. Kosikowski (38) claims significant improvements in the quality of cheese when milk is sterilized using the hydrogen peroxide-catalase system instead of being pasteurized. Commercially available catalase for the food industry is a standardized liquid enzyme system extracted from either *Aspergillus niger* or *Micrococcus lysodeikticus*.

4.4. Lipases

Cheese manufacturers use lipase catalysed hydrolysis of milk fat to enhance and accelerate cheese ripening. Lipases and esterases are enzymes that hydrolyse fatty acid esters of glycerol. They are differentiated on the basis of whether they act on soluble substrates or, immiscible substrates. The enzymes can release one, two or three fatty acids yielding diglycerides, monoglycerides and glycerol as well as free fatty acids. Different lipases have differing specifications, releasing individual fatty acids at varying rates. Lipases from *Mucor* spp. are now in widespread use in the cheese industry (1).

Shahani (39) has listed a number of microbial lipases, including those from *Candida* and *Torulopsis.*yeast species, *Rhizopus, Penicillum*, *Aspergillus, Geotrichum* and *Mucor* mould spp. and *Pseudomonas, Achromobacter*, and *Staphylococcus* bacterial species. Only a few of these lipases have found application for cheesemaking on a commercial scale. Microbial lipases can be divided into two groups based on their positional specificity. One group is non-specific and release fatty acids at random positions of the glycerol moiety. Kilara (40) reported that the second group of lipases from *Aspergillus, Mucor, Rhizopus,* and *Pseudomonas* spp. hydrolyse fatty acids located at the 1,3 positions preferentially to give free fatty acids and the di- and monoglycerides as reaction products. The lipases which have found commercial usage in cheesemaking are found in this group.

However, only limited attack on lipids occurs in the production of most cheese. Exceptions are in the blue cheese such as Blue Stilton and Danish Blue and in the manufacture of hard Italian cheese such as Romano and Parmesan, where porcine pregastric esterase is added at the curd stage, giving these cheese special flavors and properties. However, owing to the view that enzymes of animal origin should be discouraged in food processing, a good number of lipases are being investigated particularly from fungal origin and their potential in cheese ripening has been evaluated and it has been well estabilished and realised that lipase of many fungal and some bacterial origin will eventuallly take over in flavor development of various cheese processing methods. The well illustrated examples are known groups of *Aspergillus* , *Penicillium* and certain members of *Mucor* .

4.5. Proteases

The single major application of acidic proteases, particularly rennets, in the dairy industry is in cheese-making, which accounts for 10% of the total industrial enzyme market. The market is divided between animal rennets and microbial rennets in the ratio 1:1. Calf rennet is the principal commercial animal enzyme, which is extracted from the abomassum or fourth stomach of suckling calves. The predominant enzyme of standard rennet preparation is the milk coagulating enzyme, chymosin or rennin, but extracts also contain varying amounts of pepsin (6-12%) depending on the age of the animal. Pepsin also has milk-coagulating ability, although less than rennin, and greater proteolysis occurs.

The requirement to find alternatives for pepsin or rennet substitutes, relates both to the increasing scarcity of calf rennet, and to religious and ethnic regulations against animal derived enzymes. In this regard, microbial proteases are important as they have bulk availability. There are several reports but few commercial processes exist where microbial proteases are in use in the cheese industry.

Proteolysis by enzymes secreted by fungi such as *Penicillum candidum* which is encouraged to develop soft cheeses, is of central importance in which textural changes such as the ripening and softening of Camembert cheese results in a pH increase as the cheese matures. The products of proteolysis i.e. peptides and amino acids, themselves contribute flavors and provide substrates that are broken down to yield sensorially-active degradation products such as amines, thiols and thioesters .

Consequently, one of the primary strategy is the addition of exogenous proteinases to cheese curds to enhance the rate of flavor and texture development. The primary proteinase in any cheese is the coagulant, which can be added at greatest enzyme : substrate ratios. In case of calf rennet, 0.6% of added protease is retained in the curd after the subsequent mechanical dewatering stages. Retention of the protease increases with reduction in whey pH and cheese made with excessive acidity undergoes more rapid proteolysis than normal, which is due to solubilization of colloidal calcium phosphate. Since demand for enzyme is much greater than supply of chymosin calf stomach , consequently a significant part of the market is now been taken by microbial rennets from fungi of the *Mucor* family, although chymosin produced by genetic manipulation with expression of the bovine gene in micro-organisms is now available. Mucor rennet behaves differently from chymosin following clotting because it is retained by the curd at somewhat lower concentrations.

During cheese making, milk coagulation occurs in three phases:

I) Enzymic phase

II) Clotting phase and

III) Proteolytic phase

The enzymic and clotting phases for the cheese processing may be accelerated by increasing the temperature up to 45°C. At temperature below 15°C, the enzymic phase proceeds but clotting does not occur until the milk is heated and advantage can be taken of this in continuous cheese-making processes where the cheese is first cold-renneted and later coagulated continuously in special devices (41). In the proteolytic phase, further degradation by rennet of the casein fractions occurs. Processes for continuous cheese production have been described (42). Proteases other than rennet hydrolyze the curd proteins during the cheese ripening stage. Evidence has been provided which could indicate that proteases associated with *Penicillum caseiselum, P. roqueforti* and *Streptococci* all contribute to degradation of polypeptides during cheese ripening. The role of proteases from various sources on cheese ripening have also been discussed (43). Native milk protease appear to have little effect. Mesophilic starter bacteria play a significant proteolytic role, contributing both to flavor and texture development. Thermophilic starters mostly fungi also contribute to proteolysis but their action is less understood. Because of the capital equipment costs associated with cheese processing and the length of the ripening stage, enzyme mediated alternate processes for accelerating the ripening of cheese by use of added enzymes have been investigated. A range of proteases have been tested for their ability to accelerate cheese ripening and it was found that, while acid and alkaline proteases produced off-flavors, *B. subtilis* neutral protease and many fungal proteases significantly enhanced intensity without producing bitterness (44). It has been experimented and felt that proteases in combination with lipases can be used to produce strongly flavored cheese by adding the enzymes to scalded curds and then curing at 10-25°C for 1-2 month (45).

An appreciation of the contribution of proteases to cheese flavor production has led to the development of enzyme modified cheese (EMC). Here cheese flavor additives are produced by addition of specific proteases and lipases to a slurry of a natural cheese variety and incubated for several weeks under controlled condition (Fig 3). EMC pastes of Cheddar, Edam, Swiss, Provolone, Romano, Parmesan, Mozzarella, Cream dips and other products. In this respect the enzymes obtained from *A.terreus* and *A.carneus* have been found to be of immense importance.

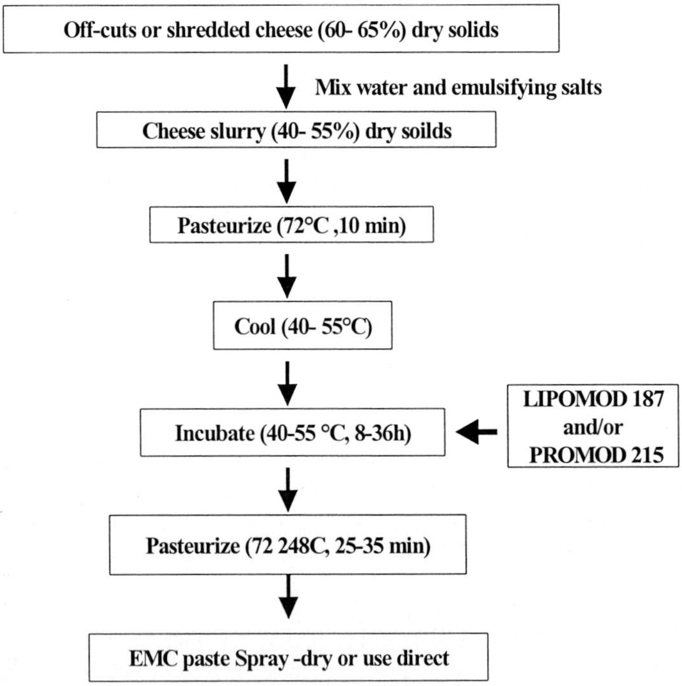

Fig:3 Outline of scheme for enzyme - modified cheese production

4.6. Enzymes in accelerated cheese ripening

The work of Kosikowski and his coworkers (41) in the early 1970s appears to be earliest report of addition of enzymes that assist in the ripening of Cheddar cheese. In these studies combinations of 41 available enzymes-acid and neutral proteinases, lipases, bacterial decarboxylases and lactases-are reported. Most of the enzymes gave little enhancement of cheese flavor, whereas others produced bitter flavors, resulting from peptides with unfavorable terminal amino acids. However, over a period, this group showed that combination of selected proteinases and peptidases together with *Mucor* coagulant could produce increased cheese flavor after one month maturation at 20°C although a higher temperature of 32°C resulted in an over-ripe, burnt flavor with secretion of fluids from the cheese. However, enzymes ripened cheeses were shown to have considerable potential for use in production of cheese where blending of highly flavored cheese with low-grade natural cheese is carried out prior to the heat-treatment used to produce the flavor stable final product .

Law and his colleagues at the Food Research Institute in Reading United Kingdom have carried out a programme of research over the past decade which has shown some of the complexities of using enzymes to develop cheese flavor (44,46-48). The type of flavor is dependent upon the incubation time as well as the source of enzyme. This is because it is reported that longer incubation with the enzymes can result in off-flavors and bitterness. Studies on the activity of acid, neutral and alkaline proteinases in developing flavor and texture in Cheddar revealed that bitterness was a problem observed with addition of many microbial enzymes. Natural proteinases were shown to have potential advantages by virtue of their instability, which limits the extent of their activity. Thus, the use of additional preparation of intracellular proteinases from cheese starter microbes has been explored to address this problem. The addition of cell-free extracts from *Lactococci* has the effect of improving the texture of cheese since use of neutral protease alone, although results in a 50% reduction of ripening time, yields cheese with a softer body that is more bitter than controls. Increasing the level of lactococcal extracts, appears to increase proteolysis but at a rate independent of increase in cheese flavor intensity, suggesting that further transformation of amino acids to sensorially important compounds is the rate-limiting step in many enzyme enhanced maturation.

Proteinase / peptidase preparations have also been described for accelerating the ripening of Dutch, Tilsit and similar cheeses (49). In this case, proteinases from *Pencillium candidum* related those used in soft cheese maturation, were used in combination with peptidases as crude extracts of cheese starter bacteria.

Lipolysis is very important in the development of strongly – flavored cheese such as Parmesan, blue cheese and Feta. In Italian cheese production crude preparations of porcine pregastric esterase, lipase from *Penicillium rowqueforti, P.candidum*, and *M.miehei* have all been employed. In the ripening of blue cheeses, *P.roqueforti* injected into the curd is central in developments of the typical peppery flavor, produced by partial β-oxidation of fatty acids in lipolysis yielding high concentration of methyl ketones. Exogenous lipase, such as preparations from an *Aspergillus* can be utilized to assist the fungus in developing flavor in less than 50% of the normal maturation time. This appears to take place partly as a result of proteinases in the *Aspergillus* enzyme but also through an apparent stimulation of fungal growth through the enhanced concentration of fatty acids in the cheese.

Enzymes are also employed to produce synthetic cheeses for food formulation application. 'Blue cheese' is a popular ingredient particularly in salad dressings and similar convenience foods in the United States. Cheese is not actually required for the product and a cheaper source of the flavor is produced by mixing free-or spray-dried whey with cream that has been incubated at 20°C for 48 hrs, with an *A. oryzae* lipase, spores of *P. roqueforti* and bacterial starter cultures. An alternative has been the isolated lipases of *P.roqueforti.*

Christian Hansen A/S in Denmark have marketed (under the trade name 'Flavorage') a preparation containing a particularly useful lipase from *A. oryzae* which has as exceptionally high specificity for C6-C8 fatty acids and a proteinase (50). This lipase releases longer-chain fatty acids than certain of the other lipases in cheese maturation enzymes, accelerating the production of the flavor of an 'aged' Cheddar. A further important factor that has made attractive the use of this lipase is that it has the valuable property of forming 200 nm micelles in aqueous media so that 94% of the enzyme added to milk is trapped in the curd, whereas 90% of the proteinases is lost into the whey.

The problem of enzyme loss into the whey, rather than the desirable retention in the curd, has attracted much attention. A typical strategy in modifying flavor development enzymes so that they are added to the milk but retained in the cheese is that developed by Olsen and his coworkers at the University of Wisconsin in Madison (51,52). Enzymes, cell free extracts and substrates for production of particular aroma components such as diacetyl and acetoin are encapsulated in milk fat globules which are added to cheese milk prior to renetting, continuing their action through the maturation period. A problem in such enzymatic actions is the need to regenerate enzyme cofactors such as NAD. This has been approached by addition of lipid-packaged cell-free extracts of two micro-organisms. *Lactococcus lactis* var. *maltigenes*, a component of cheese starter cultures which generates flavor and aroma compounds, and *Gluconobacter oxydans* which converts ethanol to acetic acid, regenerating NADH. The net result is that typical cheese volatile aroma compounds such as acetic acid, 3-methylbutanal and amino acid breakdown products such as alcohols and aldehydes are produced, in ratios that can be modified by changing the environmental conditions utilized in the cheese maturation.

5. FRUIT JUICE PROCESSING

The fruit juice industry has to deal with many intricate problems in dealing with material whose composition varies widely depending on the variety of the fruit, the year and the fruit process. The main objective of the fruit juice industry is to process as much fruits as possible at the lowest cost, while maintaining or if possible improving the organoleptic quality and the stability of the finished product. It has been realised that the fruit proteins are responsible for numerous problems encountered during processing. In this respect the first application of pectinases for apple juice clarification started in the 1930s (53). Apple is by far the most important raw material for the production of clear juice and clear concentrate. Annual production is around 54 million tons of apples from which 20% are processed mainly as clearly concentrate but a small part is processed as cloudy juice and apple sauce. The major producers of apple juice concentrate are in the United States, Poland, Argentina, India, Italy, Chile, China and Germany. Application of exogenous enzymes in the fruit processing lead to the degradation of cell wall and the selective extraction of some of their components, resulting in the creation of new types of finished products and fruit derivatives. Commercial enzymes used in fruit juice processing are produced and sold by several companies such as Gist–Brocades, Novo Nordisk, Solvay, Rohm, Quest and Danisco. The enzymes most frequently used are:

5.1. Pectinases

Pectinases obtained from *Aspergillus* species especially *A.niger* (1). These are defined and classified on the basis of their action toward the pectin. Two main groups are distinguished: pectinesterases and pectin depolymerase (54). The pectinlyase (PL) is a pectin depolymerase of endo type that has a great affinity for long methylated chains and acts by β-elimination with the formation of C4 - C5 unsaturated oligouronides (55). *Aspergillus* produces several endo-PL isoenzymes with an average pH optimum of 5.0 to 5.5. Acting on the same substrate as the PL, the pectinmethyl-esterase (PME) removes methoxyl groups from methylated pectin, decreasing at the same time the affinity of PL for this substrate. This results in the formation of methanol and lower methylated pectin, but PME has very low effect on pectin viscosity. PME from *Aspergillus* has a pH optimum around 4.0 to 4.5 and has a strong affinity for high methoxylated pectin such as apple pectin. Pectinase preparations containing numerous side activities are added to the fruit pulp for a maceration during which soluble pectin and cell walls are partially hydrolyzed, resulting in a dramatic decrease of the viscosity increasing the yield because the free-run juice volume is larger and the processing easier. Mash enzyming also increases the productivity of the plant. The pectinases induce a fast clarification of the juice (Fig 4).

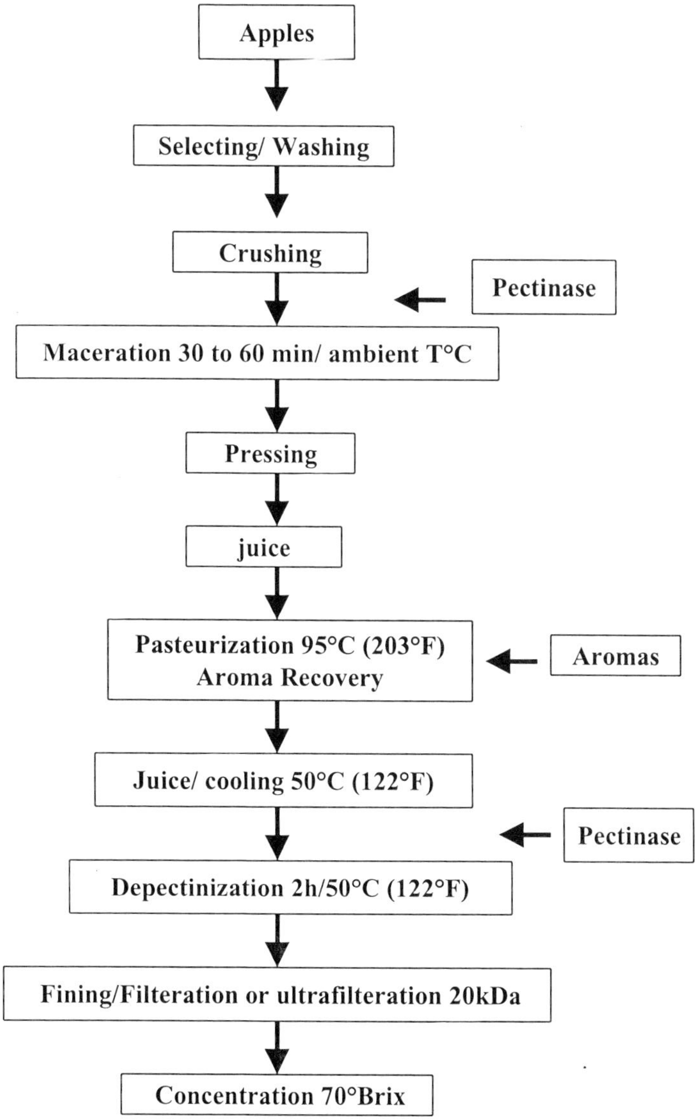

Fig: 4 A flow diagram of the process preparation for the clear apple juice concentrate

The pulp obtained after macerating enzymes are incubated for 30-60 minutes at ambient temperatures with pectinases such as pectinex ultra SP (Novo Nordisk) or Rapidase Press (Gist-Brocades) at 40-75g/tons of apples depending on fruit ripening. In the apple pulp macerating pectinases from *Aspergillus* are stable at about 75% of their optimal activity at pH 3.5-4.0. The application of exogenous pectinases from *Aspergillus* species on apple mast is a necessity because endogenous enzymes present in apple such as pectin methylesterase and polygalacturonase have too low activity to cause immediate visible effect (56,57). Apple pectin is approximately 90% methylated, hence commercial enzyme preparation must contain a high level of pectinylase or pectinmethylesterase in association with polygalacturonase and arabanases . In 1990, a new pectinase rhamnogalacturonase was obtained by another *Aspergillus* sp. which had an important role on the maceration of apple tissue (58). In 1996, new enzymes which were active toward pectic structures, had been described, among these were the rhamnogalcturonases (59) . Rapidase press (Gist - Brocades), Pectinex Ultra SP (Novo Nordisk) or Rohapect MA plus (Rohm), are sold as, mash enzyme preparations. These enzymes contain activities that are necessary to obtain high yields throughout the season. These enzymes decrease the juice viscosity resulting in an increase in its filterability, making the clarification and concentration easier. Treatments with both these enzymes allow production of clear and stable concentrate. By this treatment, the concentration of the juice by a factor of six has been achieved. This has resulted in smaller storage volume and cheaper transportation together with a better concentration stability by preventing spoilage.

5.2. Amylases

The other enzymes used in the fruit industry are the fungal acid amylases. These are used to process fruits that contain starch. This is the case of unripe apples harvested just after the crop period. . Here α-amylase is added at the gelatinization stage of starch at 75°C-80°C. If no amylase is added starch molecules can retrograde and make big aggregates that are difficult to hydrolyze and will clog the filters. The addition of fungal amylase results in the hydrolysis of starch until glucose forms. Retrogradation and postbottling haze formation thus can be prevented which results in the clarification and filterability.

Today enzyme suppliers can offer tailor-made enzyme preparations to fruit juice producers, optimically blended on the base of the fruit structure for every fruit process-citrus tropical fruit or vegetable. Numerous new enzymes have recently been discovered based on biochemical structure of the fruit cell wall. The improvement in the knowledge of fruit composition, cell wall chemistry and enzymology in parallel with better control of large scale enzyme production has made the supply of almost pure enzymes from genetically modified organisms blended in optimal composition based on fruit composition.

5.3. Naringinases

The other problem which citrus fruit juice industry is facing is that of "early" and "delayed bitterness". The former is mainly due to glycosylated flavanone – naringin and the latter are due to triterpenoid limonin. Much work has been carried out on the enzymatic debittering of grapefruit through the application of fungal 'naringinase' preparations which has rhamnosidase and β-glucosidase activities. Although the use of naringinase has been demonstrated in packed bed systems at a smaller scale but processes are yet to be developed as it is more difficult to apply these under the continuous conditions required for factory operation. The enzyme naringinase is obtained from fungi such as *Pencillium* spp. and *A.niger* (60). Limonin, a triterpenoid, an important part of the fruit can be enzymatically hydrolyzed using limonin dehydrogenase (61), obtained from microbes both bacteria and fungi still, resin based absorbents are cheaper and more convenient to remove limonin.

6. MISCELLANEOUS

In addition to the main food processing industries as described, there are many other food processing industries where microbial enzymes are used in the processing of their products. In this section we describe some miscellaneous industries where some microbial enzymes are used in the processing of food materials.

6.1. Starch and sugar industry

A number of microbial, particularly fungal enzymes are added to food stuffs to hydrolyse the starch slurries where these form useful glucose or maltose syrups and maltodextrins (partial starch hydrolysis) (2). These include bacterial and fungal amylases, glucoamylases and iso-amylases (Fig. 5)

6.1.1. Amylases

Amylases are used to produce maltose and maltotriose from liquefied starch. In food industry, two principal types of maltose syrups are produced , one contains 30-50% maltose and 6-10% glucose. It has a dextrose equivalent to 42- 49%. This is used in jams and confectionary because this enzyme is resistant to color formation. It is non-hygroscopic and does not crystallize as ready glucose syrups. The second syrup contains 30- 40% maltose and 30- 50% glucose. This has a dextrose equivalent of 63-70%. It has a high content of fermentable sugars, but is stable on storage. Hence is used in the brewing of beer and in bread making (62). It is sometimes referred to as 'brewers adjunct.' Fungal amylases used in starch and sugar industry are derived from *A.niger* and *A.oryzae* and are much less heat stable than bacterial amylases but still are widely used because they produce large amounts of maltose and maltotriose from liquified starch. The enzyme acts by hydrolysing the penultimate 1,4 -glycosidic links at the non- reducing end of the amylose or amylopectin chains liberating maltose units in a sequential fashion until the action of the enzyme is limited by an α-1-6 bond. A 10-20 dextrose equivalent (DE) syrup containing about 50% maltose is formed. Here, there is the usual relationship between enzyme concentration, substrate concentration, and the incubation time. The process is terminated by heat denaturation of the enzyme at 80-85°C, but for complete hydrolysis of starch to maltose an additional enzyme such as pullulanase is required. Maltose syrups can be manufactured using fungal α-amylase, or plant or bacterial α-amylase, to a low dextrose equivalent only. This is to minimize subsequent maltotriose formation. Maltose is then formed by the maltogenic α-amylase or β-amylase, now the syrup is heated to deactivate the enzyme and subsequently clarified, decolorized and concentrated by evaporation. Barley and soya β-amylases have been used, but because these are relatively expensive, microbial enzymes are therefore increasingly used. Unfortunately, β-amylases cannot act on the α-1,6 bonds in starch in the formation β-limit dextrins. Here the final maltose content is limited to 60%. Therefore, the combined use of β-amylase and debranching enzymes such as pullulanase have been proposed (2).

Amyloglucosidase and fungal α-amylase have also been used in combination to give highly converted syrups containing typically 30-35% glucose and 40-45% maltose. This is because both enzymes have similar pH optima. The syrups are widely used and are resistant to crystallization even at low temperatures and high solids concentrations. Microbial β-amylases are also found in *B.megaterium*, *B.circulans, B.cereus, B.polymyxa, Pseudomonas* and *Streptomyces* spp. These microbial enzymes have the advantage of being potentially cheaper and more abundant than the plant enzymes. The other enzymes being used in starch and sugar industry are as follows.

6.1.2. Glucose Isomerase

The glucose syrup can be isomerized to high fructose corn syrup (HFCS), using immobilized glucose isomerase (1,2). Production of HFCS is one of the most important commercial use of enzymes. The enzyme is obtained from variety of fungal and bacterial spp. and generally the enzyme is used in immobilized form. The syrup contains 42% fructose. This process has rapidly developed over the last dozen years, notably in the USA, and now produces millions of tonnes of HFCS product p.a. along with glucose syrups, crystalline glucose monohydrate, maltodextrins, maltose syrups, the corn gluten, oil and starch generated during the initial processing of the corn, and ethanol by fermentation of the glucose.The first re-usable glucose isomerase was produced by Takasaki & coworkers in the 1960s using *Streptomyces wedmorenis*. In 1988, Novo introduced a new form of immobilized glucose isomerase. This has a productivity of 15 t d HFCS/kg enzyme. It has an enhanced operational stability (life-times of 100-500 days) and higher activity. It is produced by *Streptomyces munnus* cells and is immobilized by procedures very similar to those used for the *B.coagulans* enzyme product.

6.1.3. Glucose oxidase

Glucose oxidase derived from *A.niger* to oxidize glucose to gluconic acid is used to prevent the browning reaction (63) which produces adverse effect on the taste and appearance of powdered egg products.

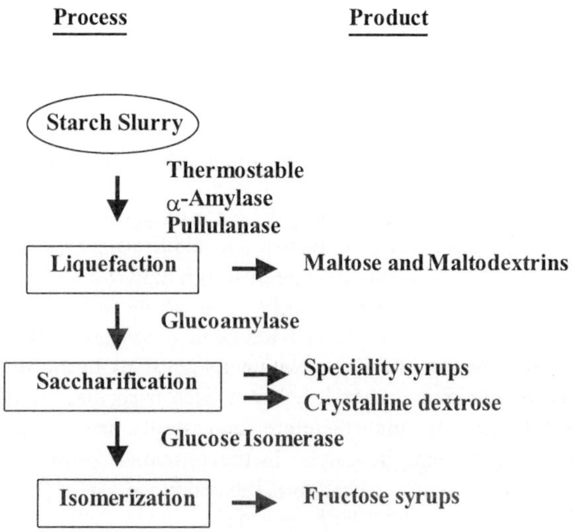

Fig: 5 Enzymatic starch conversion process

6.2. Oils and fat processing

Edible oils and fat constituents are an important part of food industry. The industry involved in the production of oils and fats are trying to develop methods by which they can extract out more amount of oils from the seeds and in this respect , a good number of enzymes as cellulases ,hemicellulases, lipases of microbial origin have beeen used successfully (64) as these enzymes perform their activity on the cellulose and hemicellulose contained in the oil seeds. The action of these enzymes acts in two ways.

1. They enhance the oil recovery.
2. Also eases the recovery procedures which enable the industry in energy saving .

Beside this, basic advantage of microbial enzymes ,particularly fungal and bacterial, these have also find applications in modifying the oils and fats to make value added products for example lipases are used for the transesterification of inexpensive oil (e.g. palm oil) to produce substitutes for cocoa butter, which is a major ingredient in chocolate and confectionery manufacture. They can also be used for the synthesis of other sturctural triglycerides, medium-long-medium-triglycerides (MLMs) which have neutropharmaceutical value. These can act as low calorie fats and as efficient food source for persons with pancreatic insufficiency and other types of mal absorbance (65). Lipases can also be used for selective and enrichment of common edible oils with polyunsaturated fatty acids (PUFA) such as eicosapentanoic acid and docosahexaenoic acid which are essential fatty acids beneficial in cardiovascular and inflammatory diseases (66).

However, sometimes the presence of lipases in certain oil seeds becomes a problem as they hydrolyse the triglycerides of oils and increase their acid value. This is undesirable as it leads to rancidity on storage . Therefore, then it becomes necessary to use certain inhibitors of lipases such as inhibitors from soy beans (67), boronic acids (68), and tetrahydrolipstatin (69).

6.3. Meat industry

The factors that influence the meat quality can be related to the spectrum of enzyme activities naturally present in the muscle tissue. The conversion of muscle into meat is a complex biochemical process, during which the tissue passes through a stage of marked rigidity called rigor mortis or more usually rigor. In the meat industry, the process of muscle softening is known as conditioning or aging. Many fungal proteinases have been shown to be revelant to the weakening of myofibrillar and connective tissue components in muscle. These proteinases have significant influence on rigor softening. Tryptase, chymase and calpain are some of the non-lysosomal proteinases and are important muscle proteinases. The main improvement in texture during conditioning arises from weakening of myofibrils, which is brought upon by proteolytic cleavage of the collagen moiety by enzymes. Apart from texture, tenderness is another important factor deciding the edible quality of meat. Of the various proteinases examined to date, crude papain has been employed most extensively as a meat tenderizer for the low cost. Another application of enzymes in the meat industry is for the recovery of useful meat from scrap, mainly bones. In this scenario, enzymes of fungal origin will take over plant enzymes in the meat processing industry in due course of time.

6.4. Enzymes for flavor production

Flavor and aroma chemicals are exceptionally bioactive molecules that exert their very characteristic taste and smell effects even when present at low concentrations. Many flavor and aroma chemicals are produced naturally in foods by the action of endogenous enzymes and/ or by naturally occuring micro-organisms such as those contributing to the tastes and smell of bread, cheese, wine, beer and many oriental products such as soy. The need to make processed foods either taste and smell characteristics redolent of traditional products creates a derived demand for natural flavors, which in turn creates a need for processes to make them. This is particularly the case for many of their volatile or heat unstable flavor chemicals during processing. It is very appropriate to use modern enzymes and microbial technologies to manufacture flavor and fragrance materials particularly as biocatalysis techniques provide the oppurtunities to carry out new reactions and especially more selective reactions not possible with conventional organic chemical synthesis.

The necessary bioprocesses to make flavor and aroma chemicals fall into four categories:

1. Traditional processes
2. Processes for low- intensity flavors
3. Processes for high-intensity flavors
4. Production of taste flavors

6.4.1. Traditional Processes: Traditional processes are used for products such as beer, wine, cheese. yogurt, and soy sauce. These generally involve microbial action on chemically complex plant- or sometimes animal-derived raw materials, with the formation of relatively low concentrations of flavor chemicals that may very well have good preservative properties as well. These can be described mostly as flavored foods produced by naturally ocurring microorganisms and/ or enzymes derived from them. For instance, intense cheese flavors can be manufactured by the hydrolysis of casein with proprietary proteases and peptidases combined with the hydrolysis of butter fat by lipases and esterases. *Aspergillus niger* and *Rhizomucor miehei* are the preferred sources for lipases for cheese making.

6.4.2. Low-intensity Flavors: Manufacturing has already been established for a range of low intensity flavoring materials such as high-fructose corn syrups, citric and acetic acids , protein hydrolysates, and cheese and cheese flavors.

6.4.3. High-intensity flavor chemicals: More recent developments include the production of high-intensity flavors, such as methylketones (blue cheese flavors) and γ-decalactone (peach and other fruit flavors), in the form of pure natural chemicals by fermentation and bioconversion processes.

6.4.4. Taste Enhancers: Taste enhancers include materials such as monosodium glutamate (MSG) and mononucleotide flavor enhancers such as guanosine monophophate (GMP) and inosine monophosphate (IMP), which are also produced as pure chemicals.

Depending on the complexity of the biochemical reactions to produce the desired products, bioflavor processes can be categorized along the lines of the following examples.

6.4.5. Single enzyme reactions: Examples of single-enzyme reactions include glutaminase, 3'-5' nucleotide diesterase (eg. from *Penicillium citrinium*), adenylic deaminase (eg. from *Brevibacterium ammoniagenes*), glycosidases for the release of terpene flavors from grape and other flavor sources, lipases and esterases for flavor ester synthesis and for resolution of racemic mixture.

6.4.6. Mixed enzyme reactions: Examples of mixed enzyme reactions include use of mixed proteases for producing protein hydrolysates flavors; for example glucosidase and nitrile lyase have been used to make benzalaldehyde from amygdalin; lipoxygenase, hydroperoxide lyase and oxidoreductases to make *cis* 3-hexenol from linoleic acid *via cis*- 3- hexenal , proteases, peptidases, lipases and esterases to make enzyme modified cheese flavors (EMCs). Recently mixed enzyme preparations have become widely used in the extraction of fruit flavors. Use of such special glycosidase combinations have been widely adapted grape fruit juice extraction (6).

7. Conclusions

By now it is well established that the enzymes will be taking over the chemical processes in the coming years as these can carry out reactions of importance much more efficiently under mild conditions. Amongst enzymes, microbial enzymes will be at the utmost top position, as production of these enzymes from microbes can be achieved much faster economically and in large quantities. The sectors of food industries such as dairy, brewing, baking, flavor development etc, are the potential places where the use of microbial enzymes particularly fungal has already established their importance, in processing and quality improvement. However, it also becomes much more essential to undertake an intensive screening programme of newer microorganisms for maximum enzyme production with novel properties, as at present not more than 5% of the microbial flora (both fungal and bacterial) has been explored. If newer organisms along with the modern technologies of molecular biology and bio-technology are combined together, will give an efficient procedure and also yield better quality products with novel properties. A table of various microbial enzymes presently used in food industry is presented as Table 4 in the last.

Table 4.

Major uses of Microbial Enzymes in Food Processing Industries

Microbial enzyme	Industry	Function
Fungal α-amylase	Bread, Baking	Dough conditioning product, proper volume of baked improved machinability
	Distillery Food	Grain mash saccharification Management of starch Containing food Wastewaters
β-amylase	Baking	Proper volume of baked product, flour malting
	Starch	Saccharification of maltose syrups
	Brewing	Barley brewing
Bacterial α-amylase	Starch Distillery	Liquefaction of starch

Table 4

Microbial enzyme	Industry	Function
	Brewing	Improved cooker performance; reduction in husklike taste low-calorie beer
Glucoamylase	Starch, Distillery	Saccharification of liquefied starch or alternative adjunct; light beer
	Brewing	Low-calorie beer
Glucose-isomerase	Starch	Production of high fructose syrup from glucose.
Pullulanase	Starch Brewing	Improved production of glucose and maltose, improved Malting
Cyclodextrin Glucosyl-transferase	Starch	Cyclodextrin production
β-glucanase	Brewing	Viscosity reduction of wort and beer
Rennet	Cheese	Milk clotting
Proteases	Bread and baking	Improved machinability
	Brewing	Chill-proofing, improved shelf life of beers
	Food	Fuctional hydrolysis of proteins, Peptone and soy sauce manufacture
	Meat	Sausage curing
	Fish	Condensed fish solids; fish waste and meat by product processing
Isoamylase	Starch	Maltose syrup prodcution
Lactase	Dairy	Overcoming lactose intolerance, manufacture of frozen mlik concentrate, whey utilization
Pectic enzymes	Fruit and vegetable juices	Mash treatment, depectinization natural cloud production; improve extraction
	Coffee and tea	Viscosity reduction of the extracts
	Wine	Clarification
Naringinase, Hesperidinase	Citrus	Juice debittering
Lipase	Oils and fats	Production of fatty acids, surfactants, emulsifiers and cocoa-butter substitute

Table 4

Microbial enzyme	Industry	Function
	Dairy	Accelerated cheese ripe;milk fat Treatment
	Meat	Milk fat modification in sausage curing
	Candy	Prevention of rancidity
Esterases	Dairy	Accelerated cheese ripening; milk fat treatment
Catalase	Dairy	Milk sterilization,peroxide decomposition
Tannase	Tea	Manufacture of instant tea
Glucose oxidase	Fish	Preventing shrimp decolourisation
	Mayonniase	Protection of water and oil emulsions
	Packaging industry	Coated film
	Fruit and vegetable juice, Beverages,wine, brewery	Flavour and colour preservative, elimination of oxygen
	Eggs	Desugaring, enhanced shelf- life, flavor and colour preservation
	Starch	Modification of fructose-glucose ratio
Mannase	Coffee	Processing of coffee beans
Cellulase and Cellobiase	Fruit and vegetable juice; wine	Enhanced extraction
Hemicellulase	Plant processing, Wine	Clarification of plant extract,enhanced extraction
Invertase	Sweetner	Production of aplha-aspartame from aspartic acid and phenylamine
α-galactosidase	Sugar	Improved sugar production from sugar beets
Koji-crude enzyme preparation	Food fermentation	Facilitating fermentation of foods

Acknowledgements

Authors acknowledge with thanks Ms. Mamta Samtani and Mr. Rajiv Chawla for preparing and making this manuscript in camera ready form. Financial support provided from the Department of Biotechnology, Government of India is also acknowledged.

REFRENCES

1. Ward O.P. In: Comprehensive Biotechnology. Vol 3 (eds:H.W.Blanch, S.Drew and D.I.C. Wang) Pergamon Press, New York, (1985) 789-818.
2. Cheetham P.S.J. In: Handbook of Enzyme Biotechnology, 3rd edn. (Ed. A.Wiseman), Ellis Horwood Ltd; Chichester (Prentice Hall, Hemel Hemptad), (1995) 419-541.
3. Rosendal. P. In: Biotechnolgy and the Food industry. (eds: P.L. Rogers and G.H. Fleet) Gordon and Breach Science Pub, (1989) 77- 93.
4. Singh. A; Kuhad R.C and Saxena R.K. In : Microbiology Today., Vol. 1 (1990) 19-27..
5. Tombs M.P. In: Biotechnology in the Food industry. Open University Press, Buckingham ,(1990) 1-46.
6. Joan Q.S. In: Encyclopedia of Bioprocess Technology. Vol. 2 (eds: M.C.Flickinger and S.W. Drew) John Wiley & Sons, Inc, New York, (1999) 947- 958.
7. Fogarty W.Mand Kelly C.T. In : Microbial Enzymes and Biotechnology 2nd edn (Eds: Fogarty W. M and Kelly C.T) . Elsevier Science Pub, (1990) 72-1218.
8. Camberlain, N; Collins, T.H and Mcdermott E. J. Food Technol, 16 (1981) 127.
9. Fox, P.F and Mulvihill, D.M. In: Advances in Cereal Science and Technology (Ed: Pomeranz, Y) American Association of Cereal Chemist, 5 (1982) 107-156.
10. H. W. Van Dam and J. D.R. Hille, Cereal Foods World. 37(1992) 245.
11. Godfrey.T. In: Industrial Enzymology. 2nd edn (eds:T.Godfrey and West.S) Macmillan Press Ltd, london (1996) 87-100.
12. J.Q.Si and Simonsen R. Proc. Int. Symp. AACC/ICC/CCOA, Beijing, November (1994) 16-19.
13. Chamber lain, N; Collins, T.H. and MeDermott, E. J. Food Technol. 16 (1981) 127.
14. Rubenthaler, G; Finnet, K.F and Pomeranz, Y. Food Technol., 19 (1965) 239-241.
15. Hebeda, R.E., Bowler, L.K. and Teague, W.M. Cerial chem., 36 (1991) 619.
16. Liener, I.E. In: Food improvements through chemical and Enzymatic modification (edn: Feeney, R.E and Whitaker, J.R.) Advs. In chem. Series. Amer. Chem. Soc., 160 (1977) 283-300.
17. R.J. Hamer, in G.Tucker and L.F. J. Woods eds., Enzymes in food process, (1991).
18. R.J.Hamer, P.L. Weegels, J.P. Marseille, and M. Kelfkens,in Y.Pomeranz ed., Wheat is unique, American Association of Cereal Chemists ,(1989) 467.
19. J.Q.Si and Goddik, Annual Meeting of the American Association of Cereal Chemists,Miami,Fla.,October (1993).
20. T.S.Jakobsen and J.Q.si, Proc. Congr. Wheat structure Biochemistry and functions, Reading, U.K.
21. Berry D.R. and Paterson A. In: Enzyme chemistry. Impact and applications (2nd Ed suckling C.J.) (1990) 306-349.
22. Reichelt, J.R. In.: Industrial enzymology, (eds: T. Godfrey and J. Reichelt) (1983) 210-220.

23. Barrett, F.F. In: Enzymes in food processing (ed. G. Reed) Academic, New York. (1975), 301-330

24. El-Dash, A.A. and J.A. Johnson. Cereal Sei. Today., 12 (1967) 282-288.

25. Beckerich, R.P and Denault,L.J. In: Enzymes and their role in Cereal Technology. (eds: J.Kruger, D.Lineback and C.E.Stauffer) American Association of Cereal Chemists, St. Paul MN (1987) 335.

26. Becher, E., Albrecht, R., Bernhard, K., Leveberger, H.G.W., Mayor, H., Mabler, R.K., Beck, C.I and Scoyy, D In food Related Enzymes (ed. J.R. Whitaker). Adv. In chem series Am. Chem. Soc. 136 (1974) 1-30.

27. Scott, D. In: Enzymes in Food Proceesing (ed. Reed, G), 2nd edition (1975) 493-517.

28. Caransa, A., Simell, M., Vaara, T. Proc. Starke Tagung Conf. Detmold, Germany (1998).

29. Woodward, J.B and Bennett, A.B. UK Patent Spec 421, (1973) 955.

30. Marshall, W.G.A., Denault, L.J., Glenister P.R and Dower J. Enzymes in brewing .Brewers Digest, 579 (1982) 14-22.

31. Sills, A.M., Russell . I and Stewart G.G. *In proceedings of the 19th congress*, European Brewery Convention IRL Press Oxford (1983) 377- 384.

32. Pentilla, M.E., Andre, L., Saloheimo, M., Lehtovaara, P and Knowles, J.K.C. Yeast 3 (1987) 175-185.

33. Dalgleish, D.G. Developments in Dairy chemistry-1. (ed. P.F. Fox), Elsevier Applied Science, London (1982) 157.

34. Scott, R. Cheese making practice (2nd edn.), Elsevier Applied Sciences London (1986).

35. Savaino, D.A. and Levitt, M.D. J. Dairy Science. 70 (1987) 397-406.

36. Tweedie, L.S. Mac Bean, R.D & Nickerson, T.A. Food Technol. Aust. 34 (1978) 57-62.

37. Smart, J.B; Crow, V.L and Thomas, T.D. J. Dairy Sci. Technol 20 (1985) 43-56.

38. Kosikowski, F.V. and Iwasaki, J. Dairy Res, 49 (1982) 137.

39. Sahani, K.M. In: Reed G (ed.), Enzymes in Food Processing. Academic Press, New York. (1974) 181.

40. Kilara, A. Process Biochem. 4 (1985) 35-45.

41. Kosikowski, F.V. and Iwasaki, T. J. Dairy Sci. 58 (1975) 963.

42. Beridge, J.J. Analyst, 77, (1952) 57-62.

43. Law, B.A. In : Food proteins, Kellogg Foundation International Symposium on Food Proteins, ed. P.F. Fox. Applied Science London (1982) 307-328.

44. Law, B.A and Wegmore, A.S. J. Dairy Res. 49 (1982) 137.

45. Godfrey, T. In : Industrial Enzymology (ed. T. Godfrey and J. Reichelt) MacMillan, Nature Press, New York (1983) 466.

46. Law, B.A. Dairy Ind. Int. 45 (1980) 15, 17, 19, 20, 22.

47. Law, B.A. and Kolstad, J. Dairy Res., 50 (1983) 519

48. Law, B.A. and Wegniore A.S. J. Dairy Res., 50 (1983) 519

49. Fox P.F. J. Dairy Sci. 72 (1989) 1379.

50. Braun, S.D. and Olson, N.F. J. Dairy Sci., 69 (1986a) 1202.

51. Braun, S.D. and Olson, N.F. J. Dairy Sci., 69(1986b) 1209.

52. A. Mehlitz, Biochem.Z. 221 (1930) 217- 231.

53. L. Rexova – Benkova and O. Markovic, Adv. Carbohydr. Chem. Biochem. 33 (1976) 323- 385.

54. A. Voragen, Ph. D, Thesis, Wageningen University, The Netherlands. 1972.

55. M. Yamasaki, T. Yasui, and K. Arima. Agric. Biol. Chem. 28 (1964) 779-787.

56. Q. Wu, M. Szabacs – Dobozi, M. Hemmats and G. Hrazdina. Plant Physiol. 102 (1993) 219-225.

57. H. Schols, M. A. Posthumus, and A. Voragen . Carbohydr. Res. 206 (1990) 117-129.

58. G. Beldman, G. Mutter, M. J. F.Searle -Van –leeuwen, L.A.M. Van den Brock, H.A. Schols and A.G. J. Voragen , in J. Visser and A.G. J. Voragen ed., Pectin and Pectinases, Elsevier Science, Amsterdam, (1996) 231-245.

59. Tsen, H.Y, Tsai, S.Y. and Yu, G.K. J. Fermentation Bioegg. 67 (1989) 186-189.

60. Vaks, B. and Lifshitz, A. J. Agric. Food. Chem. 29 (1989) 1258.

61. Maeda, H and Tsao.G.T. Process Biochem. (1979) 2-5..

62. Pilizota,V and Subaric, D. Food Technol. Biotechnol. 36(3) (1998) 219-227.

63. Christen,F.M. Biotech. Appl. Biochem. 11(1989) 249-265.

64. Babayan, V.K. Lipids 22 (1987) 417-420.

65. Babayan, V,K; Rosenau J.R. Food Technol. 45 (1991) 111-114.

66. Satouchi, K and Matsushita, S. Agric .Biol. Chem. 40 (1976) 889-897.

67. Sutton,L; Stout, J.S; Housie,L; Spencer,P and Quinn, D.M. Biochem. Biophys. Res. Commun. 134 (1986) 386-392.

68. Madvary,P; Lengsfeld, H and Wolfer,H. Biochem J. 256 (1988) 357-361.

Applied Mycology and Biotechnology
Volume 1. Agriculture and Food Production
G.G. Khachatourians and D.K. Arora (editors)

Production of organic acids and metabolites of fungi for food industry

Nirmala A. Sahasrabudhe[a] and Narendra V. Sankpal[b]

[a]Biochemical Sciences Division, National Chemical Laboratory, Pune-411008, India [b]Chemical Engineering Division

1. INTRODUCTION

Organic acids and metabolites play an important role in food processing. They serve as food ingredients or as precursors for food ingredient. The major organic acids primarily function as food acidulant. In western countries and also in orient the domestic consumption of organic acids are constantly increasing. Fungi produce a wide variety of metabolites including organic acid, vitamins, lipids, flavors, exopolysaccharides, amino acids, etc. Citric acid is the major organic acid produced by fungal fermentation and the second of all fermentation commodities following industrial ethanol. Even though the organic acids are being produced for more than one century, the new advances in the biotechnology and fermentation technology should make substantial improvements.

The present chapter, on production of organic acid and metabolites by fungi covers information on the earlier attempts to the most recent developments in this field. The cited literature may help in increasing understanding of the subject.

The economics of a fermentation process depends on the yield obtained, period of fermentation and choice of raw material. An ideal fermentation process must use raw materials which are cheap, abundant and also are amenable to convert them into a valuable metabolite in high yields during a relatively short fermentation cycle. In the manufacturing of organic acids by fermentation, serious competition to the fermentation process was faced in several instances from alternative processes based on synthesis from cheap material such as n-paraffin or other hydrocarbon byproducts of petroleum industry in last two decades. When the organic acid, which can be produced by microbial fermentation also, occurs in abundance in natural materials and industrial wastes, the chance of the fermentation processes competing with the latter are remote. For instance tartaric acid (Huang and Qian, 1990), which can be easily produced from wine-less, waste. In such cases production of these metabolites by fermentation will not be viable. In this respect the recent literature throws light on the continuous search of cheaper carbon sources such as beet and cane molasses, cane juice, dairy waste, whey and lignocellulosics for production of various metabolites and organic acids.

The majority of organic acids produced by microorganism can be readily placed into two main groups, depending on their metabolic origin and from the main metabolite sequence of aerobic organism, in the TCA cycle and glycolysis and acid arising from the oxidation of glucose (Kubicek *et al.*, 1994). The second group produced by only one or two enzymatic steps using glucose are in this respect important in biotransformation to produce organic acids. In general the biosynthetic routes to group one organic acids can utilize a variety of starting materials. Citric, itaconic, lactic, and malic acids fall into first group, while gluconic, 2-oxogluconic acids come in the second. Acetic acid should be considered a biotransformation of ethanol and is itself produced by group one process.

This review covers the biochemical and the industrial aspects of production of organic acids by fungi. This includes new developments in production processes, their applications in food industry along with the information. (Srinivasan, 1986; Bigelis and Arora, 1992; Matty, 1992; Bigelis and Shib-Perng, 1994).

2. ORGANIC ACIDS

2.1. Citric acid

Citric acid, 2-hydroxy-1,2,3,-propanetricarboxylic acid, is a major commodity chemical manufactured by industrial fungal fermentation after ethanol. Citric acid was first isolated form lemon juice by Scheele and Italians established a virtual monopoly of production in the next 100 years and the product remained expensive. Wehmer observed that citric acid occurs as a significant microbial metabolite in *Citromyces* (a fungus from citrus plant). Based on the characters, latter investigators renamed it as *Penicillium*. Currie (Currie, 1917) found a strain of *Aspergillus niger* capable of accumulating significant amounts of citric acid from a sugar medium. This work was the basis of the first industrial critic acid plant established by Chas., Pfizer and Co. in the US in 1923 and is still the largest producer of citric acid in the world.

In 1928 a plant was built in Kaznejov near Plzer in the Czech Republic, where beet molasses was used as a cheap carbon source with success. This material, however, presented difficulties due to its content of detrimental substances, like metal ions of the transition metal group. Several other plants were erected subsequently in the Soviet Union, in Germany, in Belgium and in several other countries, however, many of the production organisms and the processing technologies are proprietary. Nevertheless, articles that provide details on the citric acid production are numerous and informative (Srinivasan, 1986; Bigelis and Arora, 1992; Bigelis and Shib-Perng, 1994; Matty, 1992; 1994; Roehr *et al.,* 1996).

2.1.1. Raw materials

The critical factor in determining the use is cost of the raw material and its pretreatment. Sucrose is more suitable carbon source (Watanabe *et al.,* 1998). Other materials tested for the citrate fermentation are molasses (Gupta and Sharma, 1994), starch (Rugsaseel *et al.,* 1993; Suzuki *et al.,* 1996), cellulose hydrolysate (Sarangbin *et al.,* 1993), cocoa juice waste (El-Sharkawy *et al.,* 1996), (Shankaranand and Lonsane 1993), banana extract (Sassi *et al.,* 1991), soy whey (Khare *et al.,* 1994) and pineapple waste (Tran *et al.,* 1998).

Use of the impure materials like beet and cane molasses, crude starch, starch hydrolyzate, raw sugar needs more processing (Liu and Wang, 1997). Molasses varies in quality and their composition. The main problem with all carbohydrate sources is presence of metal ions, which are removed by the addition of sodium or potassium ferrocyanide (Haustede and Rudy, 1976). Other equivalent ferro- or ferri- cyanide salts were added either before inoculation or 8 h after inoculation. The ability of ferrocyanide salt to form complex anions has been known for many years.

The other method used for reducing the trace metal ion levels is ion exchange (Swarthout, 1966) but this is applicable when a relatively simple carbohydrate source such as glucose syrup is used, otherwise the ion exchange capacity of the resin is rapidly exhausted by other ions.

2.1.2. Production processes

Commercially citric acid is manufactured by surface-, submerged- and solid-state- or koji- fermentation processes. The medium consists of a carbohydrate supplemented with inorganic or organic nitrogen along with properly balanced mineral salts.

2.1.3. Surface fermentation

This is the original method employed on a large scale around 1920. Despite the development of more sophisticated methods, this process is still employed in existing plants, as the energy costs are lower than those for the submerged process. However, it requires more labor costs. Under present situation new plants of this type may not be economical. Many of the processes are proprietary and hence the details are kept secrete. In this process, the mycelium is grown as a surface mat in a large number of shallow trays with a capacity of 50-100 liters. The surface area of each tray is about 5meters square and the depth of medium is kept between 5-20 cm. The trays are made up of high purity aluminum or stainless steel. Trays are stacked in racks in a chamber constructed so that the process can be carried out under conditions approaching the aseptic. Liquid medium is adjusted to sugar concentration of 20-25% and poured into shallow aluminum or stainless steel pans and inoculated with the spores of the citric acid producing fungal strains. Generally, an initial pH of 3.5 is preferred since under this condition oxalate formation is minimum. In many strains, the onset of sporulation on the mycelial mat reduces or inhibits citric acid yields and hence conditions are adjusted in such a way that mycelial growth takes place for prolonged periods without initiation of sporulation. This can be achieved by raising $NH4^+$ and nonionic surfactant (Sankpal *et al.*, 1998). The fermentation is usually complete in 8-12 days. Then the solution is separated from the mycelium. In replacement cultures, fresh nutrient solution is incubated with the pre grown mycelial mat in a fed batch culture method. This reduces the lag period for subsequent cycles. Repeated use of mycelia is practiced depending upon the characteristics of the strain, particularly in relation to its tendency to show autolysis during the primary growth phase.

When molasses is used as a starting material, it is diluted to the 15% concentration of fermentable sugars. Any required treatment is carried out to remove metal ions and the medium is sterilized usually by boiling. After cooling, it is pumped into the trays within the ventilated chamber. Inoculations are either made directly using a spore suspension or introduced through the air blown over trays. For spore germination of *A. niger,* pH is kept between 5 and 7. High spore concentration in the inoculum decreases the nonproductive phase of growth and may give a filamentous growth rather that pellets (Martin and Demain, 1977).

A patent (Golubtsova *et al.,* 1988) describes a method for inoculum development for citric acid production by surface method. Instead of developing *A. niger* mycelial mat, as an inoculum, *A. niger* conidia have been mixed in 1:2 ratio with a filler, particularly malt sprouts to increase biochemical activities of the inoculum. A new approach for production of citric acid (Solinski, 1987) by surface fermentation, comprises of a series of fungal nutrient media filled trays stacked one above the other. The nutrient flows by gravity from highest to lowest tray. The last tray had a pH of 1.5-5.0. The filtrate from the last tray is then precipitated using calcium salt and separated. The medium is then enriched with appropriate nutrients. The process has worked for 85 days and based on the starting sugar produced 80% citric acid.

2.1.4. Solid state fermentation

Solid state fermentation (SSF) has impact on the environment. Various natural and industrial carbonaceous wastes can be diverted into the value-added products. It is a simple technique in which solid waste substrates such as fibers, bran etc. are utilized with a single culture or combination of cellulolytic cultures. Metabolic heat removal, humidity control, and aeration are the physical controls that improve the overall productivity. Particle size is the most important variable. Smaller particle size of the substrate can increase the productivity and reduce fermentation period considerably (Pallares *et al.,* 1996). Methanol addition stimulates the citric acid production (Hang, and Woodams 1987; de Lima *et al.,* 1995; Tran *et al.,* 1995; Tran *et al.,* 1998; Roukas, 1998). SSF can tolerate high concentrations of metal ions, (Shankaranand and Lonsane, 1994). Tray type fermentations are the widely used for SSF. Medium is sterilized by steam and the fine suspension of the spores is sprayed on the solid substrate. Culture is allowed to grow for a limited period of 10 to 15 days depending upon the degradability or rate of substrate utilization. Fungal cultures on germination secrete extracellular enzymes such as cellulases and amylases converting the substrate into monomers of sugar and further to citric acid. In some cases sugarcane-pressmud has been used as a novel substrate for production of citric acid by solid-state fermentation (Sassi, *et al.,* 1991).

2.1.5. Submerged fermentation

Submerged type fermentations are growing due to the better understanding of fungal metabolism and its positive response. About 80% of the world production of citric acid is carried out by this method. Submerged method has several advantages over surface and solid state culture such as it requires less man power, gives higher yield and productivity, low cost, less contamination, better temperature control during fermentation, handling with suspended solids etc.

2.1.6. *Aspergillus niger* fermentation

Most industrial processes for citric acid production use *A. niger*. In submerged culture, *A. niger* grows in the form of mycelial pellets under rigorously controlled environmental conditions (Kubicek and Roehr, 1986; Thomas, 1992) The medium typically contains 15-20% (w/v) decationized sugar solutions, usually consisting of cane molasses, sucrose, glucose or fructose. Either $(NH_4)SO_4$ or NH_4NO_3 serves as the nitrogen source and levels are optimized for maximum productivity. Trace metal concentrations are strictly controlled, especially the levels of iron, zinc, copper and manganese. Maintenance of low concentrations of manganese (in parts per billion) is critical. Phosphate levels must be limited as its high concentration are not necessary (Dawson and Maddox, 1989). The temperature is kept in the range 28-33°C and the pH of the fermentation medium is controlled between 1.5 and 2.8. Molasses based medium requires pH on the higher side.

A level of oxygen concentration is maintained during the entire citric acid fermentation and is critical for high yields. The fermentation time depends on the raw material and lasts for 5-8 days or may extend to 10-15 days with lower grade medium components. Under these conditions, the conversion of glucose to citric acid is remarkably efficient, ranges from 70-90% depending on the strain, purity of carbohydrate raw material and environmental conditions. Byproducts such as oxalic acid or gluconic acid are formed if growth conditions are not controlled (Kubicek and Rohr, 1986; Milsom and Meers, 1985; Milsom, 1989). During the fermentation *A. niger* exhibits a growth phase during which sugar is consumed, followed by a production phase during which little growth occurs citric acid is produced (Villelt, 1981).

Tower fermenter is more suitable as compared to stirred type fermenter. In a tower fermenter height to diameter ratio is 4:1 to 6:1. The capacity of TF can be increased 100 fold over the stirred tank fermentation (STR). Citric acid production and mycelial morphology is dependent on agitation intensity and pH. The length of the filaments has been shown to be the only parameter that could be related to citric acid production. In a reactor, shorter the filaments, the more citric acid is produced (Papagianni *et al.,* 1994).

Efficiency of citric acid accumulation by *A. niger* depends on various parameters. Among these, the type and concentration of the sugar used has most pronounced effect on acid production, while the optimal concentration of trace metals, phosphate, and nitrogen are interrelated. The suitability of various carbon sources for citric acid production by *A. niger* has been discussed earlier (Xu *et al.,* 1989). Easily metabolizable sugars such as glucose or sucrose result in high rates of acid accumulation and trigger a rise of the intracellular concentration of Fructose 2, 6 diphosphate (Kubicek-Pranz *et al.,* 1990).

Addition of nitrogen/$NH4^{+}$ ions for the growth of cells stimulate the productivity (Choe and Yoo, 1991; Yigitoglue and McNeil, 1992) and has effect on raising the level of PFK1. Controlled levels of $NH4^{+}$ during growth of *A. niger* NCIM 545 helps less contamination of gluconic acid during citric acid formation in surface culture (Sankpal, 1999).

The availability of trace metals, particularly Mn^{2+} but also Fe^{3+} and Zn^{2+}, has a significant impact on citric acid accumulation by *A. niger*. The biochemical mechanisms of action of trace elements have been studied (Roehr *et al.,* 1992). Macromolecular synthesis is disturbed due to the manganese deficiency (Hockertz *et al.,* 1989) thereby inducing increased protein degradation (Ma *et al.,* 1985; Kubicek and Roehr, 1985). Manganese deficiency influences operation of TCA cycle enzymes by increasing the levels of critical intermediates; alters cell wall biosynthesis, composition and morphology (Kisser *et al.,* 1980) changes lipid biosynthesis (Kubicek and Roehr, 1986). Citric acid accumulating strains posses lower quantities of sterols, and ergosterol is the only component (Jernejk *et al.,* 1991).

Citric acid formation increases under conditions of strong aeration. Effect of aeration on respiration of *A. niger* during citric acid production has been reviewed (Zehentgruber *et al.,* 1980; Kubicek *et al.*, 1980; Kirimura *et al.,* 1987). In addition, aeration fulfills two functions - oxygenation and heat removal. The requirement varies with the stage of growth. Initially sterile air is required at low rates to prevent contamination during the spore germination and the initial phase of growth, later heat dispersal requires rates up to 10 cubic meter medium per minute and although this does not require sterility. The air is usually filtered. The humidity of the incoming air is required to be controlled during the process, because the surface area to volume ratio leads to serious problem of evaporation losses over such an extended period of cultivation.

The accumulation of citric acid by *A. niger* is markedly influenced by the pH, where, only significant amount of citric acid accumulates at pH values below 2.5. At higher pH, oxalic and gluconic acids are formed as by products. The external pH has a small influence on the cytosolic pH. (Legrasa and Kidric 1989; Banuclas *et al.,* 1977). Observed differences in the kinetic and regulatory properties in some enzymes with the change of a single pH unit, which is in support of the fact that the small change in pH may influence fine regulation on intracellular enzymes.

2.1.7. Yeast

Earlier interest of the yeast in manufacturing of citric acid was due to its ability to grow on *n*-alkanes. Yeast species, *Candida lipolytica* or its sexual form *Yarrowia lipolytica* (Roehr *et al.*, 1996) and to some extent *Saccharomyces cerevisiae* are used for citric acid production. These cultures use various carbon sources like glucose, sucrose, molasses (Kautola, *et al.,* 1992), ethanol-containing raw materials (Finogenova *et al.,* 1997) and straight- chain *n*-paraffins (about C_9 to C_{20}). During fermentation significant amount (up to 50%) of iso-citric acid is formed. To avoid this byproduct, yeast cultures with minimum expression of aconitate hydratase are selected either by mutation, genetic engineering, addition of surfactants, (Matsumoto *et al.,* 1983) and introduction of oxygen into the fermentation medium (Matsumoto and Ichikawa, 1986).

Using glucose as a carbon source *S. lipolytica* DRL 99 showed increase biomass at higher agitation rate at the same temperature as compared to *Candida guilliermondii* Y488, *Candida lipolytica* Y1095, and *Yarrowia lipolytica* Y7576. These cultures yielded upto 0.79 g citric acid per g glucose consumed to give 78.5 gl^{-1} citric acid (Rane and Sims, 1993). *Candida guilliermondii* have been immobilized by adsorption onto sawdust and used in bubble-column reactor for the continuous production of citric acid. In nitrogen-limited medium at dilution rate of 0.21 h^{-1} productivity of 0.24-g l^{-1}.h was achieved. This was twice that observed in a batch fermenter using free culture (Tisnadjaja *et al.,* 1996). Eventhough, in membrane recycle bioreactor, *Y. lipolytica* ATCC 20346 used for repeated cycles showed reduced productivity, and yields were constant. Further minimization of substrate diversion towards cell growth is possible if the fermentation is carried out at different stages suggested (Lo *et al.,* 1997). Concentration of oxygen in the culture medium (pO2) influenced fermentation of citric acid using the mutant strain of *Y. lipolytica.* At pO2 below 20 % of air saturation the production of citric acid ceased due to decrease in the activities of several mitochondrial and cytosolic enzymes which results shifting of metabolic reactions to the synthesis of intracellular lipids (Il'chenko *et al.,* 1998).

A UV induced mutant of *Y. lipolytica* (A101-114 strain) was found to be the most suitable for citric acid production using glucose containing medium (Wojtatowicz *et al.,* 1991). A cycloheximide-resistant sake yeast mutant CY-179 is obtained by chemical mutagenesis of parental strain F7-01 with EMS. This strain produced higher total acids such as citric and malic acid than the parental strain (Suzuki *et al.,* 1997).

2.1.8. Overproduction of citric acid

Using conventional technique of fermentation of stirred vessel, productivity can be achieved up to certain level. To improve the process productivity and yield of citric acid either physical or biological parameters require modification. The overproduction of citric acid by fungi is directly applicable to the process implementation and improving process economy.

In this respect strain improvement has become the important activity. The selection of strain which would consistently yield citric acid without being sensitive to trace elements and other uncontrolled impurities in commercial raw materials would make the industrial process more easy to work and give greater consistency of result.

The Japanese workers have patented a method for growing selected strains of *A. niger* on moist wheat bran and these strains which are not influenced by the metal ions interference, produce citric acid in good yield. Apart from the strain selection, strain improvement through mutation is carried out to get strains with improved yields. Successive treatments with physical and chemical mutagens followed by testing a large number of colonies will be necessary before strains with improved performance can be isolated. In selecting strains or mutants for large-scale production, several important factors need consideration. These include stability of the strains without undergoing physiological or biochemical degeneration upon subculture for mass propagation, non-utilization of the acid formed and non-formation of other metabolic acids like gluconic, oxalic, and malic acid. Most citric acid fermentation rely on improved strains of *A. niger* that have been chosen for their high productivity and adaptability to industrial fermenters (Schreferl-Kunar *et al.,* 1989; Rugsaseel, 1993). Para-sexual recombination, diploidization and heterokaryon formations have been used for strain improvement (Kirimura *et al.,* 1988a; Kirimura *et al.,* 1988b). Traditional gene transfer approaches have been hampered by the absence of sexual cycle. The genes for pyruvate kinase and phosphofructose kinase from *A. niger* have been cloned and transformants show 20-30 fold overexpression of the enzymes (Visser, 1991).

Protoplast fusion between two citric acid-producing strains from different genetic backgrounds gives stable diploid strains. Diploid strains were treated by benomyl as a haploidizing agent, and many sergeants were obtained. Prototrophic segregants were selected and their haploidy was confirmed by their conidial size and DNA contents. Some of the prototrophic segregants were better either in shake culture or in solid culture than both the parental strains (Kirimura *et al.,* 1988a; Kirimura *et al.,* 1988b). Selection of 2-deoxyglucose-resistant mutant (Sarangbin *et al.,* 1993) having ability to grow faster than the parent strain was carried out on solid media containing 14% sucrose. Among nine mutants, four exhibited significant (>20% of control) increase of citric acid at a rate comparable to the parent strain but exhibited a faster uptake at high (14%) sucrose content (Schreferl-Kunar *et al.,* 1989). *Aspergillus niger* when exposed to UV radiation for a small period, the frequency of mutation increased with increase in the dose of radiation while the number of survivors decreased with the increase of the dose. Highest production of citric acid (49.9 g l^{-1}) was observed with *A. niger* strain E-17 as compared to parent strain which produced 29.9 g l^{-1} under the same growth conditions (Hussain *et al.,* 1987).

2.1.9. Immobilization

Immobilization of fungal mycelia is another way to make process efficient, productive, and continuous. In spite of attempts to immobilize fungal cells, still industrial scale success has not been achieved. Various kinds of solid supports have been tested for the cell immobilization of fungal mycelia, they are glass carrier (Heinrich, 1982), polyurethane foam (Lee, 1989), Polypropylene hollow fibers (Chung, 1988), cellulose carrier (Fujii, 1992; Sankpal *et al.,* 1999a; Sakurai *et al,* 1997), entrapment in Ca-alginate (Eikmeier and Rehm, 1987), (Pena Miranda and Gonzalez, 1994), polyacrylamide gels (Wongwicharn and Sumonpun, 1991; Mittal *et al.,* 1993), (Garg and Sharma, 1992), and agarose (Khare *et al.,* 1994). Various problems have been encountered during fermentation of immobilized mycelia. Further sophistication and or more care during fermentation is required.

2.1.10. Applications

Among many uses of citric acid, about 70% is used in the food industry and 10% of the in cosmetics and pharmaceuticals. The remainder is employed for diverse industrial purposes, including an increasing use in liquid wash products.

Citric acid is widely used in the food industry owing to its versatility and multifunctional nature as a food additive. It has unrestricted GRAS status. In food and beverage citric acid serves as an acidulant, preservative, pH regulator, flavor enhancer, chelating agent, stabilizer, and antioxidant. It is highly soluble, permitting use in concentrated syrups. Due to its low toxicity and medical properties it is considered as a health drink. Citric acid and other organic acids are used extensively to adjust the acid flavor of soft drinks, fruit and vegetable juices, wine and wine-based drinks, ciders, and canned fruit. Being a natural ingredient in many fruits and juices, citric acid effectively brings out flavors and blends well with flavor systems. Sodium citrate plays a similar role in beverages, especially in lemon-lime drinks. Blends of citric acid combined with malic acid, lactic acid or fumaric and in these applications rely on its ability to prevent metal catalyzed off-flavors and the deterioration of color. This important food additive also serves as an antimicrobial preservative, retarding the growth of spoilage organisms.

Citric acid lowers the pH of the food, altering processing parameters. Thus, low acid foods can be converted to high acid foods, changing the cooking time and temperature. The addition of citric acid during processing preserves the natural color of fruit and vegetable pigments. It protects the activity of ascorbic acid by chelating the trace metals that react with it, serving as an antioxidant in this manner. Citric acid also retards the action of enzymes in food, preventing their effect on color e.g., enzymatic browning, and flavor. Citric acid levels of 0.1-0.3% inhibit color and flavor deterioration during the processing of frozen fruits and vegetables during canning.

Immersion of fresh vegetables in a 1-2% citric acid bath for 30 seconds prevents browning for 2-4 h at room temp. In jam, jellies and pie filling citric acid serves as preservative. It also maintains the correct pH in the range of 3.0 to 3.5 for pectins to settle. During processing, citric acid is added as a 50% solution to permit even distribution throughout the batch. Citric acid plays a comparable role in the acidification of gelatin desserts, where it maintains the pH near the isoelectric point of gelatin, permitting proper gelation as well as tartness.

In candies, citric acid is added mainly for tartness and a pleasant softness. It also prevents graining or crystallization of sucrose in candy product. Agar-and starch containing jellies in confectioneries contain citric acid as a flavoring agent.

Sodium citrate finds wide use in dairy products. The citrate salt is added in the emulsification of processed cheeses and cheese foods. It prevents fat separation, maintains the flexibility and texture of cheese slices prevents adherence of sliced cheeses, and promotes favorable and uniform cheese-melting properties, all without changing the cheese flavor. Sodium citrate acts as a stabilizer in whipping cream products and dairy substitutes derived from vegetables. It also promotes the whippability of ice cream and custard products.

The addition of citric acid to wines and ciders is used to dissolve the tannin-iron and phosphate-iron complexes that cause turbidity. Further more, citric acid can prevent an increase in the pH of white wins and the accompanying browning reaction that sometimes occurs during storage.

The citrate derivative is suitable for use in solid fats, plants and fish products. Citric acid plays important role in seafood processing. Citric acid helps to maintain the stability and flavor of seafood products by inactivating endogenous enzymes and enhancing the action of antioxidants.

2.2. Gluconic acid (pentahydroxycaproic acid)

Gluconic acid is produced by fermentation using *A. niger* as well as electrochemical process. More than 80,000 metric tons gluconic acid is produced annually worldwide is in the form of sodium or calcium salt as glucono-d-lactone and free form. Several aspects of gluconic acid production are reviewed earlier (Bigelis and Arora, 1992; Bigelis and Shib-Perng, 1994; Roehr *et al.,* 1996) In a simple one step bioconversion reaction glucose is oxidized to glucono-d-lactone by intracellular and extracellular glucose oxidase, and glucono-d-lactone formed is hydrolyzed spontaneously or by lactonase to gluconic acid. Hydrogen peroxide formed is removed by intracellular catalase.

Although glucose is the preferred carbon source for gluconic acid production by *A. niger,* the other carbon sources such as sucrose (Rao and Panda, 1994a; Quirasco-Baruch *et al.,* 1993), beet molasses (Roukas and Harvey, 1988; Goma *et al.,* 1989; Goma and Seiller, 1987); cane molasses (Rao and Panda, 1994a; Moresi *et al.,* 1992a; Rao and Panda, 1994b), grapemust (Moresi *et al.,* 1991a; Buzzini *et al.,* 1993; Moresi *et al.,* 1992a) and starch or starch hydrolysate (Moresi *et al.,* 1991a; Takahashi, 1991) have been used. Molasses pretreatment is an essential step for fermentation to reduce various cations and anions as the synthesis of gluconic acid has been observed to influence more by cations than anions. Higher levels of metal ions like, copper, iron, zinc, manganese, calcium, and magnesium affect the yield of gluconic acid, and are removed using potassium ferrocyanide treatment (Rao and Panda, 1994b).

2.2.1. Production process

An industrial production of gluconic acid using *A. niger* employs high glucose concentrations (200-380 g l^{-1}) as a carbon source. Ammonium sulfate, di-ammonium phosphate, urea, corn-steep liquor etc. serve as nitrogen source. Addition of KH_2PO_4 and $MgSO_4$ is required along with iron and manganese as trace elements. Addition of calcium carbonate or sodium hydroxide is not only essential to keep the optimum pH for mycelial glucose oxidase but also inhibit the other acid contamination. Submerged process is the commonly used process for production of gluconic acid, but still some plants are using surface process.

In surface culture fermentation, rack holding seven large aluminum pans, one above the other were constructed. Three days old inoculum from flask was poured in 45 l medium. A thin film is observed after 24 h, which thickens rapidly. After completion of fermentation, broth was neutralized followed by mycelial separation and unreacted glucose was recycled for next batch. Eighty percent of the gluconic acid is produced by submerged fermentation in the form of sodium gluconate. *Aspergillus niger* produces gluconic acid at high level during submerged fermentation. When sufficient manganese is present in the media and other trace elements are controlled, by-product formation is negligible and final yield of gluconic acid reaches more than 95%. Conidia or actively growing mycelia are used as a standard inoculum. Spore suspension is prepared by surface culture and introduced into an aerated inoculum fermenter, which is maintained at pH 6.5 and 28-30 °C. Then the inoculum is introduced into production media and diluted approximately 10 fold with fresh medium. Unrestricted growth reduces gluconic acid production (Rao and Panda, 1994a). Efficient production of gluconic acid requires: higher glucose concentration (120-250 g l^{-1}), lower nitrogen concentration (10 mM), low phosphate concentration, presence of manganese ions (0.5 mM), pH 4.5-6.5, dissolved oxygen tension (25% or aeration 0.2 to 1 vvm) agitator speed 1000 rpm (for STR). At an optimal medium levels, pH, aeration, agitation the process completes between 24-72 h.

During fermentation, oxygen is provided in the form of air, oxygen or hydrogen peroxide (Rosenberg *et al.*, 1991; Rosenberg *et al.*, 1992a). Usually high aeration rate in fermentation disrupts the mycelial pellet. This can be prevented in airlift fermenter. Airlift fermenter requires low power as compared to bubble column reactors. The desired morphology of *A. niger*, i.e. pellet growth, is obtained under low stress (Traeger *et al.*, 1989; Trarger *et al.*, 1990).

Zipper et al., described a method to produce gluconic acid at high concentrations. Glucose is added in stages to avoid the problem of crystallization in pipes and pumps. The broth is partially neutralized to obtain mixture of sodium gluconate and gluconic acid which doesn't crystallize. (Ziffer *et al.*, 1971). In a two stage pH maintaining technique (Rosenberg *et al.*, 1992b), conversion of sucrose to gluconic acid and fructose was carried out at pH 4.5 for 12 h (for invertase action), thereafter the pH was maintained at 5.5 after 12 h for conversion of glucose to gluconic acid.

2.2.2. Overproduction of gluconic acid

Overproduction of gluconic acid is carried out by maintaining higher dissolved oxygen (DO) concentrations either by providing oxygen gas or maintaining fairly high pressure in oxygen sparged medium. *A. niger* can withstand fairly high pressure and was found more productive under high dissolved oxygen concentration. Cellular yields decreased with increasing DO (50 -150 ppm), whereas gluconate productivity was increased (Lee *et al.*, 1987; Sato, 1990; Traeger *et al.*, 1991a). In O_2 enriched air DO was raised from 30% to 100% saturation, resulted in two fold gluconic acid production rate. DO concentration was raised by increasing reactor pressure of 2 bar and 6 bar to increase the productivity (Yamashita *et al.*, 1997; Sakurai *et al.*, 1989a).

2.2.3. Genetic improvement

Strain improvement is usually performed by mutation studies or by using genetic engineering techniques such as cloning and transformation. Earlier reviews on this matter are informative (Roehr *et al.*, 1996b; Bigelis and Shib-Perng 1994).

2.2.4. Immobilized cells and enzymes

Aspergillus niger has been immobilized on different supports, like fine felt (Sakurai and Takahashi 1989a and 1989b), cellulose fiber (Matsui and Yasuda, 1989) polyurethane foam (Vassilev *et al.*, 1993); Ca-alginate beads (Moresi *et al.*, 1991b). A Japanese patent issued in 1988 describes a method of manufacturing citric and gluconic acid with *A. niger* immobilized on a high void volume cellulose support. The immobilized *A. niger* was used in a continuous process up to 26 days without any decrease in productivity (Matsui and Yasuda, 1989). High concentrations and productivity is achieved by *Aspergillus niger* immobilized on a non woven fabric for fed batch fermentation (Sakurai *et al.*, 1989a). In a new technique of surface fermentation, *A. niger* has been immobilized on a porous cellulose support (Joshi and Joshi, 1995). Sporulation encountered during fermentation is inhibited following a protocol during growth stage. Glucose solution is made to flow through the capillary of fabric, where glucose is converted to gluconic acid at interface at low pH (3.5 to 2.1). This led to higher productivity for continuous production of free gluconic acid for a period of 61 days with 140 gl^{-1} yield (Sankpal *et al.*, 1999b).

2.2.5. Applications

In addition to high solubility in water, gluconic acid has some additional activities like low corrosivity, low toxicity, form water-soluble complexes with di-, trivalent ions. It is used in the dairy industry to prevent or remove milkstone, cleaning milk cans. In beverages it prevents the calcium scaling. In food it improves mild sour taste, and forms complex with traces of heavy metals. Sodium gluconate has wide applications, mainly as a sequestering agent for calcium in the presence of sodium hydroxide.

2.3. Fumaric acid

Earlier methods of manufacturing fumaric acid involved large-scale fermentation with species of *Rhizopus* (Foster, 1954; Bigelis and Arora, 1992; Bigelis and Shib-Perng, 1994; Roehr *et al.*, 1996c). *Rhizopus nigricans, Rhizopus japonicus* and *Rhizopus arrhizus* have been extensively studied. Further work with *R. arrhizus*, indicated that a fermentation using glucose can produce fumaric acid at a yield upto 60% (Smith, 1974; Buchta, 1983). The rate of accumulation of fumaric acid can be increased by adding surfactant and vegetable oils to the medium (Goldberg and Steigiltz, 1985).

These studies made a solid foundation for further developments. Several *Rhizopus* strains produce fumaric acid from glucose or starch (Bai *et al.,* 1988) but they differ greatly in fumaric acid yields. Among 41 strains studied, *R. arrinizus* R25 was found to produce 53.5-g l^{-1} fumaric acid and 2.12 g l^{-1} of L-malate (Zhang, *et al.*, 1988).

The levels of carbohydrates, phosphate, carbonate, nitrogen, Fe, Mg and Zn and aeration greatly influenced the yield of fumaric acid in R25. The optimal medium for the production of fumaric acid consisted of 12% glucose, 0.1% urea, 0.075% corn steep liquor, 0.03% KH_2PO_4 0.02 - 0.04% $MgSO_4,7H_20$, 50 ppm $ZnSO_4,7H_20$. Several parameters have been optimized for fumaric acid production by *R. arrhizus* using cornstarch hydrolysate (Moresi *et al.,* 1992b) or cassava bagasse (Carta *et al.,* 1997).

In shake flask studies, supplementation of glucose-peptone medium with carboxy-methyl-cellulose (CMC) altered the morphology of *R. arrhizus* from clumped pellets to dispersed mycelium and increased fumaric acid production from 1.78-g l^{-1} to 5.36 g l^{-1}. In fermenter cultures at pH 5.0, mycelium was uniformly dispersed throughout the culture as aggregates of ~1cm diameter with hyphe protruding showed increased fumaric acid production in presence of CMC. Addition of carbopol either in shake flask or in fermenter did not show increase in fumaric acid production (Morrin and Ward, 1990).

In solid state fermentation, fumaric acid was produced by *R. arrhizus* ATCC 13310 from lyophilized orange peels and concentrated rectified grape must used as the carbon source. After 144h, the yield was 31.80 g/kg of the substrate. Small amounts of other organic acids were also produced (Buzzini *et al.,* 1990).

Fumaric acid production was optimized in batch fermentations, using immobilized *R. arrhizus* TKK201 -1-19 mycelium on 10% initial xylose concentration, the C:N ratio of 160 and 10.25 days retention time. The maximum fumaric acid yield increased by 3.4 times. Over 15.3 gl^{-1} acid produced using free cells (Kautola and Linco, 1989; Kautola and Linco, 1990). After the success of this experiment, several methods of immobilization were developed. The fungus was immobilized on polyurethane foam (Petruccioli *et al.,* 1996). Fumaric acid was produced using either molasses or cornstarch hydrolysate.

Immobilized *R. arrhizus* in Ca-alginate produced higher yields than cultures immobilized in other carriers such as agar, alumina, perlite or polyurethane foam

(Petruccioli and Angiani, 1995). Unconventional materials such as cork, expanded polystyrene, extended wood shavings and clay etc. were used for immobilization studies using *R. arrhizus*. Among them cork was found to be the best support to produce 24.06-g l^{-1} fumaric acid using grape must as the carbon source (Buzzini *et al.*, 1993a).

Oxygen mass transfer was improved using porous spargers (stainless steel membrane tube and porcelain tube) in a 50 L rectangular airlift loop reactor, volumetric mass transfer coefficient (KLa) was high which favored conditions for mass transfer, and also useful for forming and suspending small, well-distributed mycelial pellets of 1~2 mm of *R. arrhizus* ATCC20344 (Du Jianxin *et al.*, 1997).

In a recent method, fumaric acid is manufactured with maleic acid isomerase producing microorganisms from maleic acid (Tokumara *et al.*, 1997). Fumaric acid serves as a food acidulant and industrial chemical. It is used as a food ingredient in fruit juices, and aseptic packaged drinks, pie fillings, gelatin desserts, baked goods, jellies, jams wine egg white foams and salt substitute.

2.3.1. Applications of fumaric acid in food industry

Mostly, fumaric acid with the combination of other organic acids is used in the preparation of food additives in the form of Fe (II) fumarate. The ash of mushroom is mixed with fumaric acid and other organic acids to serve as health food, to treat hypotension, prostatitis, constipation, and vocal functions. Kamaboko, a Japanese food additive is prepared using fumaric acid, with stearic acid. It is used to form a coated food additive, which was added to the mixture of surimi, seasonings, starch, water and steamed. The product was elastic and stable at 30°C for 5 days.

Fumaric acid with other organic acids is used as mold growth-inhibitor in combination of azodicarbonamide or potassium bromate, as a food sterilizing agent in alcohol with ascorbic acid as food acidulant to preserve food from spoilage due to the growth of lactic acid bacteria or yeast or pathogenic *E. coli* O-157:H7. An antimicrobial-cleansing compound prepared for vegetables and fruits consists of fumaric acid, surfactant, xanthan gum and ethanol. It is also used as an oil-in-water emulsion type preservative.

Indirect food additives are the packaging materials used in food industries. United States Food and Drug Administration (FDA) is amending the food additive regulations to provide for the safe use of fumaric acid-modified chemicals as a component of adhesives for articles intended to contact food. Food is preserved by packing it in a gas-impermeable sealed container together with di-methyl fumarate, methyl fumarate, or di-ethyl fumarate so that food and the fumarate are not in contact with each other. For example by packaging the fumarate with a gas-permeable small bag. The method prevents food over a long time from deterioration due to bacteria, fungi, yeast etc.

2.4. Itaconic acid (Methylene succinic acid, methylenebutanedioic acid)

Kinoshita (1929) first observed itaconic acid in *A. itaconicus* but *A. terrus* showed better production ability (Kobayashi, 1967). Itaconic acid was originally known as a product of pyrolytic distillation of citric acid. The earlier work on commercial production of itaconic acid has been extensively reviewed (Lockwood, 1979; Miall, 1978; Yan and Ren, 1995; Juranyiova and Matisova, 1991).

In commercial production of itaconic acid, industrial strains of *A. terrus* are grown in aerated stainless still fermenter at pH 2-3, temperature 30-40°C for 3-7 days. The yield of itaconic acid at 10-20% initial carbohydrate concentration generally approaches 55-64% of the theoretical maximum. Itaconic acid is produced during stationary phase of the fermentation (Milsom and Meers, 1985). Several carbohydrates such as glucose (Pfeiffer *et al.*, 1952; Vasilev, *et al.*, 1992), sucrose (Kautola *et al.*, 1990), xylose (Kautala, 1990), wheat starch (Cros and Schneider, 1989), beet molasses and their juices (Nakagawa, 1990; Nakagawa *et al.*, 1991; Nakagawa 1992; Nakagawa *et al.*, 1994) have been employed. Ammonium sulfate or ammonium nitrate serves as a nitrogen source, while Ca^{++}, and Mg^{++} are essential supplements. Optimization of Cu^{++} and Zn^{++}levels is required. Some substances found in beet molasses are detrimental to itaconic acid production (Perlman and Sih, 1960; Pfeiffer *et al.*, 1952).

The process of Pfizer utilizes cheep raw material such as molasses with high sugar concentration of 100-1 80 g l^{-1}. With molasses, supplementation of Cu^{++}, Zn^{++}, and Mg^{++} is required. Disadvantages of the molasses as a whole source of carbon are that the volatile acids present in it pose inhibiting action on itaconic acid fermentation. Elaborate procedures are required to remove such volatiles from the molasses. The resulting itaconic acid is partially neutralized using lime.

Miles process utilizes decationized molasses, sugar level is made up to 125 g l^{-1}, Cu^{++} and Zn^{++} ions are added and the fermentation is carried out at pH>3.0, and 39-42°C (Batti, 1964). The yield of itaconic acid was 58% (based on sugar concentration) within 7.5 days. Addition of calcium ions (Batti and Schweiger, 1963) is beneficial for higher yield (Riviere *et al.*, 1977). Itaconic acid has been produced from various raw materials in batch and continuous process using *A. terrus* NRRL 1963. Fermentation conditions were optimized to include sucrose as the most suitable carbon source. Cell growth was prevented by phosphate limitation. Under optimal conditions the yields of itaconic acid was 60 g l^{-1} (Welter *et al.*, 1998).

Two types of fermenters, a batch stirred tank (STR) and airlift reactor (ALR) were used for itaconic acid production (Welter *et al.*, 1998). Under optimum conditions of fermentation using sucrose as a carbon source the rate of itaconic acid formation in STR was 0.8-g l^{-1}h for agitation rate of 400 rpm and aeration 0.5 vvm. In ALR, the rate of acid formation was 0.64 g l^{-1}h at an oxygen supply of 0.41 vvm. Even though ALR showed high productivity it was not suitable for large-scale production of itaconic acid in repeated batch cultures due to decrease in fluidity resulting from increase in mycelia mass. However in this method without cell recycling, repeated production of itaconic acid was stable for 4 cycles (21 days) with production rate of 0.36 g l^{-1}h (Park *et al.*, 1994). During the fermentation air is required, and stopping the aeration even for 5 min resulted in a complete cessation of itaconic acid production which can be resumed slowly after 24 h continuing of aeration. It was found that insufficient aeration damage to itaconic acid production was due to insufficient ATP production.

Apart from reactor design and fermentation conditions, productivity was increased by immobilization of the production strain and by strain improvement. *A. terrus* was immobilized on polyurethane foam (Kautola *et al.*, 1990). Immobilized *A. terrus* used in a continuous packed-bed column system and operated for 4.5 *months* to produce 328 mg itaconic acid per day per g carrier (Vasilev *et al.*, 1992) which was almost double to that obtained using free cells (Kautola *et al.*, 1989). In another study *A. terrus* mycelia entrapped in calcium beads produced less itaconic acid due to the repressed diffusion of DO to mycelia and produced brown pigment (Nakagawa *et al.*, 1994). Itaconic acid production was reduced to 60-80% of the values of free cells.

It has been possible to overproduce itaconic acid through strain improvement program. A new mutant has been isolated using a technique of concentration gradient agar plate assay. Itaconic acid production here reached from 80-g l^{-1} to 160-g l^{-1} in shake flask studies which was 1.3 times more than that of the parental strain. In another studies, heat resistant mutant KP3152 was obtained by subjecting *A. terrus* IFO6365 to NTG treatment (Karya and Fujiwara, 1994). The mutant when tested in shake culture in medium containing glucose and corn steep liquor at 40°C produced 55-g l^{-1} itaconic acid within 8 days in comparison to 10-g l^{-1} from parental strain.

Microbial manufacture of S(+)-citramalic has potential application for manufacture of styrene butadine copolymers and for lattices and emulsions.

2.5. Kojic acid (2-hydroxymethyl-5-hydroxy-γ-pyrone)

Kojic acid is produced by surface culture by members of *A. flavus* group (Samson 1992) and *A. oryzae* on koji under certain conditions (Saito, 1970). Earlier work on kojic acid has been summarized in following reviews (Miall, 1978; Kitada *et al.,* 1967; Kwak and Rhee, 1992).

Other fungal cultures include *Aspergillus tamarii* (Balint *et al.,* 1988; Chang and Chou 1996; Michalik and Horenitzky, 1992) *Aspergillus oryzae* (Ogawa *et al.,* 1995a; Ogawa *et al.,* 1995b), *Aspergillus albus* (Saegusa *et al.,* 1993) *and Penicillium puberulum* (Chang and Chou 1996). These cultures do not show specific requirement for carbon source. Various carbon sources like glucose (Michalik and Horenitzky 1988), sucrose, fructose, maltose and mixture of glucose and sucrose (Balint *et al.,* 1988; Chang and Chou, 1996) have been utilized by most of the cultures and their mutants. Ammonium salts, urea and yeast extract serve as nitrogen sources.

Fermentation was carried out either in submerged culture or surface culture. For submerged culture, pH is kept between 2.0-5.0 while for surface culture it is 6.5. During submerged fermentation aeration is necessary. Fermentation is carried out at 30°C. Addition of K^+, Mg^{++}, Fe^{++}, Mn^{++}, Zn^{++} salts is essential. Under these conditions kojic acid formation may take few hours to twenty days. In *Aspergillus*, addition of P-bromophenylisothiocyanate or iodoacetate, the inhibitors of carbon metabolism demonstrated increase in yield after 6 days. In submerged culture, the yield of kojic acid ranges between 30-35 g l^{-1} while in static fermentation it is about 17-g l^{-1}.

2.5.1. Submerged Fermentation

Cultivation characteristics of immobilized *A. oryzae* were studied. *A. oryzae* mycelia in situ grown from spores were entrapped in calcium alginate gel beads. In flask culture, kojic acid was accumulated up to of 83-g l^{-1} at this concentration the kojic acid begins to crystallize and thus, the culture was replaced with fresh medium for the next batch culture. The overall productivities of two consecutive cultivation were higher than that of free mycelial fermentation (Kwak and Rhee, 1992).

Submerged fermentation is comparatively faster and more productive than the koji process. Preferred nitrogen source is ammonium sulfate. Kojic acid accumulates under conditions of nitrogen depletion. Medium optimization was done by Box-Wilson method. For *A. oryzae,* the kojic acid production rate was 4.4-g l^{-1} (Takamizawa *et al.,* 1996). In a typical batch process, under optimized conditions of fermentations, *A. tamari* mutant produces 35.9-g l^{-1} of kojic acid (Machalik and Hortnitzky, 1988).

Effect of DOT (at 30, 50 and 80% saturation) was studied on *A. flavus* Link 44-1 in a 2 l tank fermenter at an agitation rate of 600 ppm with varying air flow. Under these conditions, there was no increase in kojic acid yield. However, the biomass was increased below 80% DOT (Ariff *et al.*, 1996). This is not in agreement with the earlier report, where at 35%DOT, the fermentation time was reduced by 35 h and kojic acid yield of 15% increase (Machalik and Hortnitzky, 1992). Higher levels of kojic acid were produced in a newly designed fermenter consisting of a cylindrical apparatus for membrane surface liquid culture. Using *A. oryzae* kojic acid was produced in a batch culture (Wakisaka *et al.*, 1998a) and in a continuous operation (Wakisaka *et al.*, 1998b; Ogawa *et al.*, 1995b).

2.5.2. Applications

Kojic acid is an inhibitor of bacteria, viruses, and fungi. It is also inhibiting the browning effect of tyrosinase in the food and cosmetic industries. To improve its lipophilic properties, *Pseudomonas cepacia* lipase and *Penicillium camembertii* lipase were used for catalyzing the esterification of kojic acid to synthesize kojic acid monolaurate and kojic acid monooleate. These products inhibited tyrosinase.

Polyphenol oxidase (PPO) of mushroom, potato, apple, white shrimp, and spiny lobster cause browning during storage which is inhibited by addition of kojic acid. Kojic acid inhibits melanosis process by interfering the uptake of oxygen required for enzymic browning. Kojic acid may be a substitute for sulfites used in food. Color of astaxanthin-containing foods e.g. salmon, trout etc. are stabilized by treating with flavanols and kojic acid. Color-developing agents and/or pigments for processed foods (e.g. ham, sausage, etc. contain kojic acid.

Discoloration of foods, e.g. vegetables and fruits, are prevented by treating with solution containing kojic acid. Fish feed containing kojic acid improved the body color of fish which inhibits the deposition of melanins on the body surface of the fish. The softening of beef was controlled by addition of kojic acid with protease.

2.6. Malic acid

Malic acid occurs naturally in several fruits and contributes significantly to the flavor, especially of unripe apples. It is an intermediate of TCA cycle. The low levels of accumulation in tissue and high cost of extraction have made its availability limited for wide spread industrial exploitation.

Malic acid is produced chemically by heating fumaric and malic acid mixture at about 200^{0}C, and pressure about 14 bar. The resulting racemic mixture is separated by ion exchange or re-crystallization. Malic acid is produced by fermentation along with fumaric acid followed by treatment with fumarate hydratase. Fermentation process of malic acid is similar to that of fumaric acid where the last step is blocked.

In China, malic acid is manufactured from fumaric acid with immobilized *Candida rugosa* cells (Zhang, 1982), strains of *Schizophyllum commune* (Tachibana and Murakami, 1973), *Paecilomyces varioti* (Takao *et al.,* 1984; Takao *et al.,* 1977), *A. falvus* , *A. paraeitycus* and *A. oryzae* (Abe *et al.,* 1962), species of *Pichia, Candida, Pullularia* and *Penicillium* (Beuchat, 1978; Rossi and Clementi, 1985), *Ustilago* spp. and *Tolyposporium* spp. (Guevarra and Tabuchi 1990).

According to the Japanese patent (Tabuchi and Takahara, 1991), L- malic acid is produced from glucose or carbohydrates by fermentation using *Ustilago* or *Tolyposporium* spp A mixed culture of *Rhizopus arrhizus* and *Proteus vulgaris* (Takao and Hotta, 1997). A strain of *Candida brumptii* is able to convert n-paraffin to malic acid (Sato *et al.*, 1977). *A. oryazae* 102 has been mutated to produce several mutants. A102-800 could produce malic acid from starch as high as 36.6-g l^{-1} (Jiang *et al.*, 1989). Among several strains of *Aspergillus, Schizophylum* and *Rhizopus,* the mutant Lm02 of *Aspergillus* sp. produced 48.37-g l^{-1} of malic acid from glucose. This strain was selected for mutagenesis. The strain mutant N1-14' showed good growth characters on glucose and produced 72.53 g l^{-1} with good conversion rate (Wu *et al.*, 1993). The yeast *Saccharomyces cerevisiae* was entrapped with polyacrylamide gel beads in presence of SDS. The immobilized cells showed 60-fold rise in conversion rate than that of free cells. Almost all fumarate was converted within 30 min of incubation (Olivera *et al.*, 1994). The *Candida rugosa* cr. 10 producing fumarase was immobilized with perlite powder and coating the resultant perlite with polyethylene alcohol for use to manufacture of malic acid from fumaric acid. The immobilized cells are mechanically strong and excellent in elasticity and retain ~ 90% of the original fumarase activity (Yang, 1992).

A mixed culture of *Rhizopus arrhizus* R 25 and *Proteus vulgaris* P produced 54.8 g l^{-1} of malic acid from 12% glucose. Similar yields were obtained in the medium containing 18% sweet potato mash (Jiang *et al.*, 1989). *Saccharomyces cerevisiae* produces malic acid in medium containing glucose and yeast nitrogen base, at pH 4.2 (Farris *et al.*, 1989). In China enzymatic method as well as direct fermentation process produce malic acid. In both cases potato flour or starch was found to be the carbon source of choice due to the low cost of production and purity of the malic acid formed (Jin *et al.*, 1994). In another studies *S. cerevisiae* SHY2- strain has been shown to produce malic acid efficiently from fumaric acid to give 109-g l^{-1} in 96 h.

Conventional yeast (Figueriredo and Carvalho, 1991) and genetically engineered strain of *S. cerevisiae* with an amplified fumarase gene (FUM1) (Neufeld *et al.*, 1991; Peleg *et al.*, 1990) have been used in immobilized systems to convert fumaric acid to malic acid. Results of the genetically modified strain are extremely encouraging. The yeast strain which bears a fumarase gene under control of the strong GAL 10 promoter, exhibit inducible expression of fumarase.

The overproducing strain is able to convert fumaric acid to malic acid with an apparent conversion value of 88% of a rate of 80.4 mmol of fumaric acid $h^{-1}g^{-1}$ cell weight. This bioconversion rate is much higher than known parameters for industrial bacterial strain (Pelag *et al.*, 1990). These findings emphasize the potential of genetically engineered microbial strain for industrial organic acid production. The mechanism and kinetics of malic acid fermentation has been studied in *A. flavus* (Pelag *et al.*, 1988) and the morphological changes associated with acid production have been described (Ward, 1967).

Malic acid plays a role similar to that of citric acid as a food ingredient as an acidulant. As an acidulant it has a tart, smooth, and long lasting flavor profile. Malic acid exhibit a synergy with aspartame permitting a 10% reduction in usage of this sweeter in beverages (Duxbury, 1986).

2.6.1. Applications

Malic acid in combination with L-carnitine, and mixture of natural plant extracts has nutritional composition that is used as medical food for debilitated state. Browning of tea beverage associated with tannin is prevented by addition of malic acid in the formulation. In combination of sugars, preparation of imitating the aroma of prunes can be made that is suitable for use in puddings, jellies, ices, drinks, special chocolates and candies, pastries, cakes, etc. Aroma and taste of the drinks and foods containing aspartame is improved with addition of tartaric acid and malic acid. The yeast producing malic acid is used in manufacture of foods and beverages such as shochu, sake, wine, beer, and bread.

Comparing the physicochemistry, organoleptic properties, taste persistence, and functional properties of malic and citric acids with various canned foods (apple preserves, sour cherry jam, strawberry jam, sour cherry juice, lemon-flavored syrup, and sour cherry syrup), it showed that malic acid could replace citric acid for use in the canning industry.

2.7. Lactic acid

Lactic acid is commonly used in food industry as a food acidulant, preservative and flavor enhancing agent. From nutritional point of view, the L form of lactic acid is the most interesting for the food industry. The human body is also adapted to assimilate this form and produces only L(-)- lactate dehydrogenase enzyme. Some *Rhizopus* spp. are capable of producing L(-) lactic acid in high concentration (Lockwood *et al.,* 1936). *Rhizopus* is the only fungus known to produce L(-) lactic acid. In the industrial production, still *Lactobacillus* spp. are used. Research on lactic acid production by *Rhizopus* has continued primarily, because of the ease of product purification, ability of the fungus to utilize complex carbohydrates and pentose sugars, and production of stereospecific pure forms of L(-) - lactic acid (Wang *et al.,* 1995).

Rhizopus spp. especially *R. arrhizus* and *R. oryzae* produce L-lactic acid from glucose, starch and molasses in the presence of $CaCO_3$ (Hang and Yu, 1989). The basic difference between the *Rhizopus* and bacterial fermentation is that the former is an aerobic process and only L(-)- lactic acid is produced, while bacterial fermentation is an anaerobic process and produces a mixture of D (-), L (+) and D L- lactate. The growth and L-lactic acid production on different carbohydrate sources by *Rhizopus arrhizus* CCM 8109 were investigated. The results indicate that only D and not the L (-) forms of xylose, fructose, galactose, mannose, glucose, maltose, sucrose, cellobiose and partially hydrolyzed starch were converted to L(-) lactic acid.

The effect of different culture conditions on the metabolism of *R. oryzae* NRRL 395 revealed that in the absence of calcium carbonate, because of the accumulation of lactic acid and fumaric acid, rapid decrease in pH resulted in a little consumption of glucose. When $CaCO_3$ was added to aerobic cultures, glucose was completely consumed and this resulted in a high production of L-lactic acid (65-g l^{-1}) along with substantial increase in the biomass. Other metabolites such as ethanol and fumaric acid were present in relatively low concentrations. It was also noticed that, in oxygen limited conditions and without $CaCO_3$, glucose consumption and metabolite production was again limited due to rapid reduction in pH. When $CaCO_3$ was added to these cultures, the fungus was capable of metabolizing practically all the glucose present and its metabolism was directed exclusively to ethanol production (Soccol *et al.,* 1994).

Rhizopus oryzae is not typically considered as organism that grows under anaerobic conditions. Yet, it doesn't possess ethanol fermentative enzymes that allow the fungus to grow for short period in the absence of oxygen (Wright *et al.,* 1996). During aerobic growth *Rhizopus oryzae* produced L(-)- lactic acid from lactate dehydrogenase mediated reduction of pyruvate, while oxygen limiting conditions yield primarily ethanol.

The recent studies (Christoper *et al.,* 1998) demonstrate that it is possible to increase lactic acid production under oxygen limited growth conditions by shifting lactic acid production away from ethanol fermentation. The mutant was selected after chemical mutagenesis on medium rich in glucose and supplemented with allyl alcohol. The adh gene associated with ethanol fermentation and glucose repression of adh genes involved in the oxidation of ethanol show enhanced expression. The enzyme assays and zymograms studies indicate that under anaerobic condition, mutant 18 had severely impaired adh activity and mutant 1-3 and 9 had higher activity than the control. It is theorized that the higher activity might confer resistance to allyl alcohol by selectively converting the toxic acrolein back to alcohol. However, mutant number 18 is unstable and unable to grow on allyl alcohol in successive generations and hence not suitable for large-scale production, still authors are trying to develop more stable mutants.

2.8. Tartaric acid

Tartaric acid (from Tamarind) is an important acidulant in Indian food. Indian production of tamarind is about 3.5 lakh tonnes per year. The fruit pulp is the richest natural source of tartaric acid (8-18%) and is the chief acidulant used in the preparation Indian food.

Conventionally tartaric acid is produced form tamarind extraction and still practiced in some industries. It is also produced from the residues of the wine making industry named as, dregs, and wine less, cream of tartar. It is further extracted using hot water or solvent extraction. L(+)- tartaric acid has better market value as compared to other acids and is called as fruit acid. L(+)-Tartaric acid was produced by cultivation of *Rhizopus validum* using 12% cis-expoxysuccinate in flask fermentation. The fermentation was carried out at 30°C for 4 days on a reciprocal shaker at 110 rpm. The accumulation of L(+) tartaric acid in the cultures was 8.16% (Huang and Qian, 1990). A recent review discusses semi-biosynthetic methods for production of cis-epoxy succinic acid, di-sodium d-tartarate, calcium, d-tartrate and d-tartaric acid, cis-epoxy succinic acid from beer stone (Liu and Xu, 1996).

2.8.1. Applications

In preparation of jelly, yogurt, ice cream, and jam etc. Addition of chitosan gives a smooth mouthful texture. For this purpose, chitosan is dissolved in water by mixing with citric acid or tartaric acid. A patented non calorigenic sweetening agent consists of sodium cyclamate, sodium saccharin, tartaric acid and other components. Drinks and foods containing aspartame, tartaric- and malic- acid improve the aroma and taste.

Low acid vegetables such as asparagus spears, green beans and the like are canned in a solution acidified with tartaric acid. Sufficient tartaric acid is added ~~to~~ the canned product so that the pH of the canned product after thermal processing remains within a range of about pH 4.1 to about 4.4. The reduced pH allows less extreme thermal processing than standard canning parameters to achieve microbiological safety to enhancing the organoleptic qualities of the food.

Diacetyl tartaric acid esters of monoglycerides (DATEM) are used in ice creams. Almonds coated with DATEM retains the texture of ice creams for longer duration, DATEM has conditioning properties and used in different shampoo formulations for grasping and combing-ability of dry hair.

2.9. Succinic acid

Succinic acid is mostly produced by chemical methods. *Saccharomyces unisporus* ferments some monosaccharides to produce succinic acid and acetic acid during ethanol fermentation (Monttanari and Zambonelle, 1997; Guidici *et al.*, 1997). Other fungal cultures include *Candida brumptii* and *Rhizopus spp.* (Sasaki *et al.*, 1970).

2.9.1. Applications

Monoglyceride derivative of succinic acid is useful as emulsifying and ameliorative agents in food, pungent taste-controlling agents for food, its oil-in-water type emulsion is markedly stabilized by lecithin and succinic acid monoglycerides which inhibit feathering, oil-off and had high organoleptic test scores when added to coffee. Succinic acid with other composition is used as an additive in meat treatment to make the meat soft, juicy, and tasty after cooking or frying. Addition of aconitic, gluconic and/or succinic acid, optionally with sclareolides, improves the organoleptic properties of foods, especially by imparting the "Umami" effect and full-mouth feel, in place of sulfur containing amino acids and their salts, pyrrolidonecarboxylic acids and their salts, and nucleotides.

A chelant selected from the succinic acid, and some other classes are provided as a bactericidal compound for use in cosmetic and food products, dioctylsodium sulfosuccinate is used as a mycotoxin inhibitor in human and animal foodstuffs.

3. FUNGAL METABOLITES

3.1. Amino Acids

Amino acids have traditionally been used as human food additives and as animal feed. In addition to the 20 common and several rare amino acids of proteins over 150 other amino acids are known to occur in different cells and tissues in either free or combined form but never in proteins. Most of these are derivatives of the a-amino acids found in proteins. Some non-protein amino acids occur in the D- configuration, such as D-glutamic acid found in the cell walls of bacteria.

Fungi and higher plants contain a variety of amino acids having very curious structure. Fermentation processes for production of amino acids have been reviewed earlier (Leuchtenberger, 1991; Hodgson, 1994; Marces *et al.,* 1994).

Almost all of the 20 amino acids used are manufactured by bacterial processes though fungal technology is used in a few cases. Among the isolated spp. *Candida, Saccharomyces, A. flavus* and *Rhizopus* have been studied for the amino acid production. Some of the glutamic acid secreting microorganisms can be easily found in nature, however this is not true for microorganisms that produce other amino acids. In general most wild strains isolated from nature cannot produce higher levels of amino acids. This is due to the tuned regulation of cellular metabolism to avoid over synthesis of microbial metabolites. In order to achieve the amino acid overproduction, frequently, it is necessary to modify the cell metabolism or the metabolic regulation of these microorganisms (Nakayama, 1973). Following examples illustrate the production of different amino acids by fungi.

Tryptophane is an essential amino acid for human beings and animals. It has many uses in food processing industry, medicine and as forage additive. Genetic engineering advances have been adapted for production of tryptophan (Wei and Wu, 1998).

In *S. cerevisiae* and its mutants (Martirnz-Force and Benitez,1994) blocking of the degradative pathway leads to amino acid accumulation of threonine, methionine and related amino acids. These studies have shown the possibility of isolation of new strains that accumulate amino acids. *Saccharomyces cerevisiae* produces amino acids within and outside the cell while fermenting glucose from molasses. Addition of NaCl has been shown to enhance the production of several enzymes inside and outside the yeast cell (Malaney *et al.,* 1991).

After 1986, the demand for D-alanine has started increasing as a medium for sweeteners. The excellent manufacturing method has been developed using *Candida* (Sato, 1999). The method is based on the ability of the culture to assimilate L-alanine from the culture medium. When such organism is cultivated in presence of DL-alanine, D-alanine of high optical purity can be manufactured. In another example immobilized *A. oryazae* pellets show excellent activity and stability for optical resolution of N-acetal-DL-alanine and produces D-alanine (Song *et al.,* 1997).

In Japan, several strains of Rhizopus have been used to prepare temphe by fermentation. Fermented temphe show high levels of free amino acids as compared to that produced from non-fermented samples (Baumann *et al.,* 1990). Fermentation of molasses by *C. utilis* spp. accumulates product rich in essential amino acids (Lou *et al.,* 1986).

L- proline has been separated from the mixture of amino acids using almost waste free method. This method is based on the ability of amino acids with a primary amino group which accompany L-proline in the reaction mixture to form schiffs's base with salicylic acid or its derivatives, immobilized on the resin surface as schiff's bases and L-proline passes from the column (Sagiyan *et al.,* 1998).

3.2. Vitamins

Vitamins are important components of food in human nutrition and are required for proper metabolism. They play a role in certain enzymatic action and are called as co-factors or co-enzymes. Till today, biological materials such as fruits and vegetables are the major sources of vitamins. Yeast biomass is another biological source used as vitamin supplement. It particularly provides water-soluble B vitamins. Recently, Niacin, Choline, Inositol (Henry *et al.,* 1999), Vitamin K (Tani, 1998) and carotenoids have been produced by fermentation using fungi in laboratory to pilot scale. However, only riboflavin and vitamin B-12 are produced at industrial scale. Industrial production of ascorbic acid involves chemical synthesis and it uses a single microbial step.

3.2.1. Vitamin B2 (Riboflavin)

Riboflavin is used as a nutritional supplement to food products. Its derivative, riboflavin 5'-phosphate is used as a food color. Many fungi and bacteria have ability to produce riboflavin (Bigelis, 1992). Cultivation of an itaconate resistant mutant of *Ashbya gossypii* on soybean oil revealed an increase of 15% isocitrate lyase activity and a 25 fold increase in riboflavin yield as compared to the wild type *A. gossypii* (Schmidt, 1996) *Eremothecium ashbuii* and *A. gossypii* fungal strains produce high levels of riboflavin. The yield upto 15-g l^{-1} obtained by *E. ashbyii* is suitable for industrial fermentation. Using glucose as carbon source, riboflavin is produced in the late growth phase after glucose has been exhausted. When grown on wheat bran and orange peel in presence of yeast or malt extract, *A. gossypii* produced 170-190 mgl^{-1} riboflavin. Addition of sunflower oil enhanced the yield (Lizama-Uc *et al.,* 1998).

Riboflavin producing *A. gossipii* is cultivated in a medium and the product is then isolated using genetic engineering technique. The endogenous ICL activity of *A. gossypii* has been simulated (Boeddecker and Seulberger, 1997). A part from *A. gossypii* the temphe mold, *Rhizopus oligosporus*, *R. arrihizus* and *R. stolonite* produce substantial amounts of riboflavin during temphe fermentation. Certain bacteria isolated from the tempeh enhanced riboflavin production by 25 folds (Keuth and Bisping, 1993).

Candida utilis QGY211 strain grows on molasses to produce mixture of vitamin B1, B2, B6, A and C along with several essential amino acids (Lou *et al.,*1986). Many yeasts are known to produce riboflavin. *Candida torulopsis* or *Candida tropicalis* utilize unrectified or rectified grape mud as a carbon source for riboflavin production. *Candidia torulopsis* also utilizes whey or glucose syrup as carbon source. On grape must, free cells of *C. torulopsis* produce more than 600 mg l^{-1} riboflavin in 5 days. *Candidia topicals* was entrapped in calcium alginate beads and used for riboflavin production either in batch or continuous process produces 350-600 mg l^{-1} riboflavin (Buzzini and Rossi, 1997). Another yeast strain *C. guilliermondii* isolated in Taiwan produces about 186 mg l^{-1} riboflavin. The mutant strain T2-4-67 derived from the wild strain T2-043 produces riboflavin upto 1563-mg l^{-1}. Under optimized conditions of medium, incubation temperature, agitation, and medium to flask volume ratio, the mutant produced 3518 mg/l riboflavin in 200 h in 5 L ml flask and 3.85 g l^{-1} in 5 l jar fermenter under increased speed of agitation (Wang *et al.,* 1994).

3.2.2. Vitamin B 6 (Niacin)

In *C. utilis* QGY211 conditions of fermentation utilizing molasses have been optimized with respect to pH, temperature, agitation, inoculum and type of fermenter and period of fermentation. Under optimized conditions the protein content reaches 50% and the product contains a mixture of several vitamins with niacin (Lou *et al.,* 1986). Several strains of *Rhizopus* play important role in temphe fermentation in solid state. All these cultures have been isolated in Indonesia during temphe production from the water soaking samples. These strains produce a mixture of several water-soluble vitamins of B group, however, final concentrations of these vitamins depends on the production strain. The thiamin content decreases during fermentation by the mold however, addition of *Citrobactor freundii and Klebsilla pneumoniae* increase B-6 production (Keuth and Bisping, 1993). A new process for the production of vitamin B-6 has been patented (Ichikawa *et al.,*1997) the process involves cultivation of *Rhizobium* spp. in a defined culture medium under aerobic condition.

3.2.3. Vitamin A

Beta carotene, a precursor of vitamin A is present in many fruits and vegetables. Some fungi have ability to synthesize ß- carotene during fermentation.

Blakeslea trispora, a heterothallic fungus of *Mucoralas* is extremely rich in mycelial ß- carotene (Goodwin, 1972). The yield of ß-carotene is as much as 1-g l^{-1} medium and is equivalent to 20 mg/g day mycelia when both sexual forms are grown together (Ninet and Renaut, 1979). In a recent new method, a crystalline carotenoid compound from *B. trispora* biomass has been described. The method involves several steps for obtaining pure crystals of ß- carotene (Sibeijn and de Peter, 1998; Sibejin *et al.,* 1998).

Rhodosporium diabovatum (VKOMY-2212) has been isolated. It produces high levels of carotenoids when cultured on medium containing variety of carbon sources. The new strain when grown on ethanol fermented rye flour or molasses for more than 72 h produced a biomass containing high levels of ß-carotene (Avicbieva, 1998). A process has been patented for the formation of intersexual heterokaryons of *P. blakesleenus* strain having nuclei of both the sexual types (Areujo *et al.,* 1992). The heterokaryon strain produced 0.5 mg of ß-carotene g^{-1} dry weight cells, but was unstable. Superproducing strain of *P. blaesleenus* has been isolated which synthesizes 25 mg of carotene/g dry mycelium (Davis, 1973; Murillo *et al.,* 1983) as against the 0.5 mg produced by the heterokaryon.

The yeast, *Phaffia rhodozyma* produces carotenoid on xylose containing medium (Parajo *et al.,* 1998a). After fermentation, the biomass was highly pigmented due to presence of carotenoid. *Candida utilis* doesn't possess an endogenous biochemical isoprenoid pathway for the synthesis of carotenoids. The central isoprenoid pathway concerned with the synthesis of prenyl lipids present in *C. utilis* is active in the synthesis of ergosterol. The enzyme in this pathway has been modified by genetic engineering. The plasmids were transformed by electroporation. A combination of ERG9 disruption and overexpression of HMGOCOA gene, yielded lycopene at 7.8-mg g^{-1} dry cell mass. The results show, how pathway modification can enhance commercial carotenoid production, (Shimada *et al.,* 1998). Gene A encoding a peptide of given sequence was engineered and introduced in yeast along with a carotenoid biosynthetic gene recovered from the culture medium of Yeast (Kirin-Brewing 1999).

3.2.4. Vitamin D

In food processing, synthetic vitamin D is usually added to dairy products such as milk, margarine, breakfast cereals and baby food as vitamin supplement. Sterols are precursors of vitamin D they are widely distributed among fungi (Rattray, 1988). Ergosterol, an important steroid has been isolated from dehydrated form of yeast biomass and converted to vitamin D by UV irradiation (Harrison, 1957). Efficient ergosterol producing yeast strains has been isolated through strain improvement (Park *et al.,* 1982). Two mutants of *S. cerevisiae* synthesizing sterols of the vitamin D group have been selected. They serve as producers of provitamin D3 (68.8 %) and provitamin D4 (95.0%). The provitamin was further converted to vitamin D4 by photo-isomerization at (300 nm) (Mikhailova, 1987). Such yeast cells may serve as alternative sources of vitamin D as food additive to the future foods.

3.3. Flavors and nucleotides

3.3.1. Flavors

Natural flavors are usually extracted from plants and animals. Production of flavors by microorganisms would offer an interesting alternative source of flavors (Welsh, 1989). In many countries flavors produced by microorganisms are considered as natural flavors. This method of production would reduce the supply problem of aromatic plant materials. Most of the studies on flavor production by microorganisms have been made on fungi (Berger *et al.*, 1987; Gallois *et al.*, 1990; Gross *et al.*, 1989; Langrand *et al.*, 1988).

Increasing demand for flavoring low caloric and processed foods and shortage of some classical plant sources have increased interest in flavor generating biosystems. Several articles have summarized progress on the microbial production of bioflavor research (Sjöström and Furia, 1968; Bigelis, 1992; Belin, 1992; Berger *et al.,* 1992; Farbood, 1991; Mateo *et al.,* 1991; Murray, 1989; Schreier, 1992; Tuite, 1992).

Industrial production of flavor nucleotides has been originated in Japan. Over 2000 tons of inosonic acid (5'-IMP) and 1000 tons of guanylic acid (5'-GMP are manufactured there every year primarily by direct fermentation with improved strains of Bacillus and *Brebacterium* spp. The IMP and GMP nucleotides and monosodium L-glutamate (MSG) are termed as "Umami" compounds in Japan, are considered as flavor enhancers (Kuninaka 1981).

Umami compounds appeared to be chemically related to substances that give rise to the four basic independent tastes, salty, sweet, sour and bitter. GMP, IMP and to small extent XMP cause the development of a fuller flavor and impart a "meaty" or "mouth filling" taste to foods, acting alone or synergistically with MSG. They give greater smoothness, body viscosity to certain foods and suppress undesirable flavors and aroma (Shimazono, 1964; Kawamuram and Kare, 1987; Sjöström and Furia, 1968). Industrial production of 5'IMP and 5'GMP began in 1961 and involved the enzymatic degradation of ribonucleic acids with nucleolytic enzymes from selected strains of *Penicillium citrinum* or *Streptomyces aureus*. The RNA is derived from yeasts (Yan *et al.*, 1998), usually improved mutant of *C. utilis* grown under conditions of maximum RNA content (i.e. 10-15% of the cell dry weight). Industrial production of 5' nucleotides has been carried out via four routes.

3.3.2 Aroma Compounds

Potential of higher fungi to form aroma chemicals has been comprehensively reviewed in several articles (Janssens, 1991; O'Connell and Kelly, 1989; Mau *et al.*, 1998a; Mau *et al.*, 1998b, Blinkovsky *et al.*, 1998a; Blinkovsky 1998b, Lelik *et al.*, 1997; Kauppinen *et al.*, 1996). *De-novo* synthesis of lipistirone ketone by the *Basidomycetes, Lepistra irina* and the feed batch conversion of octonic acid to heptanone producing fruity aroma of blue cheese are some important examples of higher fungi producing flavor compounds. Recently, several mushroom varieties such as *King Oister* mushroom, *Pleurotus eryngti* (Mau *et al.*, 1998a), five varieties of ear mushroom and Shiitake mushroom have been reported to produce volatile flavor compounds. Nonvolatile flavors such as amino and other organic acids, several nucleotides and related compounds are also produced by mushrooms (Whitfield 1998).

Yeasts are important contributions to flavor development in alcoholic beverages and the compounds produced during fermentation are many and varied. The production of flavor compounds is dependent upon the raw materials and microorganisms involved. Recently cellulose has been used as an alternative source for ethanol production (Lynd *et al.*, 1998; Gough, 1998). During brewing, quicker fermentation at high temperature or under agitated conditions tend to produce severe off flavors due to a plethora of compounds including fused oils, esters and sulfur compounds in beer. In the production of some alcoholic beverages such as vodka and gin, the ethanol produced in the fermentation is distilled to a high level of purity so other flavor compounds present in the fermentation broth don't contribute to the final flavor of the product. In other processes such as whiskey, bourbon and brandy production, the final spirit contains a large number of yeast derived flavor compounds as well as ethanol. Flavor compounds include a wide range of higher alcohols, fatty acids, esters and carbomyl compounds. The physiology and biochemistry of these flavor compounds have been reviewed (Berry *et al.*, 1987; Korhola *et al.*, 1989). The final level of flavor compounds in the wash can be influenced by growth conditions and by genetic characteristics of the yeast strain. Genetic techniques have been used to produce strains with an increased rate of formation of selected alcohols. The Yeast strain producing three times more iso-amyl alcohol and increased levels of isoamyl acetate has been produced by introducing a gene for resistance to trifluroleucine. In the modified strain, the normal feed back control by leucine had been broken down and elevated levels of isopropylmalate synthatase were obtained (Hirato and Hirol, 1991). Sake strain that produce elevated levels of ß- phenyl ethanol had been obtained by selecting the strains for resistance to the analogs of phenylalanine, *o*- and *p*-fluoro DL phenylalanine. In these strains the enzyme DAHP synthase had been released from feedback inhibition by tyrosine (Fukuda *et al.*, 1990).

Certain enzymes, due to their substrate specificity, selectivity and operation at lower temperature render enzyme as a useful catalyst for producing volatiles (Gatfield, 1988). Glycosidases are being used to liberate volatile from non-volatile glycosidic precursors. Various ß-glucosidases from yeast or filamentous fungi have been characterized regarding their substrate specificity for wine making (Darriet *et al.*, 1988). Enhancement in flavor of wine and passion fruit was further observed using immobilized endo-ß-glucosidases (Shoseyov *et al.*, 1990). Hydrolazes have been found broad applications due to their co-factor independent mode of action. A treatment of extracted vanilla pods with exo-enzyme mixture of *A. niger* led to the liberation of additional flavor compounds (Pouget *et al.*, 1990). Chymosin, a protease used in cheese making was the first enzyme produced by a heterologus recombinant *Kleuveromyces lactis* (Kugimiya *et al.*, 1992). The isolated polypeptides from *A. niger* called as flavourzyme, having aminopeptidase activity can be used as flavor improving compound (Blinkovsky *et al.*, 1998a) to prepare protein hydrolysate for premix for dough in baking ingredients (Blinkovsky *et al.*, 1998b; Kauppinen *et al.*, 1996).

Lipases of different origin such as calf porcine pancreas, *A. niger* or *Pseudomonas fluorescens* showed different specification in generating cheese flavors such as fatty acids, methyl ketone and ethyl butadnote (Lee *et al.*, 1988). The lipase-mediated release of precursor hydroxy fatty acids generates a peach like fatty odor (Berger, 1991). Reverse hydrolyses in micro-aqueous environment using isolated esterase /lipases transferred aliphatic and terpinon substrates to the corresponding carboxylic esters with fruity and fragrant odors (Langrand *et al.*, 1990). The transformed *S. cerevisiae* with alcohol acetyl transferase gene is used in the manufacture of an alcoholic beverage with an enriched ester flavor (Fuji *et al.*, 1993).

3.4. Lipids and fatty acids

Mono-, di-, and tri- glycerides are esters of glycerol containing one, two and three fatty acid chains respectively. Glycerides are the major components of storage of fats in plant and animal cells. Those that are solid at room temperature are termed as fats while those that are liquid at room temperature are termed as oils. Microbial lipid production plays a great nutritional and industrial role. The selection of the most suitable strain and optimization of conditions is very important for lipid production.

About 70% of the world's fats and oils are derived from plants. Alternative source needs to be explored to feed the growing population. Processes for the lipid production using *Candida* and *Fusarium* spp were developed during both world wars. Fat producing yeasts and molds have been thoroughly examined (Rattary, 1988; Kosugi, 1989; Ratledge, 1997) and their potential for commercialization of lipids and fatty acids has been considered. Growth yield, maintenance and lipid formation by yeast (Ykema *et al.*, 1989) and *Mortierella* have been reviewed (Suzuki *et al.*, 1988).

Research on microbial lipid production has focused on the evaluation of fungi with a high fat content. The molds *Fusarium oxysporium* (Naim and Saad, 1984; Fukuda, 1987b; *Penicillium liliacum* (Khan *et al.*, 1987); *Mortierella* spp (Yokuchi and Suzuki, 1989; Zhao and Zneng,1995; Higashiyama *et al.*, 1998; Knutzon *et al.*, 1998; Shimizu and Kobayashi 1998; Suzuki *et al.*,1998; Huang *et al.*, 1998) certain *Zygomycetes* fungi (Weete *et al.*, 1998), white rot fungus, *A. garicales* and brown rot fungus, *Aphyllophorales* (Kapich *et al.*, 1990), *Mucor* (Kikuji, 1990; Fukuda, 1987a; Nakajima and Shimauchi, 1990; Hanada and Ishikawa, 1989), *Conidiobolus* spp. (Kikuji,1990; Nakajima and Shimauchi, 1990), *Rhizopus spp.* (Immelman, 1997; Emelyanova, 1997; accumulate significant levels of fat. Many yeasts, *Candida* spp. (Kitamoto *et al.*,1990; Celligoi *et al.*, 1997; Brown *et al.*, 1989; Thaker and Yadav, 1997), *Hansenula anomala* (Shi, 1988), *Rhodosporula gracilis* (Sattur and Karanth, 1987), *Trichosporan* (Tahoun *et al.*, 1987), *Apiotrichum carvatum* (Ykema *et al.*, 1988; Ykema *et al.*, 1989; Ykema *et al.*, 1990), *Lucosporidium gelidum* (Balashova *et al.*,1987) lipomyces (Romanova *et al.*,1990), *S. cerevisiae* (Nagar-Legmann and Margalith, 1987) are also good producers of fat. In all these cases, wild strains have been isolated from various environments, but the quantities of lipids produced sufficient to warrant commercial exploitation are extremely limited.

Raw materials for lipid production include various carbohydrates such as glucose, sucrose, lactose-whey (Thaker and Yadav, 1997), molasses (Celligoi *et al.*, 1997; Emelyanova, 1997), ethyl alcohol, xylitol (Romanova *et al.*, 1988), glycerol, vegetables oils, butter fat, hydrocarbons (Suzuki and Yokochi, 1986), sulfite waster liquor etc. (Khan *et al.*,1987). The nitrogen sources include ammonium sulfate, ammonium nitrate, sodium nitrate or urea. In general ratio of C:N has pronounced effect on lipid production and it varies with cultures depending on the carbon source used (Sattur and Karanth, 1991; Sattur and Karanth, 1989;Tahoun 1987; Ykema 1988).

There is a direct correlation between the logarithmic growth phase and lipid accumulation (Emelyanova, 1997; Brown *et al.*, 1989). In *Rhodotorula* yeast, pH reflects on lipid accumulation (Johnson *et al.*,1992). When the pH increases, ergosterol production decreases. The maximum amount (60 % dry wt of biomass) of lipid was produced at pH 4.0. At pH 3.0, 5.0 and 6.0 lipid production was less than that at pH 4.0 but there is not much change observed in the fatty acid profile in the pH range of 3-6. Effect of temperature on lipid production has not been studied in wild type fungus, but a mutant in *Mortierella isabelium* shows different temperature requirement for growth and lipid accumulation. Moderate temp of 25-30°C has been found been suitable for growth while low temp ranging between 15-20°C was optimum for lipid production.

Generally mixed lipids are produced by fungi, the mixture consists of phospholipids, glycerides, free fatty acids and sterols. In many cultures, fungal lipids are rich in unsaturated fatty acids like arachodonic acid, linolenic acid and g-linolenic acid. The composition of saturated and unsaturated fatty acids is quite variable within several species of the same genus. Addition of ionic or nonionic surfactants enhances lipid production (Fukuda, 1987b). Similarly addition of NaCl and calcium pantothenate also enhance lipid production on some cases.

Mortierella isabelium mutant strain M018 has shown to accumulate large amount of lipids (Huang *et al.*, 1998) in shake flask culture and produced 135% higher lipid than the parent strain. A desaturase defective mutant, D5, also produced substantial amount of fatty acid (Kawashima *et al.*, 1997; Hiroshi *et al.*, 1997). A gamma- linolenic acid (GLA) producing mutant M6 was further mutated using UV radiation and microwave to produce M6-22 (Xing *et al.*, 1996). MM 15-1 have shown (Hiruta *et al.*, 1996) to produce more unsaturated fatty acids. During the cultivation of both pellet and filamentous forms were observed. The pellet form accumulates lipids with higher GLA content than the filamentous form.

The natural or biotechnological synthesis and hydrolysis of glyceride ester bonds play a variety of roles in relation to food flavor, quality, appearance and wholesomeness. The ability of mono- and di-glycerides to emulsify oil- water mixtures has led to their extensive use in the food and nonfood sectors.

3.5. Polysaccharides

Many exopolysaccharides (EPS) are conventionally used as food ingredients. Particularly, during the last two decades, microbial polysaccharides have been successfully explored and applied in commercial practice. (Yuen, 1974; Yalpani and Sandford, 1987). The two main commercially viable microbial polysaccharides are xanthan and dextran, which are of bacterial origin. At present industrially important fungal polysaccharides are limited to pullulan and scleroglucan.

3.5.1. Pullulan

Pullulan is a non toxic and water-soluble polysaccharide produced industrially by black yeast *Aureobasidium* (Pullularia) *pullulans* (Badr-Eldin *et al.*, 1994). It is a neutral linear homo-polysaccharide composed of maltotriose units, (i.e. three a-1,4-linked glucose molecules which are held together by α-1,6-linkages). The molecular weight of pullulan is generally in the range of 10^4- 10^6, depending on the strain and substrate, phosphate, pH and the duration of the fermentation. Glucose, fructose or certain disaccharides induce pullulan synthesis, and more than 70 % of the sugars can be converted into polymers. Cell growth precedes pullulan production and then elaboration of the exopolymer coincides with the depletion of nitrogen in the medium. A transition of *A. pullulans* from mycelial to the yeast like form which produce the polysaccharide is favored by reduced nitrogen concentration. Catley, 1973; Catley, 1979; Soldki and Cadmus, 1978; Lachake and Rale, 1994; Deshpande *et al.*,1992 have exhaustively reviewed the physiological versatility and application of pullulan.

Pullulan is tasteless and odorless. Being soluble in water and impermeable to oxygen, pullulan membranes or films are used as coating and packaging material for food (Yeun, 1974). Fresh flavor of tea is maintained in tea bags made up of pullulan laminated paper. Water-soluble packages are also used to preserve eggs and egg products. Pullulan imparts thickness of beverages, ice creams and sauces to improve emulsion stability of these products. Addition of pullulan increase quality and texture of baked food products, like meat, ham and sausages etc. It is used as food ingredient to wheat flours for preparing low caloric food like cookies, biscuits and wafers replacing amylose starch with pullulan.

3.5.2. Scleroglucan

Scleroglucan is an EPS produced by many species of the genus *Sclerotium* and related genus such as *Corticium* and *Sclerotinia*. It is a linear chain of 1,3 ß-D glucopyranose units with a single 1,6-ß linked D-glucopyranose at about every third residue in the main chain (Rogers, 1973; Sandford, 1979). *Sclerotium rolfsee* (Hallrich, 1994) and *Sclerotium glucanicum* (Taurhesias and McNeil, 1994; Wang and McNeil, 1994; Wang and McNeil, 1995a; Shelly and McNeil, 1994; Wang and McNeil, 1995b; Wang and McNeil 1995c; Wang and McNeil, 1996) are the main producers of EPS.

The rate of sucrose addition markedly influences the scleroglucon production in *Sclerotium glucanicum*. Addition of sucrose 0.084 gl^{-1}.h has showed increase EPS production by 50% (Wang and McNeil, 1994). Sucrose was added in one shot to raise the concentration of broth 50-kg /Cm, this addition was made at 54, 72 and 96 h interval. It was found that initial high concentration inhibits growth and production in *S. glucanicum*. A variety of methods have been attempted for increasing the production rate and final concentration of the biopolymer. These include nutrient limitation, control of pH, temperature, dissolved oxygen, fed batch operations, use of specialized reactors and strain improvement program (Wang and McNail, 1994; Rouv *et al.*, 1992).

In scleroglucan production sucrose is used as an energy source, ammonium sulfate, yeast extract or sodium nitrate serve as nitrogen sources. The effect of initial –C and –N concentrations on scleroglucan formation *by S. glucanium* NRRL 3006 have been examined in batch cultures containing nutrient medium, supplemented with 30 g l^{-1} sucrose as carbon source and ammonium hydroxide as nitrogen source. Incubation temperature was 27-29°C. Sucrose was added at one shot to raise the broth concentration by 15 kg/CM. The timings of addition were 54, 72 and 96 h. Increasing the initial N concentration led to increased cell mass at the expense of scleroglucan. High sugar concentrations (45 kg/CM) inhibited growth of *S. glucanicum* leading to reduced product level. However, supplementation of sucrose to the cultures after the initial growth phase, overcame this inhibition increasing scleroglucan levels by up to 80% and improving the product yield on carbon source. The clear optimum time for supplementation was 72 h, close to the point of N exhaustion.

Among the nitrogen sources used, i.e. yeast extract, ammonium sulfate and sodium nitrate high concentrations of sodium nitrate (3.0-g l^{-1}) favored growth and biomass formation but polymer production was reduced as compared with that obtained on 1.2 g l^{-1}. It is proven that, change in the yeast extract levels had effect on the fermentation than variations of sodium nitrate levels. Maximum synthesis of scleroglucan (17.4-g l^{-1}) is reached by decreasing levels of yeast extract from 0.56 to 0. 4 g l^{-1} (Hallrich, 1994).

Culture pH critically influences many aspects of the scleroglucan fermentation process. In particular fermentation of the product scleroglucan, and the unwanted formation of oxalate, but at a low pH increased diversion of the energy source to maintenance may occur. The optimum controlled pH for biomass formation is around 3.5, whereas that for scleroglucan formation is 4.5. Lowering the culture pH leads to decreased formation of oxalic acid and scleroglucan (Wang and McNeil, 1995d).

Scleroglucan and biomass formation show two distinct temperature optima, at around 28°C and 28-32°C respectively. At 28°C significant oxalate is formed. As byproduct, generally increased as the temperature declined except at 20°C, where oxalic acid accounted for half of the acidity, the remainder is accounted for by malic and fumaric acid biosynthesis. At the highest temperature (36°C) growth and biosynthetic activities were much reduced and the fungus aggregated to form dense, heavily pigmented pelleted structures. (Wang and McNeil, 1995a).

Some studies on fungal glucan have indicated that polymer synthesis may be stimulated by oxygen limitation. However, in many of the studies simple comparison was made between the normal fermentation process which rapidly becomes oxygen limited and a process where dissolved oxygen (DO) was maintained at some level above zero after the growth phase. Attempts to control DO, usually involves automatic control by increase in impeller speed. As has been shown, in scleroglucan and other fungal glucans, change in impeller speed lead to multiple effects mediated via change in bulk mixing. Average shear stress, improved pH/ temp control as well as maintenance of the target DO levels. Any one or all of these could have a pronounced effect on biopolymer synthesis, since individually each has been shown to increase glucan concentration. Dissolved oxygen neither acts as a trigger for commencement of glucan synthesis nor dose it specifically increase glucan synthesis (Wang and McNeil, 1995c).

Earlier, it was shown that the pellet form of fungus is essential for scleroglucan production (Harvey, 1984). This is somewhat at variance with recent report indicating a dispersed filamentous growth in agitated processes (Taurhesias and McNail, 1994). The morphological characters of the biomass can have a significant impact on many aspects of the process including overall broth rheology, mass and heat momentum transfer (Nienow, 1990). Increased mixing costs for the filamentous broth may directly influence the process economics. A method for increasing scleroglucan production which involves preservation of mycelia in distilled water (4-7°C). The preserved mycelium when transferred to potato-dextrose-yeast extract-agar medium produced more scleroglucan. (Farinaj *et al.*, 1996).

The scleroglucan by virtue of its structure and high molecular weight (12 x 10') has a number of interesting properties. These include marked pseudo plasticity in aqueous solution at low to moderate concentration, stability at high temperature and within a pH range of 2-12 and compatibility with a wide range of electrolyte. Scleroglucan could be widely applied in food industry as a binding, stabilizing or thickening agent (Lecaucheux, *et al.*, 1986; Bluhm *et al.*, 1982) but would have to compete with a number of existing polymers such as starch and cellulose. By understanding the mechanism an operational strategy that maximizes product formation and minimizes byproduct accumulation can be designated.

3.5.3. New polysaccharide from yeast

A new method of preparing polysaccharide involves preparing a suspension of yeast at 80-100°C and separating the polysaccharide containing solids from the suspension by centrifugation. This polysaccharide is useful as thickener and stabilizer or as an additive in the preparation of bread or baked products. Natural polysaccharide is obtained from sources that are nutritionally safe, healthy and inexpensive. Addition of polysaccharide to baked products prolongs retention of water and therefore extends the shelf life. The yeast is preferably *S. carlsbergensis, Kluyveromyces marxianus, Kluyveromyces lactis, K. fragilis or S. cerevisiae*. The suspension contains a weak acid (lactic, trichloroacetic, tartaric, glycolic or citric of pH 4-4.5 (Puglisi *et al.*, 1998).

3.6. Other metabolites

The enzymes isolated from *A. oryzae* VKPM F583 strain was heat activated at 80-90°C for 15 to 20 min. The inactive mycelium was then hydrolyzed using *A. oryaze* enzyme mixture at pH 4.5-6.0. The hydrolysate is then pasteurized at 80-90°C for 20 min. The insoluble fraction was washed and dried to obtain high quality protein hydrolysate (Rimareva and Overchenko, 1998). Renin production from Mucor was established for 575-h continuous fermentation. Lignocellulosic biomass from coconut palm was used as substrate for mushroom production.

Xylitol is used as sweetener. Lignocellulose hydrolysate was found to be suitable for production of xylitol using yeast (Parajo *et al.*, 1998). Effect of hydrolysate concentration, pH, available oxygen and culture medium supplementation, microorganism adaptation and purification of xylitol have been optimized (Pstinen *et al.*, 1998).

Candida tropical has been isolated from the sludge containing xylose or a mixture of xylose and glucose. In 45-h culture using 150-g l^{-1} xylose the cells produced 131-g l^{-1} xylitol. The rate of xylitol production decreased with xylose concentration more than 150-g l^{-1} (Oh and Kim, 1998). *Hansenula polymorpha*, yeast produces xylitol from xylose. The cells produce maximum yield at initial pH of 4.5-5.5 under agitated conditions (Sanchez *et al.*, 1998)

4. CONCLUSION

As seen from the contents of this chapter applied mycology has played a significant role in the production of organic acids and metabolites and their food applications. With the vast majority of fungi remaining to be discovered represent it is our view that we have a resource with tremendous potential. We can only expect that new explorations of fungi from new sources will continue to accelerate. With the advances in biotechnology including fermentation and downstream engineering many more metabolites of fungi should contribute to foods.

REFERENCES

Abe, S., A. Furaya, T. Saito and K. Takayama US patent No. 3063910 (1962).

Araujo, F.J.M., I.M. Calderon, I.L. Diaz and E. Olmedo. US Patent No. 4 318 987 (1982).

Ariff, A.B., M.S. Salleh, B. Ghani, M.A. Hassan, G. Rusul and M.I.A. Karim, Enzyme Microb. Technol., 19 (1996) 545.

Avicbieva, P.B., Butroa. RU Patent No. 98-435494/37.

Badr-Eldin, S.M., O.M. El-Tayeb, H.G. El-Masry, F.H.A. Mohmad and A.O. Abd El-Rehman, World J. Microbiol. Biotechnol., 10 (1994) 423.

Bai, Z., M. Jiang, H. Xie and J. Zhang. Shipin Yu Fajiao Gongye, 4 (1988) 32. (*Chem. Abs.* 110:113137).

Balashova, L.D., A.P. Belov, S.A. Zaitsev, N.B. Gradova, I. P. Bab'eva, E.G. Davidova, and E.E. Azieva. SU Patent No. 1335567 (1987).

Balint, S., J. Forsthoffer, J. Brtko and J. Dobias. Czech. Patent No. 252880 (1988).

Banuclas, M., C. Gancedo, and J.M. Gancedo. (1977) J. Biol. Chem. 252:6394.

Batti, M. and L. Schweiger. US Patent N0. 3078217 (1963).

Batti, M.A. US Patent No. 3162528 (1964).

Baumann, U., B. Bisping and H.J. Rehm, Dechema. Biotechnol. Conf. 4 Lect. DECHEMA Annu. Meet. Biotechnol. 8 (1990) 205.

Belin M., M. Bensoussan and L. Serrano-Carreon, Trends. Food Sci. Technol., 3 (1992) 11.

Berger, R.G., K. Neuhauser and F. Drawert, Biotechnol. Bioeng., 30 (1987) 987.

Berger, R.G. (Y.H. Hui eds.), Food Flavors. *In* Encyclopedia of Food Science and Technology Genetic engineering part III: pp 1313-1320, John Wiley & Sons Chichester, 1991.

Berger, R. G., F. Drawert and P. Tiefel. (1992) Naturally occuring flavors from fungi, yeasts, and bacteria. *In* Bioformation of flavors. (R.L.S. Patterson, B.V. Charlwood, G. MacLeod and A.A. Williams eds.), pp.21-32. Royal Society of Chemistry, Cambridge.

Berry D.R. and D.C. Watson (1987). Production of organoleptic compounds. *In* Yeast biotechnology (D.R. Berry, I. Russel and G.G. Stewart eds.), pp 345-368. Allen and Unwin, London.

Beuchat, L.R. (1978) Metabolites of fungi used in food processing. *In* food and Beverage Mycology, pp. 368-396. Avi. Westport.

Bigelis R. (1992) Fungal metabolites in food processing. In foods and feeds, Vol.3 (D.K. Arora and E.H. Marth eds.), pp. 415-443. Marcel and Dekker Inc. New York.

Bigelis R. and D.K. Arora. (1992) Organic acids of fungi. *In* Handbook of Applied Mycoligy. Vol. 4 (D.K. Arora, R.P. Elander and K.G. Mukerji eds.), pp. 357-376. Marcel and Dekker Inc. New York.

Bigelis R. and T. Shib-Perng. (1994) Microorgamisms for organic acid production. In Food Biotechnology (Y.H. Hui and G.G. Khachatourians eds.), pp. 239-280. VCH Pub. New York.

Blinkovsky, A., K. Brown, M.W. Rey, A. Klotz and T. Byun. WO Patent No. 9851803 (1998a).
Blinkovsky, A., K. Brown, E. Golightly, T. Byun and L.V. Kofod. WO 9851804 (1998b).
Bluhm, T.L., Y. Deslandes, R.H. Marchessault, S. Perez and M. Rinaudo, Carbohydr. Res., 100 (1982) 117.
Boeddecker, T. and H. Seulberger. WO Patent No. 9703208 (1997).
Brown, B.D.K., H. Hsu, E.G. Hammond and B.A. Glatz. (1989) J. Ferment. Bioeng. 68: 344.
Buchta, K. (1983) Organic acids of minor importance. In Biotechnology Microbial Products, Biomass, and Primary Products, Vol. 3 (H. Dellweg, eds.), pp. 467. Springer-Verlag Chemie, Weinheim.
Buzzini, P., M. Gobbetti and J. Rossi., J. Ann. Fac. Agrar. Univ. Studi Perugia. 44 (1990) 661.
Buzzini, P., M. Gobbetti and J. Rossi., Ann. Microbiol. Enzymol., 43 (1993a) 53.
Buzzini, P., M. Gobbetti, J. Rossi and M. Ribaldi., Biotechnol. Lett., 15 (1993b) 151.
Buzzini, P. and J. Rossi, Agro-Food-Ind. Hi-Tech., 8 (1997) 30.
Carta, F.S., C.R. Soccol, A.C. Prado, L.P. Ramos and L. Machado. (1997) *In* Braz. Symp. Chem. Lignins Other Wood Compon. Proc., 5th, Vol. 6. (L. P. Ramos, eds.), pp. 583-591.
Catley, B.J., J. Gen. Microbiol., 78 (1973) 33.
Catley, B.J. (1979) Pullulan synthesis by *Aurebasidium pullulans*. *In* Microbiol polysaccharides and polysaccharases (R.C.W. Berkeley, G.W. Gooday, and D.C. Ellwood, eds.), pp. 69-84. Academic Press, London.
Celligoi, M.A.P.C., D.F. Angelis and J.B. Buzato, Arq. Biol. Tecnol., 40 (1997) 693.
Chang, W. and C. Chou. (1996) Shipin Kexue. 23: 346. (*Chem. Abs.*125:219673).
Choe, J. and Y.J. Yoo, J. Ferment. Bioeng., 72 (1991) 106.
Christopher, D., N.F. Shelby and J.R. Bothust, Biotechnol Lett., 20 (1998) 191.
Chung, B.H. and H.N. Chang, Biotechnol. Bioeng., 32 (1988) 205.
Cros, P. and D. Schneider. Production of itaconic acid. Eur. Pat.No. 341112 (1989).
Currie, J.N., J. Biol. Chem., 31 (1917) 15.
Darriet, P., J.N. Boidron and D. Dubourdieu., Conn aiss. Vigne Vin, 22 (1988) 189.
Davies, B.H, Pure Appl. Chem., 35 (1973) 1.
Dawson, M.W. and S. Maddox, Biotechnol. Bioeng.,33 (1989) 1500.
de Lima, Vera Lucia, A.G., L.M. Stamford, Tania and A.M. Salgueiro, Arq. Biol. Tecnol., 38 (1995) 773.
Deshpande, M.S., V.B. Rale and J.M. Lynch, Enzyme Microbiol. Technol., 148 (1992) 514.
Du Jinanxin, N. Cao, C.S. Gong, G.T. Tsao and N. Yuan, Appl. Biochem. Biotechnol., 63 (1997) 541.
Duxbury, D.D., Food Process., 47 (1986) 42.
Eikmeier, H. and H.J. Rehm, Appl. Microbiol. Biotechnol., 26 (1987) 105.
El-Sharkawy, S.H. and M.I. Abdul Karim, Boll. Chim. Farm., 135 (1996) 176.
Emelyanova, E.V., Process Biochem., 32 (1997) 173.
Farbood, M.I., Biochem Soc. Trans., 19 (1991) 690.
Farinaj, I., F. Sineriz, O.E. Molina and N.I. Perott., Biotechnol. Tech., 10 (1996) 705.
Farris, G.A., F. Fatichenti and P. Deiana, Vigne Vin., 23 (1989) 89.
Figueriredo Z.M.B. and L.B. Carvalho, Appl. Biochem. Biotechnol., 30 (1991) 217.
Finogenova, T.V., N.V. Shishkanova, P. Straneo, and E. Moretti, Izobreteniya, 32 (1997) 268. (*Chem. Abs.*128:229422)

Foster, J.W. (1954) Fumaric acid. *In* Industrial Fermentations, Vol. 1. (L.A. Underkofier and R.A. Hickey eds.), pp. 470. Chemical Publ, New York.
Fujii, N., R. Uno, K. Yasuda and M. Sakakibara, Fukui Daigaku Kogakubu Kenkyu Hokoku, 40 (1992) 23-30.
Fujii, T., A. Iwamatsu, H. Yoshimoto, T. Minetoki, T. Bogaki and N. Nagasawa. Eur. Patent No. 574941 (1993).
Fukuda, H. Eur. Pat. No. 207475 (1987a).
Fukuda, H. Jpn. Patent No. 62006694 (1987b).
Fukuda, K., M. Watanabe and K. Asano, Agric. Biol. Chem., 54 (1990) 3151.
Gallois A., B. Goss, D. Langlois, H.E. Spinnler and P. Brunerie, Mycol. Res., 94 (1990) 494.
Garg, K. and C.B. Sharma, J. Gen. Appl. Microbiol., 38 (1992) 605.
Gatfield, I.L., Food Technol., 42 (1988) 111.
Goldberg, I. and B. Steigiltz, Biotechnol. Bioeng 27 (1985) 1067.
Golubtsova, V.M., E.Y. Shcherbakova, A.V. Korotchenko and E.S. Mints. SU Patent No. 1409658 (1988).
Goma, G. and I. Seiller. Fr. Patent No. 2 588 568 (1987).
Goma, G., I. Seiller, A.M. Bajon and J.P. Leygue. Eur. Patent No. 315496 (1989).
Goodwin, T.W., Prog. Ind. Microbiol., 11 (1972) 29.
Gough, S. and A.P. McHale, Bioprocess Eng., 19 (1998) 33.
Gross, B., A. Gallois, H.E. Spinner and D. Langlois, J. Biotechnol., 10 (1989) 303.
Guevarra, E. and T. Tabuchi, Agric. Biol. Chem., 5 (1990) 2353.
Guidici, P., C. Restuccia, C. Randazzo, V. Melia and V. Corte, Ind. Bevande., 26 (1997) 252. (*Chem. Abs.*127:133195).
Gupta, S. and C.B. Sharma, Biotechnol. Lett., 16 (1994) 599.
Hallrich, D., Biotechnol. Ferment. Physiol. Conf. Report. (1994) 4.
Hanada, K. and M. Ishikawa. Jpn. Patent No. JP 01132371 (1989).
Hang Y.D. and E.E. Woodams, Biotechnol. Lett., 9 (1987) 183.
Hang Y.D. and R.C.Yu, Biotechnol. Lett., 11 (1989) 597.
Harrison, J.S., Proc. Biochem., 3 (1957) 467-476.
Harvey, L.M. (1984) Ph.D. thesis. University of Strathelide, Glasgow, UK.
Haustede, H. and N. Rudy. US. Patent No. 3941656 (1976).
Heinrich, M. and H.J. Rehm, Euopean J. Appl. Microbiol. Biotechnol., 15 (1982) 88.
Henry, S.A., J.I. Patton, P.V. Grriac and S.D. Kohlwein. WO Patent No. 99-024123/02 (1998).
Higashiyama, K., T. Yaguchi, K. Akimoto, and S. Shimizu. WO Patent No. 9829558 (1998).
Hirato, P. and T. Hirol, Agric. Biol. Chem., 55 (1991) 919.
Hiroshi, M., Nishihara, Y. Hirano, N. Kamada, K. Akimoto, K. Konishi and S. Shimizu, Appl. Environ. Microbiol., 63 (1997) 1820.
Hiruta, O., T. Futamura, H. Takebe, A. Satoh, Y. Kamisaka, T. Yokochi, T. Nakahara and O. Suzuki, J. Ferment. Bioeng., 82 (1996) 366.
Hockertz, S., J. Plonzig and G. Auling, Appl. Microbiol. Biotechnol., 25 (1989) 590.
Hodgson, J., Bio/Technology., 12 (1994) 152.
Huang, T. and X. Qian, Gongye Weishengwu, (1990) 20 14. (*Chem. Abs.*115:6912)
Huang, J., Q. Shi, X. Zhou, Y. Lin, B. Xie and S. Wu. Weishengwuxue Tongbao., 25 (1998) 187. (*Chem. Abs.*130:123868).
Hussain, S. M.A. Akhtar, and A. Qadeer, J. Nat. Sci. Math., 27 (1987) 165.
Ichikawa, K., T. Hoshino, and M. Tazoe. Eur. Pat. No. 765938 (1997).

Il'chenko, A.P., N.V. Shishkanova, O.G. Chernyavskaya and T.V. Finogenova, Microbiology, 67 (1998) 241-244.
Immelman, M., J.C. Du Preez and S.G. Kilian, Syst. Appl. Microbiol., 20 (1997) 158.
Janssens, L., Chem. Mag. (Ghnet), 17 (1991) 37.
Jernejk, K.A., M. Cimerman, Vendramin and A. Perdh, Appl. Microbiol. Biotechnol., 332 (1991) 699.
Jiang, M., Z. Bai, J. Zhang and H. Xie, Gongye Weishengwu. 17 (1987) 1 (*Chem. Abs.*107:172366)
Jiang, M., Z. Bai, J. Zhang, H. Xie, Z. Wu, M. Xu and W. Sun, Weishengwu Xuebao, 29 (1989) 129. (*Chem. Abs.*111:22106).
Jin, Q., G. Xu and Y. Wu. Shipin Kexue. 169 (1994) 25. (*Chem. Abs.*121:132311).
Johnson, V., M. Singh, V.S. Saini, V.R. Sista and N.K. Yadav, World J. Microbiol. Biotechnol., 8 (1992) 382.
Joshi, A.P. and K. Joshi. Indian Patent Application No. 672/ DEL/95 (1995).
Juranyiova, E. and E. Matisova, Biologia, 46 (1991) 355.
Kapich, A.N., E.S. Romanovets, and S.P. Voit, Mikol, Fitopatol, 24 (1990) 51.
Karya, M. and H. Fujiwara. Jpn. Patent No. 06038774 (1994).
Kauppinen, S., J.Q. Si, T. Spendler, C. Dambmann, T. Halkier, and P.R. Oestergaard. WO Patent No. 9628542 (1996).
Kautola, H., N. Vasilev, and Y.Y. Linko, Biotechnol. Lett., 11 (1989) 313.
Kautola, H. and Y.Y. Linko, Appl. Microbiol. Biotechnol, 31 (1989) 448.
Kautola, H. and Y.Y. Linko, Appl. Biochem. Biotechnol., 24-25 (1990) 161.
Kautola, H., N. Vasilev and Y.Y. Linko, J. Biotechnol., 13 (1990) 315.
Kautola, H., Appl. Microbiol. Biotechnol., 33 (1990) 7.
Kautola, H., W. Rymowicz, Y.Y. Linko and P. Linko, Sci. Alimen., 12 (1992) 383.
Kawashima, H., N. Kamada, E. Sakuradani, S. Jareonkitmongkol, K. Akimoto and S. Shimizu, J. Am. Oil Chem. Soc., 74 (1997) 455.
Keuth, S. and B.J. Bisping, Appl. Bacteriol., 75 (1993) 427.
Khan, S.A., M. Saeed, M.K. Bhatty, M.Z. Iqbal and S. Hamid. (1987) Fett. Wiss. Technol. 89: 250-252. (*Chem. Abs.*107:114214)
Khare, S.K., K. Jha and A.P. Gandhi, Appl. Microbiol. Biotechnol., 41(1994) 571-573.
Kikuji, H. Jpn. Patent No. 02016989 (1990).
Kinoshita, K.J., Chem. Soc. Jpn. 50 (1929) 583.
Kirimura, K., Y. Hirowateri and S. Usami, Agric. Biol. Chem., 51 (1987) 1299.
Kirimura, K., I. Nakajima, S.P. Lee, S. Kawabe and S. Usami, Appl. Microbiol. Biotechnol., 27 (1988a) 504.
Kirimura, K., S.P. Lee, I. Nakajima, S. Kawabe, S. Usami, J. Ferment. Technol., 66 (1988b) 375.
Kirin-Brewing. Jpn. Patent No. 102-48-575 (1998).
Kisser, M., C.P. Kubicek and M. Roehr, Arch. Microbiol., 128 (1980) 26.
Kitada, M., G. Terui, M. Kitada and T. Fukimbara, Hakko Kogaku Zasshi., 45 (1967) 1101.
Kitamoto, D., K. Haneishi, T. Nakahara and T. Tabuchi, Agric. Biol. Chem., 54 (1990) 37.
Knutzon, D., P. Mukerji, Y. Huang, J. Thurmond, S. Chaudhary and A.E. Leonard. WO Patent No. 9846765 (1998).
Kobayashi, T., Process. Biochem., 2 (1967) 61.
Korhola M.K. Harju and M. Lehtonen, (1989). Fermentation (R.J. Piggott, R. Sharp and R.E. Duncan eds.), Longman, pp. 89 -117. Harlow Essex Publication.

Kosugi, Y. Yushi. 42 (1989) 238. (*Chem. Abs.*112:156581).

Kubicek, C.P., M. Zehentgruber, H. El-Kalak and M. Roehr, Eur. J. Appl. Microbiol. Biotechnol, 9 (1980) 101.

Kubicek, C.P. and M. Roehr, Arch. Microbiol., 141 (1985) 266.

Kubicek, C.P. and M. Rohr, Critic. Rev. Biotechnol., 3 (1986) 331.

Kubicek-Pranz, E.M., M. Morelt, M. Roehr and C.P. Kubicek, Biochem. Biophys. Acta., 1033 (1990) 250.

Kubicek, C.P., C.F.B. Witteveen, and J. Visser. (1994) Regulation of organic acid production by *Aspergilli*. *In* The Genus Aspergillus. (A. Keith and A. Powell), pp.135-1 45. Plenum Press New York.

Kugimiya, W., Y. Otani, M. Kohno and Y. Hashimoto, Agric. Biol. Chem., 56 (1992) 716.

Kuninaka, A. (1981) Taste and Flavor Enhancers. *In Flavor Research. Recent advances* (C.R. Teranishi, R.A. Flath and H. Sugisawa, eds.), pp. 305-353. Marcel Dekker, New York.

Kwak, M.Y. and J.S. Rhee., Biotechnol. Bioeng., 39 (1992) 903.

Lachake A.H. and V.B. Rale, (1994) Trends in microial production of pullulan and its novel applications in food industry. In Food Biotechnology, Microorganism. (Y.H. Hui and G.G. Khachatourians eds.), pp.589-604. VCH New York..

Langrand, G., R.C. Triantaphylides and J. Baratti, Biotechnol. Lett., 10 (1988) 549.

Langrand, G., N. Rondot, C. Triantaphylides and J. Baratti, Biotechnol. Lett., 12 (1990) 581.

Lecacheux, D., Y. Mustiere and R. Panaras, Carbohydr. Polym., 6 (1986) 477.

Legrasa, M. and J. Kidric, Appl. Microbiol. Biotechnol., 31 (1989) 453.

Lelik, L., G. Vitanyi, J. Lefler, J. Hegoczky, M. Nagy-Gasztonyi and G. Vereczkey, Acta Aliment., 26 (1997) 271.

Leuchtenberger, W. (1996). Amino acids-technical production and use. In Biotechnology 2nd Ed, Vol. 6, (Rehm, Hans-Juergen and Reed, Gerald eds.), pp. 465-502. Weinheim, Germany.

Lee, H.W., S. Sato, S. Mukataka and J. Takahashi, Hakko Kogaku Kaishi, 65 (1987) 501.

Lee, J.S., Iijima, H. Kobayashi and S.O. Yanagi, Agric. Biol. Chem. 52 (1988) 1877.

Lee, Y.H., C.W. Lee and H.N. Chang, Appl. Microbiol. Biotechnol., 30 (1989) 141.

Liu, B. and J. Xu, Huaxue Shijie. 37 (1996) 527. (*Chem. Abs.*127:94128).

Liu, P. and J. Wang, S. Yu, Fajiao Gongye, 23 (1997) 29. (*Chem. Abs.*127:345369).

Lizama-U.G., M. Grandos Baeza and V.E. Ortiz, Genetal Meet. Am. Soc. Microbiol., 401 (1998).

Lo. P.S., R. Bubbico, M. Bravi, P. Straffi and M. Moresi, Ann. Microbiol. Enzymol., 47 (1997) 1.

Lockwood, L.B., G.E. Ward and O.E. May, J. Agri Res., 53 (1936) 849.

Lockwood, L.B. (1979) Production of organic acids by fermentation. *In* Microbiol Technology, 2 ed, Vol.1 (H. J. Peppler and D. Perlman eds.), pp. 367-387. Acacemic Press New York.

Lou, C., Z. Ma, J. Pan, Z. Wu, Y. Zhang and L. Tang, Gongye Weishengwu, 16 (1986) 14.

Lynd, L.R., C.R. South, and N.H. Hanover. US Patent No. 5837506 (1998).

Ma, M., C.P. Kubicek and M. Roehr, Arch. Microbiol., 141 (1985) 266.

Malaney, G.W., R.D. Tanner and A.M. Rodrigues, Folia Microbiol., 36 (1991) 468.

Marces, M., L.M. Mateos and J.T. Martin, (1994) Microorganisms for amino acid production: *E-coli* and corgnebacteria. In Food Biotechnology (Y.H. Hui and G.G. Khachatourians (eds.), pp. 423-469.

Martin, J.F and A.L. Demain. (1977). In the Filamentous Fungi, Vol. 3 (Smith J.E. and Berry, D.R. (eds.), Edward Arnold, London.
Martinez-Force, E. and T. Benitez, Biotechnol. Prog., 10 (1994) 372.
Mateo, J.J., M. Jimenez, T. Huerta and A. Pastor, Int. J. Food Microbiol., 14 (1991) 153.
Matsui, T. and K. Yasuda. Jpn. Patent No. 01225487 (1989).
Matsumoto, T., A. Fujimaki and T. Nagata. US Patent No. 4411998 (1983).
Matsumoto, T. and Y. Ichikawa. Jpn. Patent 61219391 (1986).
Matty, M., Crit. Rev.Biotechnol., 12 (1992) 87.
Mau, J., Y. Lin, P. Chen, Y. Wu and J. Peng, J. Agric. Food Chem., 46 (1998a) 4587.
Mau, J., K. Wu, Y. Wu and Y. Lin, J. Agric. Food Chem. 46 (1998b) 4583.
Miall, L.M., Organic acids (1978) In Economic Microbiology, Vol.2 (Rose, A.H. eds.), pp-48. Academic Press, London.
Michalik, P. and R. Horenitzky. Czech. Patent No. 276815 (1992).
Michalik, P. and R. Horenitzky, Kvasny Prum. 34 (1998) 140. (*Chem. Abs.*109:127306)
Mikhailova, N.P., Z.A. Zhakovskaya, A.V. Andreev, and K.A. V'yunov, Khim. Farm. Zh. 21 (1987) 1490. (*Chem Abs.*108:110768).
Milsom, P.E. and J.L. Meers. (1985) Citric acid. *In* Comprehensive Biotechnology, Vol 3 (Blanch, H.W. Drew, S. Wang D.I. Eds.), pp. 665-680. Pergamon Press, Oxford.
Milsom, P.E. (1989) Organic acids by fermentation especially citric acid. In Food Biotechnology, Vol. 1 (R.D. King and P.S.J. Cheetham eds.), pp. 273-300. Elsevier, London.
Mittal, Y., I.M. Mishra, and B.S. Varshney, Biotechnol. Lett., 15 (1993) 41.
Montanari, G. and C. Zambonelli, Ind. Aliment, 36 (1997) 1001.
Moresi, M., E. Parente and A. Mazzatura, Appl. Microbiol. Biotechnol., 36 (1991a) 320.
Moresi, M.E. Parente, A. Ricciardi, E. Sebastiani, Chim. Ind., 73 (1991b) 648.
Moresi, M., E. Parente, A. Mazzatura and A. Ricciardi, Ann. Microbiol. Enzymol., 42 (1992a) 173.
Moresi, M., E. Parente, M. Petruccioli and F. Federici, J. Chem. Technol. Biotechnol., 54 (1992b) 283.
Morrin, M. and O.P. Ward, Mycol. Res., 94 (1990) 505.
Murray, W.D., S.J.B. Duff, and P.H. Lanthier. US Patent No. 4871669 (1988).
Murillo, F.J., L. Calderon. I. Lopez-Diaz and E. Cerda-Olmeda, Appl. Environ. Microbiol., 36 (1983) 693-642.
Nagar-Legmann, R. and P. Margalith, Appl. Microbiol. Biotechnol., 26 (1987) 49.
Naim, N. and R. Saad, Egypt. J. Bot. 27 (1984) 159-167.
Nakagawa, M., K. Kamiie, K. Ishibashi and K. Hironaka, Obihiro Chikusan Daigaku Gakujutsu Kenkyu Hokoku, Dai-Bu., 17 (1990) 7. (*Chem. Abs.*114:162386)
Nakagawa, M., K. Ishibashi and K. Hironaka, Obihiro Chikusan Daigaku Gakujutsu Kenkyu Hokoku, Dai-1Bu., 17 (1991) 123. (*Chem. Abs.*116:5269)
Nakagawa, M., Hakko Kogaku Kaishi, 70 (1992) 451. (*Chem. Abs.*118:58186)
Nakagawa, M., Y. Yashiro and T. Sato, Obihiro Chikusan Daigaku Gakujutsu Kenkyu Hokoku, Shizen Kagaku, 18 (1994) 233.
Nakajima, T, and T. Shimauchi. Jpn. Patent No. JP 02268690 (1990).
Nakayama, K., (1973) Amino acid production using microbial auxotropic mutants. In Genetics of Industrial microorganisms, Vol.1, Bacteria (Z. Venek, Z. Hostalek and J. Cudilin eds.), pp.219-248. Academic Press. New York.
Neufeld, R, J., Y. Peleg, Y.S. Rokem, O. Pins and I. Goldberg, Enz. Microbiol. Technol., 13 (1991) 991.

Nienow, A.W., Trends Biotechnol., 8 (1991) 224.
Ninet, L.S. and J. Renaut., (1979) In Microbiol Technology Microbiol Proceses, 2 ed., Vol. 2 (H.J. Peppler and D. Perlman eds.), pp-529-544. Academic Press, New York.
O' Connell, M.J. and J.M. Kelly, Gene 84 (1989) 173.
Ogawa, A., Y. Wakisaka, T. Tanaka, T. Sakiyama and K. Nakanishi, J. Ferment. Bioeng., 80 (1995a) 41.
Ogawa, A., Y. Morita, T. Tanaka, T. Sakiyama and K. Nakanishi, Biotechnol. Tech., 9 (1995b) 153.
Oh, D.K. and S.Y. Kim, Appl. Environ. Microbiol., 50 (1998) 419.
Oliveira, E.A., A.A. Costa, Z.M.B. Figueiredo and L.B. Carvalho, Appl. Biochem. Biotechnol., 47 (1994) 65.
Pallares, J., S. Rodriguez and S. Sanroman, Bioprocess Eng., 15 (1996) 31.
Papagianni, M., M. Mattey and B. Kristiansen, Biotechnol. Lett., 16 (1994) 929.
Parajo, J.C., M. Dominguez and J.M. Dominguez, Bioresource Technol., 66 (1998a) 25.
Parajo, J.C., V. Santos and M. Vazquez, Biotechnol. Bioeng., 59 (1998b) 501.
Park, Y.S., M. Itida, N. Ohta and M. Okabe, J. Ferment. Bioeng., 77 (1994) 329.
Pelag, Y.A., Barak, M.C. Scrutton and I. Goldberg, Microbiol. Biotechnol., 28 (1988) 69.
Peleg, Y., J.S. Rokem, I. Goldberg and O. Pines, Appl. Environ. Microbiol., 56 (1990) 2777.
Pena Miranda, M. and G. Gonzalez Benito, Acta Microbiol., 43 (1994) 211.
Perlman, D. and C.J. Sih, Prog. Ind. Mocrobiol., 2 (1960) 169.
Petruccioli, M. and E. Angiani, Ann. Microbiol. Enzimol., 45 (1995) 119.
Petruccioli, M., E. Angiani and F. Federici, Process Biochem., 31 (1996) 463.
Pfeiffer, V.F., C. Vojnovich and E.N. Heger, Ind. Eng. Chem., 44 (1952) 2975.
Pouget, M.P., A. Pourrat and H. Pourrat, Lebensm. Wiss. Technol., 23 (1990) 1.
Pstinen O., K. Visuri and M. Leisola, Biotechnol Tech., 12 (1998) 557.
Puglisi, P., I. Ferrero, M. Borgo and M. San. Eur. Patent No. 875580.
Quirasco-Baruch, M., F. Iturbe-Chinas, M.F. Novak and A. Lopez-Munguia, Rev. Latinoam. Microbiol., 35 (1993) 273.
Rane, K.D. and K.A. Sims, (1993) Enzyme Microb. Technol., 15 (1993) 646.
Rao, D.S. and T. Panda, Bioprocess Eng., 11 (1994a) 209.
Rao, D.S. and T. Panda, Bioprocess Eng., 10 (1994b) 99.
Ratledge, C. (1997) Biotechnology. 2nd Ed, Vol. 7. (Kleinkauf, Horst, Von Doehren, Hans eds.), pp. 133-1 97. VCH: Weinheim, Germany.
Rattray, J.B.M. Yeasts. In Microbial Lipids, Vol. 1. , pp. 555-695, Academic press, London, 1988.
Reilly, C.E. (1991) Vitamins. In Biotechnol. Food Ingredients (Goldberg, Israel, Williams, Richard A. Van Nostrand Reinhold eds.), pp. 415-432. New York.
Rimareva, L.V. and M.B. Overchenko. RU Patent No. 2104-300 (1998).
Riviere, J., M. Moss and J.M. Smith, (1917) Industrial Applications of Microbiology pp. 159-161. Surrey University Press. London.
Roehr, M., C.P. Kubicek and J. Kominek. (1992). Industrial acids and other small molecules In *Aspergillus* Biology and industrial applications. (Bennett, J.W., Kich, M.A. eds.), pp. 91 -131 Stoneham: Butterworth Heinemann.
Roehr, M. Kubicek C.P. and J. Kominek. (1996a) Citric acid. *In* A multi-volume Comprehensive Treaties: Biotechnology. Vol. 6 (H.J. Rehm and G. Reed eds.), pp. 307-345. VCH Pub. New York.
Roehr, M. Kubicek C.P. and J. Kominek. (1996b) Gluconic acid. *In* A multi-volume Comprehensive Treaties: Biotechnology. Vol. 6. (H.J. Rehm and G. Reed eds.), pp.

348- 362. VCH Pub. New York.
Roehr, M. Kubicek C.P. and J. Kominek. (1996c) Further organic acids. *In* A multi-volume Comprehensive Treaties: Biotechnology. Vol. 6. (H.J. Rehm and G. Reed eds.), pp. 364-379. VCH Pub. New York.
Rogers, N.E. (1973) Sclereglucan. *In* Industrial Gums. (Whistier, R.L. and Bemiller, J.N eds.), pp 449-511. Academic Press, New York.
Romanova, L.V., M.V. Zalashko and L.G. Grankina, Gidroliz. Lesokhim. Promst., 1 (1988) 16. (*Chem. Abs.*108:185235).
Romanova, L.V., M.V. Zalashko, I.N. Stigailo. Lipomyces starkeyi as lipid source. SU Patent No. 1541249 (1990)
Rosenberg, M., E. Sturdik, J. Gomory, S. Stanek, and R. Kacina. Czech. Patent No. 274828 (1991).
Rosenberg, M., J. Svitel, E. Sturdik and I. Rosenbergova, Bioprocess Eng., 7 (1992a) 309.
Rosenberg, M., J. Svitel, I. Rosenbergova and E. Sturdik, Acta Biotechnol. 12 (1992b) 311.
Rossi, J. and F. Clementi, Biotechnol. Lett., 7 (1985) 329.
Roukas, T. and L. Harvey, Biotechnol. Lett., 10 (1988) 289.
Roukas, T., Enzyme Microb. Technol., 24 (1998) 54.
Rouv. V., E. Gura, E. Evizewski and F. Wagner, J. Ind. Microbiol., 9 (1992) 19.
Rugsaseel, S., K. Kirimura and S. Usami, Rikogaku Kenkyusho Hokoku, Waseda Daigaku, 142 (1993) 39. (*Chem. Abs.*120:75513)
Saegusa, T., K. Ekoshi and K. Okazaki. Jpn. Patent No. 05076378 (1993).
Sagiyan, A.S., K.I. Atayan, G. Ovsepyan, A.A. Vardanyan, and A.S. Zurabyan, Khim. Zh. Arm., 51 (1998) 87. (*Chem. Abs.*130:280869)
Saito, K., Bot. Mag. Tokyo., 21 (1970) 7.
Sakurai, H. and J. Takahashi. Jpn. Patent No. 01165380 (1989a).
Sakurai, H. and J. Takahashi. Jpn. Patent 01095781 (1989b).
Sakurai, H., H.W. Lee, S. Sato, S. Mukataka and J. Takahashi, J. Ferment. Bioeng., 67 (1989c) 404.
Sakurai, A., M. Itoh, M. Sakakibara, H. Saito and M. Fujita, J. Chem. Technol. Biotechnol., 70 (1997) 157.
Samson, R.A., Current taxonomix schemes of the genus Aspergillus and its teleomorphs. *In* Aspergillus: Biology and Industrial Applications (Bennet, J.W., Klich, M.A. eds.), (1992) pp-335-390. Stoneham: Butterworth-Heinemann.
Sanchez, S., V. Bravo, E. Castro, A.J. Moyer and F. Camacho, Appl. Microbiol. Biotechnol., 50 (1998) 608.
Sandford, P.A., Adv. Carbohydr. Chem. Biochem., 36 (1979) 265.
Sankpal, N.V., A.P. Joshi and B.D. Kulkarni, (1998) Unpublished Results.
Sankpal, N.V., (1999) Ph.D. Thesis. University of Pune, Pune India.
Sankpal, N.V., A.P. Joshi and B.D. Kulkarni. Indian Patent Application No. NCL/41/99. (1999a).
Sankpal, N.V., A.P. Joshi, I.I. Sutar and B.D. Kulkarni, Process Biochem., 33/34 (1999b) 1.
Sarangbin, S., K. Kirimura and S. Usami, Appl. Microbiol. Biotechnol., 40 (1993) 206.
Sasaki, Y., S. Takao and S. Khotta, Agric. Biol. Chem., 48 (1970) 778.
Sassi, G., B. Ruggeri, V. Specchia and A. Gianetto, Bioresour. Technol., 37 (1991) 259.
Sato, S., T. Nakahara and Y. Minoda, Agric. Biol. Chem., 41 (1977) 1903.
Sato, S., Hakko Kogaku Kaishi, 68 (1990) 411. (*Chem. Abs.*113:210117)

Sato, H., Yuki Gosei Kagaku Kyokaishi., 57 (1999) 323. (*Chem. Abs.*130:295564).
Sattur, B.P. and N.G. Karanth, J. Microb. Biotechnol., 2 (1987) 116.
Sattur, A.P. and N.G. Karanth, Biotechnol. Bioeng., 34 (1989) 872.
Sattur, A.P. and N.G. Karanth, Bioprocess Eng., 6 (1991) 227.
Schmidt, G., Ber. Forschungszent. Juelich, 3260 (1996) 1. (*Chem. Abs.*126:154945)
Schreier, P., (1992) Bioflavors: An overview *In* Bioformation of Flavors. (R.L.S. Patterson, B.V. Charlwood, G. MacLeod and A.A. Williams eds.), pp. 1-20. Royal Society of Chemistry, Cambridge.
Schreferl-Kunar, G., M. Grotz, M. Roehr and C.P. Kubicek, FEMS Microbiol. Lett., 59 (1989) 297.
Shankaranand, V.S. and B.K. Lonsane, World J. Microbiol. Biotechnol., 9 (1993) 377.
Shankaranand, V.S. and B.K. Lonsane, Proc. Biochem. 29 (1994) 29.
Shelly, T. and B. McNeil, Enzyme. Microb.Technol., 16 (1994) 223.
Shi, A., Zhongguo Niangzao. 1 (1988) 25. (*Chem. Abs.*109:109154)
Shimada, H., K. Kondo, P.D. Fraser, Y. Miuray, T. Sato and N. Misawa, Appl. Envoron. Microbiol., 64 (1998) 2676.
Shimazono, H., Food Technol., 18 (1964) 36.
Shimizu, S. and M. Kobayashi, WO Patent No. 9838314 (1998).
Shishkanova, N.V., T.E. Arzumanov, V.A. Samoilenko and T.V. Finogenova, Biotekhnologiya, 1 (1998) 57. (*Chem. Abs.*129:274775)
Shoseyov, O., B.A. Bravdo, D, Siegel, A. Goldman, S. Cohen, L. Shoseyov and R. Ikan, J. Agric. Food Chem., 38 (1990) 1387.
Sibeijn M. and R.M. de Peter, WO Patent No. 9850-574 (1998a).
Sibeijn, M., R. M.de Peter and Delfl, WO 99-024127/02. (1998b).
Sjöström, L.B., (1968) T.E. Furia (ed.), pp. 493-500. The chemical Rubber Co., Cleveland.
Smith, J.E. (1974) Organic acid production by mycelial fungi. In Industrial aspects of Biochemistry (B. Spencer eds.), pp. 297-317. Elsevier, Amsterdam.
Soccol, C.R., V.I. Raimbault, World J. Microbiol and Biotechnol., 10 (1994) 433.
Soldki, M.E. and M.C. Cadmus, Adv. Appl. Microbiol., 23 (1978) 19.
Solinski, J. Pol., Patent No. 136163 (1987).
Song, Z., Z. Wang, X. Li and W. Song, Cuihua Xuebao, 18 (1997) 508. (*Chem. Abs.*128:114066).
Srinivasan, M.C., (1986) Microbial production of organic acids. *In* Selected topics in Applied Microbiology. (A. Tauro eds.). International Bioscience Publishers. Madras, India.
Suzuki, O. and T. Yokochi, WO Patent No.8604354 (1986).
Suzuki, O., T. Yokochi, K. Amano, T. Sano, S. Seto, Y. Ohtu, S. Ishida, S. Iwamoto and K. Morioka, Yukagaku, 37 (1988) 1081. (*Chem. Abs.*110:191094)
Suzuki, A., S. Sarangbin, K. Kirimura and S. Usami, J. Ferment. Bioeng., 81 (1996) 320.
Suzuki, E., A. Kuwata and M. Yasukawa, Shiken Kenkyu Hokoku, Fukushima-ken Haiteku Puraza, (1997) 55. (*Chem. Abs.*129:65384).
Suzuki, O., K. Oho, S. Shigeta, T. Aki, and K. Akimoto, WO Patent No.9839468 (1998).
Swarthout, E.J., US Patent No. 3285831 (1966).
Tabuchi, T. and T. Nakahara., Jpn. Patent No. 03180187 (1991).
Tachibana, S. and T. Murakami, J. Ferm. Technol., 51 (1973) 858.
Tahoun, M.K., Z. El-Merheb, A. Salam and A. Youssef, Biotechnol. Bioeng., 29 (1987) 358.

Takao, S., M., Tanida and H. Kuwabara, J. Ferment. Technol., 55 (1977) 196.
Takao, S. and K. Hotta, Agric. Biol. Chem., 41 (1977) 945.
Takao, S. T., M. Takahashi, M. Tanida and M. Takahashi, J. Ferment. Technol., 62 (1984) 577.
Takahashi, J., Jpn. Patent No. 3262491 (1991).
Takamizawa, K., S. Nakashima, Y. Yahashi, K.B. Kubata, T. Suzuki, K. Kawai and H. Horitsu, J. Ferment. Bioeng., 82 (1996) 414.
Tani, Y., Bitamin. 72 (1998) 639.
Taurhesias, S. and B. McNeil, Enzyme Microb. Technol, 16 (1994) 223.
Thaker, P. and N.K. Yadav, Indian J. Exp. Biol., 35 (1997) 313.
Thomas, C.R., Trends Biotechnol., 10 (1992) 343-348.
Tisnadjaja, D., N.A. Gutierrez and I.S. Maddox, Enzyme Microb. Technol., 19 (1996) 343.
Tokumaru, I.,M. Goto, M. Terasawa and H. Yukawa., Jpn.Patent No. 10257898 (1998)
Traeger, M., G.N. Qazi, U. Onken and C.L. Chopra, J. Ferment. Bioeng. 68 (1989) 112.
Traeger, M., R. Buse, G.N. Qazi, U. Onken, Annu. Meet. Biotechnol. 8th, (1990) 729.
Traeger, M., G.N. Qazi, U. Onken and C.L. Chopra, J. Chem. Technol. Biotechnol., 50 (1991a) 1.
Trager, M., D. Hollmann, R. Buse, U. Onken, J. Ferment. Bioeng., 72 (1991b) 46.
Tran, C.T. and D.A. Mitchell, Biotechnol. Lett., 17 (1995) 1107.
Tran, C.T., L.I. Sly and D.A. Mitchell, World J. Microbiol. Biotechnol., 14 (1998) 399.
Tsai, M.J. and S.J. Tsai, Shih P'in K'o Hsueh., 14 (1987) 143. (*Chem. Abs.*109:5492).
Tuite, M.F., Crit. Rev. Biotechnol., 12 (1992) 157.
Vasilev, N., H. Kautola and Y.Y. Linko, Biotechnol. Lett., 14 (1992) 201.
Vassilev, N.B., M.Ch.Vassileva and D.I. Spassova, Appl. Microbiol. Biotechnol., 39 (1993) 285-289.
Villelt, R., (1981) Citric acid. In Biotechnology for producing foods and chemicals from biomass Fermentation chemicals from Biomass, Vol. 2, pp 25. Solar Energy Research Institute, Golden, Co.
Visser, J., J. Chem. Tech. Biotechnol., 50 (1991) 111.
Wakisaka, Y., T. Segawa, K. Imamura, T. Sakiyama and K. Nakanishi, J. Ferment. Bioeng., 85 (1998a) 488.
Wakisaka, Y., T. Kiyama and K. Nakanishi., Jpn. Patent No. 10179138 (1998b).
Wang, J.J., and T. Pan, Zhongguo Nongye Huaxue Huizhi., 32 (1994) 199. (*Chem. Abs.*121:177810).
Wang, Y. and B. McNeil., Biotechnol. Lett., 16 (1994) 605.
Wang, C.W., Z. Lu and G.T. Tsao, Appl. Biochem. Biotechnol, 51/52 (1995) 57-71.
Wang, Y. and B. McNeil, Enzyme Microb. Technol., 17 (1995a) 893.
Wang, Y.and B. McNeil, J. Chem. Tech. Biotechnol., 63 (1995b) 215.
Wang, Y. and B. McNeil, Biotechnol. Lett., 17 (1995c) 257.
Wang, Y. and B. McNeil, Enzyme Microb. Technol., 17 (1995d) 124.
Wang, X., C.G. Gong and G.T. Tsao, Biotechnol. Lett., 18 (1996) 1441.
Wang, Y. and B. McNeil, Chem. Eng. Technol, 19 (1996) 143.
Ward G.E.. (1967). Production of gluconic acid, glucose oxidase, fructose and sorbose In microbiol technology. First ed. (H.J. Peppler eds.), pp. 200-331. Reinold, New York.
Watanabe, T., A. Suzuki, H. Nakagawa, K. Kirimura and S. Usami, Bioresour. Technol. 66 (1998) 271.
Weete, J.D., F. Shewmaker and S.R. Gandhi., J. Am. Oil Chem. Soc., 75 (1998) 1367.
Wei, P. and W. Wu., Yaowu Shengwu Jishu, 5 (1998) 180. (*Chem. Abs.*130:181495).

Welsh, F.W., W.D. Murray and R.E. Williams, Crit. Rev. Biotechnol, 9 (1989) 105.
Welter, K., K.Th. Willke and K.S. Vorlop, Nachwachsende Rohst.,10 (1998) 316. (*Chem. Abs.*130:167192).
Whitfield, F.B., Int. J. Food Sci. Technol., 33 (1998) 31.
Wongwicharn, A. and P. Sumonpun, Microb. Util. Renewable Resour., 7 (1991) 291.
Wojtatowicz, M., W. Rymowicz and H. Kautola, Appl. Biochem. Biotechnol., 31 (1991) 165.
Wright, B.E., A. Langarce and J. Reimers, J. Theor. Biol., 182 (1996) 453.
Wu, Q., X. Zhou, Y. Zhong and S. Chen, Zhenjun Xuebao., 12 (1993) 304. (*Chem. Abs.*121:153096).
Xing, L., H. Zhong, H. Zhou, M. Li, B. Zhang, Z. Lu, Zhenjun, Xuebao., 15 (1996) 272. (*Chem. Abs.*126:292460)
Xu, D.B., M. Roehr and C.P. Kubicek, Appl. Microbiol. Biotechnology., 32 (1989) 124.
Yalpani M. and P.A. Sandford. (1987) Commercial polysaccharides, Recent trends and development, *In* Industrial polysacharide: Genetic engineering, Structure, property relations and Applications. Progress Biotechnology. Vol-3. (M. Yalpani eds.), pp- 311, Elsevier Science Publishers.B.V. Amsterdam.
Yamashita, S., N. Fukahori, N. Baba and M. Furuta, Nippon Shokuhin Kagaku Kogaku Kaishi, 44 (1997) 164. (*Chem. Abs.*126:276415)
Yan, S. and S. Ren., Xiandai Huagong., 15 (1995) 11.
Yan, Z., M. Zhao, Z. Peng and G. Liao, Zhongguo Tiaoweipin., 7 (1998) 5. (*Chem. Abs.*130:181691,
Yang, L., CN. Patent No. 1067451 (1992).
Yigitoglue, M. and B. McNeil, Biotechnol. Lett., 14 (1992) 831.
Ykema, A., E.C. Verbree, M.M. Kater and H. Smit, Appl. Microbiol. Biotechnol., 29 (1988) 211.
Ykema, A., H.A. Bakels, R.I.G.S. Verwoert, I.H. Smit and H.W. Van Verseveld, Biotechnol. Bioeng., 34 (1989) 1268.
Ykema, A., E.C. Verbree, I.G.C. Verwoert, Ira, K.H. Van der Linden, J.J.H. Nijkamp, and H. Smit, Appl. Microbiol. Biotechnol., 33 (1990) 176.
Yokochi, T. and O. Suzuki, Yukagaku, 38 (1989) 1007. (*Chem. Abs.*112:96903)
Yuen, S., Proc. Biochem., 9 (1974) 7.
Zehentgruber, O., C.P. Kubicek and M. Roehr, FEMS Microbiol. Lett., 8 (1980) 71.
Zhang, S.Z., (1982) Industrial applications of immobilized biomaterials in china. In Enzyme engineering (I. Chibata, S. Fukai and L.B. Wingard, Jr. eds.), pp 165- 279 Plenum Press, New York.
Zhang, J., M. Jiang, Z. Bai and H. Xie., Shipin Yu Fajiao Gongye,5 (1988) 34. (*Chem. Abs.*110: 152741)
Zhao, R. and Y. Zneng, Zhenjun Xuebao., 14 (1995) 130. (*Chem. Abs.*124:115533).
Zipper, J., A.S. Gaffney, S. Rothenberg and T.J. Cairney. Br., Patent No. 1249347 (1971).

Index of Authors

Annis, S.L. 115
Archer, D.B. 73
Arora, D.K. 1

Bhatnagar, D. 165
Boysen, M.E. 267

Cary, J.W. 165
Chang, P.-K. 165

Duick, J.W. 199

Eriksson, A.R.B. 267

Gold, S.E. 199
Gowthaman, M.K. 305
Gulati, R. 353
Gupta, R. 353

Khachatourians, G.G. 1, 145
Krishna, C. 305

MacCabe, A.P. 239
Marzluf, G.A. 55

Moo-Young, M. 305

Nevalainen, K.M.H. 289

Orejas, M. 239

Panaccione, D.G. 115
Peberdy, J.F. 73
Pitt, D. 13

Ramón, D. 239
Redman, R.S. 199
Rodriguez, R.J. 199

Sahasrabudhe, N.A. 387
Sankpal, N.V. 387
Saxena, R.K. 353
Saxena, S. 353
Schnürer, J. 267

Wallis, G.L.F. 73
Weber, R.W.S. 13
Woytowich, A.E. 145

Keyword Index

Achlya 13
Actin filaments 13
Actinomucor elegans 305
Active uptake 13
Aflatoxin 1, 165
Aflatoxin genes 165
Agaricus bisporus 73, 289, 305
A. bitorquis 305
Ah-AMP1 145
AK-toxin 165
Allantoin metabolism 55
Alternaria alternata 115, 165, 267, 289
A. brassicae 115
A. brassicola 145
A. longipes 145
Alum root 145
Amanita phalloides 115
Amanitin 115
Amaranth 145
Amarathus caudatus 145
Amatoxin 115
Amino acids 387
Amplified internal competitor DNA 267
Amplified target 267
Amylases 353
Amylogluocidases 305
Anisoplia austriaca 305
Antifungal peptides 1, 145
Antimicrobial peptides 1
Apex 13
Apical body 13
Apical tip extension 13
Arabidopsis C-24 145
Aroma compounds 387
Artificial neural network 267
Aschersonia aleyodes 305
Aspergillus spp. 289
A. aegarita 305
A. awamori 73, 239
A. carneus 353
A. caronarium 305
A. ellipticus 305
A. ficuum 305
A. flavus 145, 165, 305
A. fumigatus 13, 165, 305
A. koppan 305
A. luchuensis 305
A. nidulans 55, 165, 239, 289
A. niger 73, 239, 289, 305, 353
A. niger fermentation 387
A. oryzae 239, 353
A. parasiticus 1, 165, 305
A. terreus 353
Atriptex nummularia 145
Auricularia spp. 305
Avena sativa 145

Bacillus licheniformis 353
B. megaterium 353
B. subtilis 145, 353
Baking enzymes 353
Baking industry 353
Barley 145
Barley brewing 353
Bassianolide 115
Beauveracin 115
Beauveria bassiana 115, 305
B. tenella 115, 305
Beer 353
Beta-glucuronidase 145
Biochip technologies 1
Biocontrol agents 1, 115, 305
Bioinformatics 1
Biological control 289
Biomass production 305
Bioreactor system 305
Bioremediation 305
Bipolaris maydis 145
Bjerkandera adusta 305
Blackleg 145
Blastocladiella emersonii 13
Botrytis cinerea 13
Box elements 199
Brewing first-aid kit 353
Brewing industry 353
– with adjuncts 353
Bulk enzymes 289

Calcium gradient 13
Calnexin 73
Candida albicans 13, 73
C. rugosa 305

C. utilis 305
Caniothyrium minitans 305
Carbon catabolite repression 55, 239
Carrena unicolor 305
Catalase 353
cDNA 267
Cell wall 13, 73
Cell wall associated proteases 289
Cellulases 305
Cephalosporins 165
Cephalosporium acremonium 115, 165
Cercospora beticola 145
Cereal brewing 353
Ceriporiopsis subvermispora 305
Chaetomium cellulolyticum 305
C. globosum 305
Chaperones 73
Cheese 353
Cheese production 353
Cheese ripening 353
Chitin binding domain 145
Chromosome length polymorphism (CLP) 1
cis-regulating elements 199
Citric acid 387
Clavibacter michigansis 145
Clavicepes purpurea 115
Cleonis punctiventris 305
Clitoria temateas 145
Cloned genes 73
Clustered genes 165
Cochliobolus sp. 115
C. carbonum 115, 165
C. heterostrophus 115, 165
C. vicoriae 115
Codon usage 199, 289
Colletotrichum lagenarium 145, 165
C. trucatum 305
Common balsam 145
Complementation 199
Complete DNA polymerization 267
Composting 305
Control of carbon metabolism 55
Coprinus sp. 305
C. cinereus 13, 73
Coriolus versicolor 305
Corn 145
Cotranslational pathway 73
Cotranslational translocation 13
CreA transcriptional factor 239
Ct-Amp1 145
Cut flower industry 1
Cyathus sterocoreus 305
Cyclic peptides 115, 145
Cyclosporins 115
Cytoskeleton microtubules 13
Cytosolic membrane 73

Dactylomyces crustaceous 305
Dahlia 145
Dahlia merckii 145
Dairy industry 353
Defensins 145
Degradation of PCP 305
Deoxynivalenol 267
Destruxins 115
Detection of food-borne toxigenic molds 267
– of PCR amplicons 267
Dichomitus squalens 305
Dictyloglomus thermophilum 289
Dispensable metabolic pathways 1
Dm-AMP 145
DNA intercalating dyes 267
DNA probes 1
Dolichol cycle 73
Dolichol pathway 73
Domain regulatory system 239
Down stream processing in SSF 305
Dual carbon and nitrogen control 55

Effluent disposal 305
Electrophoretic karyotyping 1
Electroporation 199
Endocytosis 13
Endomycopsis fibulinger 305
Endophytes 115
Endoplasmic Reticulum 13, 73
–, foldases 73
–, lumen chaperones 73
–, processing of N-linked glycans 73
Enniatin B 115
Entomophthora virulenta 305
Enzyme-Linked-Immunosorbent Assays (ELISA) 267
Enzymes for barley brewing 353
Enzymes for flavour production 353
Ergopeptines 115
Ergosterol 267
Ergot alkaloids 115
Ergotamine 115
Ergotism 1, 115
Ethanol utilization 55, 165
Eurotium sp. 267
Expressed sequence tags 145
Extracellular enzymes 73, 305
Extrachromosomal DNA 1

Facilitated diffusion 13
Fat processing 353
Fermented foods 305
Flammulina velutipes 305
Flavors and nucleotides 387
Flavour chemicals 353
Flax 145
Flow cytometry 289
Foldases 73
Folding of secretory proteins 73
Fomitopsis pinicola 305
Food and Agriculture Organization (FAO) 1
Food and Drug Administration (FDA) 1
Food processing 353
Food spoilage 267
Forward primer 267
Four o'clock (flower) 145
Fruit juice processing 353
Fumaric acid 387
Fuminosins 1, 165
Funalia trogii 305
Functional genomics 1
Fungal alkaloids 1, 115
Fungal Diagnosis 1
Fungal enzymes 353
Fungal expression systems 199
Fungal genomics 1
Fungal growth 13, 267
Fungal infections, humans 1
Fungal metabolites 387
Fungal peptide secondary metabolites 115
Fungal physiology 13
Fungal promoter 199
Fusarium species 1, 115
  F. avenaceum 115
  F. culmorum 145, 267, 305
  F. moniliforme 145
  F. oxysporum 115, 289, 305
  F. venenatum 289

Gamma-1- and -2-zeathionins 145
Gamma-thionins 145
GATA-factors 55
Gel electrophoresis 267
Gene cloning 199
Gene cloning strategies 199
Gene clustering 1
Gene disruption 199
Gene expression systems 199
Gene regulation of xylanase production 239
Gene structure and organization 165
Generally regarded as safe (GRAS) 1
Genetic engineering 289
Genomic approaches to fungal
  promoter analysis 199
Genomics 1
Giant saltbrush 145
Gibberella fujikuroi 165, 305
Gibberellins 165
Global positive acting regulatory protein 55
Gloephyllum sulphureus 305
Glucanase 13, 353
Glucans 13
Gluconic acid 387
Glucose isomerase 353
Glucose oxidase 353
Glycans 73
Glycosylation 13, 199
Golgi apparatus 13, 73
Golgi stacks 13
Growing apex 13
Growth patterns in SSF 305

HC-toxin 115, 165
Hemicellulases 353
Heterobasidium subvermispora 305
Heterologous gene expression in fungi 199
Heterologous protein synthesis 289
Heuchera sanguinea 145
Hevein-like peptides 145
High-throughput screening 1
History of transformation 199
Hordeum vulgare 145
Horse chestnut 145
Host specific toxins 165
Humicola sp. 305
Hydrophobicity 13, 73
Hydrophobins 1, 13, 73
Hydroxyproline-rich glycoprotein 73
Hymenoscypus ericae 305
Hyphal anastomosis 289

Immobilization of fungi 305, 387
Immobilized cells and enzymes 305, 387
Immuno-PCR 267
Immunochemical assays 267
Impatiens balsamina 145
Industrial and fuel grade ethanol 1
Industrial enzymes 353
Industrial protein and compound synthesis 199
Industrial yeasts 1
Inoculum preparation 305
International Plant Protection
  Convention (IPPC) 1
Itaconic acid 387

Knottin 145
Kojic acid 387
Kuehneromyces mutabilis 305

L-arabinofuranosidase 239
Lactase 353
Lactic acid 387
Lectins 1
Lentinus edodes 289, 305
Leptosphaeria maculans 145, 289
Lignin degradation 305
Lignocelluosic substrates 305
Linkage map of Fum loci 165
Linker sequence 267
Linum ustiatissimum 145, 289
Lipases 305, 353
Lipids and fatty acids 387
Lolitrem B 115
Lysergic acid 115
Lysosomes 13, 73

Macadamia 145
Macadamia integrifolia 145
Magnaporthe spp. 289
M. grisea 165
Malic acid 387
Meat industry 353
Mechanisms of transformation 199
Melanins 165
Melanocarpus albomyces 305
Membrane bound lipid carrier 73
Metabolic pathway 165
Metabolic regulation 55
Metarhizium anisopliae 115, 289
Mirabilis jalapa 145
Mistletoe 145
Mitochondrial protein 73
Mj-AMP1 145
Mj-AMP2 145
Molecular beacon 267
Molecular strain improvement 199
Molecular transformation 199
Monascus anka 305
M. fulginosus 305
M. purpureus 305
Mucor 353
Mucor hiemalis 305
M. meihei 305, 353
M. pusilus 305
M. silavaricus 305
Multi-copy gene expression levels 199
Multicopy episomes 199
Multicopy integration 199
Multiple mycotoxin 1
Mushroom breeding 289
Mushrooms 305
Mutagenesis 289
Mycofungicides 1, 115
Mycoherbicides 1, 115
Mycoinsecticides 1, 115
Mycotoxin 267
Mycotoxin resistant plants 1, 115

N-linked glycan structures 73
N-linked glycosylation 73
N-linked protein glycosylation 73
Naringinases 353
Nematoloma frowardii 305
Neotyphodium coenophalium 115
N. lolii 115
Neurospora crassa 13, 55, 145, 289
N. sitophila 305
niiA-niaD genes 165
Nitrate assimilation 55, 165
Nitrogen metabolic regulation 55
Nitrogen metabolite repression 239
Nonribosomally synthesized peptides 115
Norway spruce 145
Novel gene isolation 199
Nutrient utilization pathways 165

O-glycosylation 73
Oat 145
Ochratoxin 267
Oil processing 353
Oligosaccharyl transferase 73
Oomycota 13
Open reading frames (ORFs) 199
Organelle distribution 13
Organic acids 387
Overproduction of citric acid 387

PacC transcriptional factor 239
Paceliomyces fumoso-roseus 115
Packed column bioreactor 305
Pathway specific regulatory systems 239
Patulin 267
Paxilline 165
PCP-A1 145
PCR (polymerase chain reaction) 239, 289
–, for detection 267
–, internal competitor 267
–, products 267
–, quantification 267
–, –, real-time 267
PDF1.2 145

Pea defensin 145
Pectinases 305, 353
Penicillin 115, 165
Penicillium species 267, 289
P. camembertii 305
P. candidum 305, 353
P. capsulatum 305
P. chrysogenum 115, 165, 289, 305
P. cyclopium 13
P. emersonii 353
P. funiculosum 353
P. islandicum 305
P. jenthillenum 305
P. notatum 13
P. paxilli 165
P. roquefortii 305, 353
P. rugulosum 305
Peniophora lycii 305
Peptidases 353
Peptide antibiotics 145
Peptide phytotoxins 115
Peptide synthetases 115
Phanerchaete chrysosporium 13, 305
Pharbitis nil 145
Phlebia radiata 305
P. tremellosa 305
Pholioto nameko 305
Phytases 305
Phytoalacca americana 145
Picea abies 145
Pichia pastoris 73
Plant defense 145
Plant pathogenic fungi 1, 115
Plasmodiophora brassicae 145
Pleurotus sp. 305
P. eryngii 305
P. ostreatus 289, 305
P. pulmonarius 305
P. sojur-caju 305
Pn-AMP1 and Pn-AMP2 145
Pokeweed 145
Pollen coat protein 145
Polyethylene glycol–calcium chloride transformation method 199
Polymerization 267
Polypus anceps 305
Polysaccharides 387
Poria placenta 305
P. subvermispora 305
Posttranscriptional gene silencing 199
Posttranscriptional regulation 289
– of fungal gene expression 199
Posttranslational modifications 199, 289
PR-1 145
Probe 267
Production processes 305, 387
Proline utilization 165
Proteases 305, 353
Protein disulfide isomerase 73
Protein folding 289
Protein glycosylation 73
Protein localization 289
Protein processing 289
Protein secretion 73
–, physiology 73
Protein stability 199
Protein translocation 73
Proteins in the cell wall 73
Proteolytic degradation 289
Protoplast fusion 289
Protoplasts 199
Pseudomanas syringae pv. tabaci 145
Pseudomonas solanacearum 145
Pullulan 387
Purine metabolism 55
Pycnoporus sanguineus 305
Pyrenophora tritici-repentis 145
Pyricularia oryzae 145
Pythium ultimatum 13

Quinate utilization 165

Radish 145
Ralstonia solanacearum 145
Random mutagenesis 289
Raphanus sativus 145
Regulation of pathway gene expression 165
Regulatory advantages of gene clusters 165
REMI (restriction enzyme-mediated integration) 289
Research applications of gene expression systems 199
Restless transposon 1
Restriction enzyme-mediated integration (REMI) 199
Reticuloendothelial system 13, 73
Reverse primer 267
Reverse transcription 267
Reverse transcription-PCR (RT-PCR) 267
Rhizoctonia solani 145, 289
Rhizomucor miehei 353
Rhizopus arrheizus 305
R. chinensis 305
R. delemar 289
R. oligosporus 305
R. oryzae 305

RNA stability 199
Rotating drum bioreactor 305
Rs-AFP1 145
Rs-AFP2 145
Rygrass 115

Saccharomyces cerevisiae 13, 55, 73, 145, 239, 305
S. diastaticus 353
S. ubarum 353
Safety of food 1
Sanitary and Phytosanitary (SPS) Agreement 1
Saprolegnia 13
Schizosaccharomyces pombe 13, 73
Schwanniomyces castessli 353
Scleroglucan 387
Screening methods 289
Secondary metabolites 305
Secretion 13
– by the cell wall 73
Secretory pathway 73
–, genetic manipulation 73
–, organelles 73
Secretory vesicle 13
Selection of transformants 199
Sensor technology 267
Sexual crossing 289
Shared gene clusters 1
Shistosoma japonicum 239
Signal peptide sequence 73
Significance of genes clusters 165
Single nucleotide polymorphisms 1
Solid state fermentation 1, 305, 387
Somatic crossing 289
Sorghum 145
Sorghum bicolor 145
Sorghum brewing 353
Spitzenkörper 13
Sporptricum pulverulentum 305
Starch 353
Sterigmatocystein 1
Sterigmatocystin 165
Strain improvement 289
Strand displacement 267
Streptomyces cluvuligerus 305
S. rimosus 305
Strong constitutive or inducible promoters 199
Structure of the core glycan in N-glycosylation 73
Subcellular transport 199
Substrates for SSF 305
Succinic acid 387
Sugar 353
Sugar utilization 165
Sulfur controller genes 55
Sulfur regulatory circuit 55
Surface fermentation 305, 387
Sweet pepper 145

T-DNA 145
Taq DNA polymerase 267
Taralocyces spp. 305
Target, molecular beacon hybrid 267
Target DNA 267
Target:competitor ratio 267
Tartaric acid 387
Taste enhancers 353
Temperature control 305
Thermoascus aurantiacus 305
T. lanuginosus 305
Thi2.1 gene 145
Thielavia terrestris 305
Thionins 145
Tolypocladium cylindrosporum 115
T. niveum 115
Toxigenic fungal molecular probes 267
Toxigenic fungi 1, 115
Trametes versicolor 305
Trans-Golgi network 13
Transcriptional factor 239
Transcriptional gene silencing 199
Transcriptional terminators 199
Transformation efficiencies 199
Transformation strategies 199
Transformation strategies for fungi 199
Transformation vectors 199
Transgene transcription 199
Transgenic food crops 145
Translational control 199
Translational initiation signals 199
Translocation process 73
Transmembrane domains 73
Transmembrane proteins 73
Transposable elements 1
Transposition of DNA 1
Tray system 305
Trichoderma 353
T. aureaviride 305
T. hamatum 145
T. harzianum 289, 305
T. koningii 305
T. lignorum 305
T. longibrachiatum 239
T. reesei 13, 73, 239, 289, 305
T. viride 305
Trichothecenes 1, 165
Trichothecenes biosynthetic pathway genes 165

Triella fuciformis 305
Triticum aestivum 145
Tuber spp. 289
Turgor pressure 13
Tyromyces balsemeus 305

Unfolded protein response 73
Upstream ORFs 199
Uromyces fabae 13
Ustilago maydis 73
Ustulina vulgaris 305

Vacuolar nutrient storage 13
Vacuolar transport 13
Values of sales of mycology-based products 1
Vectors, autonomously replicating 199
–, integrating 199
Verticillium lecanii 115
Vesicle supply centre 13
Vesicle transport 13
Victorin 115
Viscotoxin A3 145
Viscum album 145
Vitamin A 387
Vitamin B6 (Niacin) 387
Vitamin B2 (Riboflavin) 387
Vitamin D 387
Vovariella vovacea 305

Wall bound and secreted proteins 73
Water activity 305
Water-activity (aw) 267
World Trade Organization 1

Xylan degrading enzymes 239
Xylanase 239, 305, 353

Zea mays 145
Zein 145
Zinc regulatory factors 199